Asphalts in road construction

Edited by Dr Robert N. Hunter

Thomas Telford, London

Published by Thomas Telford Publishing, Thomas Telford Ltd, 1 Heron Quay, London E14 4JD.
URL: http://www.t-telford.co.uk

Distributors got Thomas Telford books are
USA: ASCE Press, 1801 Alexander Bell Drive, Reson, VA 20191-4400, USA
Japan: Matuzen Co Ltd, Boof Department, 3–10 Nihonbashi 2-chome, Chuo-ku, Tokyo 103
Ausdtralia: DA Books and Journals, 648 Whitehorse Road, Mitcham 3132, Victoria

First published 2000

Also available from Thomas Telford Books
Bituminous mixtures in road construction. Robert N. Hunter. ISBN: 0 7277 1683 2

The cover photographs were supplied courtesy of Dynapac, Sweden

A Catalogue record for this book is available from the British Library

ISBN: 0 7277 2780 X

Typeset by MHL Typesetting Ltd, Coventry
Printed and bound in Great Britain by Redwood Books, Trowbridge, Wilts.

Foreword

The National Road Network is widely regarded as the main artery for carrying the lifeblood of the economy. With trunk roads alone valued at over £50 billions, roads are clearly a valuable asset for which good quality construction and maintenance standards are essential if a satisfactory performance is to be achieved. Standards not only dictate the value we get from the infrastructure investment, they also influence the impact highways have on the daily lives of people. Operation of trunk roads is managed by the Highways Agency which plays a key role in standard setting, working closely with equivalent bodies in Wales, Northern Ireland and Scotland, as well as with local highway authorities.

Since this book was first published in 1994, improved asphalt technology has made significant advances. Heavier trucks and super single tyres have influenced materials choices and design. The desire for longer maintenance-free life as part of whole life analysis has seen stronger roadbases emerge. Changing climatic conditions with hot summers have influenced surfacing design and, of course, there is the most important issue of road safety. Safety is the one area where the UK stands above almost any other country with the levels of fatalities continuing to remain the lowest level within international comparisons, something to which our road surfaces have made a significant contribution.

Attitudes have changed as we now see top priority given to maintaining the network in a safe and serviceable condition. For the first time, whole life costs are part of the equation which means including an allowance for disruption to road users during works. The service delivered to customers is now an important consideration for all in the highways industry.

Changes are now recognised as part of modern working lives affecting all areas of highways design, construction, management and maintenance. As a result, the asphalt being produced today has changed significantly and is now capable of meeting the most demanding performance criteria we now face. Technology has played a big part in the change particularly over the last 5 years where the pace has been more dramatic than at any other time during most professional careers. Asphalt plays an important role in the lives of everyone in the UK given that it provides the basic structure and running surface of the majority of our roads.

In recent years there has been what can only be described as a quiet revolution in the asphalt we use, something which industry and highways authorities have both played a major part in implementing. The revolution is a move to serving the customer, those people beyond the traditional supplier/client chain such as the people who live near to and use the road network we support. In serving the latest customer needs many old methods, materials and procedures have had to be replaced by more appropriate modern systems.

If the quiet revolution is taken literally it is possible to point to the use of quieter road surfaces which were a distant dream for highways engineers until recently but are now in general use. These are the most visible example of innovation and have been a huge success in addressing the very real concerns of those living near to or travelling on our busy roads. While highly textured surfaces provided very safe roads, they also generated noise. Development of the current breed of quieter surfaces has given added impetus by the change in policy of procurers towards encouraging innovation and accepting proprietary products. For many years a no-go area for public procurers, products approval schemes such as HAPAS have provided customers with the confidence they needed to accept some of the innovative new products now on the marketplace, as well as providing the industry with an incentive to invest in new ideas.

Performance specification is now a real prospect given new test methods. The Nottingham Asphalt Tester (NAT) and other equipment for measuring stiffness characteristics in conjunction with the Dynamic Cone Penetrometer and the Falling Weight Deflectometer and more traditional deflectograph surveys now enable Engineers to predict the remaining life of existing pavements more accurately. But the planned introduction of performance-related specifications will require all parties to acquire an enhanced knowledge of the behaviour of asphalts.

The use of modified binders to improve specific aspects of asphaltic behaviour is now commonplace. These include surface dressings which can more readily resist the rigours of modern traffic, high friction surfacings which are used in locations where reduced braking distances may make a contribution to safety and the new breed of thin surfacings.

Changes in road management have also had an impact on asphalt. Private finance with operator responsibility transferred to DBFO companies for a 30 year concession period have inevitably resulted in changes in attitude. Industry has looked critically at ways of maximising maintenance-free life of asphalt which, in many ways, can be achieved at little or no additional cost, just greater attention to quality and consistency. The focus is also on providing a quality service to the road users, initially driven by contract economies, but we see from Europe where customer service has driven decisions on material choice.

Finally, there is the issue of Europe. Harmonisation of standards seemed straightforward 10 years ago when work was in its infancy but is still over 3 years away for bituminous mixtures. But there have been impacts. Development of new performance related tests, greater awareness of the link between fundamental properties and ultimate performance, exposure of good ideas to European wide scrutiny have all been benefits from participating within CEN committees. These have already provided benefits to the way we design, build and operate our roads in the UK.

Advances in Asphalt Technology accompanied by the traditional basic skills have combined to give the UK the safest roads in Europe, an achievement of which all in the industry should be justifiably proud. The future for asphalt is bright. But the industry must continue to adapt to meet changing needs of customers whether local or national highway authorities, or the travelling public. This book contains details of all the major recent innovations in Asphalt Technology but does so without neglecting the fundamental elements of the subject. Contributions are made by some of the most highly respected experts, each leading in their own areas. I am sure this latest edition will be welcomed not only to the specialist in both the specifying and contracting organisations but also to students and those who have the occasional need for a reference work.

Note

Figures and other material used in the above came from the Highways Agency's website and the BRF publication *Road Fact 99*.

Graham Bowskill
Head of Pavement Engineering Group
Highways Agency

Preface

Editing this book took well over a year. Whilst editing, I was constantly reminded of the unique contribution made to highway engineering by the Transport Research Laboratory. I believe that its publications are the most significant factor in giving the UK some of the safest roads in the world and ensuring that UK engineers have the most contemporary knowledge and techniques available to them. The Highways Agency is also to be highly commended for the *Design Manual for Roads and Bridges* and the *Manual of Contract Documents for Highway Works*. There are no comparable documents in the world and they continue to reflect the most up to date approach to road design and construction. In addition, the British Standards Institution produce some magnificent Standards.

Readers who wish to keep abreast of developments in asphalt and pavement issues are advised to attend meetings of the Institute of Asphalt Technology. Despite limited resources, the Institute continues to present technical papers of the very highest quality throughout the UK. Those able to attend meetings in London will find that the Society of Chemical Industries also arranges meetings of particular merit.

In preparing the text, I was determined to produce a book which reflected the very best of UK asphalt and pavement technology. I am very grateful to the contributors who produced chapters of the very highest calibre. I am very pleased with the result and proud to be associated with the publication.

Robert N. Hunter, June 2000

Biographies of contributors

Robert N. Hunter BSc MSc PhD CEng FICE MIAT MIHT MCIArb

Robert N. Hunter was educated at Boroughmuir Senior Secondary School and Heriot-Watt University, graduating with a BSc in Civil Engineering in 1970. He has undertaken two part time Research Degrees in asphalts, resulting in the award of MSc and PhD degrees in 1983 and 1988 respectively.

Robert became a Member of the Institution of Civil Engineers in 1979 and a Fellow in 1997, a Member of the Institute of Asphalt Technology in 1983, a Member of the Institution of Highways & Transportation in 1985, an Associate of the Chartered Institute of Arbiters in 1995 and a Member in 1999. Robert is a listed Adjudicator with the Institution of Civil Engineers.

He has had over a dozen papers published in various technical journals, given many talks to learned institutions and edited and contributed to the highly successful first edition of this book *Bituminous Mixtures in Road Construction*. He also wrote a book on contractual claims, *Claims on Highway Contracts*.

Robert has been employed by local authorities for almost all his working life. Since 1972 he has been involved in road construction. Since February 1996, he has been the General Manager of the Forth Local Authority Consortium, which manages and maintains the network of motorways and other major trunk roads from the east side of Glasgow up to Stirling and down to the English border on the east side of Scotland.

He loves all aspects of asphalt technology.

Professor Stephen F. Brown DSc FREng

Steve Brown is Head of the School of Civil Engineering at the University of Nottingham and leads the Division of Pavement and Geotechnical Engineering. He served a four-year term as the University's Pro-Vice-Chancellor responsible for Research, Industry and Commerce from 1994 to 1998 and was previously Head of Civil Engineering from 1989 to 1994 and Dean of Engineering from 1992 to 1994.

Following graduation from the University of Nottingham in 1960 with a First Class Honours Degree in Civil Engineering, he spent three years with contractors and consultants in the UK and South Africa before returning to a Research Assistant post at Nottingham in 1963. He was appointed a Lecturer in 1965, Senior Lecturer in 1974, Reader in 1978 and to a personal Chair in 1983. He has been awarded the degrees of PhD and DSc by the University of Nottingham and is a Fellow of the Royal Academy of Engineering and of the Institution of Civil Engineers. In 1996 he was admitted to the City of London Livery of the Worshipful Company of Paviors.

Professor Brown's research is in the fields of pavement engineering and geotechnics in which he has published over 170 papers and attracted financial support from research councils, government and industry both in the UK and the USA. He was the British Geotechnical Society's 36th Rankine Lecturer in 1996, was awarded the James Alfred Ewing Medal of the Institution of Civil Engineers in 1997 for his contributions to engineering research and, also in 1997, was the first International Society for Asphalt Pavements' Distinguished Lecturer.

In 1985 he became a founder Director of SWK Pavement Engineering Ltd, a specialist consulting firm operating from the Nottingham Science Park and in 1993 he was appointed President of a wholly owned subsidiary based in New Jersey, USA and in 1996 as a Director of SWK Pavement Engineering Sdn Bhd in Malaysia. In 1997, the Nottingham and Malaysian companies changed their names to Scott Wilson Pavement Engineering.

Professor Brown has served on a wide range of committees within his profession, is a past chairman of the British Geotechnical Society and was an SERC Research Co-ordinator for geotechnics from 1985 to 1989. He has recently been elected as the first non-North American to serve as a Director of the US Association of Asphalt Paving Technologists.

Chris Cluett FIAT

Chris Cluett is the Business Manager for Colas Ltd, holding responsibility for Commercial Services and Marketing in the UK.

After graduating in Civil Engineering from the University of South-ampton, he commenced his career with Amey Asphalt on the construction of the M4 at Theale. During the period with Amey he became involved in airfield surfacing at Heathrow, Fairford, Greenham Common and Coltishall. He spent time in the early 1980s in the USA working on asphalt recycling and sulphur asphalt. Before leaving Amey, Chris became involved in contract management on sea defence works including the Continental Ferry Terminal at Portsmouth.

Chris joined the Colas Group as Location Manager in the South with direct responsibility for term contracting with the PSA and became involved in working with the Japanese at Toyota and Honda for the construction of major test track complexes as well as the refurbishment of the Erskine Bridge over the River Clyde using Epoxy Modified Asphalt.

Chris is a Council Member of the Institute of Asphalt Technology and has presented several papers to the Institute of Asphalt Technology, the Institution of Civil Engineers and the RHA.

Jeff Farrington

Jeff Farrington has enjoyed a career in civil engineering spanning more than 46 years, the last 35 of which have been involved in some aspect of highways. He left the Staffordshire County Council Highways Department in 1997 to set up his own consultancy specializing in trouble shooting asphalt problems on sites, advising Consulting Engineers on the better use of asphalts and training. He is a member of several British and European Standards Committees and enjoys reading asphalt books in bed.

Dr Ian M. Lancaster BSc PhD MRSC CChem MIM MInstPet

Ian Lancaster was born in Hull and educated at the University of Bradford where he gained a BSc in Chemistry in 1986 and a PhD in Polymer Science in 1989. Immediately after gaining his PhD, he joined Pilkington Glass at its Group Technical Centre in Lancashire, becoming Senior Research Scientist in 1992. He joined Nynas UK AB as Technical Manager of its Eastham Site in 1994 with responsibilities including product development, quality control and quality assurance. Ian has a particular interest in polymer modification and rheology of bitumen and has written and presented a number of publications on these subjects. He is also a Member of the Institute of Petroleum Bitumen Test Methods Committee.

John Moore

John Moore is the Technical Director of Gencor ACP Ltd, a leading manufacturer of bituminous plants. He is responsible for Sales, Estimating and Technical Publications Departments and undertakes extensive overseas business travel. He has an Honours Degree in Mathematics and is an Associate of the Institute of Quarrying with over 25 years' of experience gained in the quarrying industry. He was engaged in the draft committee for the CEN European Standards for Asphalt Plants.

Gerry Hindley FIAT MBAE

Gerry Hindley is a Bituminous Pavement Consultant running his own practice. He has over 30 years' experience of all aspects of the production, laying and compaction of all forms of asphaltic pavements and other allied highway issues.

His recent experience includes international projects funded by the European Bank for Reconstruction and Development (EBRD) and he frequently acts as an expert witness in disputes involving asphaltic pavement issues.

He is a Member of the National Council of the Institute of Asphalt Technology and a Member of the Academy of Experts.

John D. H. Gray MInstCES MIAT

John Gray is the Business Manager Contracting at Mouchel Consulting Limited, and is responsible for the management and maintenance of motorways and other major trunk roads in the UK. He has over 25 years' experience with Civil Engineering Contractors in the road construction industry and has been involved in establishing and improving financial monitoring and reporting procedures for estimating and valuation practices within contracting organisations.

John Richardson BSc CChem MRSC CEng MInstE

John Richardson graduated from the University of Strathclyde with an Honours Degree in Applied Chemistry in 1973. After a short period with the Water Board in the South of Scotland, he joined Wimpey Asphalt Ltd as a trainee Quality Engineer. He subsequently became involved at Wimpey in research and development, quality control, quality management and overseas contracting. He joined Tarmac Quarry Products Ltd in 1996 and added Pavement Engineering to his range of activities. He is now Manager Technologies.

John is Chairman of the Construction Materials Group of the Society of Chemical Industries and a Committee Member of the Greater London & Essex Branch of the Institution of Highways & Transportation. He is a Member of the Asphalt Technical Panel of the Quarry Products Association and serves on Standards Committees of both BSI and CEN. He has presented various technical papers and is a visiting lecturer of asphalt materials at the University of Birmingham.

David Rockliff BSc CEng MICE FIAT FIQ

David Rockliff is a Chartered Civil Engineer with a keen interest in pavement technology. The highways of northern England have kept him busy for most of his working life.

David joined Durham County Council as a Graduate Engineer and worked on most aspects of road design, construction and maintenance. A period in the Highways Laboratory developed an interest in asphalts as civil engineering materials which resulted in a move to Tilcon in 1988.

David now heads the team of quality system and product technology staff who support the varied operations of Tarmac Northern Limited. He is closely involved with pavement construction materials and their changing specifications.

David Whiteoak

David Whiteoak has been involved in the road construction industry in one form or another for over 25 years. After graduating from Heriot-Watt University in 1980 with a BSc Honours Degree in Civil Engineering, he joined Shell Research working in the Bitumen Laboratory at Thornton Research Centre near Chester. After five years in research, he moved to Shell Bitumen in Chertsey to work in the Technical Department and during this time he wrote the well known and highly respected *Shell Bitumen Handbook*. In 1995 he rejoined Shell Bitumen as Technical Manager after a three-year assignment in Shell International Chemical Company.

David is a member of a number of industry committees including BSI Committees on Bitumen and Surface Dressing, the Society of Chemical Industries Asphalt Sub-Committee, CEN TC 19 SC1 WG4 on Polymer

Modified Bitumens and, last but not least, the Council of the Institute of Asphalt Technology.

Dr David Woodward DPhil MPhil BSc MIHT MIAT MIEI MIQ

Dr David Woodward was appointed as Lecturer in Infrastructure Engineering in the School of the Built Environment at the University of Ulster in September 1998.

After obtaining a Combined Sciences Degree in Geology/Geography in 1982 at Ulster Polytechnic, he joined Highway Engineering Research at the University of Ulster in 1983 as a Research Assistant, became Research Officer in 1988 and Deputy Director of the Highway Engineering Research Centre (HERC) in 1995. He currently heads the Highway Engineering Research Group which is part of the Transport & Road Assessment Centre (TRAC). He was awarded a Master of Philosophy in 1988 and a Doctor of Philosophy in 1995. He has successfully co-supervised MPhil and DPhil research students. Dr Woodward has been involved in the continual development and expansion of the Highway Engineering Research Laboratory. Although its prime aim is to research highway materials, this facility also provides a research and consultancy service to the highways industry in a wide range of areas including aggregate, bitumen and bituminous mixtures.

Dr Woodward has been involved in research programmes for government agencies and private companies both nationally and internationally. With over 140 technical papers published, Dr Woodward is widely travelled and has presented research findings at national and international level ranging from local professional institutions to world conferences. His main research interest is the prediction of highway material performance with particular reference to the role of aggregate.

Acknowledgements

Robert Hunter is grateful to many people for assistance with this book including Helen Brown, Camille Cowie, Julian Cruft, Ray Diamond, John Gray and Frank Irvine of West Lothian Council, Neil Taylor, Phil Letchford and Dr Huayang Zeng of Dynapac, Paul Field of Wirtgen, Lito Achimastos of Shell Recherche SA, David Greenslade of North Lanarkshire Council, Diana Clements and Hans-Josef Kloubert of Bomag, Jack Edgar and many others who assisted with production. In particular, he would like to acknowledge the immense assistance afforded him by Ian Walsh of Babtie Laboratories, who provided whatever was requested by return on numerous occasions.

Chris Cluett would like to thank Wendy Pring and Louise Clatdon for their help in keeping him sane when deadlines had passed! Special thanks to Robin Cook, Darren Fitch, Nigel Farmer, Paul Field, Allan Newton and Terry Stagg for their patience.

Jeff Farrington wishes to acknowledge the assistance received from Derek Pearson, Martin Heslop and Ian Walsh, all of whom were kind enough to lend him some of their own photographs showing problems with surfacing materials.

Gerry Hindley would like to acknowledge the help of colleagues in the industry and in particular Martin Southall of Wrekin Construction, and David Rhodes, Brian Fern and Colin Loveday of Tarmac Heavy Building Materials UK Ltd.

John Moore would like to thank all the people who helped and supported him during the writing of his Chapter.

John Richardson would like to express his gratitude to John F. Hills for commenting on his first draft and Chris A. R. Harris for inspiring the greater use of statistical analysis.

David Rockliff would like to thank Helen Phillips for her skill and patience. The guiding spirit of the late Professor Joe Cabrera of Leeds University is also acknowledged — truly a gentleman and a scholar.

Dave Whiteoak would like to thank all those who contributed to his Chapter, in particular John Read and Colin Underwood. Last but not least, he would like to thank Robert Hunter for his contribution to the additions to the Chapter and for his immense patience when he missed deadline after deadline.

Credits

Chapter 1 was written by Dr David Woodward.

Chapter 2 was written by Dr Ian Lancaster.

Chapter 3 was written by Professor Stephen Brown.

Chapter 4 was written by David Rockliff.

Chapter 5 was written by John Moore.

Chapter 6 was written by Chris Cluett.

Chapter 7 was written by Gerry Hindley.

Chapter 8 was written by Dr Robert N. Hunter.

Chapter 9 was written by John Richardson.

Chapter 10 was written by David Whiteoak.

Chapter 11 was written by Jeff Farrington.

The estimating and quantity surveying input to Chapter 7 was provided by John Gray.

The text was edited by Dr Robert N. Hunter.

Contents

1. Assessing aggregates for use in asphalts

1.1. Preamble

This Chapter considers the assessment and role of aggregates for use in asphalts for highway construction. Any asphalt, whether hot rolled asphalt, bitumen macadam, porous asphalt, stone mastic asphalt or surface dressing is nothing more than a specified blend of aggregate and bitumen. When mixed and laid properly, a durable and long lasting structure should result. However, what properties must the aggregate have to ensure that this occurs? This Chapter considers the basic test methods available to engineers to enable them to make a choice. It also considers the factors that influence these procedures.

This background is particularly relevant as the UK highways industry is experiencing a period of change from traditional practices. This is due to reasons that include growth in both the demand and use of aggregate, environmental issues, the introduction of new mixtures such as stone mastic asphalts and thin surfacings and the harmonization of aggregate test methods and specifications within Europe. Many of these factors have only become important in the past few years. This process of change has prompted concern about how materials are assessed and how their performance is predicted.

In terms of their potential to perform (i.e. their beneficial and detrimental properties), it is important to consider how the aggregate was formed and what has happened to it since that time. This is where timescales are important. Prior to its use as a constituent in an asphalt, it should be remembered that it existed for a long time, geological time — a period of several million years. Once mixed with bitumen and laid as part of the highway structure, its behaviour should be considered in terms of engineering time, which in terms of highway construction is measured in decades.

Traditional experience and laboratory testing should indicate whether a rock that has existed for millions of years in the ground can perform during its highway design life. This is the aim of performance testing. The approach taken within this Chapter is not a simple description of test methods but rather one which seeks to introduce the concept of predicting performance and the determination of the ability of aggregate test methods to adequately predict this and thus reduce the likelihood of premature failure.

1.2. Why test aggregates?

It is important to consider why aggregates need assessment. It must be recognized that aggregates are a fundamental component of asphalts. The chosen aggregate must have certain properties to withstand the stresses imposed both at and within the road surface and sub-surface layers. The resulting performance of the constituent particles is dependent upon the manner in which they are bound together and the conditions under which they operate. Their assessment is carried out in a laboratory where a range of test methods is used to predict subsequent in-service performance.

Testing is therefore a process of prediction that provides a quantifiable indication of quality which may then be implemented in the form of specification requirements to rank different aggregates in relation to preconceived levels of performance. However, there is nothing to guarantee that a laboratory assessment of an aggregate to meet specification requirements will fully predict its potential for successful in-service performance given the changing in-service expectations of a modern road surfacing.

This is particularly true since aggregates are now being used in ways and exposed to levels of stress not experienced previously nor even anticipated in the original development of certain testing methods and specification limits. This increase in performance expectation from the finished structure prompts basic questions such as the following.

- What aggregate properties should be measured?
- Can these be quantified?
- What do current test methods measure?
- What factors influence the test method?
- Does the test method relate to actual in-service conditions?
- Does the method predict performance?
- Is the test method cheap and does it use simple equipment?

Any improvement in the prediction process of assessing aggregate performance will ultimately result in better value for money and the improved use of limited reserves of aggregate. Although performance is now common to many aspects of engineering, very little is understood about the factors that influence and control how aggregates perform in use. There have been many instances of materials failing prematurely despite having met specified requirements.

Traditionally, the immediate response has been to blame the bitumen, the surfacing contractor or the weather. Aggregates have rarely been blamed initially as the reason for failure. If premature failure occurs then it may be argued that the specification requirements and/or the testing methods used were unable to predict the suitability of the aggregate for the given set of in-service conditions, that is the prediction system which had been used had failed to perform. One of the main aims of this Chapter is to make practising engineers aware of the ability to predict aggregate performance and the limitations of existing standards and specifications.

1.3. Classifying aggregates

Geology is the study of rocks. It covers a wide spectrum, ranging from the classification of the many rock types, the study of their origins and variations in mineralogical composition to their structure and alteration over many millions of years.

Basically, the level of in-service performance of a particular rock depends on its geological properties. For example, consider the high levels of skid resistance exhibited by certain sandstones and gritstones. These materials consist, essentially, of sand grains which are held together by a cementing matrix. It is this sandpaper-like surface texture which gives the aggregate its high frictional properties.

It is advisable to ask certain questions when considering the potential suitability of a particular aggregate. These should, for example, include details of the type of rock, its composition, its chemistry, its grain size, whether it has been altered, its tendency to degrade, abrade or break under trafficking and whether bitumen will adhere to the surface. These and other questions need to be answered in order to understand fully how an aggregate will perform under the variety of conditions which may exist in a pavement.

The possible range of aggregates available for use in asphalts is potentially vast. However, if an aggregate is to be used in a particular mixture, a certain quality is required and this rules out many potential sources. Rounded gravels, for example, are rightly excluded for use in particular asphalts. Materials for use as wearing courses almost always require high values of skid resistance and this severely limits the number of potential sources. Considerable research and experience has resulted in the majority of main roads in the UK being surfaced with aggregates belonging to a single type or group, i.e. the gritstone group.

In terms of their use in an asphalt, the types of potentially available aggregate may be broken down into three main groups.

- Natural — this includes all those sources where the rock is naturally occurring in the ground and obtained by conventional blasting and quarrying, excavating or dredging from land, river, estuarine or marine deposits.
- Artificial/synthetic — these may be formed as a result of industrial processes (such as those carried out by the steel industry) or specifically made for high performance purposes, e.g. calcined bauxite used in high friction surfaces.
- Recycled — the recycling of existing asphalts is growing significantly in importance and in many countries this is now a major source of aggregate.

1.3.1. Natural rock types

Aggregate from natural rock types belongs to one of four main types — igneous, sedimentary, metamorphic or sand and gravel.

- Igneous rocks are those which have solidified from a molten state and possess a very wide range of chemical composition, grain size, texture and mode of occurrence. Certain types may have formed due to the slow cooling of large masses deep within the Earth's crust. The resulting granites and diorites are, typically, coarse grained. Other types may have been extruded as lava flows on the Earth's surface resulting in fine grained basalts.

- Sedimentary rocks are typically formed as a result of the weathering and subsequent erosion due to transport by water, ice or wind existing rocks. These processes cause deposition of these materials into stratified layers that are then consolidated or cemented by chemical action to form new rocks such as sandstone and gritstone. Sedimentary rocks may also be formed due to the chemical precipitation of minerals that were dissolved in water to form rocks such as limestone. They may also consist of organic materials, e.g. coal.
- Metamorphic rocks occur as the result of alteration of existing igneous or sedimentary rocks by heat, pressure or chemical activity and comprise a very complex range of rock types.
- The properties of sands and gravels are typically dependent on the rocks from which they were derived and their transport by water prior to deposition. Crushed gravels are generally not allowed for most asphalts in the UK, although in some other countries they are an important source of aggregate.

1.3.2. Artificial/synthetic aggregates

Artificial/synthetic aggregates have had a human involvement. They include various types of slag that are a by-product of the steel industry. The use of slags has been restricted in the UK in the past due to limited availability and problems with stability. However, in recent years their use has become more widespread particularly due to their good skid resistance characteristics.

However, the main use of synthetic aggregate in the UK has been the development of high friction surfacings (discussed in Section 10.5, Chapter 10) incorporating the use of calcined bauxite. This is a high aluminium content clay which is heated to temperatures of 1500 °C. This results in the formation of corundum crystals producing a very hard wearing aggregate. When used as a 3 mm sized aggregate, there is no other natural aggregate which can give comparable levels of skid resistance.

1.3.3. Recycled aggregates

Until recently, there was very little interest in the recycling of aggregates in the UK. However, this has changed due to a general public awareness of environmental issues and the development of techniques to produce materials which meet specified requirements. In terms of aggregate for use in pavements, most experience has been with the recyling of existing asphalts. Some specifications allow up to 50% of this material to be included with virgin constituents.

Other types of material, such as crushed concrete and demolition material, have also been considered. However, prior to the recycling of these materials, careful consideration must be given to their effect on in-service performance. The aggregate may have already been subjected to 5, 10, 15 or 30 years of use and it may not have sufficient retained strength and durability to perform satisfactorily.

1.4. Aggregate petrology — a simple classification

Many aggregates have local or traditional names which has caused considerable confusion in the past. Although many such terms are still used they must not be used in specifications or contract documents. O'Flaherty[1] has listed some of the traditional names against their designated British Standard names and these are shown in Table 1.1.

Recognizing the possible confusion that existed, the original version of BS 63[2] contained an appendix that simplified the classification of aggregates into 12 Trade Groups. Each group contained a range of rock types that was deemed by the Geological Survey to give similar levels of behaviour. (When this version of BS 63[2] was published, the Geological Survey was the senior professional geological body in the UK.) It was felt that simplistic classification was better suited for road-making purposes rather than the confusing array of names and terms used by geologists. This is shown in Table 1.2. With a few minor changes, e.g. listing andesite and basalt together, this original grouping still remains in use within the UK aggregate industry.

However, it has been found that individual sources of aggregate within each of these groupings do not perform in similar ways. Geologists have identified many different rock types that may be classified into these groups. Therefore, it is necessary to understand the importance of these basic groupings and the range of possible rock types that they may contain.

Table 1.1. Examples of traditional names for aggregates[1]

Traditional name	Appropriate group in British Standard Classification
Clinkstone	Porphyry (rarely basalt)
Cornstone	Limestone
Elvan (blue elvan)	Porphyry (basalt)
Flagstone	Gritstone
Freestone	Gritstone or limestone
Greenstone	Basalt
Hassock	Gritstone
Hornstone	Flint
Pennant	Gritstone
Rag (stone)	Limestone (rarely gritstone)
Toadstone	Basalt
Trap (rock)	Basalt
Whin (stone)	Basalt

Table 1.2. A basic classification of road aggregates[2]

Granite	Basalt (including basaltic whinstone)	Grit
Gabbro	Hornfels	Limestone
Porphyry	Schist	Flint
Andesite	Quartzite	Artificial

Table 1.3. Rock types commonly used for aggregates[3]

Petrological term	Description
Andesite	A fine grained, usually volcanic, variety of diorite
Arkose	A type of sandstone or gritstone containing over 25% feldspar
Basalt	A fine grained basic rock, similar in composition to gabbro, usually volcanic
Breccia	Rock consisting of angular, unworn rock fragments, bonded by natural cement
Chalk	A very fine grained cretaceous limestone, usually white
Chert	Cryptocrystalline silica
Conglomerate	Rock consisting of rounded pebbles bonded by natural cement
Diorite	An intermediate plutonic rock, consiting mainly of plagioclase with hornblende, augite or biotite
Dolerite	A basic rock with grain size imtermediate between that of gabbro and basalt
Dolomite	A rock or mineral composed of calcium magnesium carbonate
Flint	Cryptocrystalline silica originating as nodules or layers in chalk
Gabbro	A coarse grained, basic, plutonic rock, consisting essentially of calcic pagioclase and pyroxene, sometimes with olivine
Gneiss	A banded rock, produced by intense metamorphic conditions
Granite	An acidic, plutonic rock, consisting essentially of alkali feldspars and quartz
Granulite	A metamorphic rock with granular texture and no preferred orientation of the minerals
Greywacke	An impure type of sandstone or gritstone, composed of poorly sorted fragments of quartz, other minerals and rock; the coarser grains are usually strongly cemented in a fine matrix
Gritstone	A sandstone with coarse and, usually, angular grains
Hornfels	A thermally metamorphosed rock containing substantial amounts of rock-forming silicate minerals
Limestone	A sedimentary rock, consisting predominately of calcium carbonate
Marble	A metamorphosed limestone
Microgranite	An acidic rock with grain size intermediate between that of granite and rhyolite
Quartzite	A metamorphic rock or sedimentary rock composed almost entirely of quartz grains
Rhyolite	A fine grained or glassy acidic rock, usually volcanic
Sandstone	A sedimentary rock, composed of sand grains naturally cemented together
Schist	A metamorphic rock in which the minerals are arranged in nearly parallel bands or layers. Platy or elongate minerals such as mica or hornblende cause fissility in the rock which distinguishes it from a gneiss
Slate	A rock derived from argillaceous sediments or volcanic ash by metamorphism, characterized by cleavage planes independent of the original stratification
Syenite	An intermediate plutonic rock, consisting mainly of alkali feldspar with plagioclase, hornblende, biotite or augite
Trachyte	A fine grained, usually volcanic, variety of syenite
Tuff	Consolidated volcanic ash

In the current version of BS 812: Part 104,[3] the petrological description of natural aggregates requires that an aggregate should be identified by an appropriate petrological name selected from a given list, as shown in Table 1.3.[3] The geological age of the aggregate should also be given. For

Table 1.4. The main geological ages in chronological order

Geological age	Million years ago
Pre-Cambrian	544 to 4500?
Cambrian	505 to 544
Ordovician	440 to 505
Silurian	410 to 440
Devonian	360 to 410
Carboniferous	286 to 360
Permian	245 to 286
Triassic	208 to 245
Jurassic	146 to 208
Cretaceous	65 to 146
Tertiary	1.8 to 65

example, carboniferous limestone, silurian greywacke or tertiary basalt. A listing of the geological ages, presented in chronological order, is shown in Table 1.4.

Not all rock types make good roadstone. Some may be too soft, some may suffer from durability problems, and others may be inaccessible due to their location or difficulties in obtaining planning permission for a quarry. The range of properties possessed by one group may be different from another. It is also important to recognize that considerable variation may occur within single groups and even within individual quarries. Each of the main types of quarried rock has a different potential performance in a road.

In terms of assessing how an aggregate will perform in use, one of the main problems is variability in the composition of the quarried product. Aggregates are not uniform inert materials. They are rarely uniform and tend to be differentially affected by a range of stressing forces once quarried and used as an aggregate. It is suggested that if specifications recognized that different types of aggregate behave in different ways and have distinct types of problem, then the likelihood of premature failure could be significantly reduced.

1.5. Important studies relating petrology to aggregate properties

When considering rock types for use as aggregate in asphalts, the selection process in the UK has traditionally been influenced by the property of skid resistance. This has arisen from a history of research into the relationship between accident rates and skid resistance and has resulted in the UK demanding much higher standards of skid resistance from its aggregates than those which are required elsewhere in the world.

Two important parameters associated with skid resistance are the Polished Stone Value (PSV) and the Aggregate Abrasion Value (AAV). The PSV is a measure of the ability of an aggregate to resist polishing under the action of traffic without losing the rough surface which is presented to tyres by the individual grains, for example a gritstone will have a relatively high PSV. The AAV is a measure of the ease with which an aggregate will abrade, for example a chalk will have a relatively low AAV.

There have been a number of surveys undertaken over the last century which relate aggregate geology to its suitability as a source of roadstone. The earliest attempts to relate sources to aggregate properties have been reported by Lovegrove[4] and Lovegrove et al.[5,6] By 1929, Lovegrove had amassed data on over 400 sources of aggregate in the British Isles. Although Lovegrove could not measure skid resistance, his wet attrition apparatus was able to rank many of the aggregates which remain in use today. Unfortunately, certain sources of high PSV aggregates which enjoy substantial contemporary use did not fare well in his research as they were susceptible to high rates of wear.

Another notable investigation was carried out by Knill[7] who recognized the relationship between petrology and skid resistance. Her work showed that gritstone aggregates possessed the highest values and proposed that this was due to plucking, i.e. the progressive removal of grit-sized grains from a softer matrix resulting in a consistently rough surface. This may also support the view that high skid resistance characteristics are accompanied by poor values for other parameters, e.g. AAV.

Probably the most important period of work into understanding aggregate properties was undertaken by the Transport & Road Research Laboratory (now Transport Research Laboratory (TRL)) in the late 1960s and 1970s. This was primarily due to the introduction of standards for skid resistance and the need to identify suitable sources of aggregate to meet the prescribed limits. This work examined both natural and synthetic aggregates where significant fundamental research was carried out into understanding the factors that influence PSV characteristics.

A major investigation of arenaceous (i.e. containing sand) aggregates was carried out by Hawkes and Hosking.[8] This work reaffirmed their importance as an aggregate suitable for use in the manufacture of roadstones.

Private sector aggregate and construction interests have also actively pursued their own surveys and investigations towards identifying new sources of aggregate. An important development in the late 1980s was the creation of coastal super quarries. One of the first was Glennsanda Quarry on the west coast of Scotland. This, however, is a granite quarry whose product does not possess the levels of skid resistance which are required for the UK market. Being a coastal quarry, the scale of operations has allowed the aggregate to be sold in bulk to the south of England, Europe and America.

In the late 1980s, it became apparent that very little was known or being done outside the private sector to ascertain available resources. This was particularly true in relation to the geographical distribution and the technical properties of rock types that may meet the required specification requirements. It was recognized that this was causing major problems, particularly in demonstrating the need for additional consented reserves and for mineral planning authorities in making properly informed decisions with regard to the consent for new reserves and possible sterilization of resources.

Table 1.5. Summary of HSA aggregate distribution in the UK[9]

Location	PSV 68+	PSV 63–67.9	PSV 58–62.9	Total number
England	1	7	13	21
Wales	6	3	6	15
Scotland	0	1	19	20
Northern Ireland	0	7	14	21
Totals	**7**	**18**	**52**	**77**

In the early 1990s, the Department of Transport renewed its interest in sources of aggregates with high PSV characteristics. This led to the Travers Morgan survey in 1993, reported by Thompson *et al.*,[9] which aimed to '*collate, review and enhance the present knowledge of existing and potential sources of High Specification Aggregates (HSA) within the whole of the United Kingdom, and to assess the current and potential future supply of these materials in relation to existing and anticipated future levels of demand*'.

A summary of the number of sources which were identified in this work is shown in Table 1.5. A total of 76 sources of HSA were identified with each area of the UK having approximately the same number. However, there were only seven sources which could be classed as being of the highest grade and six of these were in Wales.

1.6. Sources of high quality aggregates in the UK

The geology of the UK includes a wide spectrum of rock types. However, in terms of their quality as aggregate for use in asphalts, many potential sources are unsuitable. Thompson *et al.*[9] found that there was an even distribution of HSA sources within the four areas of the UK. However, these are not evenly distributed and, for example, there are no sources of high quality aggregates in the south east of England where demand is greatest. Indeed, the main reserves of aggregate are centred in the south west, Midlands and north west of England, Wales, central Scotland and Northern Ireland. This necessitates expensive haulage and has provided some impetus to recycling. The following is a summary of the geological distribution of sources (after Thompson *et al.*)[9]

- *North west England*. Aggregate is quarried from greywacke sandstones of silurian/ordovician age and the Borrowdale volcanics group.
- *The Midlands*. These include igneous rocks, e.g. Nuneaton Ridge diorite, Caldecote volcanics formation, Charian volcanics sequence of pre-Cambrian age. Other sources include pre-Cambrian fluvial sandstones and indurated sandstones and mudstones of the Myton Flafs formation.
- *South west England*. These include sandstones of carboniferous and devonian age and blue elvan dolerites.
- *Wales*. These include some of the highest PSV sources in the UK and quarry dolerites, the Stapely volcanic formation, silurian and ordovician greywackes, Longmyndian Wentor group sandstones and Pennant sandstones.

- *Scotland*. Quartz dolerite from the Midland Valley sill and silurian and ordovician greywackes are the most important sources.
- *Northern Ireland*. The main sources of aggregate are found within silurian and ordovician greywackes and tertiary basalts.

1.7. Use of aggregates for asphalts around the world

A wide range of aggregates is used around the world to manufacture asphalts. However, it is beyond the scope of this Chapter to discuss these in detail, other than to mention some of the main factors involved. Obviously, ease of availability is the single most important factor in dictating usage. For example, Northern Ireland has approximately 100 hard rock quarries whilst Holland has none. Contrast Iceland, where the only available rock type is basalt, with the UK which is endowed with almost every type of rock spanning geological time. Factors such as availability, distance of source to market, climatic conditions, national specification requirements and choice of asphalts all combine to make the selection of aggregate around the world a complex subject. Europe may be broken up into a number of distinct areas.

- Possibly due to its Atlantic climate, the UK has a wide range of available rock types. Historically, great importance has been placed on skid resistance. This combination has favoured the use of high PSV gritstone and sandstone aggregates as surfacing aggregates with a minimum level of PSV specified for all other asphaltic materials.
- Scandinavian countries, with a combination of very cold winters and the need to use studded tyres, have favoured the use of very hard igneous and metamorphic aggregates to withstand these climatic and trafficking conditions.
- Mediterranean countries possess, as their main rock type, limestones with poor strength and low levels of skid resistance. Heavy traffic and warm conditions coupled with the use of limestone in dense asphalts of low texture depth and skid resistance has resulted in surfacings which are extremely dangerous when wet.
- France and Germany tend to favour the use of hard wearing aggregates in thin rut-resistant surfacing mixtures.

It can be seen from the short description above that the availability of aggregates, their properties and local conditions have affected the use and choice of asphalts across Europe. A similar picture occurs across the world. North America has a vast highway structure in which local materials are used—the import of aggregates, as in the south east of England, would be too costly. The main types of aggregate used include limestones, basalts, dolerites, granites and sandstones. Their selection has typically been assessed using the Los Angeles Abrasion Value. Unlike the UK, in the USA there was little, if any, interest in skid resistance characteristics.

One of the most economically active areas in Asia is Hong Kong. However, it has problems with its aggregate supplies. The main rock type is granite, which tends to suffer from alteration and weathering problems. Indeed, being able to secure a source of aggregate for a large construction job is a major benefit when tendering. Many countries within tropical areas suffer from deep weathering and alteration of potential aggregate sources due to their climate. In southern Australia and New Zealand, alteration of basalt and greywacke aggregates is also a problem and has prompted considerable research into the understanding of this process. Research into 'local' problems has resulted in the development of many commendable test methods which can highlight aggregates which may have problems. However, the opposite can also be the case. The use of UK or US specifications and methods in developing countries may not take due account of local materials and conditions.

Despite the geological diversity of aggregates around the world, an engineer who is aware of inherent problems associated with particular rock types and approaches the selection process by predicting the likely in-service performance should avoid costly premature failure.

1.8. Basic problems with the way that aggregates are tested

There are a number of basic problems that require consideration prior to the selection and testing of aggregate for use in an asphalt.

- It is a common view that aggregates are inert materials that are unaffected by external influences. Aggregates are rarely inert and may be subject to rapid changes over engineering time, i.e. years, months and days. For example, certain aggregates can lose 50% of their dry strength when wet. Similarly, some aggregates lose their skid resistance characteristics during the summer followed by renewal in the winter. Some lose their ability to adhere to bitumen due to moisture, or suffer crushing or abrasion under traffic.
- The reliance on a national specification which does not recognize the inter-relationship between factors such as differences in rock type, use and performance in service.
- The limitations of test methods to recreate conditions in use, e.g. the confined compressive loading of the strength tests.
- Assessment of single sizes whilst a range of aggregate sizes is typically blended together to produce an asphalt.
- The use of the 10/14 mm size for the standard strength and abrasion tests whilst skid resistance is assessed using 6/10 mm sized aggregate.
- There is no British Standard method to measure the compatibility of aggregate and bitumen and the effect of moisture on this parameter.
- New developments in surfacing materials use aggregates in ways and in conditions where there is little or no previous long term experience. For example, in terms of skid resistance the majority of sideways friction coefficient (SFC) data relates to a single group of aggregates (gritstones) used as 20 mm precoated chippings in hot-rolled asphalt (HRA) wearing courses.

The above list highlights just some of the basic problems and limitations of methods currently used to assess aggregate. The implications of these should be fully understood and accepted.

1.9.   Aggregate properties relevant to asphalts

The modern asphalt is a composite material consisting of aggregate in varying sizes and proportions held together by a bituminous binder. Despite advances in technology, its level of performance is restricted because of a lack of understanding of how the constituents react together under the conditions prevailing within a pavement. Ideally, there is an elaborate series of factors which act together to produce a successful structure.

In practice, however, the level of performance that is achieved may be a function of its least understood factor. Addressing this area is likely to have the greatest effect on overall performance. It is important that the weak link is identified. For example, there is no value in using an aggregate with a very high level of skid resistance if it abrades quickly.

Traditionally, personal experience has been the best way of selecting suitable sources of aggregate. One of the first attempts by an Irish engineer to list the required factors was undertaken by Dorman[10] who, in 1910, proposed that a roadstone should principally possess the following properties

- hardness, toughness and the ability to withstand disintegration from atmospheric and chemical action
- absence of mud or dust, and particularly very adhesive mud, which, owing to its chemical composition, injuriously affects clothing, paint, etc.
- a good foothold under all weather conditions
- sufficient binding properties to prevent it ravelling or breaking up in dry weather
- a tendency to break into a cubical shape when being converted into road metal
- consistency of texture, so as to wear evenly with a smooth, uniform surface (this seems contradictory but, given the date of publication, is believed to have meant a uniform product which contained no weathered aggregate which is likely to wear away quickly).

It is interesting to note that this list, proposed some ninety years ago by an experienced practical engineer, is basically the same as would be used today to predict performance. Nowadays, traditional experience and local knowledge remain important tools in the selection of aggregate. However, by definition, assessment on such a basis requires long periods of time and is unsuitable for the modern situation where quality aggregate is often imported considerable distances and used in different ways and exposed to in-service conditions which are dissimilar to those where it originated. Such an approach also lacks objectivity.

Accordingly, it was quickly realized that acceptance should be based on quantifiable data obtained by standardized test methods. In response, many different types of assessment have been developed and adopted throughout the world. One of the first to provide quantified data in the UK was the work of Lovegrove et al.[5] who identified the performance capability of aggregate based on the type of rock. Numerous engineers including Knight,[11] Hartley,[12] Lees and Kennedy,[13] Collis and Fox,[14] Bullas and West[15] and

Table 1.6. A summary of current Highways Agency requirements for aggregates[16]

Property	British Standard Test	Limiting value
Size	Sieve analysis (dry or wet)	
Shape *Flakiness* *Elongation*	Flakiness Index (FI) Elongation Index (EI)	<35 <35
Strength *Impact* *Crushing*	Aggregate Impact Value (AIV) Ten Per Cent Fines Value (TFV)	<30 >140 kN
Abrasion	Aggregate Abrasion Value (AAV)	<16
Polishing	Polished Stone Value (PSV)	>58
Soundness	Magnesium Sulphate Soundness Value (MSSV)	>75

Thompson *et al.*[9] have since listed the properties that should be possessed by the ideal surfacing aggregate. A summary of the current Highways Agency's requirements for aggregates[16] is shown in Table 1.6.

1.10. Current test methods for assessing aggregates

BS 63, *Sizes of Broken Stone and Chippings,*[2] was published in 1913 as the first British Standard specification for road-making aggregates. This evolved into BS 812 which was first published in 1938 and subsequently revised in 1943, 1951, 1960, 1967 and 1975. Since the mid 1980s, the 1975 Standard has been revised as individual parts containing single test methods. Many of the tests which are now found in BS 812 can trace their history back to the original 1938 Standard, although some have earlier origins. This is an important point to consider. In terms of testing aggregate there has not been any major shift in emphasis in the methods since the first edition was published in 1938. Test methods which originally assessed single pieces of material have been replaced by test samples of single size crushed rock aggregate. Until the inclusion of the Magnesium Sulphate Soundness Value (MSSV) Test[17] in 1989, the only other notable addition was the PSV test in 1960.

Despite recent revisions, there are particular problems associated with the methods that are currently specified. Examples are: the assessment of samples consisting of single size material rather than those which are graded; failure to recognize the effect of water, i.e. the dry assessment of strength and abrasion resistance; the use of 10 mm aggregate for skid resistance; and the period of three weeks which is necessary for a soundness test. Such information is fundamental to the successful prediction of aggregate performance in an asphalt.

1.11. Summary of main aggregate testing requirements

The responsibility for managing and maintaining motorways and other trunk roads in England falls to the Highways Agency. In Scotland, the equivalent organization is the Scottish Executive (formerly The Scottish Office), in Wales the Welsh Office and in Northern Ireland the Department of the Environment. The motorway and trunk road network occupies some 4.2% of the total road network in the UK. Responsibility for the balance of the network rests with Local Authorities.

The requirements of the Highways Agency etc. for construction and maintenance of their roads are contained in the *Manual of Contract Documents for Highway Works* (MCHW).[18] It consists of seven volumes, numbered volume 0 to 6 inclusive. Volume 1 constitutes the national specification and is entitled the *Specification for Highway Works* (SHW).[16] Guidance on the use of the SHW[16] is contained in the *Notes for Guidance on the Specification for Highway Works* (NGSHW).[19] A companion system dealing with design is entitled the *Design Manual for Roads and Bridges* (DMRB).[20] It consists of 15 volumes, numbered volume 1 to 15 inclusive. The most important volume for those concerned with asphalt technology is volume 7, entitled *Pavement Design and Maintenance*.[21] Most Local Authorities adopt both the MCHW[18] and the

Table 1.7. Typical testing requirements

SHW clause	Type of material	Property to be assessed	Test	Frequency of testing
901, 925	Aggregate for bituminous materials	Hardness	TFV (N) AIV (N)	Monthly Monthly
926		Durability	MSSV (N) WA (N)	1 per source as required
		Cleanness	Sieve test (%<0.075 mm) (N)	Monthly
	Crushed rock, gravel	Shape	FI (N)	Monthly
	Coarse aggregate for wearing courses	Skid resistance	PSV (N)	1 per source
		Abrasion	AAV (N)	1 per source
915, 925	Coated chippings	Grading	Grading (N)	1 per stockpile
		Binder content	Binder content (N)	1 per stockpile
		Shape	FI (N)	1 per source
		Skid resistance	PSV (N)	1 per source
		Abrasion	AAV (N)	1 per source

Notes:
All tests to be carried out by the contractor.
The frequency of testing is given for general guidance and is only indicative of the frequency that may be appropriate.
(N) indicates that UKAS (United Kingdom Accreditation Service) sampling and a test report or certificate is required.

DMRB[20] to provide standards for construction and maintenance of roads under their care. All of these documents are essential references for anyone involved in road construction in the UK. The MCHW[18] and the DMRB[20] must be read in combination as information is often spread between both systems.

The MCHW[18] requires contractors to carry out certain tests on aggregates. A summary of the main requirements is shown in Table 1.7.

The NGSHW[19] contains a list of tests and suggested frequencies. The contents of Table 1.7 are based on this list. These tests may be carried out on behalf of the contractor by a testing laboratory, manufacturer or supplier.

It should be noted that the SHW[16] states that the specified testing is not exhaustive and other tests may be required, giving the engineer the opportunity to introduce these if he sees fit. The frequency of testing is given as a general guidance to that which may be appropriate depending on contract size, location, time for completion and the existence and nature of Quality Assurance (QA) systems.

It is the policy to require the use of testing laboratories accredited for certain tests by the United Kingdom Accreditation Service (UKAS). (The United Kingdom Accreditation Service (UKAS) is the UK national accreditation body including the national accreditation of testing laboratories in the UK.)

Again, a fundamental problem with these testing requirements is that many of them relate to the amount of material that arrives on site, which often varies on a daily basis. If the aim is to predict performance, quick simple methods such as visual comparison with acceptable samples of aggregate would improve this problem.

1.12. Main test methods used to assess the properties of aggregates

In the UK, the test methods which are typically specified for the assessment of aggregates are found in BS 812. A summary of the main test methods used is shown in Table 1.8. This indicates the name of the test method, a common abbreviation, the property measured, the appropriate Part of BS 812 and typical specification limits.

1.12.1. Sampling

Sampling is a very important aspect of assessing aggregate. The most thorough testing regimes can only report on the material which is tested. If the sample is not representative then the testing is irrelevant. Sampling of asphalts and their constituents including aggregates is discussed in Chapter 9. As demonstrated within Table 1.7, many of the testing requirements are for source approval, that is test certificates that are representative of the source of aggregate to be used are provided by the supplier. However, in terms of performance in service it is very important to consider variability within a given source and the need to ensure that the data provided relate to the aggregate which is actually being supplied to the site. For example, consider the values of PSV as shown in Fig. 1.1

Table 1.8.   Summary of main BS 812 test methods used to assess aggregates

Test method	Abbreviation	Property measured	BS 812 method	Typical limit
Sampling	—	Representative sample	Part 102	—
Sieve analysis	SA	Size distribution	Part 103 Section 103.1	Depends on use
Flakiness Index	FI	% flaky aggregate	Part 105 Section 105.1	<25%
Elongation Index	EI	% elongated aggregate	Part 105 Section 105.2	<25%
Aggregate Impact Value	AIV	Resistance to sudden impact	Part 112	<30%
Ten Per Cent Fines Value	TFV	Resistance to compressive loading	Part 111	>160 kN (dry)
Aggregate Abrasion Value	AAV	Resistance to dry abrasion	Part 113	<10, <12, <14, <16
Polished Stone Value	PSV	Skid resistance	Part 114	50, 55, 60, 65, 68+, 70+
Magnesium Sulphate Soundness Value	MSSV	Soundness	Part 121	>75%

within four different greywacke quarries. It is apparent that a considerable range occurs within each source. It is quite possible that other aggregate parameters will also display similar variations.

This potential for variability within a single source must be acknowledged. It illustrates the importance of a sampling procedure which provides a true indication of the quarried product. Further information on sampling is included in Section 9.2 of Chapter 9.

1.12.2.   Size distribution

Determination of the aggregate grading is carried out in accordance with BS 812: Part 103, Section 103.1.[22] This is a fundamental property which governs how an aggregate will perform. In simple terms, any asphalt is a particular grading with a specified amount of a particular bitumen.

Figure 1.2 shows the variation in grading for a number of different asphaltic materials. These range from asphalts such as HRA and stone mastic asphalt (SMA) to the predominantly single size surface dressing chippings. The 10/14 mm and 6/10 mm single sizes which are assessed in the laboratory are also shown. As can be seen, the laboratory tests are

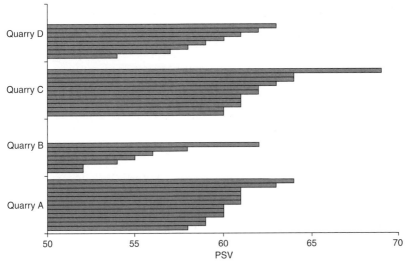

Fig. 1.1.   Range of PSVs from bulk samples collected at four greywacke quarries

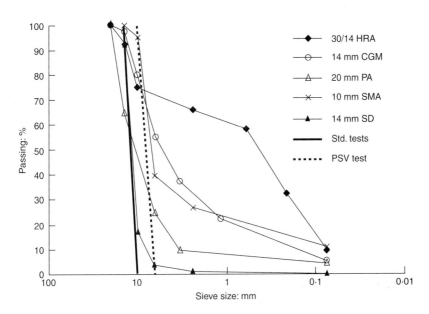

*Fig. 1.2.   Variation in aggregate size. 30/14 HRA, Column 21 hot rolled asphalt;
14 mm CGM, 14 mm close graded wearing course macadam; 20 mm PA, 20 mm
porous asphalt; 10 mm SMA, 10 mm stone mastic asphalt; 14 mm SD, 14 mm
surface dressing chippings; std. tests, 10/14 mm aggregate size assessed for
standard tests; PSV test, 6/10 mm size assessed for PSV test method*

based on only a small range of the entire grading and the finer aggregate sizes are not considered. This bias may be reflected in a result which does not truly represent the material under test.

1.12.3.   Relative density and water absorption
BS 812: Part 2[23] includes three different ways of describing aggregate density — oven dried (odRD), saturated surface dry (ssdRD) and apparent (appRD). Water absorption (WA) is expressed as the difference in mass before and after drying at $105\,°C \pm 5\,°C$ for 24 hours. Several methods are specified and use either a wire basket, gas jar or pycnometer to hold the sample. The choice of container depends on the grading of the test sample.

The gas jar method is used for 10/14 mm crushed rock aggregate. Duplicate test samples of 1000 g are washed to remove dust and immersed in water in a gas jar at $20\,°C \pm 5\,°C$ for $24\,h \pm \frac{1}{2}h$. Trapped air is removed by gentle agitation soon after immersion and at the end of the soaking period. The container is overfilled and a glass plate slid over the mouth, ensuring that no air is trapped. The outside of the container is dried and weighed (mass *B*). The container is emptied, refilled with water sliding the glass plate into position as before, dried and reweighed (mass *C*).

The aggregate is placed on a dry cloth, gently surface dried so as to have a damp appearance, and weighed (mass *A*). It is then heated in an oven at $105\,°C \pm 5\,°C$ for $24\,h \pm \frac{1}{2}h$, cooled in an airtight container and weighed (mass *D*). The four masses obtained are used to determine the following values.

$$\text{Relative density on an oven dried basis} = \frac{D}{A - (B - C)} \qquad (1.1)$$

$$\text{Relative density on a saturated surface dry basis} = \frac{A}{A - (B - C)} \quad (1.2)$$

$$\text{Apparent relative density} = \frac{D}{D - (B - C)} \qquad (1.3)$$

$$\text{Water absorption} = \frac{100(A - D)}{D} \qquad (1.4)$$

Relative density is a commonly used parameter in highway engineering. Its value is often specified as a requirement when dealing with asphalts and their mix design since it relates to the volume of an aggregate and hence the area of road which is covered per unit volume of material. Relative density on a saturated surface dry basis is also used in the calculation of the AAV to adjust for the loss of volume.

High values of WA are generally believed to be indicative of durability problems such as frost susceptibility or soundness. In the SHW,[16] an aggregate source should have a MSSV >75%. However, in recognition of the time taken to provide an MSSV result, the WA test is specified as a means of routine testing with a maximum permissible limit of 2%. Values greater than this necessitate the aggregate being assessed by means of the MSSV Test.[17] However, based on 296 sets of data for a wide range of rock types, Woodward[24] produced the equation

$$MSSV = 100.23 - 9.431 \, WA \; (R = 0.8) \tag{1.5}$$

where R is the Correlation Coefficient.

Equation (1.5) relates the two properties but it was found that using a 2% limit for WA was appropriate for the superseded limit of 82% for the MSSV and not the current maximum of 75%. When other test methods were correlated it was found that all of the predicted values varied depending on rock type and that the dry strength test limits (usually specified in the SHW[16] were too low if they were to relate to a WA of 2%.

1.12.4.   Cleanliness

Dust can cling to larger sizes of aggregate and the cleanliness of an aggregate is a measure of this parameter. Typically, it is determined as that proportion which is less than a specified value (usually 0.075 mm or 0.063 mm). It is a very significant parameter since the presence of dust will inhibit binder sticking to the surface of the aggregate. It is important in a wide range of aggregate uses ranging from surface dressing chippings to asphalts. If the bitumen does not fully adhere to the aggregate, water may then penetrate the mixture or the aggregate and cause the material to fail either immediately or with the passage of time.

Not only is the quantity of dust important, but its properties must also be considered. It is possible to have relatively inert types of dust which do not react to the presence of moisture. Alternatively, certain types of dust, particularly those containing expansive clay minerals are very damaging even if present in small amounts. These will expand in the presence of moisture and physically break the aggregate/bitumen bond causing failure of a surface dressing or an asphalt.

There is no method prescribed in BS 812 that may be used to highlight harmful types of dust. However, it has been found that the Methylene Blue Value (MBV) Test can be used to measure this property.[25] Different versions of the method are available. One of the simplest has now been adopted by the American International Slurry Seal Association.[26] In this test, 1 g of dust is added to 30 ml of water. Increasing amounts of methylene blue are added. The amount of dye which is added is directly proportional to the detrimental effect of the clay minerals which are present. Another version of this method is being considered as a European Standard.

1.12.5.   Shape

The shape of an aggregate particle is important as it affects both testing in the laboratory and its performance in service. The need to classify shape has long been recognized.[27–29] The last of these papers described the development of a method to quantify shape on the basis of five categories: rounded, sub-rounded, curvilinear, sub-angular and angular. Zingg[30] described four shape classifications based on the measurement of length, breadth and thickness. These compared

- thickness to breadth — a flatness ratio — and
- breadth to length — an elongation ratio.

Using values for these ratios of 0.66, four basic shapes were proposed, i.e. discs, equi-dimensional, bladed and rod-like. In 1936, Markwick[31] developed a method which was to become the British Standard method for classifying chipping shape. This proposed that shape could be defined in terms of thickness and length ratios which were independent of the size of the material. This method was considered to be adequate for practical purposes. Gauges were developed to classify particles according to their length and thickness. These enabled a graded sample of aggregate to be classified into a number of groups composed of stones which were broadly similar in shape. Since then, Markwick's original method[31] has changed relatively little although the BS method is considered inadequate by some authorities on the grounds that it does not give a true indication of aggregate shape. Lees[32] proposed a method for the complex shape of crushed rock particles. This proposed that shape parameters should be independent of size and take the form of ratios on the three axes — greatest dimension, intermediate diameter and least diameter. The following is a brief summary of the effects of shape on the testing of aggregate in the laboratory.

- As the proportions of flaky and elongated chippings increase so the strength of aggregate decreases. Strength is measured by means of the Aggregate Impact Value (AIV) Test,[33] the Aggregate Crushing Value (ACV) Test[34] and the Ten Per Cent Fines Value (TFV) Test[35] (all discussed in Section 1.12.6 below). Inclusion of flaky and/or elongated chippings in a sample may give strengths which are 50% of the values which would have been obtained if these fractions had been removed.
- In order to give an impression of apparent strength, a supplier may be tempted to select samples which have lower values of flakiness or elongation than is actually the case.
- Both the PSV and AAV Test[36,37] methods require that, prior to testing, the sample is deflaked on the next largest sieve size, i.e. the size above the nominal stone size of the sample. This may enhance the apparent quality of the test sample.

The above explains why aggregate testing may result in a prediction of performance that is not actually achieved in service. Engineers should be aware of the influence of shape on performance. The following summarizes the key issues.

- A cubic shape is required to ensure texture depth in HRA and surface dressings. However, such chippings may fail prematurely due to inadequate embedment. This may be overcome by increasing the rate of spray of the bitumen or by a double chipping or a racking in system which locks the cubical chippings in place.

- Chippings which are flaky and elongated will have a greater tendency to crush under the roller or due to trafficking. In a mixture with a high stone content such as SMA or a thin surfacing, this tendency to crush will also be a problem due to the stressing at points of stone-to-stone contact.

Clearly, the shape of an aggregate has an important influence on performance. There is a substantial case for improving the way that shape is measured and any such changes will have a significant influence on the routine methods of aggregate assessment both in the laboratory and in service.

1.12.6. Strength

Most methods used to assess strength have a common ancestry. The original British Standard tests on roadstone — crushing strength, attrition, abrasion and impact methods — were designed to evaluate stone setts or water-bound macadam trafficked by horse drawn vehicles having wheels with steel rims. These original methods have changed little despite the fact that modern pavements are subjected to a type of traffic which is considerably different to that experienced in Victorian England.

Current methods of assessing strength are quite simple and are contained in several Parts of BS 812 (see Table 1.8 for details of the main tests). All these methods require a sample of dry aggregate of size 10/14 mm and express a value of strength based on determining the amount of fine aggregate produced after impact or crushing. There are three parameters which are typically obtained

- Aggregate Impact Value (AIV)
- Aggregate Crushing Value (ACV)
- Ten Per Cent Fines Value (TFV).

Aggregate Impact Value

The AIV gives a relative measure of the resistance of an aggregate to sudden shock or impact. This type of testing dates from the Page Impact Test used to assess rock cores and first included in BS 812 in 1938.[38] This method was modified in the mid-1940s and became adopted as the AIV now found in BS 812: Part 112.[33] The main development in the Stewart Impact Test was a change from a single prepared core sample to a test sample of crushed rock aggregate particles which were held in a rigid container. It quickly became apparent that by assessing crushed aggregate using this portable equipment, a much better indication of quality could be predicted. Indeed, the basic Stewart Impact Test has changed relatively little over the last 50 years.

In the BS 812: Part 112[33] method, a test sample of single size dry 10/14 mm aggregate is placed in a steel container and subjected to 15 blows from a 13.5 kg weight dropped from a standard height. The resistance to sudden impact is assessed by sieving the test sample using a 2.36 mm test sieve and determining the mass of material passing as a percentage of the original sample weight. Two samples are assessed and the AIV expressed as the mean of the two results.

Although the AIV method has been in BS 812 for many years, it has generally not been used as a national specification requirement. Where it has been used, a value of <25 has typically been regarded as a suitable requirement. However, following the work of Bullas and West,[15] the AIV test was included in the SHW[16] with the acceptable maximum limit lowered to 30 for aggregates in asphalts.

Aggregate Crushing Value
The ACV gives a relative measure of the resistance of an aggregate to crushing under a gradually applied compressive load. The early British Crushing Strength Test[38] used an aggregate core which was similar to that employed in the Page Impact Test. This was subjected to a direct compression load applied to both ends of the sample. The value of the load which caused initial cracking and that which caused crushing to occur were noted. However, numerous problems existed due to the variation in possible strength in relation to planes of weakness. This resulted in the adoption of a method which assessed samples of crushed rock aggregate in sizes which were similar to those used in road construction. In Germany a crushing test using a steel cylinder containing crushed aggregate had been adopted as early as 1935. In the 1943 revision of BS 812, a British version appeared alongside the testing of stone cylinders as the method to evaluate the mechanical strength of coarse aggregate. This was, in effect, the ACV Test. The British method expressed the result as a percentage of fines produced. In the 1951 revision of BS 812, the ACV Test on prepared cores had been dropped. The remaining version of the ACV Test has remained relatively unchanged since that time. The British method was developed by Markwick and Shergold.[39] Their work gave a valuable insight into the testing of aggregate strength and showed the following.

- The smaller the aggregate size, the stronger it is.
- The range of test values varied with type of rock, e.g. 10 to 20 for hard igneous rocks, quartzite, flints and sandstone; 20 to 30 for limestone; and greater than 25 for slag.
- Drying prior to testing increased the value of strength obtained. This was found to be dependent on the type of rock with very little difference occurring for different varieties of limestone to more than 40% for some igneous rocks.

In the current BS 812: Part 110[34] method, a test sample of single size dry 10/14 mm aggregate is placed in a steel container and subjected to slowly applied compressive load of 400 kN for a period of 10 min. The resistance to crushing is assessed by sieving the test sample through a 2.36 mm test sieve and determining the amount of material which passes as a percentage of the original mass of the sample. Two samples are assessed with the ACV expressed as the mean of the two results. There is no requirement in the current SHW[16] for an aggregate to possess a particular minimum value of ACV. However, a value of <25 is typically regarded as being appropriate.

Ten Per Cent Fines Value
The TFV is the load, expressed in kN, required to produce ten per cent of fine material when subjected to a gradually applied compressive load. Traditionally, the AIV and ACV Test methods were used to assess the strength of British aggregates. However, occasionally, anomalous results occurred during the testing of weaker aggregates. This problem was investigated by Shergold and Hosking[40] and resulted in the development of the TFV Test which was included in the 1960 version of BS 812.

The equipment used in the TFV Test[35] is similar to that which is used to ascertain the ACV. A test sample of single size dry 10/14 mm aggregate is placed in a steel container. However, the TFV Test[35] differs in that the load applied during testing is not constant but varies according to the expected strength of the aggregate. Thus, a weak aggregate is subjected to a lower load than that which is applied to a strong aggregate. The reason for this approach is due to the finding that, during ACV testing, low-strength aggregates absorbed the load and thus, did not give a true indication of their strength Experimentation established that if the load was varied with the aim of producing 10% fine material (i.e. that which passes a 2.36 mm sieve) then this gave a better indication of strength. The load to be applied is first determined using the AIV of the aggregate from the equation

$$\text{required load} = \frac{4000}{\text{AIV}} \text{ kN} \tag{1.6}$$

This load is then used to assess the aggregate sample using the ACV apparatus. The percentage of fine material produced is determined using a 2.36 mm test sieve and, for the load used, should be within the range 7.5 – 12.5%. If it is outside this range then the applied load is adjusted accordingly. The TFV is determined using the equation

$$\text{TFV} = \frac{14 \times \text{applied load}}{(\text{percentage of fines} + 4)} \text{ kN} \tag{1.7}$$

It is possible to have results ranging from 400 kN for the strongest aggregates to 20 kN or even less for the weakest materials. The TFV Test[35] is frequently used as the main specification requirement for aggregate strength. The SHW[16] specifies that aggregate for use in asphalts must have a dry value >140 kN whilst unbound sub-base aggregates must have a soaked value of >50 kN. These different methods frequently cause misunderstandings.

Figure 1.3 shows the differences in TFV levels which are returned on wet and dry samples of the same aggregate from a number of different sources. Wet aggregate is substantially weaker than the dry equivalent.

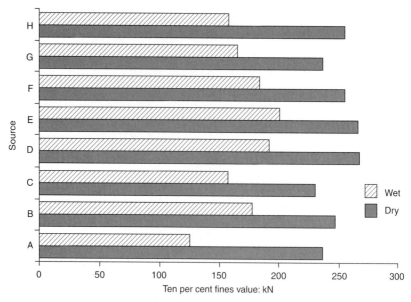

Fig. 1.3.   Wet TFV strength loss for limestone aggregates

engineers must be aware that an event as common as a shower of rain may affect the performance of an aggregate. This phenomenon does not feature in the SHW[16] yet the data shown in Fig. 1.3 demonstrate the necessity of considering this aspect of wet strength loss. This is particularly worthy of note given the popularity of porous or semi-porous asphalts.

Aggregate Abrasion Value
The AAV is a measure of the resistance of an aggregate to surface wear by dry abrasion. Aggregate which is exposed at the road surface must be resistant to wear caused by trafficking. Texture depth is a measure of the roughness of a wearing course and is required to facilitate the removal of water. The higher the texture depth, the lesser the distance required for a vehicle to brake in wet conditions. Some aggregates exhibit good texture depth characteristics initially, only to abrade away under traffic. The possession and retention of texture depth has, in the UK, long been recognized as being crucially important to the performance of a surfacing.

In the UK, the assessment of aggregate abrasion has changed little over the last 100 years. Dry abrasion resistance has been determined using a grinding lap and abrasive. The original Dorry Abrasion Test assessed two cylinders of rock, similar to those of the Page Impact and Crushing Tests. After weighing, they were pressed against a cast steel disc with a force of $250 \, \text{g/cm}^2$ whilst rotating in a horizontal plane at 28 r/min. This continued for 1000 revolutions. Crushed quartz was initially used as the abrasive. This was later standardized to Leighton Buzzard silica sand. After testing, the amount of material abraded was determined. The Coefficient of Hardness was then calculated from the formula

$$\text{Coefficient of Hardness} = 20 - \frac{\text{loss in weight}}{3} \qquad (1.8)$$

Investigation of this original Dorry Abrasion or Hardness Test resulted in the Aggregate Abrasion Value (AAV) Test method being developed in 1949. Although this method employed a similar abrasion machine the test was carried out on aggregate chippings rather than cores. In the 1951 revision of BS 812, the Dorry Abrasion Test was superseded by the AAV Test.

The Road Research Laboratory (now TRL) began studying the performance of road surfacing aggregates during the early 1950s. The main aim of this work was to establish the relationship between the skidding resistance of a surface and the PSV of the aggregate used. It was found that the skidding resistance was related to the durability of the aggregate exposed at the surface. This was highlighted in a full scale road trial at West Wycombe in 1955,[41] which looked at the performance of 13 surface dressing aggregates. One of these aggregates had an AAV of 17 and wore away rapidly, resulting in poor skid resistance characteristics. This subsequently led to the inclusion of requirements in specifications for maximum levels of AAV in roadstone aggregates.

In the current AAV Test,[37] the sample preparation is complex requiring the deflaking of a sample of single size 10/14 mm chippings using a 20/14 mm flaky sieve. It has been reported by Woodward[24] that this typically excludes a large percentage of the original bulk sample which was supplied. It was found that for a range of aggregates, only 15% to 30% of the required size and shape remained available for making AAV moulds. This may have significant implications for heterogeneous aggregates. An example would be a high PSV greywacke containing a high percentage of finer grained shale. As shale typically crushes to produce a flaky chipping which is often significantly weaker than the coarser grained greywacke component, its removal during the deflaking process may have a considerable effect on the obtained AAV result. According to Woodward,[24] this phenomenon was not specific to the type of rock but appeared to affect all of those which were assessed. Using the deflaked aggregate, two flat resin-backed specimens are made. Each is made with approximately 23 correctly orientated particles. These are weighed and subjected to 500 rotations on a horizontal steel grinding lap using Leighton Buzzard sharp sand as an abrasive. The AAV is calculated from the mass loss using the formula

$$\text{AAV} = 3 \times \frac{\text{loss in mass}}{\text{saturated dry relative density}} \qquad (1.9)$$

The requirements for AAV in roadstone aggregates are contained in a Part of Volume 7[21] of the DMRB[20] designated HD 36[42] and these are shown in Table 1.9. It can be seen that for increasing numbers of commercial vehicles, the requirements for aggregate abrasion are raised. This is because the aggregate is required to resist the abrading action of

Table 1.9.   Maximum AAV of chippings, or coarse aggregates in unchipped surfaces, for new wearing courses[42]

Traffic in commercial vehicles per lane per day at design life	Under 250	251–1000	1001–1750*	1751–2500	2501–3250	Over 3250
Maximum AAV for chippings	14	12	12	10	10	10
Maximum AAV for aggregate in coated macadam wearing courses	16	16	14	14	12	12
Coarse aggregate for use in porous asphalt wearing courses	12					

* For lightly trafficked roads carrying less than 1750 commercial vehicles per lane per day, aggregate of higher AAV may be used where experience has shown that satisfactory performance is achieved by aggregate from a particular source.

greater numbers of vehicles tyres. More information on AAV can be found in Section 4.10.2 of Chapter 4.

Polished Stone Value

Microtexture is the roughness of a particular type of aggregate. The efficiency of braking by vehicles at low speeds (<50 km/h) is substantially affected by the microtexture of an aggregate. Under the action of traffic, the texture of the aggregate reduces and the resistance to this wear mechanism is measured by means of the Polished Stone Value (PSV) Test.[36] In the UK, this is regarded as one of the most important properties possessed by an aggregate. The current method for assessing PSV is contained in BS 812: Part 114.[36] British interest in skid resistance developed along two separate but connected paths:

• the routine measurement of existing roads and
• the laboratory assessment of an aggregate.

The ability to measure these two parameters has subsequently provided the basis for current skidding standards and has influenced the specification of other properties. A method to measure the skid resistance of a road surface existed before a laboratory system was developed. Early laboratory assessment considered single chippings.

The present day PSV Test[36] evolved from work carried out in the early 1950s by Maclean and Shergold,[43] who started investigating aggregate skid resistance using an accelerated polishing machine. This was the first satisfactory method of quantitatively assessing the polishing resistance of different aggregates. It was originally designed as a research tool to study the different factors that cause wear in a surface dressing. Initial results clearly demonstrated that the machine showed promise in simulating the polishing action of traffic and indicated the important role that grit plays in the test. It was also found that the rate of polishing could be accelerated by inducing slip, which increases stress between the tyre and the aggregate. The PSV Test was first introduced into BS 812 in 1960. Since then, a number of changes have been made and the current version[36] includes revisions to improve the repeatability and reproducibility of the method:

- the use of a solid tyre instead of a pneumatic tyre (during the test, the air inside the pneumatic tyre heated up resulting in an increase in tyre pressure which in turn affected contact areas and rates of polish)
- a new control stone was introduced to replace the original source (Enderby Quarry) which had ceased the production of aggregate
- the introduction of a second set of tests.

The PSV Test[36] is the only standard test that does not employ 10/14 mm aggregate. Instead, it uses single size cubic aggregate that passes a 10 mm sieve and deflaked using a 10/14 mm flaky sieve. This process can have a significant influence in determining how much aggregate remains for testing. Similar to findings in relation to the preparation of samples for the AAV Test,[37] Woodward[24] found that as little as 4–30% of the original 10 mm sample remained for testing — hence, the concern that the material which is used for testing is not truly representative of the aggregate. Such anxieties are exacerbated by the fact that the establishment of the PSV of a 14 mm or 20 mm aggregate is based on the testing of a 10 mm sample.

In the test, approximately 35 to 50 of the deflaked cubic aggregate chippings are placed flat surface down in a curved mould and a resin is poured into the mould to hold the aggregate in place, thus forming a test specimen. Four test specimens are made for a particular aggregate. The PSV test is in two parts.

- 14 test specimens are placed on the wheel of an Accelerated Polishing Machine (see Fig. 1.4). These specimens consist of six pairs of samples for different aggregates plus two control specimens. The test specimens are then subjected to simulated wet polishing conditions for three hours using a coarse emery abrasive and a further three hours using a fine emery abrasive.
- The degree of polishing of each test specimen is then determined using the Portable Skid Resistance Tester (see Fig. 1.5). The test is then repeated. The PSV of each aggregate is then calculated using the average values of the four test specimens.

The PSV is calculated from

$$PSV = S + 52.5 - C \tag{1.10}$$

where S is the mean of the four test specimens and C is the mean of the four control test specimens. HD 36[42] specifies the minimum PSV of chippings, or coarse aggregates in unchipped surfaces for new wearing courses and the relevant requirements are reproduced in Table 1.10.

Fig. 1.4.   The Accelerated Polishing Machine. Photograph courtesy of Wessex Engineering & Metalcraft Ltd

Most motorways require the use of aggregates with a PSV of 60 or 65. There are many quarries in the British Isles producing aggregates of these values. Sites which are slightly more critical require aggregates with PSVs of 68+. Very few sources in the UK can consistently produce aggregates which exhibit such values (there are a few in Cumbria and in

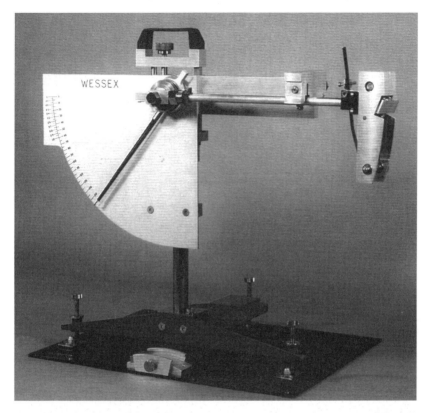

Fig. 1.5.   The Portable Skid Resistance Tester. Photograph courtesy of Wessex Engineering & Metalcraft Ltd

Wales). The high cost of haulage can add significantly to the cost of surfacing works. A PSV of 70+ is required for the situations deemed by HD 36[42] to be the most critical in accident terms. This can only be achieved by means of high friction surfacings such as Shellgrip. Experience has shown that such surfacings are very effective in reducing accidents on sites where traffic levels and skidding risk are high. Their use is controlled by Clause 924 of the SHW[16] and they are discussed further in Section 10.5 of Chapter 10.

The result of this specification is that a single type of aggregate now dominates the surfacing of UK highways.[9] This rock type is given the general term gritstone and includes types such as greywacke, sandstone and volcanic tuff. Within the UK, there are many sources of these rock types. However, they do not occur where demand is greatest, in the south east of England. Rather, the geological distribution limits sources to areas such as South Wales, north west England and Northern Ireland.

High PSV aggregate is therefore a scarce, high quality and valuable product. The situation is not assisted by the introduction of thin surfacings which use higher proportions of aggregates that possess better PSV

Table 1.10. Minimum PSV of chippings or coarse aggregates in unchipped surfaces, for new wearing coarses[42]

Site definitions	Traffic at design life: Commercial vehicles/lane/day									
	0–250	251–500	501–750	751–1000	1001–2000	2001–3000	3001–4000	4001–5000	5001–6000	Over 6000
Motorway (mainline), dual carriageways (non-event)	50	50	50	50	50	55	60	60	65	65
Motorway mainline, 300 m approaches to off-slip roads	50	50	50	55	55	60	60	65	65	65
Single carriageways (non-event) dual carriageways approaches to minor junctions	50	50	50	55	60	65	65	54	65	68+
Single carriageways minor junctions, approaches to and across major junctions, gradients 5–10%>50 m (dual, downhill only), bends <250 m radius >40 mph	55	60	60	65	65	68	68+	68+	68+	70+
Gradients >50 m long >10%	60	68+	68+	70+	70+	70+	70+	70+	70+	70+
Approaches to roundabouts, traffic signals, pedestrian crossings, railway level crossings and similar	68+	68+	68+	70+	70+	70+	70+	70+	70+	70+
Roundabouts*	50–70+	55–70+	60–70+	60–70+	60–70+	65–70+	65–70+			
Bends <100 m*	55–70+	60–70+	60–70+	65–70+	65–70+	65–70+	65–70+			

Notes:
1. Where '68+' material is listed, none of the three most recent results from consecutive tests relating to the aggregate to be supplied shall fall below 68.
2. Throughout this table '70+' means that specialized high skid resistance surfacings complying with MCHW1 Clause 924 will be required.
3. For site categories marked *, a range is given and the PSV should be chosen on the basis of local experience of material performance. In the presence of other information, the highest values should be used.

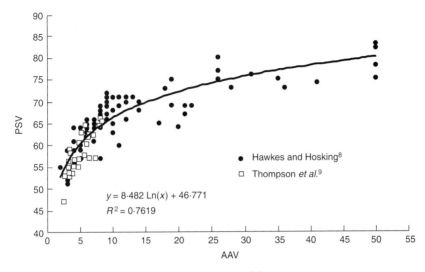

$$y = 8\cdot482 \, \text{Ln}(x) + 46\cdot771$$
$$R^2 = 0\cdot7619$$

Fig. 1.6. Relationship between PSV and AAV[8,9]

characteristics than most other asphalts. However, modern road surfacings should be safe, economical and long lasting. Aggregates which are now specified are premium products which attract high prices. Being a natural material, its useful life has natural performance limits. Yet the performance expectations have never been higher. In addition, aggregates are now being used in mixtures such as thin surfacings, stone mastic asphalt and porous asphalt where there is little traditional experience.

The requirement to maintain performance over time suggests that PSV must be considered in relation to other properties. For example, Fig. 1.6 plots reported values of PSV and AAV reported by Hawkes and Hosking[8] and Thompson *et al.*[9] for a wide range of high PSV Gritstone aggregates. It appears that small increases in PSV correspond to large increases in AAV for higher PSV aggregates, i.e. high PSV aggregate is likely to abrade more quickly than lesser PSV aggregate. Woodward[24] has found this to be the case for other test properties where PSV is gained at the expense of every other test property. Once again, it can be seen that careful consideration is necessary if aggregates are to be assessed properly and such assessment should be based on a range of tests. Further information on PSV can be found in Section 4.10.2 of Chapter 4.

Soundness
Some aggregates, thought to be suitable in terms of their strength, have been found to fail in use. Soundness is a measure of the durability of an aggregate in service. The Magnesium Sulphate Soundness Value (MSSV) Test[17] is used to measure this parameter. A sample of aggregate is subjected to repeated cycles of immersion in a saturated solution of magnesium sulphate followed by oven drying.

The term clean, hard and durable has traditionally been used to describe aggregate suitable for surfacing use. Historically, in the UK, durability has not been accorded the importance warranted by this parameter. This is substantiated by the fact that when the MSSV Test[17] was first included in BS 812 in 1989, it was for source approval only. The method dates from early in the nineteenth century and was included as a requirement in countries throughout the world for many years but features in few modern specifications. In contrast to many other countries, relatively little work has been carried out in the UK into the assessment of aggregate durability.

One of the earliest references to the use of a sulphate test was by the French engineer Brard.[44] The test, described in 1828, originated as a method to recreate the disruptive freezing conditions experienced by stone used for building. As freezing cabinets were not available at the time, this was simulated by soaking aggregate in different types of sulphate salts and assessing the effects of the resulting expansion as it crystallized when drying. A number of salts were tried but sodium sulphate and magnesium sulphate offered the greatest potential. The bulk of the development work on the sulphate soundness test was carried out in the USA.

The standard UK MSSV Test[17] assesses a sample of 10/14 mm aggregate. Test samples are placed in a basket and subjected to five cycles of soaking in a magnesium sulphate solution for about 17 hours followed by oven drying for 8 hours. During the soaking phase, salt solution penetrates any planes of weakness that are present within the aggregate. During the drying phase, the growth of salt crystals within these weaknesses results in internal stressing that induces disruption or breaking of the material. The degree of breaking is determined by sieving the test sample after grading using a 10 mm sieve and calculating the percentage of 10/14 mm aggregate remaining.

The current SHW[16] requirement is that >75% of the mass remains after testing. This was based on work carried out by Bullas and West[15] who assessed acceptable aggregate for use in bitumen macadam roadbase. Table 1.11 summarizes results obtained by Woodward[24] for a wide range of rock types of differing quality. It can be seen that the majority of aggregates assessed would meet the specified limit of >75% retained.

Problems may occur in aggregates containing small amounts of contaminants such as shale or badly weathered basalt. Despite the fact that the proportion is low, these may fail quite rapidly in the MSSV Test.[17] In-service, rapid deterioration may accelerate the failure

Table 1.11.   Basic MSSV statistics[24]

Group	Number	Mean	Min.	Max.	Range	Lower quartile
All	292	84.3	0.0	100.0	100.0	83.5
Basalt	142	77.3	0.0	100.0	100.0	74.1
Gritstone	112	90.8	0.0	99.2	99.2	89.1
Limestone	19	87.0	30.9	98.7	67.8	82.0

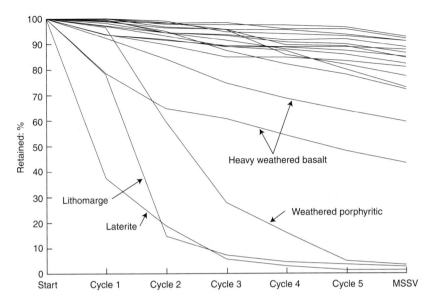

Fig. 1.7. The progressive breakdown of basalt aggregates during the five cycle MSSV Test[24]

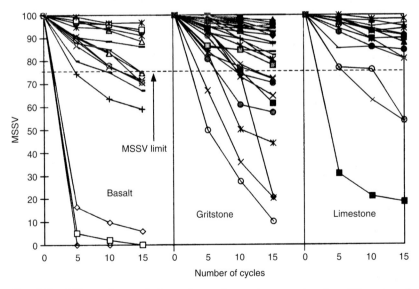

Fig. 1.8. Effect of extended numbers of soundness cycles for different rock types[24]

mechanism within the pavement. An example of data[24] obtained from samples of tertiary basalt aggregate is shown in Fig. 1.7. It shows the progressive breakdown during the five soaking cycles and demonstrates the suitability of the MSSV Test[17] for the determination of this tendency. Indeed, it was found that the MSSV Test[17] was one of the few BS 812 test methods that could highlight the presence of this weathered material in a heterogeneous mixture.

Another example[24] of MSSV Test[17] data is shown in Fig. 1.8. This indicates that many aggregates continue to fail if the testing continues beyond the standard five cycles. In terms of performance, it may be worth simulating the environment in which the aggregate is to be used by extending the number of cycles during testing. Examples would be materials which are subjected to increased cycles of wetting/drying, freezing/thawing or the application of de-icing salts.

1.13. Proposed European norms

The testing of aggregates has been under review for the last decade as part of the European harmonization process. Within Member States of the European Community, there are considerable variations in the geology and the methods of processing, storage and transportation of both aggregates and asphalts. In order to avoid problems, it is vital that appropriate test methods and specifications are chosen carefully.

Individual countries were originally asked to propose test methods to assess a range of aggregate properties including skid resistance, abrasion, fragmentation, soundness, freeze/thaw and thermal shock. However, there was no centralized research programme to study the suitability of these methods for predicting performance. Many countries have little, if any, experience with the full range of methods now being considered.

Involvement by the UK was initially poor and resulted in only a small number of British Standard test methods being proposed for consideration. As a result, the UK aggregate industry has, in general, limited practical experience with the test methods which are being proposed by other member states. Even within Europe, there is still no general consensus of applicable limits as few have been used outside of the country of origin. A listing of the main aggregate European Standards is shown in Table 1.12.

1.13.1.   Selection of test methods

The harmonization process is the responsibility of the Comité Européen de Normalisation (CEN) and is undertaken by CEN's Technical Committees (TCs). Aggregate is the responsibility of TC154 and is sub-divided into six Sub-Committees (SCs) with SC6 dealing with test methods. These are further divided into Task Groups (TGs) which consider individual test methods. The approach adopted was to produce a single test method for each specification parameter. Thereafter, research programmes would consider these methods and make appropriate revisions or develop new methods, as necessary. In 1990, the estimated target date for acceptance was 1992. However, progress was poor, with the methods only being decided in the late 1990s. The main parameters and methods are shown in Table 1.13.

Table 1.12.   The main European Standards relating to aggregates

Property	European Standard
Polishing resistance	BRITISH STANDARDS INSTITUTION. Tests for Mechanical and Physical Properties of Aggregates, Determination of the Polished Stone Value, BS EN 1097-8, 1999. BSI, London.
Abrasion	BRITISH STANDARDS INSTITUTION. Tests for Mechanical and Physical Properties of Aggregates, Determination of the Resistance to Wear (micro-Deval), Part 1, BS EN 1097-1, 1999. BSI, London.
Fragmentation/impact	BRITISH STANDARDS INSTITUTION. Tests for Mechanical and Physical Properties of Aggregates, Methods for the Determination of Resistance to Fragmentation, Part 2, BS EN 1097-2, 1998. BSI, London.
Soundness	BRITISH STANDARDS INSTITUTION. Tests for Thermal and Weathering Properties, Magnesiun Sulfate Test, Part 2, BS EN 1367-1, 1998. BSI, London.
Thermal shock	COMITÉ EUROPÉEN DE NORMALISATION. Determination of Resistance to Thermal Shock of Aggregates or the Production of Hot Bituminuos Mixtures and Surface Dressing Aggregates, CEN/TC154/SC6, Draft prEN 1367-5, 1996.

Table 1.13.   Aggregate test methods proposed as European Standards

Property	Proposed test method	Country of origin
Polishing resistance	Polished Stone Value	United Kingdom
Abrasion	Aggregate Abrasion Value	United Kingdom
	Micro-Deval	France
Fragmentation	Los Angeles Test	Originally USA but now used throughout Europe
Impact	Schlagversuch Impact Test	Germany
Soundness	Magnesium Sulphate Soundness Value	United Kingdom
Freeze/thaw	Freeze/Thaw Test without salt	Germany
	Freeze/Thaw Test with salt	Iceland
Thermal shock	Heating to 700 °C	Holland/Norway

Resistance to polishing
The need to improve safety is fundamental to the process of European harmonization. Measurement of the resistance of an aggregate to polishing will be assessed using the BS 812 test method, i.e. the PSV with few changes to the British method[36] (recently published as BS EN 1097-8[45]). Until the process of harmonization started, most mainland European countries paid little, if any, attention to the skid resistance characteristics of wearing courses. Many Member States considered other properties such as strength and durability to be more important. Others, such as the Mediterranean countries, only have poor quality limestones available and, accordingly, may have great difficulty in producing aggregates with a PSV of 50 or more.

Abrasion
There are two methods which have been considered to assess abrasion characteristics. The UK AAV Test[37] assesses dry resistance to mass loss resulting from abrasion by sand on a grinding lap. The second method is the French wet Micro-Deval (MDE) Test (published as BS EN 1097-1[46]). This determines the mass loss of an aggregate sample as it tumbles in a steel container for 2 hours with 10 mm ball bearings and water. The French classification of aggregate specifies an MDE of <10 for surfacing aggregate (Norme Française[47]).

Table 1.14 shows basic AAV and MDE statistics for the main types of aggregate available in Northern Ireland.[24] The majority which were assessed, as shown by the upper quartile value, would meet the requirements of the dry AAV Test.[37] However, many would find it difficult to meet the French MDE requirement. This reflects the difference in opinion between the two countries as to which parameters are important when assessing the suitability of a particular aggregate for use in an asphalt. The French believe in a hard aggregate that does not abrade, particularly when wet. In contrast, the British require a high PSV aggregate where skid resistance is paramount albeit at the expense of wet abrasion resistance.

The MDE Test[46] method has probably the biggest implication for any aggregate source in the UK, particularly those which possess higher

Table 1.14.   Basic AAV and MDE statistics

Group	Number of tests	Mean	Min.	Max.	Upper quartile	Range
AAV Test statistics						
Basalt	127	8.29	1.60	33.31	9.10	31.70
Gritstone	161	6.95	2.43	38.40	7.83	35.97
Limestone	55	9.34	1.70	40.75	11.00	39.05
MDE Test statistics						
Basalt	49	38.87	7.29	98.80	41.31	91.51
Gritstone	66	21.69	5.88	67.20	27.49	61.32
Limestone	29	16.27	5.30	68.08	17.47	62.78

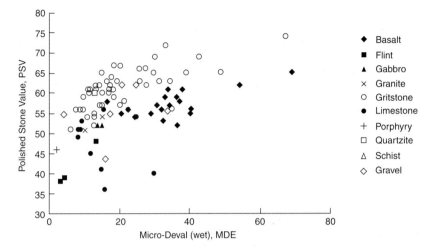

Fig. 1.9. PSV and MDE Test results for different rock types[24]

values of PSV. Figure 1.9 shows a plot of PSV and MDE for UK aggregates. If the French limit of <10 MDE is used, this would exclude the majority of high PSV aggregates currently in use in the UK. Figure 1.9 also demonstrates that small increases in PSV are accompanied by large reductions in the ability to withstand wet abrasion. This suggests that by selecting a high PSV aggregate, there is a risk that this will result in beneficial short performance but with long-term costs due to premature resurfacing works.

Fragmentation
Fragmentation will be measured by the Los Angeles (LA) Test.[48] Originally developed in the USA in 1916, the method is now used in many European countries and elsewhere. Marwick and Shergold[39] considered it as a possible British Standard and showed that it had almost numerical agreement with ACV[34] results. Despite this historical interest, experience of the LA Test[48] in the UK remains minimal.

The European method assesses single size 10/14 mm aggregate. Work by Woodward[24] has shown a close correlation to ACV[34] and AIV[33] Test methods. Any aggregate currently meeting British requirements for ACV or AIV should have reasonably comparable results from LA Test[48] method. Failure to meet requirements for fragmentation is not expected to be a problem for UK aggregate sources.

Impact
Resistance to impact will be assessed by the German Schlagversuch Impact Test (SZ).[48] Work carried out in Germany by Woodside and Woodward[49] has shown close correlation between this method and the AIV Test.[33] A comparison for a range of Northern Ireland greywacke and basalt aggregates is shown in Fig. 1.10.

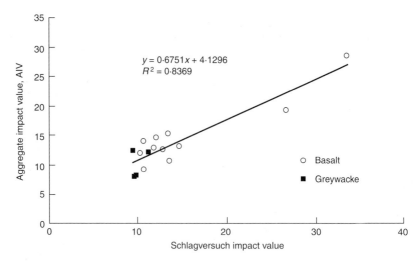

Fig. 1.10.   German SZ and UK AIV Test results[49]

Experience has shown that any aggregate which currently meets UK requirements for AIV should have reasonably comparable results from the SZ Test[48] method. Although the method is used extensively in Germany, few other European laboratories have the equipment due to its high cost and so the method is likely to be restricted in its use within the different European Member States.

Soundness
Aggregate soundness assessment will be based closely on the UK MSSV Test[17] method. This involves five cycles of soaking in a magnesium sulphate salt solution followed by oven drying to determine the susceptibility of an aggregate to failure which is related to Soundness. The resulting Soundness Value[50] differs from the UK MSSV Test[17] in that the percentage of material passing the end test sieve is used to express the soundness of an aggregate. It is suggested that the Soundness Test[50] is one of the few methods that can adequately highlight the presence of such contaminants. The MDE Test[46] is the only other CEN method which has also been found to have this capability. The MDE Test[46] has the advantage of taking only two days to complete whereas the Soundness Test[50] takes some three weeks. In practical terms, this makes the MDE Test[46] very attractive for the prediction of aggregate performance.

Freeze/thaw
Within CEN, the initial proposed testing methods did not include the MSSV Test[17] to predict soundness. The only method initially proposed to measure such a property was the German Freeze/Thaw Test.[51] The proposed method consists of ten freeze/thaw cycles of ±20 °C with the aggregate contained in water (CEN/TC154/SC6[52]). The experience of Woodside and Woodward[49] suggests that this method has limited

capabilities in respect of predicting the effects of the presence of small proportions of contaminants. What is more important is the cost of the equipment and the time taken to produce a result. As a result, a second Freeze/Thaw Test has been proposed which involves the addition of salt to the water. Experience in Iceland[53] with a wide range of basalt aggregates, resulted in the failure of numerous samples which had previously passed the German requirement.

Thermal shock
A test to assess the effects of thermal shock has recently been introduced for consideration (prEN 1367-5[54]). Immediately prior to mixing with bitumen to produce an asphalt, aggregates are heated to elevated temperatures within a short time period. During this process, certain types of aggregate may experience thermal shock. The proposed method will assess the propensity of an aggregate to suffer a reduction in strength as a result of rapid heating. The test[54] involves heating a sample of aggregate to 700 °C for three minutes and comparing the strength before and after heating. Thermal shock causes instantaneous failure of any aggregate which is susceptible to mechanical strain such as shale particles in a greywacke or weathered basalt. Research has shown that failure may be particularly violent if moisture is present within the aggregate.

The determination of bitumen content of asphalts is routinely carried out in the UK by means of the test prescribed in BS 598: Part 102.[55] Another test which also determines the bitumen content is the Ignition Test.[56] Unfortunately it also produces thermal shock in the sample of aggregate. The Test[56] uses a furnace operating at 540 °C to burn off the bitumen present in an asphalt. The bitumen content is determined as the loss of mass expressed as a percentage. Examination of samples following testing has shown that this method may cause failure resulting from the effects of thermal shock. Should this occur, the resulting change may cause the mixture to fail on grading whereas bitumen assessment by the traditional solvent method[55] would have suggested a compliant sample. Determination of bitumen content by the solvent method[55] is being withdrawn as an acceptable test method in many countries around the world. The Ignition Test[56] is now in the late stages of becoming a British Standard.

Figure 1.11 shows the amount of break-up which occurred for a range of 10/14 mm sized chippings when subjected to heating to 580 °C. After exposure to heat, the amount of break-up was calculated as the percentage of material which was <10 mm in size. It can be seen from Fig. 1.11 that this was up to 18%. The testing has indicated that, while the use of the Ignition Test[56] may be attractive on health and safety grounds, it may also be inducing thermal shock within certain types of aggregate which have traditionally been regarded as acceptable both in use and in their laboratory analysis using solvent extraction methods.

1.14. Summary
This Chapter has considered a range of British Standard and proposed European test methods which are used to assess the properties of

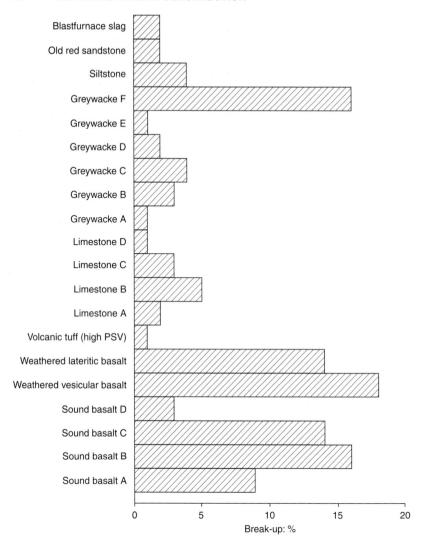

Fig. 1.11.   Percentage break-up for a range of aggregates subjected to heating to 580 °C

aggregates. It has sought to demonstrate that it is more important to be aware of the factors that affect each method and its ability to predict performance than simply to be familiar with the methods. Many of the methods were developed to determine the suitability of aggregates for horsedrawn traffic. Although it was developed more recently, the PSV test is almost 50 years old and has not fundamentally changed during this time. Many of the tests assess samples of single size dry aggregates and do not take into consideration that, in use, a variety of sizes is blended together. More importantly, the simple effect of rain is not assessed.

Engineers who have to choose an aggregate for a particular purpose must be aware of these limitations, in particular, the single most important aggregate property, skid resistance. All research shows that only a few aggregate sources can provide very high levels of PSV and that small increases in very high PSV materials are gained at the expense of every other test property.

In the future, engineers cannot continue to select aggregates on the basis of current specifications and test methods — rather the approach must be in terms of predicting performance in carriageway pavements.

References

1. O'FLAHERTY C. A. *Highways*. Edward Arnold, London, 1988, 3rd edn.
2. BRITISH STANDARDS INSTITUTION. *Sizes of Broken Stone and Chippings*. BSI, London, 1913, BS 63.
3. BRITISH STANDARDS INSTITUTION. *Testing Aggregates, Method for Qualitative and Quantitative Petrographic Examination of Aggregates*. BSI, London, 1994, BS 812: Part 104.
4. LOVEGROVE E. J. *The Surveyor and Municipal and County Engineer*. 15 April 1910, **37**, 451–455.
5. LOVEGROVE E. J., J. S. FLETT and J. A. HOWE. Road Making Stones, Attrition Tests in the Light of Petrology. *Surveyor*, 1906.
6. LOVEGROVE E. J., J. A. HOWE and J. S. FLETT. *Attrition Tests of British Road-Stones, Memoirs of the Geological Survey and Museum of Practical Geology*. HMSO, London, 1929.
7. KNILL D. C. Petrological Aspects of the Polishing of Natural Roadstones. *Applied Chemistry*, 1960, **10**, 28–35.
8. HAWKES J. R. and J. R. HOSKING. *British Arenaceous Rocks for Skid-Resistant Road Surfacings*. TRL, Crowthorne, 1972, LR 488.
9. THOMPSON A., J. R. GREIG and J. SHAW. *High Specification Aggregates for Road Surfacing Materials*. Department of the Environment, London. Travers Morgan Ltd, East Grinstead, 1993, Technical Report.
10. DORMAN R. H. *Road Metal, First Irish Roads Congress, Dublin*, 1910, 145–151.
11. KNIGHT B. H. *Road Aggregates, their Use and Testing, The Roadmakers Library*. Edward Arnold, London, 1935, **III**.
12. HARTLEY A. A Review of the Geological Factors Influencing the Mechanical Properties of Road Surface Aggregates. *Quarterly Journal of Engineering Geology*, 1974, **7**, 69–100.
13. LEES G. and C. K. KENNEDY. Quality, Shape and Degradation of Aggregates. *Quarterly Journal of Engineering Geology*, 1975, **8**, 193–209.
14. COLLIS L. and R. A. FOX (eds). *Aggregates, Sand Gravel and Crushed Rock, Aggregates for Construction Purposes*, Engineering Geology Special Publication No 1. The Geological Society, London, 1985.
15. BULLAS J. C. and G. WEST. *Specifying Clean, Hard and Durable Aggregate for Bitumen Macadam Roadbase*. TRL, Crowthorne, 1991, RR 284.
16. HIGHWAYS AGENCY *et al. Manual of Contract Documents for Highway Works, Specification for Highway Works*. TSO, London, 1998, **1**.
17. BRITISH STANDARDS INSTITUTION. *Testing Aggregates, Method for Determination of Soundness*. BSI, London, 1989, BS 812: Part 121.
18. HIGHWAYS AGENCY *et al. Manual of Contract Documents for Highway Works*. TSO, London, 1998, **0–6**.

19. HIGHWAYS AGENCY *et al. Manual of Contract Documents for Highway Works, Notes for Guidance on the Specification for Highway Works.* TSO, London, 1998, **2**.

20. HIGHWAYS AGENCY *et al. Design Manual for Roads and Bridges.* TSO, London, various dates, **1–15**.

21. HIGHWAYS AGENCY *et al. Design Manual for Road and Bridges, Pavement Design and Maintenance.* TSO, London, various dates, **7**.

22. BRITISH STANDARDS INSTITUTION. *Testing Aggregates, Method for Determination of Particle Size Distribution, Sieve Tests.* BSI, London, 1985, BS 812: Part 103, Section 103.1.

23. BRITISH STANDARDS INSTITUTION. *Testing Aggregates, Methods for Determination of Density.* BSI, London, 1995, BS 812: Part 2.

24. WOODWARD W. D. H. *Laboratory Prediction of Surfacing Aggregate Performance.* University of Ulster, PhD thesis. Jordanstown, 1995.

25. WOODSIDE A. R. and W. D. H. WOODWARD. *Use of the Methylene Blue Test to Assess the Soundness of Aggregate.* University of Ulster, Jordanstown, 1984, Internal Report.

26. INTERNATIONAL SLURRY SEAL ASSOCIATION. *Test Method for Determination of Methylene Blue Adsorption Value (MBV) of Mineral Aggregate Fillers and Fines,* ISSA, Washington DC, Proposed February 1989, Technical Bulletin No. 145.

27. SORBY H. C. On the Structure and Origin of Non-Calcareous Rocks. *Proc. of the Geological Society, 1880,* The Geological Society, London, **36**, 46–92.

28. WENTWORTH C. K. *The Shape of Beach Pebbles.* Geological Survey Professional Paper, 1922, **131**, 75–102.

29. TESTER A. C. The Measurement of the Shapes of Rock Particles. *Journal of Sedimentary Petrology,* 1931. **1**, 3–11.

30. ZINGG T. H. A Contribution to the Analysis of Coarse Gravel. *Schweizerische Mineralogische und Petrographische Mitteilungen,* 1935, **15**, 133–140.

31. MARKWICK A. H. D. *The Shape of Road Aggregate and its Testing,* Department of Scientific and Industrial Research. TSO, London, 1936, RRL Road Research Bulletin No. 2.

32. LEES G. The Measurement of Particle Elongation and Flakiness, A Critical Review of British Standard and other Methods. *Magazine of Concrete Research,* 1964, **16**, No. 49, 225–230.

33. BRITISH STANDARDS INSTITUTION. *Testing Aggregates, Method for Determination of Aggregate Impact Value (AIV).* BSI, London, 1990, BS 812: Part 112.

34. BRITISH STANDARDS INSTITUTION. *Testing Aggregates, Method for Determination of Aggregate Crushing Value (ACV).* BSI, London, 1990, BS 812: Part 110.

35. BRITISH STANDARDS INSTITUTION. *Testing Aggregates, Methods for Determination of Ten Per Cent Fines Value (TFV).* BSI, London, 1990, BS 812: Part 111.

36. BRITISH STANDARDS INSTITUTION. *Testing Aggregates, Method for Determination of Polished-Stone Value.* BSI, London, 1989, BS 812: Part 114.

37. BRITISH STANDARDS INSTITUTION. *Testing Aggregates, Method for Determination of Aggregate Abrasion Value (AAV),* BSI, London, 1990, BS 812: Part 113.

38. BRITISH STANDARDS INSTITUTION. *Sampling and Testing of Mineral Aggregates, Sands and Fillers.* BSI, London, 1938, BS 812.
39. MARKWICK A. H. D. and SHERGOLD F. A. The Aggregate Crushing Test. *Journal of the Institution of Civil Engineers* 1945, **24**, No. 6, 125–133.
40. SHERGOLD F. A. and J. R. HOSKING. A New Method for Evaluating the Strength of Roadstone with Particular Reference to the Weaker Types used in Road Bases. *Roads and Road Construction*, 1959, **37**, No. 438, 1964–67.
41. HOSKING J. R. *Road Aggregates and Skidding.* TRL, Crowthorne, 1992.
42. HIGHWAYS AGENCY *et al. Design Manual for Roads and Bridges, Pavement Design and Maintenance, Surfacing Materials for New and Maintenance Construction.* TSO, London, 1999, 7.5.1, HD 36/99.
43. MACLEAN D. J. and F. A. SHERGOLD. *The Polishing of Roadstone in Relation to the Resistance to Skidding of Bituminous Road Surfacings*, Department of Scientific and Industrial Research. HMSO, London, 1958, Road Research Technical Paper No. 43.
44. BRARD M. Sur le Procede Propose par M Brard Pour Reconnaitre Immediatement les Pierres Qui Ne Peuvent Pas Resister a la Gelee, et que l'on Designe Ordinairement par les Noms de Pierres Gelives ou Pierres Gelisses. *Anales de Chimie et de Physique*, 1828, **38**, 160–170.
45. BRITISH STANDARDS INSTITUTION. *Tests for Mechanical and Physical Properties of Aggregates, Determination of the Polished Stone Value.* BSI, London, 2000, BS EN 1097-8.
46. BRITISH STANDARDS INSTITUTION. *Tests for Mechanical and Physical Properties of Aggregates, Determination of the Resistance to Wear (Micro-Deval).* BSI, London, 1996, BS EN 1097-1, Part 1.
47. NORME FRANÇAISE. *Aggregates — Characteristics of Aggregates Intended for Road Works.* 1982, NF P 18-321.
48. BRITISH STANDARDS INSTITUTION. *Tests for Mechanical and Physical Properties of Aggregates, Methods for the Determination of Resistance to Fragmentation.* BSI, London, 1998, BS EN 1097-2, Part 2.
49. WOODSIDE A. R. and W. D. H. WOODWARD. *Report on Visit to the BASt, Germany to Test Northern Ireland Aggregates using Proposed CEN Test Methods.* University of Ulster, Jordanstown, 1993, Internal Report.
50. BRITISH STANDARDS INSTITUTION. *Tests for Thermal and Weathering Properties, Magnesium Sulfate Test*, BSI, London, 1998, EN 1367-2, Part 2.
51. DIN. *Testing of Natural Stone, Freeze-Thaw Cyclic Test, Methods A to Q*, DIN 52 104, Part 1, 1982.
52. COMITÉ EUROPÉEN DE NORMALISATION. *Tests for Thermal and Weathering Properties of Aggregates, Determination of Resistance to Freezing and Thawing*, Part 1. CEN/TC154/SC6. 1996, Draft prEN 1367-1.
53. PETURSSON, P. *Final Report on Frost Resistance Test on Aggregates, Intercomparison of a New Nordtest Method, A Base for CEN Standardisation.* Icelandic Building Research Institute, 1996, Report No. 96-18.
54. COMITÉ EUROPÉEN DE NORMALISATION. *Determination of Resistance to Thermal Shock of Aggregates for the Production of Hot Bituminous Mixtures and Surface Dressing Aggregates*, CEN/TC154/SC6. 1996, Draft prEN 1367-5.
55. BRITISH STANDARDS INSTITUTION. *Sampling and Examination of*

Bituminous Mixtures for Roads and other Paved Areas, Analytical Test Methods, BSI, London, 1996, BS 598: Part 102.

56. BRITISH STANDARDS INSTITUTION. *Method for the Analysis of Bituminous Mixures by Ignition, British Standard Draft for Development*, DD ABH. BSI, London, May 1997.

2. Bitumens

2.1. Preamble

The *Concise Oxford Dictionary* gives the primary definition of bitumen as 'any of various tarlike mixtures of hydrocarbons derived from petroleum naturally or by distillation and used for road surfacing and roofing'. Most lay persons still use the words 'tar' and 'tarmacadam' when in fact neither is still produced in any significant quantity. Tar has been classified as carcinogenic. Bitumen is now the binder for asphalts and this Chapter considers bitumens in detail. There are sections on its nature, production, specification and rheology. Nowadays, bitumens can be modified by many techniques and these are considered. The various emulsified forms are explained along with foamed bitumen which is enjoying increasing popularity. Trinidad Lake Asphalt, gilsonite and rock asphalt enjoy limited use but are of interest for specialist purposes. Tar is described briefly. Finally, the important issues of health and safety are discussed.

Many punningly describe bitumen technology as a 'black art'. This Chapter demonstrates that such a description is inappropriate and underlines the view that a knowledge of bitumen is vital for anyone involved in road construction and maintenance.

2.2. The nature of bitumen

Bitumen is perhaps best described as a complex mixture of components with various chemical structures. The majority of these structures are composed of carbon and hydrogen only and are termed hydrocarbons. However, in addition to hydrocarbons there are a number of other structures containing heteroatoms, i.e. atoms other than hydrogen and carbon, such as oxygen, sulphur and nitrogen. It is the complex arrangement of the hydrocarbon molecules and those molecules containing heteroatoms which gives bitumen its unique balance of properties.

Several models have been used to describe the structure of bitumen but by far the most common is the Micellar Model.[1] (A micelle is an aggregate of molecules in a colloidal solution.) This model describes the structure of bitumen as a colloidal dispersion of 'asphaltenes' in 'maltenes' which is stabilized by 'resins'. (A colloid is a mixture which consists of large single molecules usually dispersed through a second substance.) In order to understand this classical model[1] it thus becomes necessary to further subdivide the hydrocarbon and heteroatom structures referred to above into distinct chemical groups. The individual compounds are then classified as either saturates, aromatics, resins or asphaltenes (referred to as the 'SARA' classification).[2] It is possible to obtain relatively pure samples of each of these components using a combination of precipitation, filtration and column chromatography. In order to understand the properties of bitumen, it is necessary to understand the properties of the individual components. These are described below.

2.2.1. Saturates

These are straight and branched chain molecules consisting of carbon and hydrogen only. They are termed saturates because they contain almost exclusively single carbon–carbon or carbon–hydrogen bonds although there may be some aromatic and naphthenic ring structures present (see Fig. 2.1).

Since they are composed of only carbon and hydrogen, the saturates are non-polar and show no great affinity for each other. When extracted from bitumen, the saturates appear as a viscous, white to straw coloured liquid with a molecular weight in the range 300–2000. The saturates constitute between 5% and 20% of the total bitumen structure.

Due to the chemical nature of saturate molecules there is very little inter-molecular attraction. The absence of polar species means that very few hydrogen bonds can form and consequently the weak van der Waal's forces predominate. (Van der Waal's forces are forces of attraction between uncharged molecules.) Consequently, the saturates contribute very little to the stiffness of bitumen. They do, however, play an important part in defining the overall properties.

2.2.2. Aromatics

These are low molecular weight compounds comprising ring and chain structures and form the major part of the dispersing medium for the asphaltenes. Unsaturated ring structures predominate in the overall aromatic structure which can contribute up to 65% of the total bitumen structure. When purified, the aromatics appear as viscous brown liquids with a molecular weight in the range 300–2000.

2.2.3. Resins

The resins are highly polar, predominantly hydrocarbon molecules with a significantly higher concentration of heteroatoms than the other species which are present in bitumen. They have been found to contain both acidic and basic groups which means that possibilities exist for hydrogen bonding and strong inter-molecular and intra-molecular attraction. When purified, the resins appear as black or dark brown solids or semi-solids with molecular weights in the range 500–50 000. The polar nature of the resins gives the bitumen its adhesive properties and, to a certain extent, controls the degree of sol- and gel-type structures. (Sol is an abbreviation of solution and is used to describe a fluid colloidal suspension of a solid in

Typical structure of an asphaltene molecule
The R groups may be aromatic or naphthenic

Typical structure of an aromatic molecule
The R groups may be other carbon ring structures

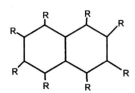

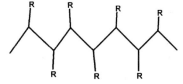

Some typical saturate molecules
The R groups may be carbon rings or chains

Fig. 2.1.   Typical structures of some of the molecules found in bitumen[2]

a liquid whereas a gel is a semi-solid colloidal suspension of a solid dispersed in a liquid.)

2.2.4.   Asphaltenes

Asphaltenes are arguably the species with the highest molecular weights within the bitumen structure. They are generally described as being very polar molecules containing a high concentration of aromatic ring structures. When purified, the asphaltenes appear as solid black particles with a gritty or brittle feel. Depending on the method of purification and analysis, the asphaltenes have been found to have molecular weights ranging between 600 and 300 000. However, the majority of data suggests that the predominant species in the asphaltene structure have molecular weights in the range 1000–100 000. The asphaltene content, which may range between 5% and 25%, has an enormous effect on the overall properties of the bitumen. Bitumens with a high asphaltene content will have higher softening points, higher viscosities and lower penetrations than those with low asphaltene contents. The presence of heteroatoms and polar groups within the asphaltene structure means that there is a very high possibility for hydrogen bonding. Consequently, it is not only the asphaltene content that influences the structure of the bitumen but also the degree of interaction and bonding between the asphaltene molecules themselves.

The classical Micellar Model[1] states that the asphaltenes form tightly packed structures or clusters within a matrix of aromatics and saturates. The resins act as stabilizers to prevent the asphaltenes from separating. The extent of dispersion of the asphaltenes within the matrix will determine the overall properties of the bitumen.[3–5] A high concentration of tightly packed asphaltenes tends to give the bitumen a gel-type structure whereas a low concentration of well dispersed asphaltenes gives a sol-type structure (Fig. 2.2).

The main supporting evidence for the Micellar Model[1] comes from X-ray scattering experiments.[6–8] However, given the chemical nature of the molecules in bitumen, it is not surprising that some scattering takes place. Several authors have attempted to prove the existence of asphaltene micelles using techniques including: size exclusion chromatography,[9,10] electron microscopy[11,12] and nuclear magnetic resonance (NMR).[13] Thus far the evidence is at best conflicting and consequently a better model for describing the structure of bitumen has been sought.

As part of the USA's Strategic Highways Research Program (SHRP), attempts were made to separate the micelles or, if this was not possible, to separate the individual components in bitumen using other techniques. It was also recognized that there were a number of possible interactions between the bitumen molecules including hydrogen bonding and acid-base interaction. Evidence for hydrogen bonding was gathered by Barbour and Petersen[14] using infrared spectroscopy.

Acid–base interactions were established using Ion-Exchange Chromatography (IEC). Although IEC had been used previously to characterize bitumen, the SHRP experiments identified that there were 'Amphoteric' species in bitumen which contained both acidic and basic groups.[9] These had very low viscosities and consequently it was concluded that the viscosity of the bitumen was controlled mainly by the acidic and basic species. (Viscosity is defined as the resistance to flow and its measurement is discussed in Section 2.6.)

One appealing method for describing the structure of bitumen is the solubility parameter approach. In essence, this method involves determining the 'attractiveness' or otherwise of individual molecules to each other using a series of titrations. One of the most useful facets of this approach is that it allows the prediction of the stability of bitumen and

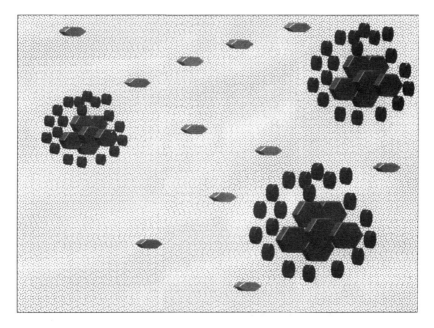

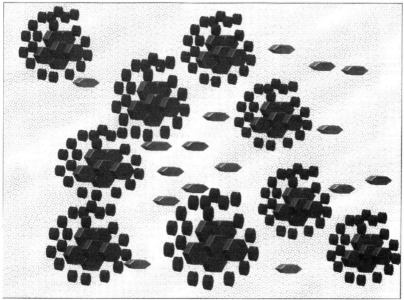

Fig. 2.2.   (a) Schematic representation of a sol-type bitumen showing well dispersed asphaltene clusters surrounding and stabilised by resin molecules. The clusters are then dispersed in a matrix of saturates and aromatics.[1] (b) Schematic representation of a gel-type bitumen showing tightly packed asphaltene clusters[1]

Polymer–bitumen blends through common theories. Ageing phenomena can also be explained in a more scientific fashion than is possible with the Micellar Model.[1]

2.3.   Physical properties of bitumen

In order to interpret the physical properties of bitumen and the effect of temperature on those properties, it is necessary to understand the model for the structure of bitumen. By returning to first principles and using the concept of bitumen as a dispersion of large, polar clusters of species within a relatively simple non-polar matrix, it becomes easier to interpret the effects of temperature.

At low temperatures, the overall structure is 'frozen', i.e. the molecules are unable to move over each other and the bitumen has a high viscosity. As the temperature increases, some of the molecules begin to move or flow and some of the intermolecular bonds may be broken down. As the rate of movement of the molecules increases, the viscosity decreases until, at high temperatures, the bitumen behaves as a liquid. Thus, a further classification criterion for bitumen can be set based on its susceptibility to temperature. At low temperatures, bitumen resembles a solid, perhaps even brittle, material but at higher temperatures it becomes a liquid.[†] This transition is fully reversible and consequently bitumen may be classified as a thermoplastic material. The temperature range over which the transformation from solid to liquid occurs is governed largely by the relative concentrations of the four components of the bitumen and the degree of interaction between those components.

A common tool for representing this change in physical form is the Bitumen Test Data Chart (BTDC) developed by Heukelom.[15] The BTDC allows penetration, softening point, viscosity and Fraass point to be plotted as a function of temperature on a single chart (Fig. 2.3).

The temperature scale on the chart is linear, the scale for penetration is logarithmic and the viscosity scale has been devised specially so that penetration grade bitumens give straight lines. It is important to have a knowledge of the viscosity of bitumen at certain key points during the manufacture and laying of a mixture, particularly during coating of the aggregate and during compaction of the mixture. The BTDC allows this information to be plotted on a single chart and also allows the optimum ranges of viscosity to be plotted. For example, if the viscosity of the binder is too high during mixing then it is unlikely that all of the aggregate will be evenly coated and a poor mixture will result. Conversely, if the viscosity is too low then there is a chance that the binder will drain from the aggregate during storage and transportation. This can be a significant problem with open graded asphalts but is of less importance with dense mixtures. It has been established that the ideal coating viscosity is around $0.2\,Pa\,s$ (1 poise (P) $\equiv 0.1\,Pa\,s$).[16]

[†] It is not strictly correct to describe bitumen as a solid at low temperature as there will always be an element of viscous flow. Rather like a glass, bitumen at low temperatures is best described as a supercooled liquid.

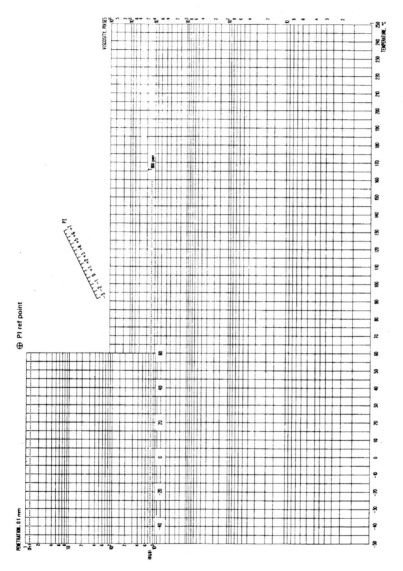

Fig. 2.3.   The Bitumen Test Data Chart (BTDC) developed by Heukelom to illustrate the basic rheological properties of bitumen[15]

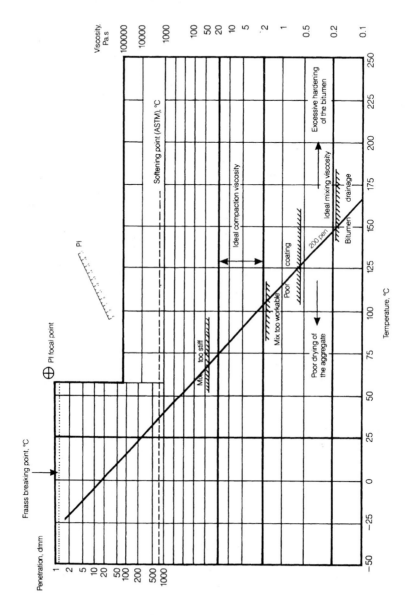

Fig. 2.4. A Bitumen Test Data Chart showing 'ideal' bitumen viscosities for mixing and compaction of dense macadam[16]

The viscosity of the binder will also have a significant contribution to the compactability of an asphalt. A binder with a high viscosity can result in a reduction in workability and poor compaction. Binders with low viscosities can produce mixtures which are excessively mobile and push out in front of the roller. It is therefore important to be able to control the viscosity of a binder within a practical temperature range to allow effective compaction. The ideal binder viscosity during compaction is between 2 and 20 Pa s. The BTDC allows bitumen suppliers to advise clients of the appropriate mixing and compaction temperatures to ensure optimal performance of the mix (Fig. 2.4). However, the BTDC also allows the estimation of temperature susceptibility of a binder and generic classification of the type of bitumen.[16]

Although temperature has the greatest effect on the properties of the bitumen, it is also important to consider the effects of time. At room temperature, the molecules within the bitumen are in constant, albeit rather slow, motion. Assuming that some of these molecules may be polar in nature, then there is the possibility for hydrogen bonding to occur if two or more of these polar molecules approach each other. When this bonding occurs a new, larger structure is created. The speed of movement of this large structure through the bitumen is considerably reduced. Evidence for this structure formation comes from 'physical hardening' experiments at low temperature.[17] Physical hardening is revealed by an increase in stiffness of the bitumen over a period of time.

2.4. Production of bitumen

Although there are a vast number of different crude oils currently available, only a small percentage of these are directly suitable for the manufacture of bitumen. It has been estimated that the total number of crude oils available at present exceeds 1500, with less than 100 of these being suitable for bitumen production. The reason for this is the widely

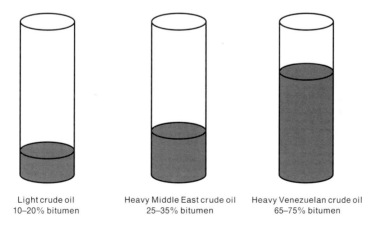

Light crude oil
10–20% bitumen

Heavy Middle East crude oil
25–35% bitumen

Heavy Venezuelan crude oil
65–75% bitumen

Fig. 2.5. Typical composition of some crude oils

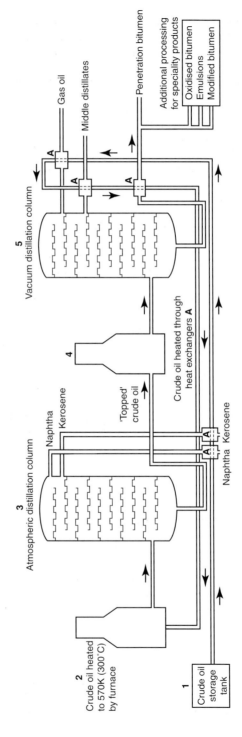

Fig. 2.6.   Schematic illustration of a bitumen production plant

varying chemical composition and bitumen content of these crudes. Most bitumen manufacture in the UK is from Middle Eastern, Venezuelan or Mexican crudes with a small percentage of North Sea crude being used. The typical composition of some crudes is shown in Fig. 2.5.

The production of bitumen from crude oil is essentially a two stage distillation process. In the first stage, the crude is heated to approximately 300 °C and fractionally distilled under atmospheric pressure (Fig. 2.6).

Steam may be injected at the base of the column to aid the distillation process. This atmospheric distillation effectively 'tops' the crude by removing the lightest (i.e. lowest boiling point) fractions such as naphtha and kerosene (atmospheric gas oil). The product at the end of the atmospheric distillation is termed 'long residue' and, if tested, would have a penetration in the range 2000–3000. (Penetration is described in Section 2.5.)

From the atmospheric column, the long residue is passed to a vacuum column where it is heated further (typically to 350 °C) under a significantly reduced pressure. This combination of heat and reduced pressure causes the heavier (higher boiling point) fractions such as diesel (vacuum gas oil), process oil and lubricating oil to evaporate from the long residue. At the completion of the vacuum distillation, the product is termed 'short residue' or 'penetration grade bitumen'. The severity of the distillation conditions and, hence, the concentration of volatile fractions removed from the long residue will determine the grade of bitumen produced.

The bitumen may then be passed to a 'blowing unit' or air-rectification plant where it is heated while air is passed through it. This process significantly alters the chemical composition of the bitumen, particularly the carbon/hydrogen ratio. The blowing process also changes the overall structure of the bitumen from sol to gel. In terms of physical properties, the blowing process raises the softening point and viscosity of the bitumen while reducing the penetration. This process produces the oxidized and hard grade bitumens used for roofing and other industrial applications.

2.5. Specification of bitumen

2.5.1. Current UK practice

Bitumen grades are specified primarily in terms of their needle penetration (referred to as 'pen'). Soft bitumens have high pen numbers and hard bitumens have low pen numbers. Within specifications however, there are certain other criteria which must be met including softening point, solubility and resistance to hardening. Descriptions of these test methods, which form the basis for most bitumen specifications, now follow.

The Penetration Test

The Penetration Test[18] involves subjecting a sample of bitumen to needle penetration under specified conditions of time, temperature and load (Fig. 2.7). In most cases, the test is carried out at 25 °C with a load of 100 g for

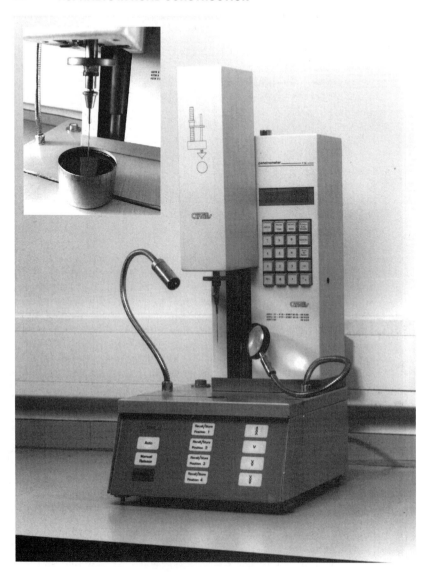

Fig. 2.7.   The Penetration Test. This test is carried out under water

a duration of 5 s. Low-temperature penetration is also measured occasionally with the usual conditions being 5 °C with a load of 200 g for 60 s. The depth of penetration of the needle into the surface of the bitumen is measured in tenths of a millimetre (dmm); hence 200 pen bitumen has a needle penetration of 200 dmm at 25 °C when measured with a load of 100 g for 5 s. It is common practice to describe the penetration of a particular bitumen without the unit, i.e. 100 pen means a penetration of 100 dmm.

2.8. The Softening Point Test.[19] *The photograph shows a disc of bitumen which is beginning to soften. The test is complete when the bitumen and balls touch the bottom plate*

The Softening Point Test

The Softening Point Test[19] involves preparing discs of bitumen and determining the temperature at which these discs are unable to support a standard metal ball (Fig. 2.8). At the softening point, all unmodified bitumens have approximately equal penetration (800 dmm) and viscosity (1200 Pa s). The softening point is therefore an approximation of the equi-viscous temperature (EVT) for bitumen.

It is possible to obtain an estimate for the temperature susceptibiltiy of the bitumen from a knowledge of the penetration and softening point. Equations developed by Pfeiffer and Van Doormal[20] demonstrate that a Penetration Index (PI), which estimates the effect of temperature on the properties of the bitumen, can be determined. The higher the PI, the more temperature susceptible will be the bitumen. The PI has traditionally been used as part of the specification for bitumen, particularly in mainland Europe.

Ageing Tests

The most common methods for determining the ageing characteristics of bitumen are the Thin Film Oven Test (TFOT)[21] and the Rolling Thin Film Oven Test (RTFOT).[22] The first of these methods involves placing a thin film of bitumen into a sample pan which is heated at 163 °C for 5 h whilst being slowly rotated. The RTFOT[22] involves placing the sample into a glass bottle which is rotated (Fig. 2.9). Air is blown into each bottle as it rotates at 163 °C for a period of 85 minutes. Both of these test methods give an indication of the ageing processes which occur during production of an asphalt.

There have been several attempts to simulate the ageing processes which occur during the service life of a pavement, none of which have met with absolute success. The American Superpave[TM] specifications[23] suggest the use of the RTFOT[22] followed by use of the pressure ageing

Fig. 2.9.   The apparatus used to carry out the Rolling Thin Film Oven Test (RTFOT) which simulates ageing of bitumen[22]

vessel (PAV) as a means of estimating ageing in service. A modified version of this procedure (termed HiPat — High Pressure Ageing Test) was recently used by the Refined Bitumen Association (the Trade Association of bitumen producers) to estimate the ageing that would be experienced by a binder in service.[24] Ageing Indices were 4–12 times higher for HiPat aged samples compared to those extracted from a road surface after 10 years. (The Ageing Index is the ratio of complex modulus before and after ageing, complex modulus is defined in Section 2.6.1.) Given that the naturally aged sample was obtained from high quality hot rolled asphalt (HRA), the results are deemed acceptable. The HiPat technique is still being evaluated, however it is likely to become a useful tool for the prediction of binder ageing.

There are a number of mechanisms involved in the ageing of bitumen,[25] but the two most important are loss of volatile components (saturates, aromatics) and chemical reaction of the bitumen with oxygen (oxidation). Both of these processes lead to a decrease in penetration, an increase in softening point and increase in viscosity.

It is important to have an accurate measure of the resistance to ageing of bitumen as this will influence its storage, mixing and end performance (Fig. 2.10). Bitumen must be stored at a temperature which facilitates easy pumping around a coating plant. When mixed with the aggregate, it must have a sufficiently low viscosity to evenly coat all of the aggregate, including the fines and filler. The storage and mixing temperatures

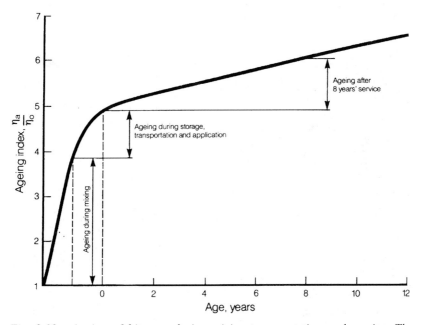

Fig. 2.10. Ageing of bitumen during mixing transportation and service. The 'Ageing Index' is expressed as the ratio of viscosity of the 'aged' bitumen compared to its initial viscosity

normally lie in the range 130–190 °C for penetration grade bitumen.[26,27] Under static storage in a closed tank, there is very little chance of oxygen entering the bitumen and reacting with it. The loss of volatiles is also controlled since the tank is closed. However, when the bitumen is pumped to and from storage tanks there is the possibility for hardening to occur, particularly if the return to the tank is via a splash-back through the vapour space above the surface of the bitumen. Under these conditions, there is a significantly increased possibility for oxygen to meet and react with the bitumen.

Similarly, when bitumen is mixed with aggregate in a coating plant there is a very high possibility of both loss of volatiles and reaction with oxygen due to the high temperatures involved and the very thin bitumen film. The oxidation continues during the transportation of mixed material to site and its subsequent passage through the paver.

As the mixed material cools, the rate of oxidation decreases significantly. For dense mixtures such as HRA with low voids contents, there is little further change in chemical composition beneath the surface once the mixture has cooled. However, more open graded materials with higher voids content allow oxygen to permeate into the mixture, penetrate the bitumen and react with it. Therefore, open mixtures can be more susceptible to atmospheric oxidation and hardening. However, the use of modified bitumens (see Section 2.7) giving increased film thicknesses can significantly reduce the oxidation rate.

The rate of oxidation of bitumen is doubled for every 10 °C increase in temperature above 100 °C. Therefore, to minimize hardening during storage, mixing, transportation and compaction, it is essential that the temperatures are as low as possible, commensurate with the practicalities involved. Although it is very difficult to quantify the effects of oxidation on the physical properties of bitumen due to the different operating conditions in each coating plant, it is generally accepted that during production of coated material the bitumen will 'drop a grade'.[28,29] Therefore, in very general terms, a nominal 50 pen binder can harden to around 35 pen after mixing.

The chemical reactions occurring during oxidation lead to a significant increase in the concentration of asphaltenes and resins. The degree of interaction between individual asphaltene molecules also increases and this is revealed by an apparent increase in molecular weight. The concentration of aromatics is also significantly reduced. The saturates, due to their relatively inert chemical nature, are generally unaffected by oxidation.

The Solubility Test
The purity of a sample of bitumen can be determined using the Solubility Test.[30] This involves dissolving a small amount of bitumen in a chlorinated solvent and filtering the solution. Any insoluble material is retained on the filter. Penetration grade bitumen should have a solubility in excess of 99.5%.

The tests described above are of only limited usefulness in defining the properties of bitumen. Several other properties must also be measured in

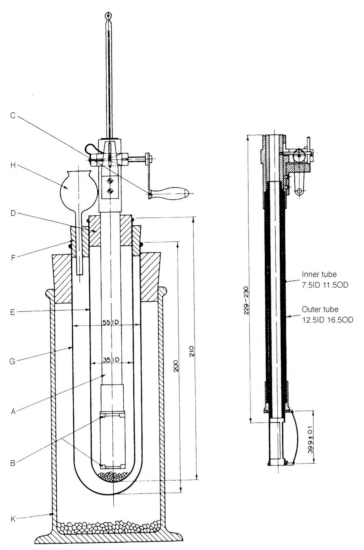

Fig. 2.11. The apparatus used to carry out the Fraass Breaking Point Test.[31]
This test determines the brittle point of a thin film of bitumen applied to a thin
plate which is flexed as the sample is cooled. All dimensions are in millimetres

order to fully understand the properties of each bitumen grade and its
relation to final performance.

The Fraass Breaking Point Test
The Fraass Breaking Point Test[31] is used to determine the low-temperature
properties of bitumen. This test involves placing a thin film of bitumen
onto a metal plate which is flexed whilst being cooled (Fig. 2.11). The

Fraass Breaking Temperature is the point at which the sample cracks. The reproducibility of the Fraass Test is not good and accurate sample preparation is essential if good quality results are to be obtained.

The difficulties in obtaining reproducible Fraass data have led to the development of other methods for obtaining low-temperature properties of bitumen. The simplest of these methods is the low-temperature penetration method which was described above.

British Standard specifications
The British Standard specifications for bitumens used in road construction are given in BS 3690: Part 1.[32] This document was originally published in 1963 and has seen a number of revisions since its first publication. Figure 2.12 shows the specifications for bitumen as defined in BS 3690: Part 1.[32] The grades of bitumen most commonly used for road construction are 50, 70, 100 and 200 pen but the stiffer grades 15, 25 and 35 pen are starting to be used more.

Recent research[33,34] into the performance of pavements has concluded that the stiffness of the roadbase has a significant effect on the performance of the pavement. As a result, there has been a tendency to move away from the use of 50 or 100 pen bitumen to 15, 25 or 35 pen bitumen in roadbases. However, this practice is not recommended for thinner pavements where fatigue is an issue (fatigue is discussed in Chapters 3 and 4). The softer grades of 300 pen and 450 pen bitumen are not commonly used for construction. They tend to be used to produce mixtures that can be stored and subsequently laid cold for patching, called 'storage grade macadams'.

BS 3690: Part 1[32] also lists three grades of cutback bitumen (i.e. bitumen containing a quantity of relatively light distillate) which are used for the production of storage grade macadam or for surface dressing. Surface dressing is discussed in detail in Chapter 10.

2.5.2. European specifications
The specification of bitumens has been under review for the last decade as part of the European harmonisation process. It is the responsibility of the Comité Européen de Normalisation (CEN). The process has taken a significant amount of time to come to fruition. However, the harmonized CEN specification will be adopted by June 2000 by each Member State except the UK who will adopt in 2002.

There are several differences in the CEN specifications for bitumen and those described in BS 3690: Part 1.[32] The most notable is the change in the definition of the grades.[35] Whereas BS 3690: Part 1[32] denotes each grade by the mid-point of the penetration specification, the CEN grades are specified by their limits. In effect, this means the 50 pen as specified in BS 3690: Part 1 becomes 40/60 under CEN. The grades themselves are also different. The current UK grade of 15 pen is removed from the proposed specification completely with the hardest grade being 20/30. In addition, 100 pen is replaced by two grades — 70/100 and 100/150. An additional specification for 'hard

BS 3690 Table 1. Properties for penetration grade bitumen

Property	Grade									
	15 pen	25 pen	35 pen	40 HD	50 pen	70 pen	100 pen	200 pen	300 pen	450 pen
Penetration at 25°C (dmm)	10±5	25±5	35±7	40±10	50±10	70±10	100±20	200±30	300±45	450±65
Softening point (°C)	63–76	57–69	52–64	58–68	47–58	44–54	41–51	33–42	30–39	25–34
Maximum loss in mass (%)	0.1	0.2	0.2	0.2	0.2	0.2	0.5	0.5	1.0	1.0
Maximum drop in penetration (%)	20	20	20	20	20	20	20	20	25	25
Solubility in trichloroethylene (%)	99.5	99.5	99.5	99.5	99.5	99.5	99.5	99.5	99.5	99.5

BS 3690 Table 2. Properties of cutback bitumen

Property	Grade		
	50 seconds	100 seconds	200 seconds
STV at 40°C (10 mm cup)	50±10	100±20	200±40
Maximum distillate at 225°C (%)	1	1	1
Maximum distillate at 360°C (%)	8–14	6–12	4–10
Pen of residue from distillation to 360°C (dmm)	100–350	100–350	100–350
Solubility in trichlorethylene (%)	99.5	99.5	99.5

Fig. 2.12. Tables 1 and 2 of BS 3690 for penetration grade and cutback bitumen[32]

CEN Table 1. Specifications for paving grade bitumens with penetrations from 20 dmm to 330 dmm

Property	Grade								
	20/30	30/45	35/50	40/60	50/70	70/100	100/150	160/220	250/330
Penetration at 25 °C (dmm)	20–30	30–45	35–50	40–60	50–70	70–100	100–150	160–220	250–330
Softening point (°C)	55–63	52–60	50–58	48–56	46–54	43–51	39–47	35–43	30–38
Maximum change in mass (%)*	0.5	0.5	0.5	0.5	0.5	0.8	0.8	1.0	1.0
Minimum retained penetration (%)*	55	53	53	50	50	46	43	37	35
Minimum softening point after hardening (°C)*	57	54	52	49	48	45	41	37	32
Minimum flash point (°C)	240	240	240	230	230	230	230	220	220
Minimum solubility (%)	99	99	99	99	99	99	99	99	99

* After hardening. Several methods of hardening are acceptable, however only the RTFOT is to be used for reference purposes

Fig. 2.13. Tables 1, 2 and 3 of the harmonized European specification for paving grade bitumen — mandatory properties

CEN Table 2. Specifications for paving grade bitumens with penetrations from 250 dmm to 900 dmm

Property	Grade			
	250/330	330/430	500/650	650/900
Penetration at 15 °C (dmm)	70–130	90–170	140–260	180–360
Dynamic viscosity at 60 °C (Pa s)	18	12	7	4.5
Kinematic viscosity at 135 °C (mm^2/s)	100	85	65	50
Maximum change in mass (%)	1	1	1.5	1.5
Maximum viscosity ratio at 60 °C	4	4	4	4
Minimum flash point (°C)	180	180	180	180
Minimum solubility (%)	99	99	99	99

CEN Table 3. Specification for paving grade bitumens — soft bitumens: grades specified by viscosity

Property	Grade			
	V1500	V3000	V6000	V1200
Kinematic viscosity at 60 °C (mm^2/s)	1000–2000	2000–4000	4000–8000	8000–16000
Minimum flash point (°C)	160	160	180	180
Minimum solubility (%)	99	99	99	99
Maximum change in mass (%)	2	1.7	1.4	1

Fig. 2.13. Continued

paving bitumens', which includes 10/20 and 15/25, is currently under review.

The CEN specification is divided into three tables (Fig. 2.13), the first of which is similar, but not identical to Table 1 of BS 3690: Part 1.[32] Table 2 gives the specifications for the softer grades characterized by penetration and Table 3 gives the specification for very soft grades characterized by the viscosity at 60 °C.

Another key difference in the organization of the CEN specifications is the presence of a series of 'national' conditions for various properties which are applicable in some, but not all, the Member States. The mandatory specifications for penetration, softening point, resistance to ageing and solubility are retained and a specification for flash point has been added.

In addition to these mandatory requirements, there are several tests which may be used including wax content, viscosity (dynamic and kinematic), Fraass breaking point and increase in softening point or penetration index after ageing. Only the requirement to measure kinematic viscosity at 135 °C is applicable to the UK.

2.5.3. Performance specifications

Although the adoption of the CEN specifications represents a significant step forward for the bitumen industry, it is important to remember that, with the exception of the inclusion of a small number of additional tests, the specifying method is relatively unchanged. The grades of bitumen produced will continue to be characterized primarily by their penetration and softening point. Some authorities have suggested that this is no longer sufficient and that better methods are required to prescribe bitumen and to determine its performance within pavements. The impetus towards performance specifications has emanated from the USA following the Strategic Highways Research Program (SHRP). The SHRP was established by the US Congress in 1987 as a five year, $150M programme to improve the performance, durability and safety of its highways. As such, the SHRP produced performance specifications that employed the use of non-standard methods. These specifications became known as Superpave[TM] (Superior Performing Asphalt Pavements)[36] specifications and are revolutionary in their abandonment of the penetration/softening point approach.

The first issue that SHRP addressed was that asphalt, or rather the bitumen, is subject to a vast change in temperature from mixing, laying and compaction to end use. It also established that the pavement, depending on its location, may undergo relatively large changes in temperature. Secondly, the SHRP recognized that the binder would change during its service life and determined ageing criteria. Thirdly, the SHRP established that rheological characterization of the binder was of fundamental importance to the structural design of the road. The methods used to establish these fundamental properties are outlined below.

Determination of high temperature properties
Measurement of the high-temperature properties of a bitumen gives an indication of its ease of handling at a coating plant. The SHRP uses a rotational viscometer to measure the viscosity of bitumen at elevated temperatures (Fig. 2.14).

The use of rotational viscometers is not, however, a development instigated by SHRP. ASTM Method D4402,[37] which describes the use of a Brookfield viscometer and thermosel to measure the viscosity of bitumen at elevated temperatures, was first published in 1984. The SHRP procedure does, however, set various criteria under which the determination should be made.

Under SHRP, the determination of viscosity is carried out using a Brookfield viscometer and thermosel at 135 °C. The maximum allowable viscosity is 3000 cPs ($\equiv$3 Pa s). In practice, it is desirable to measure the viscosity over a range of temperatures and shear rates so that an indication of binder behaviour during mixing and compaction can be obtained.

Determination of low temperature properties
As discussed in Section 2.5.1, the Fraass Breaking Point Test[31] is used to characterize the low-temperature behaviour of bitumen. However, the difficulties in obtaining reliable Fraass data coupled with the drive to produce valid engineering data led to the development of the bending Beam Rheometer (BBR).[38] In this test, a narrow beam of bitumen is prepared and subjected to a Three-Point Bending Test at low temperature (Fig. 2.15).

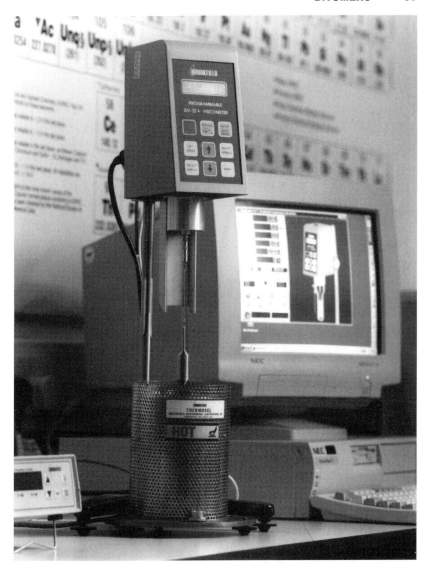

Fig. 2.14.   A computer-controlled Brookfield viscometer used to determine the high-temperature properties of bitumen[37]

The creep stiffness (*S*) and the rate of change of creep with time (*m*-value) are then determined. The SHRP procedure requires that this test is carried out at the minimum service temperature of the road for binders which have been subjected to ageing in a pressure ageing vessel (PAV). As such, it gives an indication of the stiffness of the binder after some years in service. The Superpave[TM] Specification[36] states that the creep stiffness should not exceed 300 MPa and the *m*-value should be ≤0.3.

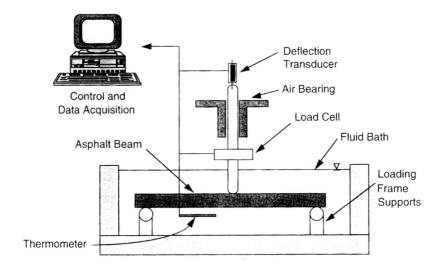

Fig. 2.15.   Principle of operation of the bending beam rheometer (BBR)[38]

Although the BBR[38] gives a very good indication of the low temperature properties of the binder, it does not always reflect in service performance. For example, some binders with creep stiffnesses >300 MPa have been found to perform very well in service. The reason for this is that some binders have a greater capacity for elongation than others, despite their high stiffness values. Consequently, the BBR study can be supplemented with information gathered via the Direct Tension Test (DTT).[39] In essence the DTT is similar to the classical determination of ductility but is carried out at much lower temperatures and with smaller samples. A 'dog bone' of material is prepared and stretched at a slow, constant rate until it fractures. The strain to failure at a particular temperature is then reported. The DTT[39] is carried out on materials which have been subjected to ageing in the PAV.

Intermediate temperature properties
The properties of the binder between the two extremes of mixing temperature and brittle point are determined using the Dynamic Shear Rheometer (DSR) (Fig. 2.16).

A more detailed description of rheology is given in Section 2.6 below but, in essence, the Rheometer measures the elastic and viscous nature of the bitumen across a range of temperatures. The SHRP established precise links between the rheological behaviour of the bitumen and the end performance of a pavement. In particular, rheological parameters were established for fatigue and rutting tendency. The rheological performance of the binder as prepared and after ageing in the RTFOT and PAV are required for specification.

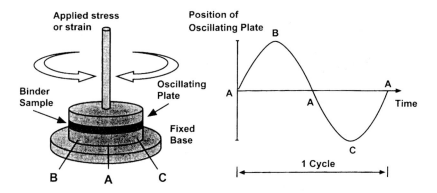

Fig. 2.16. Principle of operation of the dynamic shear rheometer (DSR)[36]

The ultimate goal of the SHRP was to produce a new set of specifications based entirely on performance indicators. The new specification proposed a series of paving grades with required levels of performance based on the upper and lower pavement temperatures. Although precise figures are not available at present, it is estimated that the savings due to the implementation of SHRP have offset the original expenditure several times over. The SHRP has not, however, remained static since its first findings were published and a significant amount of additional work has been undertaken into the effects of storage and physical hardening on end performance.

2.6. Rheology

2.6.1. Terminology and theory

Rheology is defined as the study of the deformation and flow of matter.[40] In general terms, rheology explains the behaviour of materials when subjected to a stress. There are two extremes of behaviour for any material[41] — elastic and viscous. If a stress is applied to a perfectly elastic material then the resultant strain appears immediately. This can be visualized as follows. A load is applied to the material and the material responds immediately by deforming. When the load is removed, the material recovers its original form instantaneously. Viscous materials behave differently as the resultant strain does not appear until the stress is removed. In this case, when the load is applied the material will show an initial resistance to the load but will then flow. When the load is removed, there is no recovery of the initial form.

If the stress is continually applied and removed in a cyclic fashion then, for elastic materials, the stress and strain are said to be 'in phase' whereas for viscous materials the stress and strain are completely 'out of phase' (Fig. 2.17). Another way of expressing this is to state that the strain lags behind the stress by an angle of 90° for a viscous material whereas for elastic materials the angle is 0°.

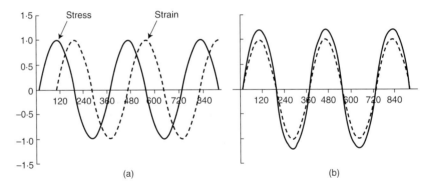

Fig. 2.17. The two extremes of rheological behaviour[41] *(a) A perfectly viscous material; stress and strain are out of phase by 90°. (b) A perfectly elastic material; stress and strain are in phase, the phase angle is 0°*

Real materials exhibit behaviour which is somewhere between these two extremes. These materials are described as 'viscoelastic' and the stress and strain are out of phase by an angle delta (δ) which will have a value between 0° and 90°.

The relationship of stress to strain is known as the modulus. For real materials, the modulus consists of two components which represent the relative magnitudes of the in-phase and out-of-phase components of the applied stress. These components are related by the equation

$$E^* = E' + iE'' \tag{2.1}$$

where $i = \sqrt{-1}$, E^* is known as the complex modulus; E' the storage or elastic modulus and E'' the loss or viscous modulus. This terminology applies for measurements carried out in tension or flexure.[42] However, most rheological measurements on bitumen are carried out using parallel plate geometry (Fig. 2.16) which applies shear to the sample. The values of E in Equation (2.1) are then substituted by G, giving

$$G^* = G' + iG'' \tag{2.2}$$

The basic terminology for the components remains the same, however for absolute accuracy the word 'shear' should be included. Hence G^* becomes the complex shear modulus. The shear storage modulus (G') is best defined as a measure of the amount of energy which a sample can store (and subsequently release) when subjected to a stress. The shear loss modulus (G'') represents the amount of energy lost by the sample through internal motion and heat. A simple representation of the relationship of the moduli is given in Fig. 2.18.

G' is occasionally described as the real component of the complex modulus and G'' the imaginary part. The ratio of loss modulus to storage modulus gives an indication of the relative contributions of each component and also gives the tangent of the phase angle

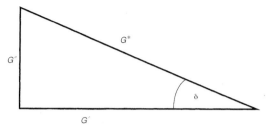

Fig. 2.18. A simple representation of the relationship of moduli: tan δ = loss, tangent = G″/G′

$$\tan \delta = G''/G' \tag{2.3}$$

The magnitude of each of the components will be influenced by the temperature, frequency and loading time, or stress, to which the sample is subjected. Rheological measurements thus represent an excellent means of characterizing the performance of bitumen. These rheological measurements are usually carried out using a DSR which allows temperature, frequency and stress to be varied individually.

2.6.2. Rheological tests and data manipulation

The most common test is to hold the frequency and stress constant while varying the temperature. These tests are called 'temperature sweeps' or 'isochrones' (Fig. 2.19) and are useful in predicting the maximum and minimum service temperatures for a bitumen.

The second common test is to hold the temperature and stress constant while varying the frequency. These tests are called 'frequency sweeps' or 'isotherms' (Fig. 2.20) and are useful in determining the effect of frequency on the modulus of the sample.

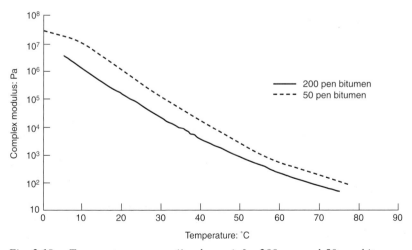

Fig. 2.19. Temperature sweeps (isochrones) for 200 pen and 50 pen bitumen

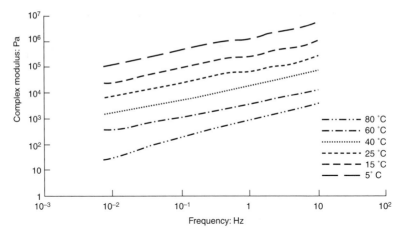

Fig. 2.20.   Typical frequency sweeps (Isotherms)

Bitumen shows a lower modulus when measured at low frequency than it does at high frequency. This explains why asphalt pavements can be susceptible to permanent deformation (i.e. rutting) when subjected to slow moving (i.e. low frequency) traffic. It is usual to carry out frequency sweeps over a range of temperatures so that an overall picture of the response of the bitumen is obtained.

The third test involves holding the temperature and frequency constant, while varying the stress. Although this test can be useful in determining the effects of traffic loading on the bitumen, it is normally used to determine the 'linear viscoelastic region' of the sample (Fig. 2.21). Truly

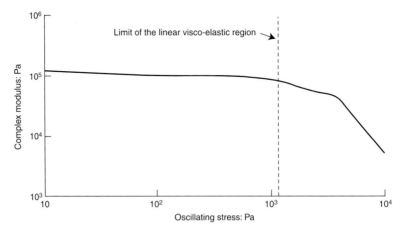

Fig. 2.21.   A typical sress or amplitude sweep used to determine the linear viscoelastic region (LVER) of bitumen. Measurements outside the LVER do not produce reliable data

accurate measurements of the modulus of bitumen can only be obtained when working in the linear viscoelastic region.

Although temperature and frequency sweeps give a good indication of the performance of the bitumen, it is often useful to express both sets of data on a single axis. There is a generally accepted principle in rheological characterization which states that the effects of time and temperature are related. This Time/Temperature Superposition Principle[43] involves performing a series of frequency sweeps at discrete temperatures and then shifting each of the curves until they lie on the same plane as the curve at a reference temperature (usually 25 °C). This allows the prediction of the properties of bitumen at very high or very low frequencies[44] which may be outside the measurement range of the instrument. When shifted together, the data form is known as a 'mastercurve' (Fig. 2.22).

It is important that data manipulation is carried out very carefully, particularly with polymer modified bitumens which can show significant deviations from the mastercurve due to non-linear effects.[45] Hence, the Time/Temperature Superposition Principle does not apply to polymer modified bitumens, which are discussed in detail in Section 2.7.1.

It is also useful to plot the relative contributions of the storage and loss modulus on a single graph. This 'Black diagram' (Fig. 2.23) gives a good view of which type of behaviour, elastic or viscous, predominates over a particular temperature range. The Black diagram can also be used to predict whether or not the binder will behave in a linear viscoelastic manner.

Care must be taken in the preparation of the sample to ensure that the measurements are accurate. In particular, the volume of the sample should be sufficient to just fill the gap between the plates. Excess material at the

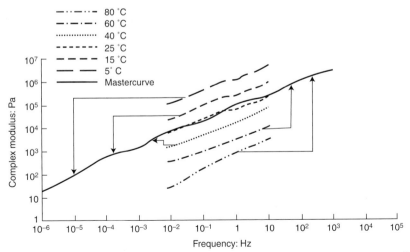

Fig. 2.22.   Construction of a mastercurve from a series of frequency sweeps.[43] The arrows indicate the application of the shift factor

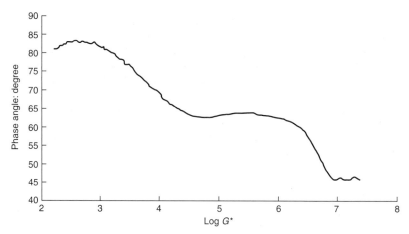

Fig. 2.23. A typical Black diagram showing the relationship of complex modulus (G^*) and phase angle (δ)

edges of the sample can lead to the generation of inaccurate data. The two most common methods of overcoming edge effects are the preparation of discs of bitumen of a similar size to the test plates and loading of precise volumes of material onto the lower test plate.

The thermal history of a sample can have a significant effect on its properties.[46,47] The phenomenon of steric and physical hardening, in which a sample can become stiffer over a relatively short period of time, should be taken into consideration (steric is defined as relating to the spatial arrangements of atoms in a molecule). Therefore, it is extremely important that samples are subjected, wherever possible, to the same thermal history before testing to ensure that the measurements are both accurate and reproducible.

It is generally accepted that the effects of steric hardening can be overcome by heating the sample to approximately 100 °C above its softening point.

It is also important that the geometry of the sample is correct. Bitumen undergoes enormous changes in modulus with temperature and not all sample geometries are suitable for measuring this change. The standard geometry employed for measuring the bitumen rheology uses a 25 mm plate and a sample thickness of 1 mm.[48] However, under certain conditions, usually at low temperature, the bitumen can become so stiff that the instrument detects the compliance of the measuring geometry. Following an extensive period of research in the SHRP it was found that these compliance effects could generally be overcome using smaller test assemblies and larger gap sizes. Low-temperature properties are determined using 8 mm plates with gap sizes of 1–2 mm. Consequently, whenever the results of a rheological experiment are presented it is important to include the conditions under which the experiment was carried out.

Although the foregoing has highlighted some of the problems associated with rheological measurements on bitumen it also explains, in part, some of the reasons why these measurements can be so useful. As the rheometer has a variable gap size, it is possible to determine the modulus of a sample at a range of thicknesses and therefore predict its behaviour in a mixture. In practice, it is very difficult to do this as the film of bitumen around an aggregate particle is typically less than 15 μm thick.

The rheometer also allows a study of the interaction between binders and fillers.[49] It has been established that there is a degree of synergy between certain binders and aggregates which can lead to significant increases in G^* that cannot be explained through simple compositional effects.

2.6.3. Relationship of rheological parameters to mix performance

One of the most important results from the SHRP was the establishment of a correlation between rheological measurements and end performance data.[50] The SHRP established that there were two rheological parameters which could be directly related to the tendency of a pavement to show rutting and fatigue cracking.

It was established that the rutting tendency of a pavement was influenced by the ratio of the complex modulus to the phase angle. The rutting parameter, $G^*/\sin\delta$, was thus established. It was also found that rutting is primarily influenced by temperature and consequently the rutting parameter was set at a minimum of 1 kPa at the specified temperature. For aged (i.e. after RTFOT and/or PAV) binders, this value rises to a minimum of 2.2 kPa. In order to maximise the rutting parameter, high values of G^* and low values of δ are required. The Superpave[TM] specification[36] thus promotes the use of relatively stiff, elastic binders to reduce rutting.

The parameter established for fatigue cracking is the product of the complex modulus and phase angle, $G^*\sin\delta$. This parameter is equivalent to the loss modulus G''. As it was perceived that fatigue cracking is a long-term process occurring at relatively low pavement temperatures, the fatigue parameter is determined using binder which has been subjected to ageing under RTFOT and in the PAV. The maximum permissible value of $G^*\sin\delta$ is 5000 kPa at the relevant test temperature. In order to reduce the fatigue parameter, low values of G^* and δ are required. The Superpave[TM] specification[36] promotes the use of elastic but compliant binders.

Recent developments in the SHRP have identified that the rutting parameter ($G^*/\sin\delta$) may not be entirely applicable to polymer modified bitumens. A better correlation may be achieved with low shear viscosity.[51]

2.7. Bitumen modification

As traffic volumes and axle loads have continued to increase over the years so have the demands placed on the bitumens used in road construction. It could be argued that the bitumen in a highway makes the single most important contribution to the performance of the road. The supporting evidence for this statement comes from an appreciation of the key performance requirements for any bituminous binder used in asphalts. In summary, the binder must

- be flexible enough to absorb traffic stresses and prevent fatigue cracking
- be cohesive enough to bind the minerals together
- be adhesive enough to prevent mineral erosion (fretting)
- have a broad in-service temperature range.

As the traffic volumes and loads have increased, it has been found that unmodified (sometimes referred to as 'straight run') bitumen has been less capable of satisfying all of these requirements. Hence the need to develop bitumens with a higher level of performance.

There are many methods of describing the improvements that are necessary for unmodified bitumens to improve their performance characteristics. One of the best is based on consideration of the rheological elements established in Section 2.6.

It has been established in this Chapter that bitumen is a thermoplastic, viscoelastic material and therefore susceptible to the effects of temperature, loading stress and frequency. In summary, bitumen will become softer as the temperature increases, the frequency of loading decreases or the loading stress increases. This apparent 'softness' at low frequency and high stress is reflected in an increased tendency for a pavement to exhibit rutting under heavy, slow-moving traffic.

2.7.1. Polymer modification

It seems logical that a reduction in the temperature susceptibility of the pavement should lead to improved performance and service life. The most common method of reducing the temperature susceptibility of bitumen is by the addition of polymers, this has the following effects[52,53]

- an increase in softening point
- a decrease in penetration
- suppression of the Fraass breaking point
- an increase in viscosity.

However, the benefits of polymer modification do not end with a reduction in temperature susceptibility; improvements in flexibility, workability, cohesion, ductility and toughness[53] may also occur. Over the years, many different polymers have been assessed and found to improve the performance of bitumens. The various types are shown Table 2.1. Despite this, relatively few have been exploited commercially.

Thermoplastic (plastomer) modifiers

A general distinction can be drawn between the two major classes of polymer used to modify bitumen. Thermoplastic polymers, often referred to as plastomers, have a similar although less dramatic temperature susceptibility to bitumen in that they are hard at low temperature and fluid at high temperatures. Thermoplastic polymers tend to influence the penetration of the bitumen more than the softening point. The softening

Table 2.1.   Types of bitumen modifier

Categories of modifier	Examples of generic types
Thermosetting polymers	Epoxy resin Polyurethane resin Acrylic resin
Elastomeric polymers	Natural rubber Vulcanized (tyre) rubber Styrene–butadiene–styrene (SBS) block copolymer Styrene–butadiene–rubber (SBR) Ethylene–propylene–diene terpolymer (EPDM) Isobutene–isoprene copolymer (IIR)
Thermoplastic polymers	Ethylene vinyl acetate (EVA) Ethylene methyl acrylate (EMA) Ethylene butyl acylate (EBA) Polyethylene (PE) Polyvinyl chloride (PVC) Polystyrene (PS)
Chemical modifiers and extenders	Organo-manganese/cobalt compound (Chemcrete^{TM}) Sulphur Lignin
Fibres	Cellulose Alumino-magnesium silicate Glass fibre Asbestos Polyester Polypropylene
Anti-stripping	Organic amines Amides
Natural binders	Trinidad Lake Asphalt (TLA) Gilsonite Rock asphalt
Fillers	Carbon black Fly ash Lime Hydrated lime

point increases as the concentration of the polymer in the bitumen rises until a maximum softening point is reached. Beyond this concentration there is no increase in the softening point even at significantly higher polymer contents.

Ethylene vinyl acetate (EVA) copolymer has the longest history of the thermoplastic polymers and has enjoyed the greatest amount of commercial success. EVA was originally used to increase the stiffness of HRA and hence improve its performance.[54] However, an unexpected additional benefit of the modification was improved workability at low temperatures. EVA is a random copolymer; the molecules of ethylene and vinyl acetate are randomly distributed along the polymer chain. However, within this random structure there is a degree of association between the polymer chains. Ethylene-rich domains in the polymer tend to segregate and associate with each other. These ethylene-rich domains are semi-crystalline in nature and improve the high temperature properties of bitumen. The vinyl acetate domains are more flexible and improve the low temperature properties of bitumen.

A number of EVA grades are available and are classified according to their melt flow index (MFI) and vinyl acetate content. The MFI is an empirical measure of viscosity and gives an indication of the molecular weight of the polymer. The higher the MFI, the lower is the molecular weight. Historically, the most common grade of EVA used for bitumen modification in the UK has been 18/150 which has a vinyl acetate content of 18% and an MFI of 150. Other grades of EVA have been used successfully in bitumen modification, including 33/45, 24/03 and 20/20.

Elastomer modifiers
Elastomeric polymers have also been very successfully exploited in bitumen modification where they introduce elastic recovery into the modified binder.[55] This property of elasticity is generated by the structure of the polymer and can be a result of entropic (driven by structure and disorder) or enthalpic (driven by energy) considerations depending on the polymer concerned. There is a wide range of elastomeric polymers available, as shown in Table 2.1. Several of these have been exploited commercially and found to make significant improvements to the properties of the bitumen. The random elastomers such as natural rubber, SBR and chloroprene (Fig. 2.24) generate their elasticity primarily from

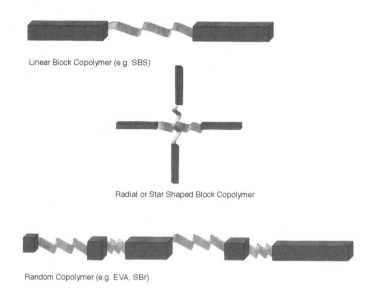

Linear Block Copolymer (e.g. SBS)

Radial or Star Shaped Block Copolymer

Random Copolymer (e.g. EVA, SBr)

Fig. 2.24. Typical polymer structures. The red structures represent 'hard' domains such as styrene. The grey regions represent 'flexible' domains, such as butadiene entropic effects. In other words, the lowest energy configuration for a random elastomer is coiled and entangled. When stretched, a degree of order is introduced into the structure and the polymer's natural tendency is to resist this structuring. When the stretching force is removed, the chains collapse back onto each other and the free energy is minimized again.

Although random elastomers have been used with some success in bitumen modification,[56] their use is now somewhat overshadowed by the block copolymers. A typical example of the use of a random elastomeric polymer is the inclusion of natural rubber latex in porous asphalt wearing course.[57] The effect of the modification is primarily to increase the viscosity, minimize binder drainage and increase the bitumen film thickness on the aggregate. However, increased durability is also noted. ('Binder drainage' is a measure of the ability of a mixture to hold bitumen.)

Block copolymers were first reported in 1968 and are produced by the successive polymerization of two or more monomers.[58] (Monomers are molecules which can be polymerized.) The key to the commercialization of block copolymers came with the development of anionic catalysts which gave precisely defined blocks of each monomer. The most common block copolymers used in bitumen modification are styrene–butadiene–styrene (SBS) and styrene–isoprene–styrene (SIS). Using the polymerization processes outlined above, it is possible to produce both linear and radial (or star shaped) polymers.

The elastic properties of block copolymers are influenced by both entropic and enthalpic considerations. The natural tendency for the chains to seek the lowest energy conformation still exists but there is also an inherent structure to these molecules. As each chain is ordered, there is a natural tendency for blocks of similar structure to associate with each other. This leads to the generation of 'hard' and 'soft' regions within the structure. The polymer will then stretch to the limits dictated by the hard blocks.

For SBS copolymers, the styrene domains constitute the hard blocks and affect the high temperature properties of the bitumen. The butadiene soft blocks are very flexible and affect the low temperature properties. These inherent properties of rigidity and flexibility explain, in part, why SBS copolymers have seen such great exploitation in bitumen modification. However, it is this 'structuring property' which has such a great influence on the behaviour of SBS modified bitumen since, generally speaking, there is always a semblance of rigid and flexible structure in the bitumen regardless of the temperature. Although SBS copolymers tend to show their greatest influence on the softening point of the blend, there are also significant improvements in cohesion, ductility and brittle point.

The overall properties of a polymer modified bitumen (PMB) are not only dependent on the correct selection of polymer type and concentration but also the effectiveness of dispersion of the polymer. This is governed by: the overall structure of the bitumen, the processing, and type and concentration of the polymer.[59]

2.7.2. Storage stability

The above discussion illustrates some of the difficulties which may be encountered when preparing PMBs. As has been shown, bitumen is a very complex blend of chemical species consisting of a dispersion of asphaltenes in a matrix of maltenes. The addition of a high molecular weight polymer to this system will cause a disturbance of the overall structure. The polymer and the asphaltenes will then compete for the stabilizing and solubilizing components of the maltene phase. If there is an insufficient quantity of the maltene phase then separation of the polymer and the asphaltenes can occur and the system is described as being 'incompatible'. Evidence for incompatibility comes from a number of sources including fluorescence microscopy, in which an incompatible system is seen as a coarse dispersion. However the most common test for incompatibility is the Hot Storage Test[60] in which a sample of PMB is placed in a suitable container and stored at elevated temperature in an oven for up to seven days. At the end of this period, the sample is withdrawn and the top and bottom sections separated and tested, as shown in Fig. 2.25. Incompatibility may be revealed by a difference in penetration, softening point, viscosity or rheology between the upper and lower portions.

SBS tends to rise to the surface of an incompatible mix whilst EVA sinks to the bottom. The implications of incompatibility are that the binder may become heterogeneous during storage, e.g. in the case of SBS, a polymer-rich phase may form at the top of the storage tank while a polymer-lean phase may form at the bottom. Ultimately, this means that, during production, mixtures could be practically unmodified (and therefore not of the required performance) at the start and heavily modified (i.e. impossible to compact) at the end.

The Hot Storage Test[60] is usually carried out over a period of seven days at 160 °C with the binder being described as 'storage stable' if the

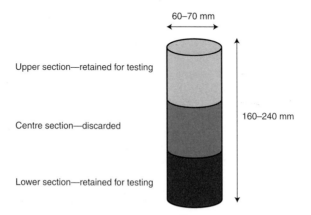

60–70 mm

Upper section—retained for testing

Centre section—discarded

Lower section—retained for testing

160–240 mm

Fig. 2.25.   The Hot Storage Stability Test.[60] The test may be performed in a purpose-built vessel with drain valves or in a beer can which is subsequently cooled and sectioned

difference in softening point between the upper and lower portions of the sample is less than 5 °C. However, it has been realised that binders which are stable in terms of softening point may not be stable in penetration terms. The most recent draft of the BS Storage Stability Test seeks to address this by including an expression for 'penetration stability'.

In practice, it is possible to use binders which are not storage stable as long as they are treated with special care to prevent the separation of the polymer. This usually involves the use of stirred tanks or extended circulation regimes. There is also evidence that binders which are found to be unstable using the conventional Hot Storage Test[60] may be stable in practical storage due to the presence of convection currents. Methods to simulate this are currently being explored.

There are a number of factors that must be taken into consideration in defining storage stability, including the effects of temperature and time. The key requirement for a binder should be that it is storage stable at the recommended storage temperature for the period over which it is normally expected to be stored.

As discussed above, it is not common for polymers and bitumen to be inherently compatible with each other. Therefore, it is usual to modify the bitumen, the polymer or both to ensure that a storage stable system results.

2.7.3. Production considerations for polymer modified binders

The processing of the polymer bitumen blend will also have a profound effect on the effectiveness of the dispersion and therefore the ultimate properties of the product. The degree of processing required is influenced by the type and concentration of the polymer.

Thermoplastic polymers such as EVA tend to melt on addition to hot bitumen and therefore require only low shear to generate an adequate dispersion. Elastomeric polymers such as SBS require much more work to generate adequate dispersion. There are two reasons for this. Firstly the polymer, because of its structure, does not melt into the bitumen. The first stage in preparing SBS/bitumen blends is a swelling of the styrene domains by the bitumen. The swollen polymer must then be subjected to high shear if an adequate dispersion is to be achieved. The production variables of time, temperature and shear must be carefully balanced to ensure that the polymer is well dispersed in the bitumen without degradation and loss of properties.[61]

Finally, the molecular weight and concentration of the polymer have an effect on the ease of processing and compatibility.[62] High molecular weight polymers, regardless of their chemical type, tend to be more difficult to disperse than those with low molecular weight. Generally, the higher the concentration of polymer, the more difficult will be the dispersion. Although the effects of polymer modification of bitumen are readily observed through the classical tests of penetration, softening point and brittle point, a better understanding of the effects of the polymer is gained through tensile testing and rheological experiments.[63–67]

Fig. 2.26.   The Ductility Test for determining the basic tensile properties of bitumen.[68] *The photograph illustrates the superior performance of polymer modified bitumen (upper sample) in comparison to unmodified bitumen*

2.7.4.   The effects of polymer modification

The benefits of polymer modification are dramatically illustrated by Ductility Tests at low temperature (Fig. 2.26). Unmodified bitumen tends to show a low elongation to break and no elastic recovery. Elastomer modified bitumens, however, exhibit a much greater elongation to break and a significant recovery.

The Toughness/Tenacity Test[68] also gives a good indication of the benefits of polymer modification. This test involves pulling a spherical probe from the surface of the bitumen (Fig. 2.27) at high speed and measuring the load and elongation to break. PMBs have a greater initial peak (indicating resistance to stress) and a longer elongation to break. The Toughness/Tenacity Test is useful in screening different binders for performance. However, significant problems with reproducibility have meant that, realistically, it cannot be used for specification purposes.

Polymer modification tends to alter the gradient of the complex modulus when plotted against temperature (Fig. 2.28). In comparison to unmodified bitumen, PMBs show a reduction in modulus at low temperature and an increase in modulus at higher temperature. Thus, the two prevailing distress mechanisms for a pavement — fatigue cracking and rutting — are reduced. The extent of this reduction in temperature susceptibility will be affected not only by the type and concentration of the polymer but also by the source and grade of the bitumen. The

Fig. 2.27. The Toughness Tenacity Test for determining the tensile properties of bitumen. The spherical probe is pulled from the sample at 1000 mm/min

rheological effects of modifying bitumen with a thermoplastic polymer such as EVA are particularly evident at lower temperatures as significant reductions in modulus, compared to unmodified bitumen, can be achieved.

With elastomer modified bitumens, the effects are noted both at high and low temperature. At low temperature, the modulus is significantly reduced due to the presence of the soft portion of the elastomer. This

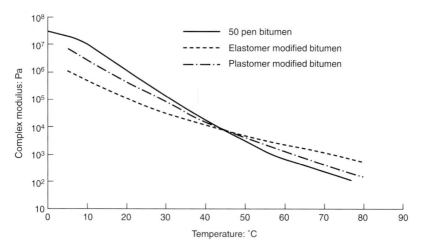

Fig. 2.28.   The effect of polymer modification on rheology. Note the reduction in modulus at low temperature and increase in modulus at high temperature

effect is more pronounced if a relatively soft grade of base bitumen is used. At higher temperatures, the modulus is significantly greater due to the presence of the hard blocks.

It is often useful to investigate the effect of polymer modification on the phase angle, which indicates the degree of elasticity of the binder. The lower the phase angle, the more elastic is the binder and hence the tendency to exhibit rutting in a pavement is reduced.

It is also useful to determine the temperature at which the binder changes from displaying predominantly elastic to predominantly plastic behaviour. The temperature at which this occurs can give an indication of the rutting tendency of the pavement. Polymer modification of the bitumen, particularly with elastomers tends to move this crossover point to higher temperatures.

Polymer modification also leads to a significant improvement in the cohesion of the binder as determined by the Vialit Pendulum Test[69] (Fig. 2.29). This test involves placing a thin film of binder between two cubes and measuring the energy required to remove the upper block. The maximum impact energy is usually significantly increased by polymer modification, as is the overall energy across the entire temperature range. The Vialit Pendulum Test[69] is of greatest significance in situations where aggregate is placed in direct contact with traffic stresses, for example in surface dressings and the chippings in HRA wearing courses. Its significance for other materials has yet to be fully evaluated.

Thus far, we have concentrated mainly on the effects of bitumen modification on the properties of the bitumen itself. However, to fully understand the benefits of bitumen modification, it is necessary to investigate the effects of the modified binder on the mixture.[70] The effects of polymer modification are usually best appreciated at the extremes of the in-service temperature range of the road. Table 2.2 illustrates the

Fig. 2.29.   The Vialit Pendulum Test for determining the cohesion of bitumen.[69]
The photograph shows the pendulum swinging down before impacting the sample
on the left

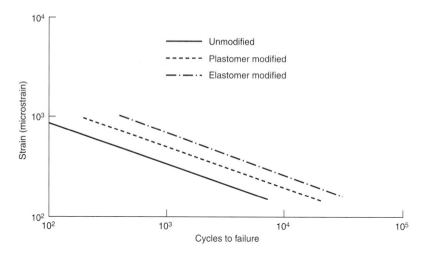

Fig. 2.30. The improvements in fatigue characteristics of asphalt mixtures obtained using polymer modified binders

reduction of wheel tracking rate and total rut depth achieved using modified binders in a HRA wearing course. Figure 2.30 shows the improvement in fatigue properties achieved using modified binders in asphalt mixtures.

To date, modified binders have primarily been used in wearing course applications. However, it is possible that the future will see modified binders being incorporated into basecourses to improve the durability of the road.[71] Permeable basecourse materials required for new variants of porous asphalt are also likely to require modification to ensure the required performance levels are met.

2.7.5. Other bitumen modifiers
Thermosetting polymers
Although thermoplastic polymers have seen the greatest exploitation in the modification of bitumen for roads, thermosetting polymers also have an important role to play. Thermosetting polymers, once cured, are unaffected by temperature and consequently require special handling techniques to render them suitable for road construction or maintenance. The difficulty in handling thermoset modified bitumens comes from the fact that the polymer has to be created 'in situ' usually by the reaction of

Table 2.2. Effect of polymer modification on wheel tracking rate of HRA

Binder	Wheel Tracking Rate at 45 °C: mm/h
50 pen	3.2
100 pen	9
50 pen + EVA	0.5
200 pen + SBS	0.6

two or more components. Only when the two components have reacted to form a cross-linked polymer network will the benefits of modification be realized.

The main thermosetting polymer used in bitumen modification is epoxy, although some polyurethane is used. Epoxy is supplied as a two-component system of resin and hardener which is mixed together in the bitumen. The resin and hardener ultimately react together to produce a very stiff cross-linked polymer. As this is a reactive system, the relative concentrations of resin and hardener and the temperature must be very carefully controlled to ensure that the mixture remains workable for a sufficient time to allow compaction.

Despite the difficulties in handling epoxy modified binders, some significant benefits can be achieved, including increased stiffness and reduced rutting characteristics. The cross-linked polymer also imparts fuel resistance properties to the bitumen.

Thermosetting polymers can also be used as binders for the high-friction surfacings used on the approaches to traffic lights, roundabouts, pedestrian crossings and the like. These systems employ very specialized equipment that mixes together a resin modified stream and a hardener modified stream immediately before application to the road surface. A very hard aggregate such as calcined bauxite is then applied, which results in a road surface of exceptional skid resistance. These materials are discussed in some detail in Chapter 10.

Waxes

It is also possible to add 'waxy' materials to bitumen to improve its workability. Care has to be taken both in the choice of modifier and its concentration to ensure that the benefits gained by modification are not outweighed by the reduction of adhesion to the aggregate. This class of modifier has been mainly used in stone mastic asphalts to improve workability and compactibility.

Chemical modifiers

The addition of polymers to bitumen should be seen as primarily a physical modification as, generally, there is very little chemical reaction between the components. There are, however, other modifiers which improve the properties of bitumen by chemically reacting with its components.

Sulphur, as a chemical modifier, has perhaps the longest history and greatest exploitation both in the UK and the USA.[72,73] The addition of sulphur to bitumen in high concentrations (12–18%) can lead to significant modification of the final properties of the mixture. The nature of the interaction between sulphur and bitumen is not clear, however, it is evident that some of the sulphur will chemically react with the bitumen components. The remaining, unreacted, material forms a separate phase, the properties of which depend on the particular temperature. At mixing and laying temperatures, the unreacted sulphur acts as a plasticizer and improves the workability of the mixture. At lower temperature, the sulphur recrystallizes and produces a stiffer mixture.

The benefits of sulphur modified bitumen in terms of improved performance at relatively modest cost are, however, outweighed by environmental considerations. At temperatures above 150 °C, a reaction occurs between the bitumen and sulphur which produces asphaltenes and releases significant quantities of toxic hydrogen sulphide gas. Although there have been several attempts to improve temperature control and thus minimize the possibilities of this reaction taking place, the risk of exposure of operatives remains high. Significant concerns have also been raised about emissions during the recycling of sulphur asphalt. Consequently, there is very little sulphur modified bitumen used in road construction.

A number of other chemically reactive modifiers are available which claim to enhance the properties of bitumen. The most famous, by far, is a combination of organo-manganese, cobalt and copper species in an oil carrier sold under the trade name 'Chemcrete'^{TM}. As with sulphur, the precise role of the compounds is uncertain although it is believed that they catalyse the reaction between bitumen and oxygen. This reaction leads to the formation of stable 'macrostructures' within the bitumen which increases its strength and viscosity. The degree of enhancement is influenced by the concentration of the manganese and the source of bitumen and, consequently, the process is very sensitive to these factors. Extensive trials were undertaken with Chemcrete in the early 1980s, the results of which were variable. In some cases, the reaction between the bitumen and the manganese led to excessive hardening of the bitumen and embrittlement of the pavement whilst in others, the reaction did not proceed effectively and rutting occurred.[74,75]

Adhesion agents
Agents may be added to bitumen to improve the adhesion of the bitumen to the aggregate. In many cases, adhesion is not a problem. However, in the presence of water, the bond between the bitumen and the aggregate can be reduced to the extent that the bitumen loses contact with the stone. This process is known as 'stripping'. The tendency of a binder to strip is influenced by its viscosity and the nature of the aggregate. High-viscosity binders tend to be more resistant to stripping although the aggregate may be more difficult to coat initially.

Certain minerals are known to reduce stripping, for example hydrated lime. It is also possible to reduce the stripping tendency of the binder by incorporating surface-active agents which enhance the physical bond between the bitumen and aggregate. These agents are usually fatty amines and are incorporated at levels up to 1% in the binder. It is believed that the amine groups orient themselves toward the aggregate while the fatty group remains in the bitumen. A chemical bridge is thus formed between the aggregate and the binder. A limitation of the fatty amines is that they tend to be relatively unstable at bitumen storage temperature and can rapidly become deactivated. Other materials have been developed to overcome this stability issue.

Miscellaneous modifiers

A range of other materials may be used to modify bitumen although the distinction between binder and mix modifier becomes increasingly vague. These include fibres which, although added to the mixture, tend to influence the binder properties, particularly the viscosity.

Natural bitumens such as Trinidad Lake Asphalt (TLA) or gilsonite may be incorporated to improve stiffness and stability.[76] Pulverized tyre rubber has also been used to increase binder viscosity in porous asphalt[77–79]. An unexpected side effect of this modification was that the noise characteristics were improved. Carbon black and fly ash have been added to bitumen in partial replacement of fillers.[79] Pulverized glass has been used to replace a portion of the filler in a HRA wearing course, where it was also found to improve the reflectivity of the surface.[79]

2.8.   Emulsions and cutbacks

It was stated in Section 2.3, that bitumen viscosity should be around 0.2 Pa s in order to coat the aggregate effectively.[16] In normal hot-mix asphalt production, this is achieved by heating the bitumen and aggregate to appropriate temperatures. However, there are other methods of influencing the viscosity which do not involve the application of heat. Firstly, a volatile diluent can be added to the bitumen to produce what is called a 'cutback'. Secondly, a bitumen emulsion can be manufactured. These processes are described below.

2.8.1.   Cutback bitumen

Cutback bitumens are produced by the addition of a volatile diluent or flux to penetration grade bitumen. BS 3690: Part 1[32] lists three grades of cutback (50, 100 and 200 secs). They are specified in terms of their viscosity which is measured using the standard tar viscometer (STV). The higher the number, the more viscous the material. The two main areas of application for cutback bitumen are surface dressing and storage grade macadam mixtures. In either case, the performance of the cutback is entirely dependent on the rate of evaporation of the diluent. Commonly used diluents for cutbacks include kerosene, gas oil and white spirit. The use of creosote is now almost obsolete due to health and safety considerations.

The use of cutback in storage grade macadams is primarily to overcome the difficulties associated with producing small quantities of hot material which cool quickly and, thus, cannot be laid. Traditional uses for these mixtures include temporary trench reinstatements and patching. The use of these fluxed materials is, however, becoming less popular as the demand grows for reinstatements which are completed in one visit.

Cutbacks are also used in surface dressings where they provide instantaneous grip to dry aggregate. However, safety, environmental considerations and the vulnerability of surface dressings using cutback bitumens to wet weather during their early life has led to a significant reduction in their use.

2.8.2. Bitumen emulsion

Production

The second method of reducing the viscosity of bitumen is to produce an emulsion. An emulsion is defined as a stable dispersion of two or more immiscible liquids.[80,81] Bitumen emulsions represent a particular class of oil-in-water emulsions in which the oil phase has a relatively high viscosity. Obviously, bitumen and water will not form an emulsion merely by mixing the two components together. A surfactant is needed to achieve a stable emulsion. (A surfactant is a substance which reduces surface tension.) The surfactant must allow the bitumen particles to remain in suspension in the water and prevent the particles approaching each other and coalescing.

Bitumen emulsions are normally produced by dispersing hot bitumen in water containing a surfactant.[82] The aqueous phase (or soap) may also contain stabilizers and other additives to improve emulsion quality. The dispersion of the bitumen in the aqueous phase is achieved using a Colloid mill (see Fig. 2.31) which comprises a high speed rotor inside a fixed stator. The high speed of the rotor produces high shear inside the mill and produces very small droplets of bitumen. The size of the droplets produced is dependent on a number of variables including bitumen viscosity, rotor–stator gap, rotor speed and the concentration of the surfactant. It is normal for particles in the range 1–20 μm to be produced. The surfactant keeps the particles dispersed and prevents them from coalescing

Surfactants (often called 'emulsifiers') generally have two chemical components which have different affinities for oil and water. This usually comprises a hydrophilic 'water loving' charged head and a hydrophobic 'water hating' tail. During emulsification, the hydrophobic portion of the surfactant will embed itself into the bitumen, as depicted in Fig. 2.32, while the charged hydrophilic head orients itself towards the water.

The charge on the hydrophilic head of the surfactant may be either positive or negative depending on its chemical nature. 'Traditional' surfactants such as sodium dodecyl sulphate (SDS) have a negatively charged head and thus produce bitumen particles with a negative charge. If subjected to an electrical field the negatively charged particles would migrate to the positively charged anode. Emulsions of this type are termed 'anionic' and were among the first to be manufactured.

Other surfactants, such as fatty amines, produce particles which are positively charged and, consequently, migrate to the cathode. These emulsions are thus termed 'cationic' and are in far greater use today than anionic emulsions. A third class of surfactants exist which produces emulsions without any charge. These are termed 'non-ionic' emulsions.

Before a surfactant can become active it must be reacted with a species to produce the charged head. For anionic emulsions this usually involves reacting the surfactant with a basic material such as sodium hydroxide

$$RCOOH + NaOH \rightarrow RCOO^- + Na^+ + H_2O$$

where R is a hydrocarbon chain.

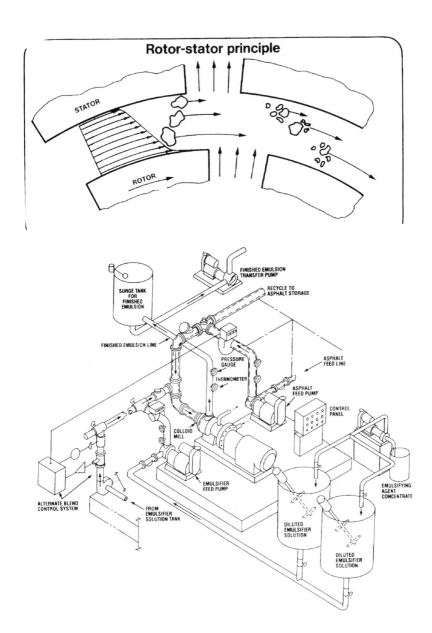

Fig. 2.31.   Schematic diagram of a typical bitumen emulsion plant.[82] Inset shows
the rotor–stator principle

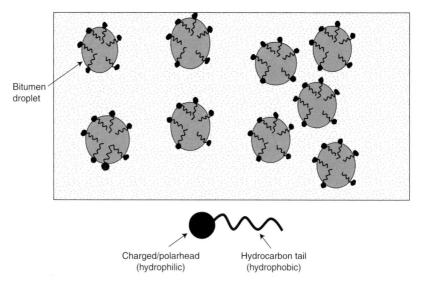

Bitumen droplet

Charged/polarhead
(hydrophilic)

Hydrocarbon tail
(hydrophobic)

Fig. 2.32. Typical emulsifier structure and the orientation of emulsifiers in a bitumen emulsion. For anionic emulsions the head has a negative charge. Cationic emulsions have a positive charge

For cationic emulsions, the reaction is with an acidic species such as hydrochloric acid

$$RNH_2 + HCl \rightarrow RNH_3^+ + Cl^-$$

The above establishes the mechanism of reducing the viscosity of the bitumen so that it may be handled readily. However, it is important that the original properties of the bitumen are recovered through the loss of water (and/or reaction with the aggregate) and coalescence of the particles. This process is known as 'breaking' and is illustrated in Fig. 2.33.

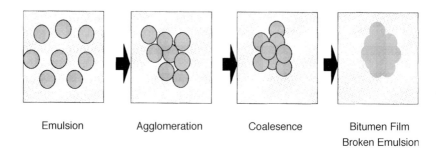

Emulsion   Agglomeration   Coalesence   Bitumen Film
                                                              Broken Emulsion

Fig. 2.33. The processes involved in 'breaking' of a bitumen emulsion[88]

Classification and performance

The British Standard specification for emulsions is BS 434: Part 1,[83] which classifies bitumen emulsions according to their chemical type, rate of break and bitumen content.

The classification works as follows

- chemical type — anionic (A) or cationic (K)
- rate of break — 1, rapid; 2, medium; 3, stable/slow; 4, slow
- bitumen content — expressed as a percentage of the total.

Thus K1-70 is a cationic, rapid breaking emulsion with nominally 70% bitumen. A2-57 is an anionic, medium setting emulsion containing nominally 57% bitumen. There are moves to harmonize the bitumen emulsion specifications throughout europe in a similar fashion to those for bitumen.

The key performance requirements for bitumen emulsions are viscosity, stability and rate of break.[84-86] The viscosity, which determines the ease of handling, is influenced by the bitumen content, emulsifier loading and particle size. For bitumen contents up to approximately 60%, there is very little effect on viscosity. However, increasing the bitumen content above 65% has a significant effect on the viscosity. If the bitumen content is increased beyond 75% there is a significant chance of the emulsion 'inverting' and becoming solid.

The balance between stability and rate of break of an emulsion is critical in ensuring its performance. The emulsion should have sufficient stability to allow storage and transportation but should not be so stable that it does not break in use. The rate of break of an emulsion dictates its end use.[87] For example, emulsions used in surface dressing need to have a rapid rate of break so that there is a quick build up in bond strength between the aggregate and the binder. Consequently, K1 emulsions are used. Conversely, emulsions used for slurry seals and similar mixtures need a much lower rate of break so that the aggregate and the binder become intimately mixed. K3 emulsions are used for slurry sealing. A peculiar balance of properties has to be struck for emulsions used to produce storage grade macadams. During mixing, it is important that the emulsion mixes intimately with the aggregate without significant breaking. When stored, the emulsion should not drain from the mixture nor should the emulsion break completely and prevent further working. Finally, at compaction, the emulsion should break quickly so that the mixture achieves a rapid build up of strength.

The breaking of a cationic emulsion is usually initiated by a chemical reaction between the positively charged emulsion and the negatively charged aggregate.[88,89] This type of reaction is much less likely to occur with anionic emulsions where the breaking process is governed almost entirely by the evaporation of water. The absence of any chemical initiation of the breaking process means that anionic emulsions can be very sensitive to climatic conditions.

Developments in emulsion technology have been mirrored by developments in binder technology. Polymer modified bitumens with

Table 2.3.   Typical uses of bitumen emulsions

Process	Emulsion(s) used
Surface dressing	K1-70/K1-60/proprietary polymer modified
Tack coats	K1-40/K1-60/proprietary polymer modified
Slurry seal	K3-60/A4
Micro-surfacing	Proprietary
Concrete curing	K1-40/K1-60/A1-55/A1-60
Patching	K1-60/A1-60/A1-55
Land remediation	Proprietary
Retread/recyling	A2-50/K2
Reinstatements	K4 proprietary
Soil stabilization	K1-40/A1-40

improved cohesion, toughness and reduced temperature susceptibility[90–92] are now more widely used.

It is important that emulsions are protected from low temperatures as this can cause settlement and, in the worst cases, breaking. If the product is drummed, it is important that the drums are rolled regularly to prevent settlement of the bitumen. It is also important that cationic and anionic emulsions are not allowed to mix as rapid coalescence and breaking will occur.

Bitumen emulsions are now used in a wide variety of pavement construction and maintenance techniques[93,94] as shown in Table 2.3. Increasing awareness of environmental responsibility and the need for preventive maintenance mean that this is likely to continue with increasing importance.

2.8.3.   Testing of bitumen emulsions and cutback bitumens

The British Standard specifications for bitumen emulsion and cutback bitumen are BS 434: Part 1[83] and BS 3690: Part 1[32] respectively, although both will be replaced by harmonised CEN specifications. Both standards specify the measurement of viscosity using flow-cup viscometers which determine the volume of material flowing through the orifice for a specified period of time. The viscosity of emulsions with high bitumen contents, such as K1-70, is determined using a Redwood II viscometer at 85 °C[83] and is measured in seconds. This measurement gives an indication of how well an emulsion will spray. The viscosity of all other emulsions is determined using the Engler viscometer[95] at 20 °C with the results being reported in degrees (ratio of the flow time of the emulsion to that of water). Cutback viscosity is determined using the standard tar viscometer[96] at 40 °C with the results reported in seconds.

Bitumen emulsions complying with BS 434: Part 1[83] have a base grade in the range 70–300 pen while the base grade for cutbacks complying with BS 3690: Part 1[32] is in the range 100–350 pen.

It is very important that the bitumen content of these products can be controlled and determined accurately since variations in bitumen content could lead to inconsistencies in use and ultimately unsatisfactory end performance. The bitumen content of emulsions is measured by performing a Dean & Stark Distillation[97] which uses a light solvent such as xylene to azeotropically distil the water from the emulsion. (An azeotrope is a mixture of liquids in which the boiling point remains constant during distillation, at a given pressure, without change in composition.) This is a comparatively lengthy procedure and great care must be taken to avoid carry-over of the bitumen during the distillation. Several other methods for the determination of bitumen content have been proposed, all of which have some limitations in terms of reproducibility or the time taken to recover the sample. The bitumen content of cutback bitumen is determined by distillation to constant weight.[98] It is important to remember that the properties of the base grade after recovery may differ from those of the binder before manufacture of the emulsion or the cutback.

The rate of break of an emulsion, which defines its end use, is normally determined by mixing with a filler such as silica flour or cement. Filler is added to a sample of cool emulsion in small quantities until the emulsion becomes unstable and the sample becomes semi-solid. The 'Break Index' is then reported as the quantity of added filler.[99] The Break Index Test[99] can be performed manually or automatically and the results obtained may vary significantly between operators and with different grades of filler. At present, the Break Index does not form part of the bitumen emulsion specifications in the UK but it is useful for production control. The Break Index gives an indication of the rate of break of an emulsion in service. However, varying climatic conditions on site mean that correlation between this test[99] and end performance is difficult.

2.8.4. Performance specifications for emulsions and cutbacks

Specifications for emulsions and cutback have developed considerably over the last few years, with a significant amount of work focussing on the production of performance-based specifications. Draft end performance specifications now exist in the UK for surface dressing,[100] micro-surfacing[101] and carriageway slurry seal.[102]

These 'end performance' specifications differ from traditional 'recipe' specifications in that fundamental information on the binder and proposed aggregate must be submitted for approval before the contract is awarded. Typical end performance specifications require information on the rheology and cohesion of the binder, its compatibility with specified aggregates and any limiting climatic conditions. In the case of surface dressing, it is usual to include minimum levels of skid resistance and texture depth of the finished surface.

The aim of end performance specifications is to improve the rational design of traditional maintenance techniques and minimize the risk of failure. As such, they should be viewed as a welcome step forward for the bitumen industry.

2.9. Foamed bitumen

Two methods of reducing the viscosity of bitumen to a level suitable for coating aggregate were discussed in detail in Section 2.8. However, a third method exists which is enjoying increased popularity. This method involves producing a bitumen foam by the controlled introduction of a small amount of water into hot bitumen. The foam produced has a very high surface area and extremely low viscosity, making it ideal for coating aggregate.

The foam-mix process is not a new concept. Csanyi[103] originally developed the process at Iowa State University in 1956. The first trials with foam-mix were carried out in Iowa and Arizona between 1957 and 1960. Adoption of Csanyi's process was, however, limited due to its reliance on the injection of high pressure steam into hot bitumen to produce the foam. Mobil (Australia) refined the process by using cold water to generate the foam and was granted a patent in 1968.

The foam-mix process was originally developed as a means of stabilizing marginal aggregates but has since evolved into a widely accepted maintenance and construction technique. There are several reasons for this. Firstly, the process is very adaptable as it can be used in static mixing plants or specially constructed 'in situ' production units. There are also significant energy savings in comparison to conventional hot-mix production as the aggregate does not require heating or drying. Mixed material can be stockpiled for a period before being used; this results in minimal wastage.

The foam-mix process is equally applicable to the coating of virgin and recycled aggregates which explains its increasing popularity for 'in situ' recycling of roads. The process is also seen as being less susceptible to the effects of weather in its early life than other cold mix processes.[104] As the drive for more energy efficient, environmentally friendly construction continues, significantly increased exploitation of foam-mix technology is likely.[105,106]

2.10. Natural bitumen

Although the bitumen derived from crude oil is by far the most common material used in road construction today, there are several sources of natural bituminous material around the world. Although they do not find substantial use at present, these materials can offer significant benefits when blended in the correct proportions with petroleum bitumen.

2.10.1. Trinidad Lake Asphalt (TLA)

Approximately one kilometre from the sea in the southern half of the Caribbean island of Trinidad lies one of the world's largest deposits of natural bitumen. Covering an area of just under 100 acres and with a depth of around 90 m, the 'pitch lake' contains an estimated 10–15 million tonnes of natural bitumen. The material is excavated mechanically from the lake and contains a significant proportion of water, organic and mineral matter.

The organic matter is removed by passing the product though coarse screens and the water removed by heating to 160 °C. The 'refined' material is termed 'Trinidad Epuré' and has the following composition

- bituminous material, 54%
- mineral matter, 36%
- organic matter, 10%.

With a penetration of 2 and a softening point of around 95 °C, the Epuré is too stiff to be used alone and is therefore commonly blended with an equal quantity of 200 pen bitumen. The blended material has excellent natural weathering characteristics[107,108] and durability.

2.10.2. Gilsonite

The state of Utah in the mid-west of the USA holds another of the world's large deposits of natural bitumen. The material was discovered in the 1860s and was first exploited by Samuel H. Gilson in 1880 as a waterproofing agent for timber. Gilsonite is found below ground in vertical deposits which are mined by hand. The product is exceptionally hard[109] (zero pen) with a softening point between 115 and 190 °C. Blending gilsonite with bitumen dramatically reduces the penetration and increases the softening point.[110] Due to the nature of the mining process, gilsonite is relatively expensive and is therefore not in common use in paving materials. It is, however, used to improve the softening point and stiffness of mastic asphalt used as a bridge and roof waterproofing material.

2.10.3. Rock asphalt

Found mainly in France, Italy and Switzerland, rock asphalt contains approximately 12% bitumen in combination with limestone or sandstone. Rock asphalt was used in early road construction where it was crushed and heated. Nowadays, the use of rock asphalt has all but ceased.

2.11. Tar

The terms tar and bitumen are often used to describe the same material due their similarity in appearance and properties. However, there are very significant differences between the two materials. Whereas bitumen is generally refined by the distillation of crude oil, tar is a by-product of the destructive distillation of solid fuel.

The specification for road tar is given in BS 76,[111] which defines eight grades according to their equi-viscous temperature (EVT). The more viscous tars are used on more heavily trafficked roads. Tar offers increased adhesion to aggregate and imparts a high resistance to attack by fuel spillage compared to bitumen.

The residue from this destructive distillation is termed 'pitch' and, when combined with light oils produces refined tar. The temperatures involved in the destructive distillation of solid fuel are very high (700–1200 °C) in comparison to those which are used to produce bitumen (≈350 °C) and consequently a significant degree of cracking occurs. This cracking can lead to the production of significant concentrations of carcinogenic polycyclic aromatic hydrocarbons.[112] Tar is now classified as a carcinogen[113] (a substance which causes cancer) and, consequently, its use has declined dramatically in recent years.

2.12. Health and safety aspects

The last few years have seen a significant increase in the awareness of the health and safety aspects associated with the use of bitumen. The bitumen industry's representative organizations have all published or updated guidance documents[114–117] on the safe handling of bitumen and bitumen-based products over the last few years.

It has long been recognized that the primary risk associated with the use of bitumen based products is of burns. It is, therefore, imperative that the appropriate Personal Protective Equipment (PPE) is worn at all times.

It is also recognized that exposure of operatives to the effects of bitumen fume and bitumen solutions is an area of some concern. Before addressing the validity of this concern, it is necessary to appreciate the composition of bitumen (in solid and diluted form) and bitumen fume. It was established in Section 2.2 that bitumen is a complex mix of hydrocarbons containing a significant proportion of ring structures. Some of these fall into the class of compounds known generally as polycyclic aromatic hydrocarbons (PCAs). It has long been known that a number of PCAs exhibit biological activity and are carcinogenic. However, a number of studies have shown that the concentration of these PCAs in bitumen is very low.[112] Consequently, the carcinogenicity of bitumen is low, however it is important that appropriate control measures are in place to mediate exposure.

The composition of bitumen fume is complex but can be described as consisting of organic vapours, condensates and aerosols[118] in combination with non-hydrocarbon particulate matter. The composition of the fume is usually expressed in terms of the total particulate matter and the benzene (or in some cases cyclohexane) soluble matter. A study by Concawe[119] concluded that as the particulate matter could arise from the aggregate, more representative figures on the composition of bitumen fume could be gained by determining the benzene soluble matter (BSM). The conclusions of this report were that the BSM could constitute between 50% and 80% of the total particulates and that exposure of workers to bitumen fume was generally well below the UK's recommended exposure limit of 5 mg/m^3. The study also looked at the PCA content of the fume and found this to be extremely low.

A further study by Concawe[120] considered the acute, chronic and toxic effects of bitumen and bitumen fume. This study further confirmed the exceptionally low levels of carcinogenic PCAs in bitumen compared to coal tar. The overall conclusions reinforced those of the earlier study in that the carcinogenicity of bitumen was found to be low.

There have been attempts to correlate the quantity of fume associated with the manufacture and laying of asphalts. Badriyha et al.[121] concluded that the greatest generation of fume was during loading and paving due to the increased turbulence of the mixture during these operations. This study also simulated the emissions from a placed pavement and found them to be an order of magnitude lower than those reported during paving.

2.13. Summary

Over the last few years there have been significant developments in the field of bitumen technology. The overall structure of bitumen is now more fully understood and hence the properties of this unique engineering substance can be better predicted. There have also been significant advances in test methods which have led to a more fundamental understanding of the performance of asphalts.

The adoption of the Superpave^{TM} specifications in the United States has led to a step change in the way that bitumen is specified, with penetration and softening point no longer being seen as fundamentally important. Increasing knowledge of the demands placed on bitumen in service has led to an increase in use of polymer modified materials to improve end performance and reduce the possibility of failures. Increasing awareness of environmental issues has seen an increase in recycling, which has seen further developments in emulsion and foam-mix technology.

It was said at the beginning of this Chapter that bitumen technology was seen by many to be a black art. It is more correct to say that bitumen technology is now 'state of the art' and will continue to develop to ensure that asphalt remains a high performance construction material.

References

1. GIRDLER R. B. Constitution of Asphaltenes & Related Studies. *Proc. Assoc. Asphalt Paving Tech.*, 1965, **34**, 45. Journal of the Association of Asphalt Paving Technologists, Seattle.
2. AMERICAN SOCIETY FOR TESTING and MATERIALS. Standard Test Method for Separation of Asphalt into Four Fractions. *Annual Book of ASTM Standards 1997.* ASTM, Philadelphia, 1997, **04.03**, D4124-97.
3. PAULI A. T. and J. F. BRANTHAVER. Rheological and Compositional Definitions as they Relate to the Colloidal Model of Asphalt & Residue, Proc Symposium on Stability and Compatibility of Fuel Oils & Heavy Ends. *Proc. 217th National ACS Meeting, March 1999.* American Chemical Society, Washington DC.
4. BONEMAZZI F. and C. GIAVARINI. Shifting the Bitumen Structure from Sol to Gel. *Journal of Petroleum Science & Engineering*, 1999, **22**, 17–24.
5. JAIN P. K., U. C. GUPTA and H. SINGH. Refining Processes to make High Performance Bitumen. *Proc. of Petro Tech-97, New Delhi*, 1997, 33–40.
6. STORM D. A. and E. Y. SHEU. Characterisation of Colloidal Asphaltenic Particles in Heavy Oil. *Fuel*, 1995, **74**, No. 8, 1140–45.
7. STORM D. A., E. Y. SHEU and DETAR. Macrostructure of Asphaltenes in Vacuum Residue by Small-Angle X-Ray Scattering. *Fuel*, 1993, **72**, No. 7, 977–81.

8. BARRÉ L., D. ESPINAT, E. ROSENBERG and M. SCARSELLA. Colloidal Structure of Heavy Crudes and Asphaltene Solutions. *Revue de l'Institut Français du Pétrol*, **52**, No. 2, 161–175.

9. STRATEGIC HIGHWAYS RESEARCH PROGRAMME. *Binder Characterisation & Evaluation, Chemistry*, SHRP-A-386, **2**.

10. DOMIN M. *et al.* A Comparative Study of Bitumen Molecular Weight Distributions. *Energy & Fuels*, 1999, **13**, 552–57.

11. LI J. E. *et al.* Colloidal Structure of Three Chinese Petroleum Vacuum Residues. *Fuel*, 1996, **75**, No. 8, 1025–29.

12. ROZEVELD S. J. *et al.* Network Morphology of Straight and Polymer Modified Asphalt Cements, *Microscopy Research and Techniques*, 1997, **38**, pp. 529–43.

13. FONNESBECK J. E. *Structural Elucidation of Asphalt using NMR Spectroscopy and GPC Fractionation with Laser Desorption Mass Spectroscopy.* University of Montana, PhD thesis, 1997.

14. BARBOUR R. V. and J. C. PETERSEN. Molecular Interactions of Asphalt, An Infra-Red Study of the Hydrogen-Bonding Basicity of Asphalt. *Analytical Chemistry*, 1974, **46**, No. 2.

15. HEUKELOM W. An Improved Method of Characterising Asphaltic Bitumen with the Aid of their Mechanical Properties. *Proc. Assoc. Asphalt Paving Tech.*, 1973, **42**, 62–98.

16. WHITEOAK C. D. *The Shell Bitumen Handbook*. Shell Bitumen, Chertsey, 1990, 70–71.

17. BAHIA H. U. Isothermal Low-Temperature Physical Hardening of Asphalt Binders. *Proc. Int. Symposium on the Chemistry of Bitumens, Rome, 1991.*

18. BRITISH STANDARDS INSTITUTION. *Methods of Test for Petroleum and its Products, Determination of Needle Penetration of Bituminous Materials.* BSI, London, 1993, BS 2000: Part 49.

19. BRITISH STANDARDS INSTITUTION. *Methods of Test for Petroleum and its Products, Determination of Softening Point of Bitumen, Ring & Ball Method.* BSI, London, 1993, BS 2000: Part 58.

20. PFEIFFER J. P. H. and P. M. VAN DOORMAAL. The Rheological Properties of Asphaltic Bitumens. *Journal of the Institute of Petroleum*, 1936, **22**, 414–40.

21. BRITISH STANDARDS INSTITUTION. *Methods of Test for Petroleum and its Products, Determination of Loss on Heating of Bitumen and Flux Oil.* BSI, London, 1993, BS 2000: Part 45.

22. AMERICAN SOCIETY FOR TESTING AND MATERIALS. Standard Test Method for Effect of Heat and Air on a Moving Film of Asphalt (Rolling Thin-Film Oven Test). *Annual Book of ASTM Standards.* ASTM, Philadelphia, 1995, **04.03**, D2872-88.

23. ASPHALT INSTITUTE. *Superpave Performance Graded Asphalt Binder Specification and Testing*, Superpave Series No. 1. AI, Kentucky, 1997, 15–19.

24. HAYTON B. *et al.* Long Term Ageing of Bituminous Binders. *Proc. EUROBITUME Workshop, Luxembourg, 1999.*

25. HERRINGTON P. R. and Y. WU. Effect of Inter-Molecular Association on Bitumen Oxidation. *Petroleum Science and Technology*, 1999, **17**, Nos 3–4, 291–318.

26. BRITISH STANDARDS INSTITUTION. *Hot Rolled Asphalt for Roads*

and other Paved Areas, Specification for Constituent Materials and Asphalt Mixtures. BSI, London, 1992, BS 594: Part 1.

27. BRITISH STANDARDS INSTITUTION. *Coated Macadam for Roads and other Paved Areas, Specification for Constituent Materials and for Mixtures*. BSI, London, 1993, BS 4987: Part 1.

28. WHITEOAK C. D. *The Shell Bitumen Handbook*. Shell Bitumen, Chertsey, 1990, 124.

29. FORDYCE D. (R. N. HUNTER ed.). *Bituminous Mixtures in Road Construction*. Thomas Telford Publishing, London, 1994, 26.

30. BRITISH STANDARDS INSTITUTION. *Methods of Test for Petroleum and its Products, Solubility of Bituminous Binders*. BSI, London, 1993, BS 2000: Part 47.

31. INSTITUTE OF PETROLEUM. *Standard Test Methods for Analysis and Testing of Petroleum & Related Products, Determination of the Breaking Point of Bitumen–Fraass Method*. IP, London, 1998, IP80/87.

32. BRITISH STANDARDS INSTITUTION. *Bitumens for Building and Civil Engineering, Specification for Bitumens for Roads and other Paved Areas*. BSI, London, 1989, BS 3690: Part 1.

33. NUNN M. E. and T. SMITH. *Road Trials of High Modulus Base for Heavily Trafficked Roads*. TRL, Crowthorne, 1997, TRL 231.

34. NUNN M. E. *et al. Design of Long-Life Flexible Pavements for Heavy Traffic*. TRL, Crowthorne, 1997, TRL 250.

35. COMITÉ EUROPÉEN DE NORMALISATION. *Bitumen & Bituminous Binders–Specification for Paving Grade Bitumen*. CEN, Berlin, 1999, prEN 12591.

36. ASPHALT INSTITUTE. *Superpave Performance Graded Asphalt Binder Specification and Testing*, Superpave Series No. 1. AI, Kentucky, 1997, 62–5.

37. AMERICAN SOCIETY FOR TESTING AND MATERIALS. *Viscosity Determinations of Unfilled Asphalts using the Brookfield Thermosel Apparatus*. ASTM, Philadelphia, D4402-84.

38. BAHIA H. U., D. A. ANDERSON and D. W. CHRISTENSEN. The Bending Beam Rheometer: A Simple Device for Measuring Low-Temperature Rheology of Asphalt Binders. *Proc. Assoc. Asphalt Paving Tech.*, 1991, **61**, 117–35.

39. AMERICAN ASSOCIATION OF STATE HIGHWAY OFFICIALS. *Test Method for Determining the Fracture Properties of Asphalt Binder in Direct Tension*, Test Method TP3. AASHTO, Washington DC.

40. WALKER P. M. B. (ed.). *Chambers Science & Technology Dictionary*. Chambers, London, 1991, 765.

41. BARNES H. A., J. F. HUTTON and K. WALTERS. *Introduction to Rheology*. Elsevier, Barking, 1989.

42. EUROBITUME. *Glossary of Rheological Terms*. European Bitumen Association, Brussels.

43. WILLIAMS M. L., R. F. LANDEL and J. D. FERRY. The Temperature Dependence of Relaxation Mechanisms in Amorphous Polymers and other Glass Forming Liquids. *Journal of American Chemical Society*, 1955, **77**, 3701–7.

44. CHRISTENSEN D. W. and D. A. ANDERSON. Interpretation of Dynamic Mechanical Test Data for Paving Grade Asphalt Cements. *Proc. Assoc. Asphalt Paving Tech.*, 1992, **61**, 67–115.

45. LESUEUR D. *et al.* A Structure-Related Model to Describe Asphalt Linear Visco-Elasticity. *Journal of Rheology*, 1996, **40**, No. 5, 813–36.
46. CLAUDY P. M. *et al.* Thermal Behaviour of Asphalt Cements. *Thermochimica Acta*, 1998, **324**, 203–13.
47. PLANCHE J. P. Using Thermal Analysis Methods to Better Understand Asphalt Rheology. *Thermochimica Acta*, 1998, **324**, 223–27.
48. ASPHALT INSTITUTE. *Superpave Performance Graded Asphalt Binder Specification and Testing*, Superpave Series No. 1. AI, Kentucky, 1997, 23.
49. SHASHIDHAR N. and B. H. CHOLLAR. Suspension Rheology of Asphalt-Filler Systems. *Proc. Society of Rheology, 68th Annual Meeting, Galveston, Texas*, Feb. 1997.
50. HICKS R. G. *et al.* Validation of SHRP Binder Specification through Mix Testing. *Proc. Assoc. Asphalt Paving Tech.*, 1993, **62**, 565–609.
51. ROBERTUS C. and M. PHILLIPS. *Binder Rheology and Asphaltic Pavement Deformation – The Zero Shear Viscosity Concept*. EuroBitume and EuroAsphalt Congress. European Bitumen Association, Brussels, 1996, Paper E & E5.134.
52. LEWANDOWSKI L. H. Polymer Modification of Paving Asphalt Binders. *Rubber Chemistry and Technology*, 1994, **67**, 447–80.
53. WARDLAW K. R. and S. SHULER (eds). *Polymer Modified Asphalt Binders*. ASTM, Philadelphia, 1992, STP1108.
54. NICHOLLS J. C. *Generic Types of Binder Modifier*. SCI, London, 1993, Lecture Papers 0035.
55. BONEMAZZI *et al.* Correlation between the Properties of Polymers and Polymer Modified Bitumens. *EuroBitume & EurAsphalt Conference, Strasbourg*. European Bitumen Association, Brussels, 1996.
56. SZATKOWSKI W. S. *Resistance to Cracking of Rubberised Asphalt: Full Scale Experiment on the Trunk Road A6 in Leicestershire*. TRL, Crowthorne, 1970, LR308.
57. HIGHWAYS AGENCY *et al. Manual of Contract Documents for Highway Works, Specification for Highway Works*. TSO, London, 1998, 1, Cl. 938.
58. HENDERSON J. F. and M. SZWARC. The Use of Living Polymers in the Preparation of Polymer Structures of Controlled Architecture. *Reviews in Macromolecular Chemistry*, 1968, **3**, 317–401.
59. COLLINS J. H., M. G. BOULDIN, R. GELLES and A. BERKER. Improved Performance of Paving Asphalt by Polymer Modification. *Journal of the Assoc. Asphalt Paving Tech.*, 1991, **60**, 43–79.
60. INSTITUTE OF PETROLEUM. *Determination of the Penetration Stability and Softening Point Stability of Stored Modified Bituminous Binders*. IP, London, 1999, IP PM CH/99.
61. BRÛLÉ B., Y. BRION and A. TANGUY. Paving Asphalt Polymer Blends: Relationships between Composition, Structure and Properties. *Proc. Assoc. Asphalt Paving Tech.*, 1988, **57**, 41–61.
62. LU X. *Fundamental Studies on Styrene-Butadiene-Styrene Polymer Modified Road Bitumens*. Royal Institute of Technology, Licenciate thesis, Stockholm, 1996.
63. FAWCETT A. H., A. MCNALLY and G. MCNALLY. *Polymer Bitumen Blends*. 216–17.
64. LU X, U. ISACSSON and J. EKBLAD. Rheological Properties of SBS,

EVA and EBA Modified Bitumens. *Materials and Structures*, 1999, **32**, 131–9.

65. GAHVARI F. and I. L. AL-QADI. Dynamic Mechanical Properties of SBR Modified Asphalt. *Proc. 4th Materials Engineering Conference, Materials for the New Millenium.* Americam Society of Civil Engineering, Washington DC, Nov. 1996, 133–43.

66. HERRINGTON P. R., Y. WU and M. C. FORBES. Rheological Modification of Bitumen with Maleic Anhydride and Dicarboxylic Acids. *Fuel*, 1998, **78**, 101–10.

67. BRÛLÉ B. and M. MAZE. Application of SHRP Binder Tests to the Characterisation of Polymer Modified Bitumens. *Journal of the Assoc. Asphalt Paving Tech.*, 1995, **64**, 307–92.

68. INSTITUTE OF PETROLEUM. *Standard Test Methods for Analysis and Testing of Petroleum & Related Products, Determination of Tensile Properties of Bituminous Binders–Toughness Tenacity Method*, Proposed Method BQ/94. IP, London.

69. COMITÉ EUROPÉEN DE NORMALISATION. *Bitumen and Bituminous Binders, Determination of Cohesion of Bituminous Binders for Surface Dressings.* CEN, Brussels, 1999, Draft prEN 13588.

70. DENNING J. H. and J. CARSWELL. *Improvements in Rolled Asphalt Surfacings by the Addition of Organic Polymers.* TRL, Crowthorne, 1981, LR989.

71. CARSWELL J. *An Assessment of Bituminous Basecourse and Roadbase Materials containing EVA and Sulphur.* TRL, Crowthorne, 1986, LR92.

72. PETROSSI U., P. L. BOCCA and P. PACOR. Reactions & Technological Properties of Sulfur-Treated Asphalt. *Industrial & Engineering Chemistry Product Research & Development*, 1972, **11**, No. 2, 214–19.

73. LEE D. Modification of Asphalt and Asphalt Paving Mixtures by Sulfur Additives. *Industrial & Engineering Chemistry Product Research & Development*, 1975, **14**, No. 3, 171–7.

74. DAINES M. E., J. CARSWELL and D. M. COLWILL. *Assessment of 'Chem-Crete' as an Additive for Binders for Wearing Courses and Roadbases.* TRL, Crowthorne, 1985, RR54.

75. NICHOLLS J. C. *A Review of the Effects of using Chemcrete in Bituminous Materials.* TRL, Crowthorne, 1990, TR271.

76. VANELSTRAETE A., L. FRANCKEN and R. REYNAERT. Influence of Binder Properties on the Performance of Asphalt Mixes. *EuroBitume & EurAsphalt Conference, Strasbourg.* European Bitumen Association, Brussels, 1996.

77. MORRISON G. R., H. HEDMARK and S. A. M. HESP. Elastic-Steric Stabilisation of Polyethylene-Asphalt Emulsions by using Low Molecular Weight Polybutadiene and Devulcanised Rubber Tyre. *Colloid & Polymer Science*, 1994, **272**, 375–84.

78. PELTONEN P. A New Method for using Waste Rubber Powders in Bitumen. *EuroBitume & EurAsphalt Conference, Strasbourg.* European Bitumen Association, Brussels, 1996.

79. SOCIETY OF CHEMICAL INDUSTRY. Waste and Industrial Products in Asphalt. *SCI/IAT/IHT Joint Symposium, London.* SCI/IAT/IHT, London, October 1995.

80. SHAW D. J. *Introduction to Colloid and Surface Chemistry.* Butterworths, London, 1980.

81. EVERETT D. H. *Basic Principles of Colloid Science*. Royal Society of Chemistry, London, 1988.

82. AMERICAN SOCIETY FOR TESTING AND MATERIALS. *A Basic Asphalt Emulsion Manual*. ASTM, Philadelphia, MS-19.

83. BRITISH STANDARDS INSTITUTION. *Bitumen Road Emulsions (Anionic and Cationic), Specification for Bitumen Road Emulsions*, BSI, London, 1984, BS 434: Part 1.

84. DURAND G. *Influence of Emulsion Size Gradation on the Properties of Emulsions*. *Proc. 1996 AEMA Symposium on Asphalt Emulsion Technology, Phoenix, Arizona, 1996*, 217–51.

85. BOUSSAD N. and T. MARTIN. Emulsifier Content and Particle Size Distribution: Two Key Parameters for the Management of Bituminous Emulsion Performance. *EuroBitume & EurAsphalt Congress, Strasbourg*. European Bitumen Association, Brussels, May 1996, Paper E&E.6.159.

86. CHAVAROT P. and P. CHENEVIERE. Four Key Parameters to Control Bitumen Emulsion Viscosity. *Proc. 1st World Emulsion Congress, Paris, October 1993*, 3-10-124.

87. BRITISH STANDARDS INSTITUTION. *Bitumen Road Emulsions (Anionic and Cationic), Code of Practice for use of Bitumen Road Emulsions*, BSI, London, 1984, BS 434: Part 2.

88. SCOTT J. A. N. *A General Description of the Breaking Process of Cationic Emulsion in Contact with Mineral Aggregate, Theory and Practice of Emulsion Technology*. Academic Press, New York, 1976, 179–201.

89. SALAGER J. L. *et al.* Breaking of an Asphalt Emulsion on Mineral Aggregate: Phenomenology, Modelling and Optimisation. *Proc. 2nd World Emulsion Congress, Bordeaux, September 1998*, 3-3-096.

90. LYCETT C. Development and Performance of Polymer-Modified Emulsion Binders for Surface Dressing. *SCI Symposium on Modified Bitumen, London, 1993*.

91. SERFASS J-P. and A. JOLY. SBS-Modified Bitumen Emulsions for Surface Dressing–Properties and Behaviour. *Proc. 1st World Emulsion Congress, Paris, October 1993*, 4-11-112.

92. CARSWELL J. *The Testing and Performance of Surface Dressing Binders for Heavily Trafficked Roads*. TRL, Crowthorne, 1994, PR12.

93. STROUP-GARDINER M. and P. S. KANDHAL (eds). *Flexible Pavement Rehabilitation & Measurement*. ASTM, Philadelphia, 1998, ASTM STP1348.

94. MUNCY H. W. *Asphalt Emulsions*. ASTM, Philadelphia, 1990, ASTM STP1079.

95. INSTITUTE OF PETROLEUM. *Standard Test Methods for Analysis and Testing of Petroleum and Related Products, Determination of the Viscosity of Bitumen Emulsions–Engler Method*. IP, London, 1998, IP212/82.

96. INSTITUTE OF PETROLEUM. *Standard Test Methods for Analysis and Testing of Petroleum and Related Products, Determination of the Viscosity of Cut-Back Bitumen*. IP, London, 1998, IP72/86.

97. INSTITUTE OF PETROLEUM. *Standard Test Methods for Analysis and Testing of Petroleum and Related Products, Determination of Water Content of Bitumen Emulsions–Distillation Method*. IP, London, 1998, IP291/92.

98. INSTITUTE OF PETROLEUM. *Standard Test Methods for Analysis and Testing of Petroleum and Related Products, Determination of Distillation*

Characteristics of Cut-Back Bitumen. IP, London, 1998, IP27/74.

99. COMITÉ EUROPÉEN DE NORMALISATION. *Petroleum Products–Bitumen and Bituminous Binders, Determination of the Breaking Value of Cationic Bitumen Emulsions, Mineral Filler Method*. CEN, Berlin, 1997, prEN 13075-1.

100. HIGHWAYS AGENCY *et al. Manual of Contract Documents for Highway Works*. TSO, London, 1998, **1–2**, Cl. 922.

101. HIGHWAYS AGENCY *et al. Manual of Contract Documents for Highway Works*. TSO, London, 1998, **1–2**, Cl. 927.

102. HIGHWAYS AGENCY *et al. Manual of Contract Documents for Highway Works*. TSO, London, 1998, **1–2**, Cl. 918.

103. CSANYI L H. *Bituminous Mixes Prepared with Foamed Asphalt*. Iowa State University, Iowa, 1960, **189**.

104. COUNTY SURVEYORS' SOCIETY. *Permanent Cold Lay Bituminous Materials, Overview Report*. CSS, 1997, ENG/1-97.

105. BROWN D. C. *Wisconsin Demo Explores In-Place Asphalt Rehabilitation, Roads & Bridges*. Roads and Bridges Online 1997, 28.

106. http://foamasph.csir.co.za:81/

107. LAKE ASPHALT OF TRINIDAD & TOBAGO LTD. *Trinidad Lake Asphalt in Highway Surfacing*. 1978.

108. SHIELDS C. F. *The Effect of Flux Oil and Trinidad Lake Asphalt Content on the Photo-Oxidation of Asphaltic Cements*. TRL, Crowthorne, 1967, LR120.

109. AMERICAN GILSONITE COMPANY. *Gilsonite Information Bulletin*.

110. http://pioneerAsphalt.com/gilsonit.htm

111. BRITISH STANDARDS INSTITUTION. *Specification for Tars for Road Purposes*. BSI, London, 1974, BS 76.

112. WALLCAVE *et al.* Skin Tumourigenesis in Mice by Petroleum Asphalts and Coal Tar Pitches of Known Polynuclear Aromatic Hydrocarbon Content. *Toxicology and Applied Pharmacology*, 1971, **18**, 41–52.

113. HEALTH & SAFETY COMMISSION. *Approved Supply List, Information Approved for the Classification and Labelling of Substances and Preparations Dangerous for Supply*, (CHIP98). HSC, London, 1998, 4th edn.

114. INSTITUTE OF PETROLEUM. *Model Code of Safe Practice in the Petroleum Industry, Bitumen Safety Code*. IP, London, 1990, 3rd edn, Part 11.

115. REFINED BITUMEN ASSOCIATION LTD. *Code of Practice for the Safe Delivery of Bitumen Products*. RBA, London, 1996.

116. REFINED BITUMEN ASSOCIATION LTD. *Safe Handling of Petroleum Bitumens*. RBA, London, 1998, Technical Bulletin No. 7.

117. ROAD EMULSION ASSOCIATION LTD. *Code of Good Practice for the Use and Safety of Mobile Storage Tanks*. REAL, Clacton, Essex, 1996.

118. ROGGE W. F. *et al.* Sources of Fine Organic Aerosol, Hot Asphalt Roofing Tar Pot Fumes. *Environmental Science and Technology*, 1997, 7, **31**, No. 10, 2726–30.

119. CONCAWE. *Review of Bitumen Fume Exposures and Guidance on Measurement*, The Hague, Netherlands, 1984, Report No. 6/84.

120. CONCAWE. *Bitumens and Bitumen Derivatives*, Product Dossier No. 92/104. The Hague, Netherlands, 1992.

121. BADRIYHA B. N. *et al.* Emission of Volatile and Semi-Volatile Organic Compounds from Hot Asphalts in Laboratory Scale and Field Studies. *Conference on Toxic and Related Air Pollutants, Triangle Park, North Carolina, USA.* Environmental Science and Technology, 1996.

3. Functions and properties of road layers

3.1. Preamble

Road construction involves various combinations of layers between the surface, in contact with vehicle tyres, and the ground over which the road is built. Figure 3.1[1] shows a selection of these combinations, which range from simple unsurfaced compacted aggregate pavements for haul roads or low volume roads in developing countries to the thick multi-layer bituminous or concrete roads used for heavily trafficked routes in the developed world. This Chapter concentrates on the typical British asphalt construction, shown in Fig. 3.2, which includes the surfacing (i.e. the wearing course plus the basecourse) supported by a combination of basecourse and roadbase, all of which are asphaltic.

Flexible composite construction involves a lower roadbase of cement bound material below the bituminous layers. These bound layers are placed over a foundation, usually built from unbound aggregates, placed on the subgrade soil (simply called the subgrade) which may be natural ground in a cutting or compacted fill in an embankment. The top of the soil may be covered by a capping layer (simply called capping) to provide the pavement formation.

The design of pavements for trunk roads including motorways in the UK is described in the Highways Agency's *Design Manual for Roads and Bridges* (DMRB)[2] whilst materials and workmanship are specified in the companion document, the *Manual of Contract Documents for Highway Works* (MCHW).[3]

Each of the layers in the pavement has a specific role and its properties need to relate to that role. These must be examined in the context of pavement design, since layer thicknesses and the mechanical properties of the layers are intimately linked. Consequently, although Chapter 4 deals with the design of roads, it is necessary in this Chapter to provide, at least, a philosophical design framework within which to discuss the functions and properties of road layers.

3.2. The ideal pavement

Figure 3.3[4] shows the concept of an 'ideal' pavement and highlights the key roles of the main component layers. The wearing course is essentially a cosmetic treatment in that it is not required to necessarily have a structural role. However, the surface must provide the appropriate

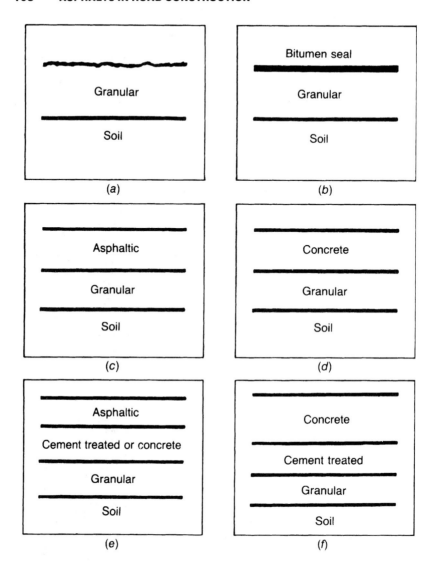

Fig. 3.1.   Typical pavement constructions[1]

conditions for safe and comfortable contact with vehicle tyres. It must also fulfil a waterproofing function since moisture in pavements is generally undesirable as it can cause various types of damage to the lower layers. The particular case of porous asphalt wearing course is an important exception to this requirement and is discussed in Section 3.7.2.

The main structural layer of the pavement is the roadbase (Fig. 3.2), which is usually covered by a basecourse provided as a practical expedient in order to achieve more accurate finished levels than is

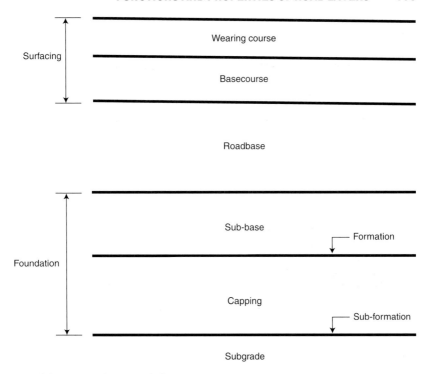

Fig. 3.2.   Typical UK asphalt pavements

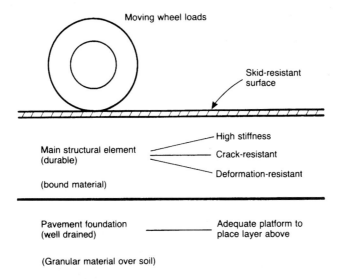

Fig. 3.3.   The 'ideal' pavement[4]

possible with the thick lift construction used for roadbase layers. The combination of basecourse and roadbase must have the ability to spread the wheel load so that the underlying layers are not overstressed. Damage, through cracking or permanent deformation within the layer itself, must also be prevented. These concepts are discussed in more detail in Section 3.10.1 below.

The bituminous layers are supported by a pavement foundation which must be adequate to carry construction traffic and the paving operation. It is most highly stressed during the construction phase. Since the foundation usually consists of granular materials and soils, the principles of soil mechanics apply. Application of these principles reveals the need for good drainage to depress the water table and prevent a build up of water in the granular layers.

3.3. Design concepts

Figure 3.4[4] illustrates the load spreading principle in pavement engineering. The level of stress induced in the supporting layers depends on the elastic stiffness of the roadbase for a given layer thickness. Elastic stiffness is defined as the ratio of stress to strain and may be regarded as analogous to Young's modulus of elasticity. Figure 3.5[4] shows how the shear stress in the foundation decreases as the elastic stiffness of the roadbase increases.

However, this decrease in shear stress is accompanied by an increase in the tensile stress induced at the bottom of the roadbase. Clearly, then, there is a possibility that in providing good protection to the foundation, the roadbase itself may crack. This effect is a very real one, since repeated loading by traffic can cause failure through cracking at a tensile stress level much lower than that which the material can sustain under a single load application. This is known as fatigue cracking. Figure 3.6[1] shows the characteristic relationship between the tensile stress in the bituminous material and the number of load applications to failure by cracking. It can

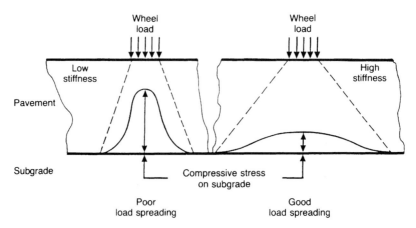

Fig. 3.4. The load spreading principle[4]

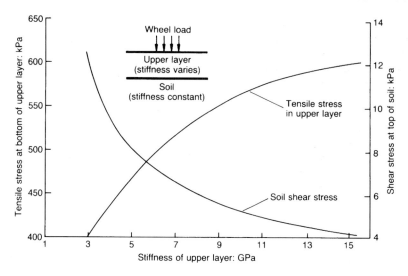

Fig. 3.5.   Influence of asphalt layer stiffness on critical stresses[4]

be noted that the sustainable stress decreases as the number of load applications increases.

The interaction between the elastic stiffness of the roadbase, the shear stress in the foundation and the tensile stress in the roadbase provides the essential key to pavement design. Other related parameters are the roadbase thickness and its composition, the latter dictating the stiffness and fatigue cracking characteristics which are exhibited in practice.

Research carried out by the Transport Research Laboratory (TRL) in recent years,[5] focusing on heavy duty pavements (often called long life pavements), has concluded that fatigue cracking initiated at the bottom of

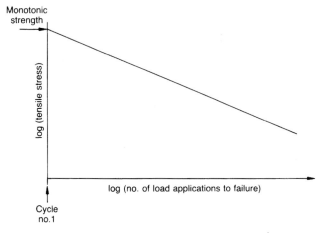

Fig. 3.6.   Characteristic fatigue relationship for an asphalt[1]

the roadbase does not occur in practice when the bituminous layer thickness exceeds 160 mm. Furthermore, such cracking as does occur appears to propagate downwards from the surface. This matter is still under investigation at a research level to develop a full understanding of the detailed mechanisms.

Contemporary thinking in other countries, notably France and the US, still supports the concept of classical fatigue cracking starting at the bottom of the asphalt layer.[6,7] In the US, particularly, thinner asphalt layers tend to be used in practice and this probably explains their approach. Since the majority of the UK road network falls outside the motorway and trunk road system, where the heaviest duty pavements are used, pavement design principles should still include the possibility of classical fatigue cracking.

Figure 3.7 illustrates the possible locations for initiation of fatigue cracking. Although, theoretically, the tensile stress and strain at the surface are lower than at the bottom of the roadbase, conditions at the surface are conducive to crack initiation. These conditions include bitumen which may oxidize and harden, thus becoming brittle, and the presence of micro-cracks from the rolling operation and surface texture which assist the initiation of fatigue cracking.

The pavement foundation has a dual function. It must provide short-term service as a haul road and construction platform and long term service as a support to the asphaltic construction above. For this reason, the concept of designing a flexible pavement in two stages has been developed in the UK.[2,8]

Stage one involves designing the foundation, which has to carry a limited number of heavy wheel loads. The design considerations are shown in Fig. 3.8. It is essential that overstressing and plastic (i.e.

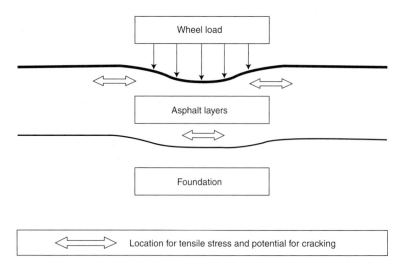

Fig. 3.7. Locations of crack initiation

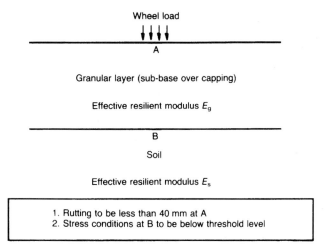

Fig. 3.8.   Design considerations for a pavement foundation

permanent) deformation in the soil are prevented. This is done by ensuring that the combination of thickness and elastic stiffness of the granular material placed above it is adequate. In present UK practice, this granular layer usually consists of two parts. A lower quality 'capping' is placed over the soil and forms part of the earthworks. This is covered by a better quality crushed rock sub-base which is commonly 150 mm thick but can be as thick as 350 mm.[9] The capping layer is not required when the assessed California Bearing Ratio (CBR) for the soil exceeds 15%. Sub-base is not required on 'hard rock subgrades that are intact or, if granular, would have a laboratory CBR of 30%, and which do not have a high water table'.[9] The CBR and other relevant properties of foundation layers are discussed in Section 3.5. The design principles remain the same whether the granular layer is in two parts or only one. In addition to protecting the soil from overstressing, the layer itself must not develop serious ruts under the action of construction traffic. For practical purposes, the rut depth should not exceed 40 mm.

Stage two of the design process involves selection of the correct combination of asphalt stiffness and thickness to ensure that the foundation is not overstressed and that the layer itself does not crack. In addition, the roadbase, basecourse and wearing course combination must not develop permanent deformation which leads to wheel track rutting. This is most effectively dealt with by correct mixture design and attention to quality control on site, particularly in regard to compaction, including the temperature of materials at the time of compaction.

3.4.   The subgrade

Although the subgrade is not formally a pavement layer, its properties must be fully understood in order to design and construct a satisfactory pavement over it. The soil will either be in 'cut' or 'fill' and its geological stress history will influence the mechanical properties exhibited once the pavement is constructed. This matter is discussed in detail by Brown.[10]

In order to properly understand soil characteristics, the principle of effective stress must be appreciated. Figure 3.9 illustrates this. The simple equation involved is:

$$\sigma' = \sigma - u \tag{3.1}$$

where

σ' = effective stress

σ = total stress

u = pore water pressure.

In simple terms, effective stress is the normal stress carried by the particle-to-particle contacts. Since soil is a frictional material, its strength depends on this normal stress. By reference to Fig. 3.10, it can be seen that the shear stress (τ) to cause one particle to slip relative to another is given by

$$\tau = \sigma' \mu \tag{3.2}$$

where μ is the coefficient of friction between particles. Since soil particles interlock and interact in a complex three dimensional manner, the effective coefficient of friction is expressed as

$$\mu = \tan \phi' \tag{3.3}$$

where ϕ' is known as the angle of shearing resistance with respect to effective stresses. The shear strength is, therefore, expressed as

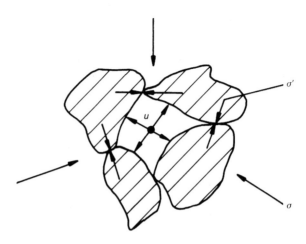

Fig. 3.9. Principle of effective stress

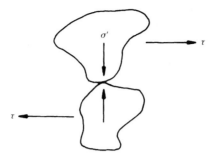

Fig. 3.10.   Interparticle friction

$$\tau = \sigma' \tan \phi' \tag{3.4}$$

Effective stress always has to be determined by Equation (3.1). The total stress is caused by the externally applied load while the pore pressure depends on the hydraulic conditions in the soil at the point concerned. Figure 3.11[1] illustrates the principle.

For an element of soil at depth z below formation level when the water table (the natural level of water in the soil) is at depth h, the total stress (σ) is given by $\sigma = \sigma_p + \gamma z$, where σ_p is the stress caused by the weight of the pavement layers on the subgrade and γ is the unit weight of the soil.

The pore water pressure (u) at depth z is given by $u = \gamma_w(z - h)$ where $\gamma_w = $ unit weight of water. Hence, the effective stress (σ'), by equation (3.1), is

$$\sigma' = \sigma_P + \gamma z - \gamma_w(z - h) \tag{3.5}$$

For pavement construction, interest is generally centred on the soil near formation level (top of the subgrade) which will be above the water table. For clays, water can be held in the soil pores under capillary action above the water table and saturated conditions can apply for some height above

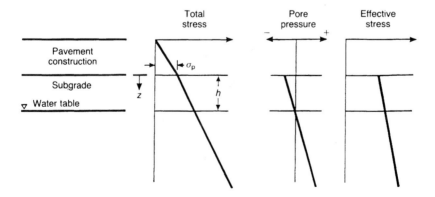

Fig. 3.11.   In situ stresses caused by self-weight[1]

the water table. In these circumstances, the pore water pressure will be a suction and Equation (3.5) becomes

$$\sigma' = \sigma_p + \gamma z + \gamma_w(h - z) \tag{3.6}$$

Soil strength, by Equation (3.4) increases with effective stress. It follows then, from Equation (3.6), that a lower water table is desirable (i.e. a large value of h).

The effective stress at formation level is often equated to the soil suction (S). This parameter has been used for many years in the UK for determination of design CBR values, on the basis of extensive research by TRL.[11] The CBR of a soil is an empirical parameter based on a test which measures the resistance to penetration of a standard plunger into a compacted specimen of the soil contained in a standard mould.[12]

Failure in the subgrade is defined, for pavement design purposes, in terms of the permanent (plastic) deformation which can develop if the shear stress, repeatedly applied by wheel loading, is too high. The concept of a threshold level for this shear stress has been developed. Below this critical value, negligible plastic strains will accumulate. This provides a useful basis for design since the granular layer thickness/stiffness combination must be sufficient to prevent the threshold stress in the subgrade being exceeded under the action of construction traffic. Further detail on this matter has been presented by Brown.[10]

The resilient modulus of the soil is a mechanical property which is important for pavement design, since the stress induced in the upper layers, as well as in the soil itself, will depend on it. It is significant for both stages in the design process and for interpretation of deflection testing results when pavements are subsequently being evaluated in service using the Falling Weight Deflectometer[13] (discussed in detail in Section 3.10.2).

The general definition of resilient modulus is the quotient of applied stress pulse (due to wheel loading) and the recoverable (resilient) strain. More specifically, for soils, the resilient modulus (E_r) may be estimated from

$$E_r = C + A\sigma' - Bq_r \tag{3.7}$$

where q_r is the shear stress pulse, σ' is the effective stress (defined above) and, A, B and C are constraints which depend on the soil type.[10]

Equation (3.7) is important since it demonstrates that a particular soil does not have a unique value of resilient modulus but that the value is influenced by the stress conditions. These include both the static effective stress (σ), due to overburden and water table position, and the shear stress caused by traffic loading (q_r). The implication of Equation (3.7) is that soil stiffness increases with depth. For pavement design purposes, a value appropriate to formation level is normally used. However, for pavement evaluation purposes, the profile with depth is very important.[13]

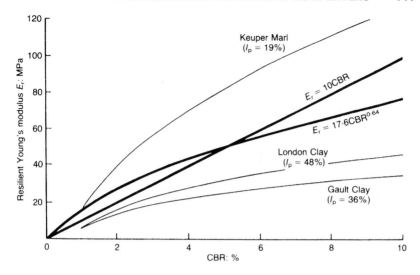

Fig. 3.12. Relationships between resilient modulus and CBR of soils[4]

The CBR of a soil is not directly comparable with the resilient modulus and this matter has been discussed elsewhere.[10] A unique relationship between the two parameters does not exist. However, for typical stress conditions, Fig. 3.12[4] shows relationships for three different soil types, together with empirical equations often used for design.[14]

3.5. The foundation

In one of the documents which controls the design of trunk roads including motorways,[9] the foundation is specified as the capping (if present) plus the sub-base. The foundation is usually constructed of granular materials. The main difference between granular materials (crushed rock or gravel) and clays lies in the contrasting particle sizes. The basic principles of soil mechanics still apply. Granular materials will, generally, exist in a partially saturated state within the pavement, unless drainage provision is completely ineffective.

Equation (3.4) for shear strength is equally applicable for granular materials and the concept of a threshold stress also appears to hold. In this case, it is defined as 70% of the shear strength.[1]

The stress dependence of resilient modulus for granular materials is very marked. The influence of effective stress (σ') is more important than that of the shear stress. Since overburden stresses are very low in the granular layer, the effective stress induced by wheel loading becomes significant. A simple equation for resilient modulus in granular materials is

$$E_\mathrm{r} = K\sigma'^D \tag{3.8}$$

where K and D are materials constants.

For pavements with asphaltic roadbases, the non-linear resilient properties of the sub-base do not present a major problem in design. Experimental and theoretical studies[15] have shown that, for a good quality crushed rock sub-base, the effective in situ resilient modulus is about 100 MPa. This figure will decrease if saturated conditions are approached. Higher figures can arise if the layer has self-cementing properties, such as are exhibited by the softer limestones. Materials used in capping will generally have resilient moduli of about 80 MPa. Laboratory testing[16] has clearly demonstrated that there is no real relationship between CBR and resilient modulus for granular materials.

Compaction is an important factor in the construction of granular layers. High density will increase strength and, hence, resistance to rutting, since these are related as noted above. If the material has a high fines content, particularly the proportion less than 75 microns, then it will be particularly susceptible to weakening by water ingress. In a highly compacted state, such materials have a greater affinity for water as the small particle sizes and void spaces increase suction forces. Conversely, drainage becomes difficult as permeability will be low.

A drainage blanket of high-permeability material can be useful because of the problems which water can cause in soils and granular materials. The 'ideal pavement foundation' incorporating such a layer is illustrated in Fig. 3.13.[1] The high permeability can be achieved through control of the aggregate grading. It is very important, however, that the layer is properly compacted. Geotextiles can usefully be employed to separate the soil from the granular layer to minimize contamination which would reduce permeability and, if severe, shear strength too. In the US, several states use open-textured asphaltic materials to form stable drainage blankets below compacted dense granular sub-bases while, in the UK, 'no fines' concrete has been tried. Clearly, drainage blankets must allow water to escape to the side drains and not vice versa.

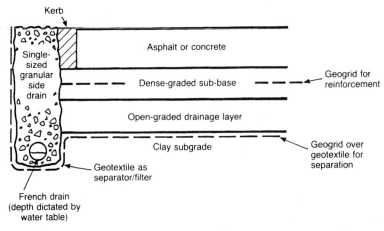

Fig. 3.13.   The 'ideal' pavement foundation[1]

In view of this latter possibility, these layers have not been used very extensively in the UK.

Reinforcing grids of high tensile polymeric material can be effective in granular layers by increasing resistance to rutting from construction traffic. It is important that such grids are placed relatively near the surface of the layer where the shear stresses causing rutting are highest. A rule of thumb suggests a depth equal to half the width of the tyre contact patch (say 150 mm). These grids are effective when they interlock with the aggregate particles. The correct aperture size must, therefore, be selected to match the coarse aggregate dimensions or vice versa.

3.6. Bituminous roadbases and basecourses

The combination of roadbase and basecourse provides the main structural layer in the road (see Figs 3.2 and 3.3[4]). The essential requirements are that the material should offer the following mechanical properties.

- *High elastic stiffness.* This is required to ensure good load spreading ability as illustrated in Figs 3.4[4] and 3.5.[4]
- *High fatigue strength.* This is required to prevent the initiation and propagation of cracks due to repeated loading by traffic.
- *High resistance to permanent deformation.* This is required to ensure that these layers do not contribute significantly to surface rutting. The shear stresses induced by wheel loads, which are the main cause of rutting, are most severe in the top 100 mm to 150 mm of the pavement, so it is the wearing course and basecourse which need to exhibit the highest deformation resistance. In addition to these mechanical properties, the materials must be durable, which implies that they are resistant to the effects of air and water.

In current British practice, most roadbase and basecourse materials are specified in British Standards[17,18] using a recipe approach. These provide an aggregate grading envelope, a grade of bitumen and its quantity in the mixture. No mechanical properties are measured. The success of the materials in practice is based on past experience and good workmanship in construction.

Although asphalts are all essentially combinations of aggregate and bitumen, with some air in the voids following compaction, two generic types of mixture have emerged from historical developments in the UK. Dense bitumen macadam (DBM) is a continuously graded material with a relatively low binder content and, traditionally, a relatively soft binder. Hot rolled asphalt (HRA) is a gap graded material with higher binder content and harder binder. The grading curves for typical examples of these materials are compared in Fig. 3.14.

The elastic stiffness of a mixture is influenced by its volumetric proportions and, in particular, the quantity and stiffness of the bitumen which is used. DBM and HRA roadbases tend to have similar elastic stiffness values when properly compacted, since the small amount of soft binder in the DBM has the same effect as the larger amount of hard binder in the HRA. Resistance to fatigue cracking is provided by the volume and

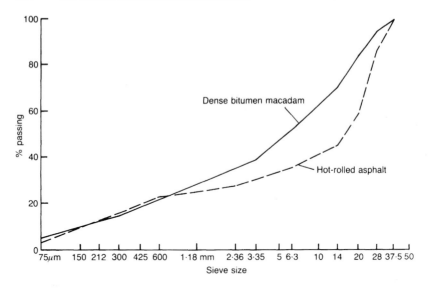

Fig. 3.14.   Grading curves for typical hot rolled asphalt and dense bitumen macadam

hardness of the binder. As HRA has more binder of a harder grade, it offers much better resistance to cracking than DBM.

Resistance to permanent deformation depends, principally, on the aggregate grading and the particle characteristics. Minerals which are rough and angular when crushed offer better resistance than smooth, rounded materials. A low but adequate binder content is needed. Consequently, DBM resists permanent deformation better than HRA. Figure 3.15[19] summarizes the relative permanent deformation characteristics of DBM (continuously graded) and HRA (gap graded).

The mechanical properties of the two traditional British Standard materials have been improved in recent years by simple adjustments to the mix formulations. DBM50 incorporates a harder grade of binder (50 pen) and heavy duty macadam (HDM) has a higher filler and binder content, together with the stiffer bitumen.

In practice, the importance of good compaction has been recognized and the Percentage Refusal Density Test[20] is now required for roadbase and basecourse bitumen macadams.[3] This allows field compaction to be expressed as a percentage of the maximum density which the particular material can achieve in a 150 mm diameter mould with a vibrating hammer.

Durability and compactability are important properties of an asphalt. Durability refers to the resistance offered by the material to environmental damage caused by air and water. It is mainly influenced by the binder film thickness on the aggregate relative to the air void content in the mixture. Durability should usually be adequate for the dense mixtures used in roadbases and basecourses with air void contents between 2% and 8%.

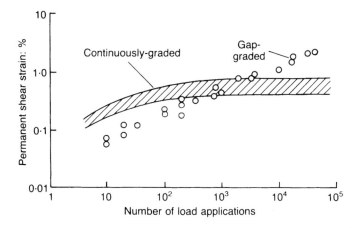

Fig. 3.15.   Permanent deformation characteristics of typical asphalts[19]

Test protocols have been suggested to assess durability through the resistance to water damage and to oxidative ageing based on the Bitutest project conducted at the University of Nottingham between 1991 and 1994 in conjunction with both sides of the industry.[21] In order to assess resistance to water damage, which can cause stripping of binder from the aggregate, the compacted mixture is subjected to a soaking regime in warm water and the change to its stiffness modulus is measured. Assessment of the effects of long-term ageing is measured by exposing the compacted mixture to an elevated temperature in a forced-draft oven.

The compactability of a mixture is not presently the subject of a recognized test in the UK. However the gyratory compactor, the principle of which is depicted in Fig. 3.16, has been used for this purpose in France for many years and has been adopted in the US as part of the Superpave mixture design method.[22]

It will be apparent from the foregoing discussion that the mechanical properties of a roadbase or basecourse mixture depend on several variables. The objectives of good mixture design are to proportion the constituent materials and control the site operation so that satisfactory mechanical properties are realized. This requires a two stage process involving compaction of trial mixtures in the laboratory while varying the key parameters of aggregate grading, degree of compaction and binder content. The subsidiary variables of aggregate type and binder grade can also be studied.

A chart such as that shown in Fig. 3.17[23] can be used to establish which formulations are likely to produce satisfactory mixtures. This is a plot of void content (V_V) against voids in the mineral aggregate (VMA) showing lines of equal binder content by volume (V_B). (Note that VMA $= V_V + V_B$). The target box is shown for $V_V = 3\%$ to 7% and VMA $= 14\%$ to 17%, while V_B should be greater than 8% to avoid the possibility of fatigue cracking. A full discussion of these criteria is given in published

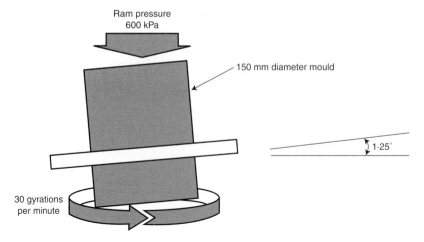

Fig. 3.16.   Gyratory compaction[22]

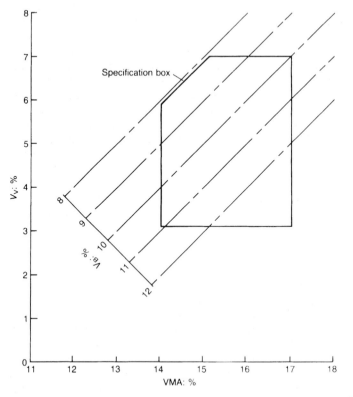

Fig. 3.17.   Typical volumetric proportions chart for mixture design[23]

texts.[23,24] Those mixture formulations which fall in the target area are taken forward for tests of their mechanical properties. These should include elastic stiffness modulus and resistance to permanent deformation.

These performance-based mechanical properties are now widely measured in mixture design procedures and are increasingly featuring in the end-product specifications developed by the Highways Agency for motorways and other trunk roads.[3]

3.7. Wearing courses

The wearing course has to provide resistance to the effects of repeated loading by tyres and to the effects of the environment. In addition, it must offer adequate skid resistance in wet weather as well as comfortable vehicle ride. It must also be resistant to rutting and to cracking. It is also desirable that the wearing course is impermeable, except in the case of porous asphalt which is discussed in Section 3.7.2.

This demanding list of requirements results in practical compromises in terms of the material types which are used. These are now summarized.

3.7.1. Hot rolled asphalt (HRA)

This is a gap graded material with less coarse aggregate than the basecourse version shown in Fig. 3.14. In fact, it is essentially a bitumen/ fine aggregate/filler mortar into which some coarse aggregate is placed. The mechanical properties are dominated by those of the mortar. This material has been extensively used as the wearing course on major roads in the UK, though its use has recently declined as new materials have been introduced (discussed later in Section 3.7.5). It provides a durable layer with good resistance to cracking and one which is relatively easy to compact. It is specified in BS 594: Part 1.[17] The coarse aggregate content is low (typically 30%) which results in the compacted mixture having a smooth surface. Accordingly, the skid resistance is inadequate and pre-coated chippings are rolled into the surface at the time of laying to correct this deficiency.

In Scotland, HRA wearing course remains the preferred wearing course on trunk roads including motorways but, since 1999 thin surfacings have been the preferred option in England and Wales.[25] Since 1999 in Northern Ireland, HRA wearing course and thin surfacings are the preferred permitted options.[25]

3.7.2. Porous asphalt (PA)

This is a uniformly graded material which is designed to provide large air voids so that water can drain to the verges within the layer thickness. If the wearing course is to be effective, the basecourse below must be waterproof and the PA must have the ability to retain its open textured properties with time. Thick binder films are required to resist water damage and ageing of the binder. In use, this material minimizes vehicle spray, provides a quiet ride and lower rolling resistance to traffic than dense mixtures. It is often specified for environmental reasons but stone mastic asphalt (SMA) (see Section 3.7.4 below) and specialist thin surfacings are generally favoured in current UK practice. There have been high profile instances where a PA wearing course has failed early in its life. The Highways Agency does not recommend the use of a PA made with a normal 100 pen binder at traffic levels above 6000 commercial vehicles per day.[26]

3.7.3. Asphaltic concrete and dense bitumen macadam (DBM)
These are continuously graded mixtures similar in principle to the DBMs used in roadbases and basecourses but with smaller maximum particle sizes. Asphaltic concrete tends to have a slightly denser grading and is used for road surfaces throughout the world with the exception of the UK. DBM wearing course is specified in BS 4987: Part I[18] but it is generally only used on minor roads. It is more difficult to meet UK skid resistance standards with DBMs than with HRA, SMA or PA. This problem can be resolved by providing a separate surface treatment but doing so generally makes DBM economically unattractive.

3.7.4. Stone mastic asphalt (SMA)
This wearing course material was pioneered in Germany and Scandinavia and is now widely used in the UK. SMA has a coarse, aggregate skeleton, like PA, but the voids are filled with a fine aggregate/filler/bitumen mortar. In mixtures using penetration grade bitumens, fibres are added to hold the bitumen within the mixture (to prevent 'binder drainage'). Where a polymer modified bitumen is used, there is generally no need for fibres. (See Section 2.7 of Chapter 2 for more information on this topic.) SMA is a gap-graded material with good resistance to rutting and high durability. It differs from HRA in that the mortar is designed to just fill the voids in the coarse aggregate whereas, in HRA, coarse aggregate is introduced into the mortar and does not provide a continuous stone matrix. The higher stone content HRAs, however, are rather similar to SMA but are not widely used as wearing courses in the UK, being preferred for roadbase and basecourse construction.

3.7.5. Thin surfacings
A variety of thin and what were called ultra thin surfacings (nowadays, the tendency is to use the term 'thin surfacings' for both thin and ultra thin surfacings) have been introduced in recent years, principally as a result of development work concentrated in France. These materials vary in their detailed constituents but usually have an aggregate grading similar to SMA and often incorporate a polymer modified bitumen. They may be used over a high stiffness roadbase and basecourse or used for resurfacing of existing pavements. For heavy duty pavements (i.e. those designed to have a useful life of forty years), the maintenance philosophy is one of minimum lane occupancy, which only allows time for replacement of the wearing course to these 'long life' pavement structures. The new generation of thin surfacings allows this to be conveniently achieved.

3.8. Comparison of wearing course characteristics

The various generic mixture types described above can be compared with respect to their mechanical properties and durability characteristics by reference to Fig. 3.18. This shows, in principle, how low stone content HRA, asphaltic concrete, SMA and PA mixtures mobilize resistance to loading by traffic.

Asphaltic concrete (Fig. 3.18(a)) presents something of a compromise when well designed, since the dense aggregate grading can offer good resistance to the shear stresses which cause rutting, while an adequate binder content will provide reasonable resistance to the tensile stresses which cause cracking. In general, the role of the aggregate dominates. DBMs tend to have less dense gradings and properties which, therefore, tend towards good rutting resistance and away from good crack resistance.

HRA (Fig. 3.18(b)) offers particularly good resistance to cracking through the binder rich mortar between the coarse aggregate particles. This also provides good durability but the lack of coarse aggregate content inhibits resistance to rutting.

SMA and PA are shown in the same diagram (Fig. 3.18(c)) to emphasise the dominant role of the coarse aggregate. In both cases, well coated stone is used. In PA, the void space remains available for drainage

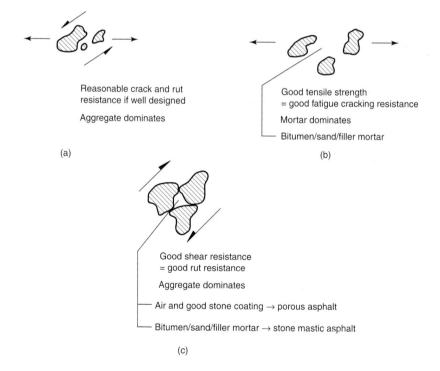

Fig. 3.18. Characteristics of wearing course materials

of water, whilst in SMA, the space is occupied by a fine aggregate/filler/bitumen/fibre mortar. Both materials offer good rutting resistance through the coarse aggregate content. The tensile strength of PA is low whilst that of SMA is probably adequate but little mechanical testing data have been reported to date.

3.9.   Skid resistance

The principles involved in providing good resistance to vehicle skidding in wet weather are well established and are covered in some detail in one of the publications[27] which forms part of the standard DMRB.[2]

Low-speed skid resistance relies on adequate surface texture for the aggregate particles exposed on the pavement surface. This is referred to as 'microtexture'. Suitable mineral types can provide resistance to the polishing action of traffic in the long term, so that adequate microtexture is preserved. This is quantified by the Polished Stone Value (PSV) Test.[28]

At higher speeds, the removal of bulk water from the tyre/stone interface is the major requirement so that good contact is maintained. This is achieved by adequate 'macrotexture' measured in terms of the texture depth of the wearing course combined with deep treads in tyres. Recent work at TRL by Roe *et al*[29] has demonstrated that adequate texture depth does reduce accident risk at speeds in excess of 64 km/hr. When PA is used, the bulk water is removed by drainage within the surfacing layer thickness so the concept of texture depth is not relevant.

3.10.   Measuring the mechanical properties of road layers

3.10.1.   The Nottingham Asphalt Tester

Development of the equipment

Research at the University of Nottingham and elsewhere in the late 1970s and early 1980s concerned with mechanical properties of typical UK asphalts[30] identified the need to develop an appropriate method of mixture design[23] and associated practical, low cost test apparatus to measure the key properties.

The Nottingham Asphalt Tester (NAT), shown in Fig. 3.19, emerged from this work.[31] Its development was based on a substantial background of research which used quite complex apparatus to obtain accurate data from tests which were designed to closely simulate field loading conditions. Such tests were, however, inappropriate for routine application and practice. Development of the NAT also drew on experience in the US and involved careful attention to those practical details which were considered as being most important in evolving a reliable and accurate test regime which could be used in contractual situations.

The NAT uses a pneumatic loading system mounted on a rigid test frame which is manufactured from stainless steel, as shown in Fig. 3.20. The height of the crosshead is adjustable to accommodate specimens of different lengths and diameters. Load is measured with a precision strain-gauged load cell, deformations are measured with linear variable

Fig. 3.19.   The Nottingham Asphalt Tester (NAT)

differential transformers (LVDTs) and there is an option for temperature measurement with thermocouples. It is possible to carry out Indirect Tensile Tests and Direct Compression Tests by using different sub-systems in the test frame, as shown in Figs 3.20(b) and (c). The test frame operates within a temperature-controlled cabinet. Data acquisition and control are carried out using a conventional personal computer, a state of the art digital interface unit and user friendly software, which is available to operate within the Windows environment. A number of different tests can be carried out using a range of sub-systems and these provide the means of measuring and assessing stiffness modulus, resistance to permanent deformation and resistance to fatigue cracking, the first two of which are covered by British Standard Drafts for Development[32,33] whilst the third is under consideration.

The Indirect Tensile Test regime was adopted in the US to provide a simple way of measuring the elastic stiffness of asphalts. The test principle is illustrated in Fig. 3.21 and is a development of the Brazilian Test used to determine the tensile strength of concrete under monotonic loading, as in Fig. 3.21(c). The test can be used to determine stiffness modulus and fatigue strength by applying lower and repeated loads, as in Figs 3.21(a) and 3.21(b). Application of the load is through steel strips machined to fit the curved surface of the test specimen. The deformation of the specimen is measured across a horizontal or vertical diameter, depending on which property is being measured.

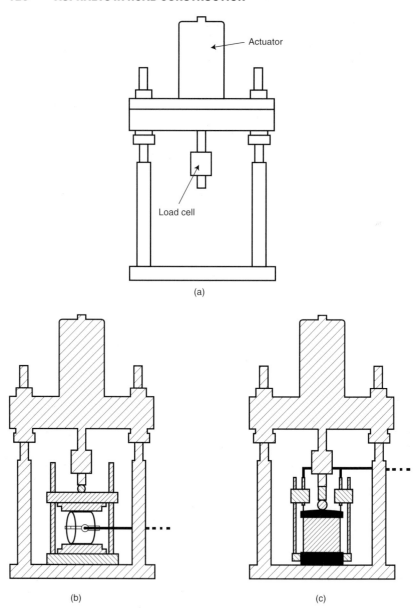

Fig. 3.20. NAT test arrangements[21]

The test has three principal advantages over alternatives. It allows cylindrical specimens, usually either 100 mm or 150 mm in diameter, to be tested. These may be prepared by compaction in a mould or by coring from site or a test slab. The test specimen does not need to be very long, typically 30 mm to 80 mm, which is convenient when coring from

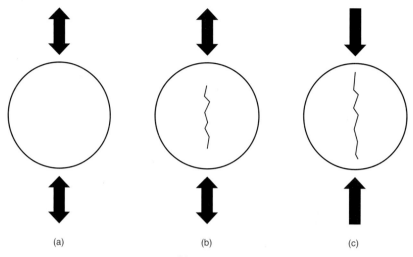

(a)   (b)   (c)

Fig. 3.21.   Indirect Tensile Tests.[21] *(a) Repeated load stiffness, (b) fatigue, (c) monotonic splitting*

pavements. The load to be applied is lower than that required, for example, in a Direct Compression Test. The third advantage is that the test offers a convenient means of applying tensile stress under compressive loading. The tension of the specimen is generated indirectly on the vertical diameter.

Early criticisms about the accuracy of the repeated load version of this test have been overcome through improved instrumentation and attention to the detail of test conditions. This allows accurate measurements to be recorded and ensures that the material is tested under conditions in which essentially elastic response is obtained.

Figure 3.21(a) shows the repeated load Indirect Tensile Stiffness Modulus (ITSM) Test in which only a few load applications are applied and the specimen suffers no significant damage. Figure 3.21(b) involves the same test regime but repeated loading is continued until failure by cracking is achieved. This is the Indirect Tensile Fatigue Test (ITFT).

Measurement of stiffness modulus
Stiffness modulus is the ratio of stress to strain under uniaxial loading conditions and is analogous to Young's modulus of elasticity. Since asphalts are sensitive to temperature and loading time, this alternative terminology is used since stiffness modulus is also a function of these parameters and is, therefore, not a constant for a particular material. Stiffness modulus is measured using the ITSM Test.[33] A load pulse is applied along the vertical diameter of the specimen as shown in Fig. 3.21(a) and the resultant peak transient deformation along the horizontal diameter is accurately measured to better than 1 micron. The stiffness modulus is a function of load, deformation, specimen dimensions and

Poisson's ratio. This last parameter is normally assumed to be 0.35 which is a representative value for most asphalts.

Measurement of fatigue strength

Most of the experimental procedures for characterizing the fatigue strength of asphalts have previously been quite time consuming, involving 15 to 20 test specimens and sophisticated apparatus. The ITFT, shown in Fig. 3.21(b), has proved to be a quick and reliable procedure with potential use in practice. The experimental arrangement is similar in principle to that used for the ITSM test except that the specimen thickness should be between 30 mm and 50 mm. Modifications have been made so that the crosshead is fitted with linear bearings to maintain a horizontal position at all times during testing as shown in Fig. 3.22.

Typical results of ITFT on a continuously graded dense bitumen macadam are shown in Fig. 3.23. These are expressed as a relationship

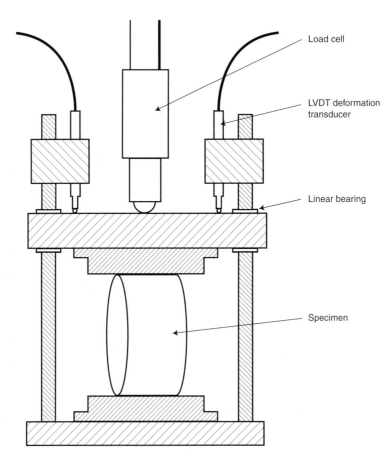

Load cell

LVDT deformation transducer

Linear bearing

Specimen

Fig. 3.22.   ITFT arrangements[21]

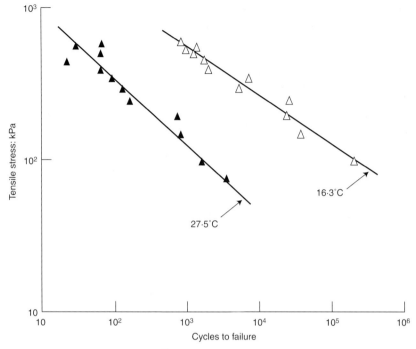

Fig. 3.23.   Typical ITFT data[21]

between tensile microstrain and the number of cycles to failure. Tests were carried out at two temperatures and it can be seen that there are two distinct linear relationships when the results are plotted using logarithmic scales.

Initial tensile strains are determined using a technique based on an analysis of the stress distribution within the specimen and stiffness moduli measured using the ITSM Test. This enables the results to be plotted in the conventional manner, in terms of tensile strain as shown in Fig. 3.24. On this basis, the results tend to conform to a single linear relationship, which is typical of controlled-stress fatigue tests carried out in other types of equipment.

Figure 3.24 compares results for the ITFT with those from a Bending Test carried out on specimens cut to a trapezoidal shape and tested as cantilevers. This method has been extensively used in research and more routinely in French practice.[34] The results are seen to be consistent. It has been found that sufficient tests to define a fatigue relationship can be carried out with the ITFT in a few hours. This represents a very significant advance in cutting the cost of this type of testing and is, therefore, of considerable interest to practical highway laboratories. Initial reproducibility trials involving ten laboratories have provided encouraging results.[35]

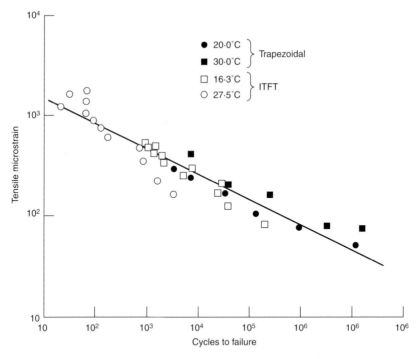

Fig. 3.24.   Comparison of results for the ITFT with those from a Bending Test[21]

Measurement of resistance to permanent deformation
Accumulated permanent deformation under repeated loading leads to wheel track rutting. The implementation of a suitable test for assessing resistance to such deformation is probably the most important requirement for performance-based specifications. This is because it is affected by a wide range of mixture parameters, not least those associated with the aggregate. It was for these reasons that the uniaxial static creep test was introduced in the 1970s.[36] It is now recognized that repeated loading is a necessary requirement and, hence, the Repeated Load Axial (RLA) Test has been developed at Nottingham. The test configuration is shown in Fig. 3.20(c) and usually involves a test specimen which is 70 mm in length and either 100 mm or 150 mm in diameter. A 100 kPa axial stress is applied in 1 second square waves pulses with 1 second rest periods. The test is conducted at either 30 °C or 40 °C and lasts for 3600 load cycles which takes 2 h. An initial conditioning stress of 10 kPa is applied for 10 minutes. Axial deformation is monitored by a pair of displacement transducers mounted above the loading platen. RLA test results are expressed as accumulated permanent axial strain against number of load cycles. A version of the test, which allows the specimen to be subjected to some confining stress, has been developed and is known as the Confined Repeated Load Axial (CRLA)[37] Test. This facility is considered very

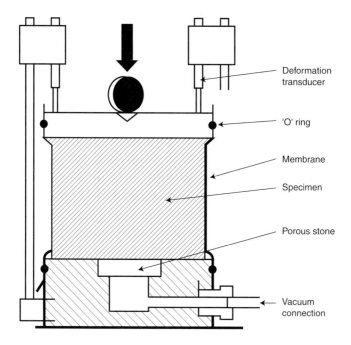

Fig. 3.25. Confined Repeated Load Axial Test[37]

important, particularly for mixtures such as SMA which rely on aggregate interlock to resist permanent deformation. Such interlock can only be properly mobilized when some confinement is provided to simulate field conditions. Figure 3.25 shows the CRLA apparatus indicating how the confining stress is applied through an internal partial vacuum. Some typical results of RLA tests are shown in Fig. 3.26.

3.10.2.   The Falling Weight Deflectometer

Whilst the NAT is helpful for testing small specimens in the laboratory, the Falling Weight Deflectometer (FWD) has become the equivalent tool for field measurements. The device is shown in Fig. 3.27 and the principles of operation are shown in Fig. 3.28.

The FWD was originally invented in France and further developed in Scandinavia where the equipment is now manufactured. Its use for pavement evaluation has been summarised by Brown *et al.*[38]

The FWD applies a load pulse to the pavement surface which is simulative of traffic moving at about 30 km/hr. The resulting deflected shape of the surface, known as the 'deflection bowl' is measured using a set of geophones mounted at various radial positions up to approximately 2 m from the centre of the loading platen, as shown in Fig. 3.28. The geophones measure velocity, which is electronically integrated to determine deflection. The equipment is trailer mounted, computer controlled and easy to use. Some 200 measurements can be taken in a

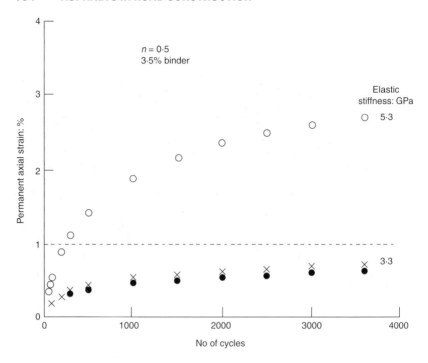

Fig. 3.26.   Typical RLA Test results

working day, depending on the nature of the site. Tests can be performed on partially constructed pavements or completed construction either to assess 'as built' condition or for evaluation of pavements in service. The deflectograph has been routinely used in the UK for many years. The advantages of the FWD over the deflectograph[39] are that it obtains a very accurate measurement of the complete deflection bowl for subsequent structural analysis and does so much more quickly than the deflectograph. The deflectograph, on the other hand, simply records the central, maximum, deflection under very slow loading conditions; typically 3 km/h. The interpretation of deflectograph data is entirely empirical, based on accumulated experience relating central deflection to potential pavement life to critical or failure conditions.[40] Figure 3.29 shows the key differences indicating two very different structural responses with the same central deflection. Unlike the FWD, the deflectograph cannot distinguish between these two situations.

There are two levels of interpretation for FWD site data. Figure 3.30 shows a typical set of key deflection parameters from a length of road. D_1 is the central deflection, beneath the centre of the load, D_7 is the smallest deflection measured at the largest radial distance (typically 1.8 m) whilst D_1-D_4 is a deflection difference. D_1 is a measure of the overall response of the layered pavement structure, D_7 is a function of the subgrade stiffness, independent of the other layers, and D_1-D_4 is a function of the

Fig. 3.27. The Falling Weight Deflectometer. Photograph courtesy of Scott Wilson Pavement Engineering Ltd

roadbase stiffness. The reasons for these characteristics are illustrated in Fig. 3.31, which shows that a change in subgrade stiffness affects the whole deflection bowl, whilst a change in roadbase stiffness only affects the shape of the bowl towards its centre. The shape is quantified by the parameter D_1-D_4.

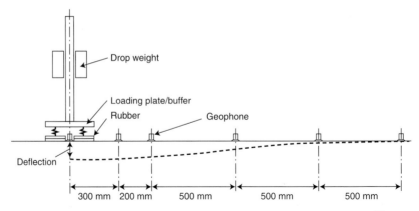

Fig. 3.28.   Principles of operation of the Falling Weight Deflectometer[16]

The second level of interpretation is 'back analysis'.[13,38] This involves the use of structural analysis for the layered pavement system in an iterative procedure until the computed deflection bowl matches the measured one. The layer stiffnesses associated with this matching condition are then identified. These are known as 'effective stiffnesses'. This term is used since back-analysis assumes each layer is coherent and uniform. Hence, any damage will result in a decrease in the effective value of stiffness. In order to carry out back-analysis and to make reliable engineering judgements about the state of the layers, it is essential to have details of the material types and thicknesses in the existing pavement and these are normally obtained by coring or, preferably, by excavating test pits typically 1 m square. This allows a judgement to be made about the structural integrity of the layers by comparing the back-analysed effective stiffness with 'as-new' values to be expected from the particular

Fig. 3.29. Key differences between the deflectograph and the FWD

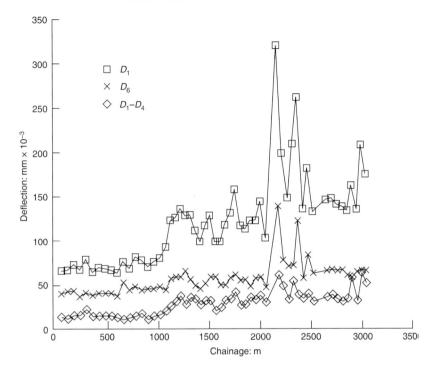

Fig. 3.30. Typical deflection profile for a length of road

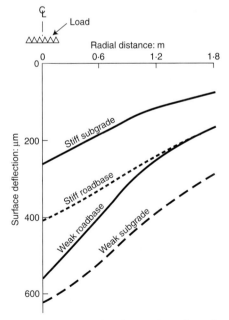

Fig. 3.31. Sensitivity of deflection bowls to subgrade and roadbase stiffness[38]

Table 3.1. Effective stiffness values for pavement layers in various states[41]

Material	Effective stiffness: GPa		
	Poor integrity throughout	Some deterioration	Good integrity
Bituminous <250 mm	<2	2–4	>4
Bituminous >250 mm	<4	4–7	>7
Lean concrete	<10	10–15	>15
Pavement quality	<20	20–30	>30

material.[41] Typical values are given in Table 3.1 for materials in various states from sound to badly damaged. The damage is generally the result of cracking in bound materials, while moisture ingress causes the deterioration in unbound granular layers and the subgrade.

It is possible to compare stiffness values obtained from ITSM tests in the NAT carried out on field cores with back analysed values from FWD data. However, the different conditions associated with each method of test need to be recognized and, for some of these, adjustments are required before legitimate comparisons can be made. The key points are as follows.

- ITSM Tests are on coherent cores representing a small volume of the pavement layer. The FWD value is based on the response of a large volume which may include damaged sections. This is why the term 'effective stiffness' is used.
- The temperature and loading time for the NAT and the FWD are generally different so adjustments must be made when dealing with bituminous layers.

Figure 3.32 shows a comparison of FWD and NAT stiffnesses for asphalts from three pavement sections involving heavy duty macadam (HDM), dense bitumen macadam (DBM) roadbase and recycled material bound with a foamed bitumen technique. In all cases, appropriate adjustments have been made. It will be noted that the FWD values are generally smaller than those from the NAT for the reasons outlined above.

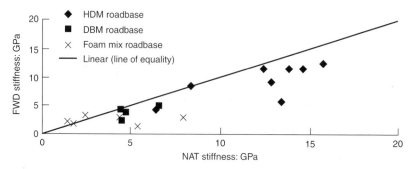

Fig. 3.32. Comparison of FWD and NAT stiffness for asphalts from from three pavement sections

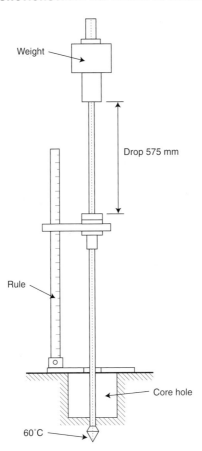

Weight

Drop 575 mm

Rule

Core hole

60°C

Fig.3.33.   The Dynamic Cone Penetrometer[43]

3.10.3.   The Dynamic Cone Penetrometer

This simple device, illustrated in Fig. 3.33, has been in use for many years in various countries following its introduction to pavement evaluation procedures in South Africa.[42] A standard cone is hammered through the unbound layers of a pavement using a standard effort from a falling weight. These lower layers of the pavement are typically accessed through a core hole. The number of blows per unit of penetration are recorded and relate to the shear strength of the material. Calibrations allow either this parameter, or the CBR, to be determined.

Figure 3.34 shows a typical profile of data indicating how these characteristics change on penetrating from one layer to the next. The change of slope is very useful for determining the thicknesses of unbound layers in pavement evaluation work and complements the extraction of cores from the bound layers above.

Use of the Dynamic Cone Penetrometer (DCP) combined with NAT testing on cores and FWD surveys with back analysis form the basis for

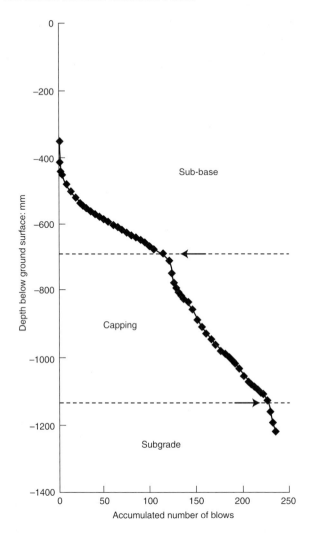

Fig. 3.34.   Typical profile of data from the Dynamic Cone Penetrometer[43]

modern structural evaluation of pavements. The procedures have been summarized by Brown.[43]

3.11. Summary

This Chapter has described the significant mechanical properties of materials used to construct asphalt pavements, from the soil to the running surface. This has been done in the context of the role played by each layer of material within the pavement. The basic mechanics of pavements, therefore, provided a backdrop to the discussion of material properties. This discussion emphasized the need to understand soil mechanics principles for subgrades, capping and sub-bases and asphalt mechanics for the asphaltic layers above.

The concepts of an 'ideal pavement' and an 'ideal pavement foundation' were discussed in order to highlight key features considered desirable for the materials in the various layers. These are mainly mechanical properties, which are being gradually introduced into 'end product' or 'performance-based' specifications.

For wearing courses, properties other than those to resist wheel loading were considered. These include the provision of adequate skid resistance and good durability to resist deterioration resulting from environmental factors.

Convenient modern techniques were described to measure the stiffness of pavement layers both in the laboratory and the field. The Nottingham Asphalt Tester is a relatively simple piece of laboratory equipment introduced to the industry in order that the mechanical properties of asphalts can be determined in the laboratory with ease. The Falling Weight Deflectometer and the Dynamic Cone Penetrometer are used to obtain field measurements, either during construction or, more usually, for structural evaluation purposes.

References

1. BROWN S. F. and E. T. SELIG (M. P. O'REILLY and S. F. BROWN (eds)). *The Design of Pavement and Rail Track Foundations, Cyclic Loading of Soils: from Theory to Design*. Blackie, London, 1991, 249–305.
2. HIGHWAYS AGENCY *et al. Design Manual for Roads and Bridges, Pavement Design and Maintenance*. TSO, London, various dates, **7**.
3. HIGHWAYS AGENCY *et al. Manual of Contract Documents for Highway Works*. TSO, London, 1998, **0–6**.
4. BROWN S. F. and R. D. BARKSDALE. Pavement Design and Materials. *Proc. 6th Int. Conf. Structural Design of Asphalt Pavements, University of Michigan, 1987*, **2**, 118–48.
5. NUNN M. E. Structural Design of Long-Life Flexible Roads for Heavy Traffic. *Proc. of ICE (Transp.)*, 1998, **129**, 3, 126–33.
6. CORTÉ J-F. and M-T. GOUX. *Design of Pavement Structures: The French Technical Guide*, Transportation Research Record 1589. Transportation Research Board, Washington DC, 116–24.
7. TAYEBALI A. A., J. A. DEACON and C. L. MONISMITH. Development and Evaluation of Surrogate Fatigue Models for SHRP A-003A Abridged Mix Design Procedure. *Journal of the Assoc. Asphalt Paving Tech.*, 1995, **64**, 340–60.
8. BROWN S. F. and A. R. DAWSON. Two-Stage Approach to Asphalt Pavement Design. *Proc. 7th Int. Conf. on Asphalt Pavements, Nottingham, 1992*. **1**, 16–34.
9. HIGHWAYS AGENCY *et al. Design Manual for Roads and Bridges, Pavement Design and Maintenance, Foundations*. HMSO, London, 1994, 7.2.2, HD 25/94.
10. BROWN S. F. Soil Mechanics in Pavement Engineering, 36th Rankine Lecture. *Géotechnique*, 1996, **46**, No. 3, 383–426.
11. POWELL W. D., J. F. POTTER, H. C. MAYHEW and M. E. NUNN. *The Structural Design of Bituminous Roads*. TRL, Crowthorne, 1984, LR 1132.

12. BRITISH STANDARDS INSTITUTION. *Methods of Test for Soils for Civil Engineering Purposes, Compaction-Related Tests.* BSI, London, 1990, BS 1377: Part 4.

13. BROWN S. F., W. S. TAM and J. M. BRUNTON. Development of an Analytical Method for the Structural Evaluation of Pavements. *Proc. 2nd Int. Conf. on the Bearing Capacity of Roads and Airfields, Plymouth, 1986,* **1**, 267–76.

14. BROWN S. F., M. P. O'REILLY and S. C. LOACH. The Relationship between California Bearing Ratio and Elastic Stiffness for Compacted Clays. *Ground Engineering,* 1990, **23**, No. 8, 27–31.

15. BROWN S. F. and J. W. PAPPIN. *Modelling of Granular Materials in Pavements.* Transportation Research Board, Washington DC, 1985, Transportation Research Record 1022, 45–51.

16. SWEERE G. T. H. *Unbound Granular Bases for Roads.* Delft University of Technology, Delft, 1990.

17. BRITISH STANDARDS INSTITUTION. *Hot Rolled Asphalt for Roads and other Paved Areas, Specification for Constituent Materials and Asphalt Mixtures.* BSI, London, 1992, BS 594: Part 1.

18. BRITISH STANDARDS INSTITUTION. *Coated Macadam for Roads and other Paved Areas, Specification for Constituent Materials and for Mixtures.* BSI, London, 1993, BS 4987: Part 1.

19. BROWN S. F. and K. E. COOPER. Improved Asphalt Mixes for Heavily Trafficked Roads. *Proc. 4th Conf. Asphalt Pavements for Southern Africa. Cape Town, 1984,* 545–60.

20. BRITISH STANDARDS INSTITUTION. *Sampling and Examination of Bituminous Mixtures for Roads and other Paved Areas, Methods of Test for the Determination of Density and Compaction.* BSI, London, 1989, BS 598: Part 104.

21. BROWN S. F. Practical Test Procedures for Mechanical Properties of Bituminous Materials. *Proc. of ICE (Transp.),* 1995, **111**, 289–97.

22. ASPHALT INSTITUTE. *Superpave Level 1 Mix Design,* Superpave Series No. 2 (SP-2). AI, Kentucky, 1994.

23. COOPER K. E., S. F. BROWN, J. N. PRESTON and F. M. L. AKEROYD. *Development of a Practical Method for the Design of Hot Mix Asphalt.* Transportation Research Board, Washington, DC, 1991, Transportation Research Record 1317, 42–51.

24. BROWN S. F., J. N. PRESTON and K. E. COOPER. Application of New Concepts in asphalt mix design. *Journal of the Assoc. Asphalt Paving Tech.,* 1991, **60**, 264–86.

25. HIGHWAYS AGENCY *et al. Design Manual for Roads and Bridges, Pavement Design and Maintenance, Surfacing Materials for New and Maintenance Construction.* TSO, London, 1999, 7.5.1, HD 36/99.

26. HIGHWAYS AGENCY *et al. Design Manual for Roads and Bridges, Pavement Design and Maintenance, Bituminous Surfacing Materials and Techniques.* TSO, London, 1999, 7.5.2, HD 37/99.

27. HIGHWAYS AGENCY *et al. Design Manual for Roads and Bridges, Pavement Design and Maintenance, Skidding Resistance.* TSO, London, 1994, 7.3.1, HD 28/94.

28. BRITISH STANDARDS INSTITUTION. *Testing Aggregates, Method for the Determination of Polished Stone Value.* BSI, London, 1989, BS 812: Part 114.

29. ROE P. G., D. C. WEBSTER and G. WEST. *The Relation between the*

Surface Texture of Roads and Accidents. TRL, Crowthorne, 1991, RR 296.
30. BROWN S. F., K. E. COOPER and G. R. POOLEY. Improved Road Bases for Longer Pavement Life. *Proc. 3rd Eurobitume Symposium> The Hague, 1985*, **1**, 217–22.
31. COOPER K. E. and S. F. BROWN. Development of a Simple Apparatus for the Measurement of the Mechanical Properties of Asphalt Mixes. *Proc. Eurobitume Symposium, Madrid, 1989*, 494–8.
32. BRITISH STANDARDS INSTITUTION. *Methods for Determination of the Indirect Tensile Stiffness Modulus of Bituminous Mixtures.* BSI, London, (Final draft March 1997), DD 213.
33. BRITISH STANDARDS INSTITUTION. *Method for Determining Resistance to Permanent Deformation of Bituminous Mixtures Subject to Unconfined Dynamic Loading.* BSI, London, 1996, DD 226.
34. BONNOT J. *Asphalt Aggregate Mixtures*, Transportation Research Board, Washington DC, 1986, Transportation Research Record 1986, 42–51.
35. READ J. M. and S. F. BROWN. Practical Evaluation of Fatigue Strength for Bituminous Paving Mixtures. *Proc. Eurobitume/Euroasphalt Congress, Strasbourg, 1996.*
36. HILLS J. F. The Creep of Asphalt Mixes. *Institute of Petroleum*, 1973, **59**, No. 570.
37. BROWN S. F. and T. V. SCHOLZ. Permanent Deformation Characteristics of Porous Asphalt Determined in the Confined Repeated Load Axial Test. *Highways and Transportation*, 1998, **45**, No. 12, 7–10.
38. BROWN S. F., W. S. TAM and J. M. BRUNTON. Structural Evaluation and Overlay Design: Analysis and Implementation. *Proc. 6th Int. Conf. on the Structural Design of Asphalt Pavements. University of Michigan, 1987*, **1**, 1013–28.
39. KENNEDY C. K., P. FEVRE and C. S. CLARKE. *Pavement Deflection, Equipment for Measurement in the United Kingdom.* TRL, Crowthorne, 1978, LR 834.
40. KENNEDY C. K. and N. W. LISTER. *Prediction of Pavement Performance and Design of Overlays.* TRL, Crowthorne, 1978, LR 833.
41. HIGHWAYS AGENCY *et al. Design Manual for Roads and Bridges, Pavement Design and Maintenance, Structural Assessment Methods.* HMSO, London, 1994, 7.3.2, HD 29/94.
42. KLEYN E. G. and P. F. SAVAGE. The Application of the Pavement Dynamic Cone Penetrometer to Determine Bearing Properties and Performance of Road Pavements. *Proc. Int. Symp. on Bearing Capacity of Roads and Airfields, Trondheim, 1982.*
43. BROWN S. F. Achievements and Challenges in Asphalt Pavement Engineering, Keynote Lecture. *Proc. 8th Int. Conf. on Asphalt Pavements. Seattle, 1997*, **3**, 19–41.

4. Design and maintenance of asphalt pavements

4.1. Preamble

This Chapter describes how roads are designed in the UK. In fact, the approach which is described is used in many parts of the world, particularly those locations where UK consultants operate. The Highways Agency (HA), the Scottish Executive (formerly The Scottish Office), the Welsh Office and the Department of the Environment for Northern Ireland are the executive agencies of national government in the UK which have responsibility for the maintenance of trunk roads including motorways. Much of their duties are undertaken on their behalf, usually under contract, by local authorities or private sector consultants or contractors. These agencies have requirements for design and maintenance of the trunk road network and these are contained in the *Design Manual for Roads and Bridges*[1] (DMRB). Most highway authorities for non-trunk roads (i.e. local authorities) also adopt the DMRB for relevant operations.

The documents which constitute the DMRB are prepared, generally, by the HA in conjunction with the Scottish Executive, the Welsh Office and the Department of the Environment for Northern Ireland. Most elements of the system operate throughout the UK, although some apply only to one particular country. Scotland has, for some matters for example, tended to issue memoranda which apply solely to that country.

Currently, the DMRB is published as 15 volumes, some of which spread over several binders. In total, the system occupies over 20 binders and costs in excess of £1000 to purchase. The system is published by The Stationery Office (TSO) formerly Her Majesty's Stationery Office (HMSO). TSO operates an update service which automatically issues new documents and these often take the form of amendments to particular existing Parts.

The areas covered by particular volumes are

- Volumes 1, 2 and 3   Structures
- Volume 4   Geotechnics and Drainage
- Volume 5   Assessment and Preparation of Road Schemes
- Volume 6   Road Geometry
- Volume 7   Pavement Design and Maintenance

- Volume 8   Traffic Signs and Lighting
- Volume 9   Traffic Control and Communications
- Volumes 10 and 11   Environmental Issues
- Volume 12   Traffic Appraisal
- Volumes 13, 14 and 15 Economic Assessment

Volume 7 deals with asphalt (and concrete) technology applied to roads. The contents related to asphaltic pavement design and related issues are discussed in this Chapter.

The contents of the DMRB are frequently updated and users are advised to ensure that reference is made to that version which applies to a particular situation or contract or is the latest available, as appropriate. The effective construction, maintenance and management of roads requires engineers to be familiar not only with the design process per se but also with the myriad of other activities associated with asphalt technology. DMRB Volume 7[2] commendably reflects that view with sections on design and construction, pavement assessment, pavement maintenance and the various surfacing materials. This Chapter considers each of these areas from a practical standpoint.

The current road design method is essentially empirical (i.e. based on experience rather than calculation) but there is no doubt that design will, in future, be based entirely on computational methods. Accordingly, a number of the current analytical methods (sometimes called 'mechanistic methods') are described.

4.2. Introduction

Probably there is no form of construction that is subject to such varying conditions as the structure forming the surface of a road. These words were written by Francis Wood in June 1912, in his textbook on modern road construction.

Wood was the Borough Surveyor of Fulham in south west London, a practising engineer adapting the design principles of Telford and Macadam to cater for increasing traffic loads and the growth of the motor vehicle. In 1912, most of the main routes in Fulham were paved with 5 inch softwood blocks with a design life of between twelve and twenty years. The arrival of the motor vehicle was another step change in the economic structure of Great Britain for Wood and his colleagues. In the middle of the eighteenth century, the UK was essentially agricultural. One hundred years later, it was predominantly a series of industrial communities support by a transport infrastructure.

The early road engineers were charged with building roads to open up trading routes. In the North Yorkshire village of Spofforth lies the grave of John Metcalfe (1717–1810) known as Blind Jack of Knaresborough. Metcalfe built his first turnpike (i.e. toll) road in 1765, linking Knaresborough and Harrogate to the Great North Road at Boroughbridge. He went on to build over 180 miles of turnpike in Yorkshire, Lancashire and Derbyshire. His methods were said to be highly individual. He built a road over a deep bog using stones and gravel supported by a bed of heather and his bridge at Boroughbridge is said to be made of stones taken from the nearby Roman road.

Thomas Telford (1757–1834) was dubbed the 'Colossus of Roads' by his poet friend Robert Southey and became the first President of the Institution of Civil Engineers in 1820. Telford was responsible for 920 miles of new roads to the north of his Caledonian Canal which runs through the Great Glen of Scotland from Inverness to Fort William. Telford developed the principle of two-layer construction — a hand-placed foundation of large stones topped with a surfacing of smaller stones and a layer of gravel. His fellow countryman John Loudon Macadam preferred a lighter form of construction which was more suited to the high construction rates needed to develop the turnpike network. macadam claimed that a 'certain resilience in the surface was an advantage and easier on the horses'.

The Victorian civil and municipal engineers developed, refined and combined the principles of Macadam and Telford. They also compromised if material was not available at a reasonable price. However, whilst the iron tyres of wagons and carts compacted the surface of Telford's roads, the pneumatic tyre sucked up the grit and turned it into mud or clouds of dust. Francis Wood and his Edwardian colleagues faced new challenges in the design and maintenance of their road pavements. This Chapter aims to help the present day engineer respond to the challenges of the twenty-first century.

4.3. Volume 7 of the Design Manual for Roads and Bridges

Current UK pavement design and structural maintenance practice has been developed by a combination of practical experience, laboratory research and full scale road trials. Structural design standards for roads were first set out in Road Note 29 which was published in three editions in 1960, 1965 and 1970.[3] It used the observed performance of a number of experimental constructions within the public road network as a basis for design curves relating traffic loading to subgrade strength. Ever growing traffic levels exposed weaknesses in the empirical approach adopted in Road Note 29.[3] This led to the publication by Transport and Road Research Laboratory (TRRL, since 1992 TRL) of Laboratory Report LR 1132[4] in 1984, which blended empirical knowledge with theoretical concepts of pavement design.

Much of the recent research into asphalt pavements has been carried out for the HA by the TRL with support from Local Highway Authorities and the principal trade associations, the Quarry Products Association (QPA, which was until 1997 the British Aggregate Construction Materials Industries, BACMI) and the Refined Bitumen Association (RBA, the trade association of bitumen manufacturers). Research findings have been translated into best practice as departmental standards and advice notes for many years. Since 1994, this advice has been consolidated into Volume 7 of the DMRB.[2]

The use of Volume 7 is mandatory for all works under the direct control of the overseeing departments — shorthand for the HA, the

Table 4.1.   Main contents of Volume 7 of the Design Manual for Roads and Bridges

Sections	Parts
1. Preambles	1. General Information (HD 23[5])
	2. Technical Information (HD 35[6])
2. Pavement Design and	1. Traffic Assessment (HD 24[7])
Construction	2. Foundations (HD 25[8])
	3. Pavement Design (HD 26[9])
	4. Pavement Construction Methods (HD 27[10])
3. Pavement Maintenance	1. Skidding Resistance (HD 28[11])
Assessment	2. Non-Destructive Assessment Methods (HD 29[12])
	3. Maintenance Assessment Procedure (HD 30[13])
	4. Use and Limitations of Ground Penetrating Radar
	for Pavement Assessment (HD 72[14])
4. Pavement Maintenance	1. Maintenance of Bituminous Roads (HD 31[15])
Methods	2. Maintenance of Concrete Roads (HD 32[16])
5. Surfacing and Surfacing	1. Surfacing Materials for New Maintenance
Materials	Construction (HD 36[17])
	2. Bituminous Surfacing Materials and Techniques
	(HD 37[18])
	3. Concrete Surfacing and Materials (HD 38[19])

Scottish Executive, the Welsh Office and the Department of the Environment for Northern Ireland. Local Highway Authorities and private sector clients accept the DMRB as a 'best practice' document and use it as a basis for most highway-related construction and maintenance contracts.

The objective of Volume 7 is to provide an instruction manual. It is made up of five sections which are broken down into Parts (fifteen in total), as detailed in Table 4.1. Each Part of Volume 7 is designed to be amended and updated as a free standing document. Parts are interlinked, as shown in Fig. 4.1. Parts 2, 3 and 4 reflect the lifecycle of a road pavement — design, construct, assess then maintain.

Each individual Part is designated by the year in which it was published or updated, although some smaller changes do not necessarily result in a change of designation, e.g. HD 29/94[12] contains amendments dated November 1994 and May 1999.

The design of all works commissioned by the overseeing organizations must comply with paragraphs in Volume 7 that are in highlighted boxes. Designers are encouraged to consider the advice and guidance contained in the remainder of the text.

4.4.   Overview of pavement structure, type and performance

An asphalt pavement is a multi-layer structure which protects the underlying soils from the effects of traffic and climate both during construction and for the duration of the working life of the road. Chapter 2 of HD 23[5] provides the background to the various layers and Chapter 4 provides a glossary of terms. The main layers of a flexible pavement are illustrated in Fig. 4.2.

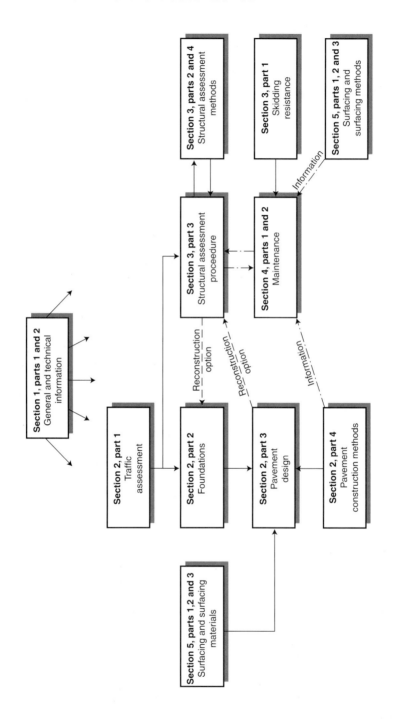

Fig. 4.1.   Use of Volume 7

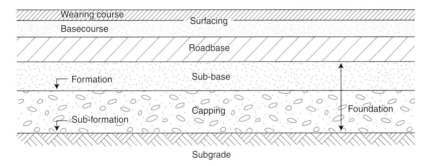

Fig. 4.2. Typical asphalt pavement[5]

The pavement foundation is made up of the subgrade soil, the capping layer (if used) and the sub-base layer. Together they form the platform for the key structural layers. Although the foundation is often constructed of low cost materials, its importance must not be underestimated. The key to Telford's success as a highway engineer was the foundation of his roads. He took the trouble to have a foundation of large stones placed by hand and packed with smaller aggregates. Telford also recognized that excess water in the structure of a road would be detrimental to its long term performance.

Design of the foundation (see Section 4.6) strives to ensure a minimum level of support for the construction of the asphalt layers whatever the characteristics of the underlying soil. The sub-base layers also have a significant structural contribution in thin asphalt pavements. The foundation cannot be considered as impermeable. Any water within the layers must be able to drain away somewhere or their strength will be adversely affected. The foundation also protects the subgrade soils from winter frosts.

The roadbase (often simply called the 'base') is the main structural layer which distributes applied loads and ensures that the foundation and subgrade are not overstressed. Roadbase asphalts must be both stiff and resistant to fatigue cracking. A good understanding of the applied traffic loading (see Section 4.5) and the characteristics and capability of roadbase asphalts is the key to cost effective design of a long-lasting asphalt pavement.

The surfacing is traditionally made up of two layers — the basecourse and the wearing course. The basecourse is also known as the binder course. It may be omitted from some heavy duty pavements and its role is generally to ensure an even surface for laying the wearing course. In maintenance works, the basecourse over the existing surface may vary in thickness. In such circumstances it is often termed a regulating course. When used below a porous surfacing, the basecourse is often required to resist the ingress of water into the pavement.

The wearing course forms the durable and skid-resistant surface used by the vehicle tyre. It must present an even ride, withstand deformation, resist cracking and (usually) keep water out.

Table 4.2. Pavement types

Pavement type	Roadbase	Surfacing
Flexible	Asphalt	Asphalt
Flexible composite	Cement bound	Asphalt
Rigid	Both layers concrete — may be a single layer	Concrete
Rigid composite	Continously reinforced concrete	Asphalt

Whatever the layer, the material chosen by the designer must be capable of being laid and compacted in all but the worst of weather conditions.

4.4.1. Pavement types
HD 23[5] defines four main types of pavement, based on combinations of flexible asphalt and rigid concrete layers. The terminology used is summarized in Table 4.2. There is at least one UK example of a hybrid pavement which uses asphalt roadbase below a continuously reinforced concrete pavement to solve a very difficult maintenance problem on a motorway built over very weak ground.

4.4.2. Pavement performance
The stresses and strains in an asphalt pavement are illustrated in Fig. 4.3. Pavements do not fail suddenly, they gradually deteriorate to a level of

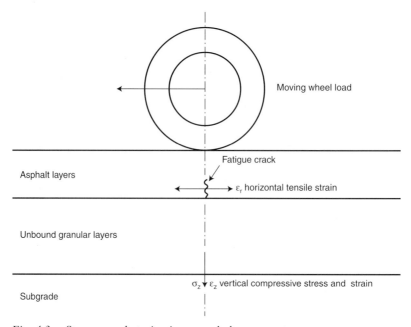

Fig. 4.3. Stresses and strains in an asphalt pavement

serviceability which may be defined as failure. Chapter 5 of HD 26[9] describes four main phases of structural deterioration in a flexible pavement

- *stabilizing* — strength generally increases as equilibrium conditions are established in a new or strengthened pavement
- *stability* — changes are very slow and can be predicted with some confidence
- *investigatory phase* — behaviour becomes unpredictable, performance needs to be monitored if future maintenance is to be cost effective
- *failure* — the pavement requires reconstruction to restore its condition (but perhaps not immediately).

The assessment of maintenance treatments (see Section 4.12) is an important element of highway engineering at a time when maintenance of the existing infrastructure is often more important than the building of new roads.

4.4.3.   HD 35 Technical Information
HD 35[6] provides general advice to the designer. Areas considered include conservation techniques and the appropriate use of aggregates, whatever their source. The greatest contribution to sustainable construction is effective initial design, but, HD 35[6] reminds users that the *Specification for Highway Works* (SHW)[20] permits a wide range of reclaimed and secondary aggregates to be used in unbound layers. Slags and reclaimed asphalts can also be used in new asphalts.

4.5.   Traffic assessment
In the early years of the twentieth century, Francis Wood and his fellow highway engineers were working hard to enumerate the damaging effects of the traffic at the time and to formulate design rules. The Highways Act of 1878 imposed a limit of 14 tons on the total weight of a steam-powered road locomotive. The Motor Car (Locomotives) Act of 1903 also imposed a limit of 8 tons on the weight carried by an individual axle.

Over the course of the twentieth century, economic prosperity and industrial development triggered massive growth in the total vehicle numbers, the loads on individual axles and the average number of axles in each commercial vehicle. HD 24[7] contains the findings of the latest research on pavement wear and traffic growth. Data on axle configurations, vehicle weights and vehicle numbers are used to calculate total traffic loading over the selected design life.

It is important to note that the empirically derived data in HD 24[7] are valid for traffic conditions in the UK. Many countries have higher permitted axle loads and gross vehicle weights so traffic assessments from other sources may not be comparable.

HD 24[7] sets out two methods of traffic assessment — a simple method for new roads which uses default national data applied to traffic flows estimated using traffic prediction techniques and a more detailed calculation for maintenance work which uses actual traffic data for the existing road.

4.5.1.   Pavement wear factors

The damaging effect of a vehicle is related to the load imposed by each of its axles. The concept of the 'standard axle' was an outcome of the American Association of State Highway and Transportation Officials (AASHO) Road Pavement Tests[21] carried out in the USA in 1959 and 1960. TRL Report LR 833[22] defines the standard axle in the UK as having a force of 80 kN (about 8.16 tonnes).

The damaging effect of a specific axle can be related to the standard axle by a fourth power law

$$\text{(Wear/axle)} \propto \text{(axle load)}^4 \tag{4.1}$$

The fourth power is actually an average value. Various studies have derived values between 3.2 and 5.6 for different vehicle configurations.[7] The relationship is illustrated in Table 4.3, which also defines the vehicle classifications and groupings used by HD 24.[7]

It should be noted that the values in Table 4.3 reflect the average weight of vehicles in service. Vehicles on the highway network are not loaded to the maximum permitted gross vehicle weight at all times. If a pavement is to be designed for a specific industrial use by vehicles of known axle weight, a site-specific calculation can be used. The values in Table 4.4 have been derived using the fourth power rule.

Two other factors which influence the wear factor are as follows

- As the flow of heavy goods vehicles increases, a greater proportion of them are likely to be in Group OGV2.
- On dual carriageways, the proportion of commercial vehicles using the outer lane(s) increases as traffic increases, as illustrated in Table 4.5.

Table 4.3.   Pavement wear factors for UK traffic [7]

Vehicle classification	Wear factor
Cars and vans below 1.5 t unladen weight	Negligible
Group: PSV Buses and coaches	1.3
Group: OGV1 (above 1.5 t unladen weight) 2 axle rigid 3 axle rigid 3 axle articulated Mean value OGV1 + PSV	 0.34 1.70 0.65 0.6
Group: OGV2 4 axle rigid 4 axle articulated 5 axle articulated Mean value OGV2	 3.0 2.6 3.5 3.0

Table 4.4.   *Wear factors related to axle load*

Axle load: t	Wear factor: sa
6.0	0.29
8.0	0.92
10.0	2.3
12.0	4.7
14.0	8.7
16.0	15
18.0	24

sa = standard axles
t = tonnes

Table 4.5.   *Percentage of commercial vehicles in the most heavily trafficked lane*[7]

Commercial vehicles per day	Percentage in lane 1
200	97
500	95
1 500	92
2 500	88
5 000	79
10 000	65

All lanes, including the hard shoulder, are constructed to the same design thickness to allow for traffic management during future maintenance works. In complex junctions, it may be appropriate to consider the actual flows in individual lanes.

4.5.2.   *The standard method of traffic assessment*

The standard method described in Chapter 2 of HD 24[7] is a simplified method for new roads which uses default national data applied to traffic estimated using traffic prediction techniques. The traffic flow prediction is done on the basis of the Annual Average Daily Flow (AADF). The following parameters are extracted from the traffic prediction data for structural design of a pavement

- commercial vehicles per day in one direction (cv/d)
- proportion of large Other Goods Vehicles (% OGV2).

The designer can select one of four traffic prediction charts depending upon

- single or dual carriageway;
- staged construction (20 years) or long (indeterminate) life.

The characteristics of the charts are illustrated in Table 4.6.

The lower design traffic values for a dual carriageway compared with a single carriageway with the same flow reflects the use of lanes other than the nearside lane. The design traffic increases in a non linear way down the columns, reflecting the greater proportion of heavier vehicles at higher flows. This can be seen by the curved nature of the lines in the HD 24[7]

Table 4.6.   Illustration of design traffic flows (msa)[7]

Traffic	Staged construction		Long life	
(Standard % of OGV2)	Single	Dual	Single	Dual
200	2.6	2.1	7.5	7.0
500	9.5	9.0	29	27
1000	24	23	75	60
1 500	40	38	more than 80	
2 000	60	52	more than 80	

charts. These charts also allow higher than usual proportions of Group OGV2 vehicles to be included in the design traffic assessment.

The non-linear increase in design traffic between staged and long life traffic reflects traffic growth assumptions and planned changes to permitted axle loading and gross vehicle weights. Increase in traffic is related to general growth in the national economy.

The concept of long life design and an associated minimum pavement thickness (discussed later in Section 4.7) means that the top and bottom of the design traffic assessment curves are not so important. HD 26[9] pavement design curves have a minimum design traffic level of 1.0 msa, which means that the calculations are not critical below about 100 cv/d with low percentages of class OGV2, levels which are typical of many housing estate and quiet rural roads. The current upper limit of pavement thickness is equivalent to 80 msa. So, if the traffic flow is above 3200 cv/d for staged construction and 1300 cv/d for a long life design, the accuracy of the traffic calculation is, again, not critical.

4.5.3.   The structural assessment and maintenance method
The more detailed calculation protocol described in Chapter 3 of HD 24[7] provides for a more accurate assessment of traffic loading because it is based on actual traffic data for an existing road. In particular, the national growth rate may not be relevant to the local situation. The calculated past growth rate may be a better basis for the prediction of future traffic levels provided reliable data are available for a period of at least ten years.

The calculation protocol assesses the traffic loading to date, usually since the opening of the road or since the last major structural maintenance treatment was undertaken, and predicts future traffic over the required residual design life or maintenance treatment life. Charts allow the derivation of growth factors for past and future traffic. Chapter 3 of HD 24[7] also includes a *pro forma* calculation sheet and examples of calculations.

4.5.4.   Heavy duty asphalt pavements
The empirical traffic loading assessments are based on typical tyre pressures for heavy commercial vehicles of about 0.5 MPa. Variations in tyre pressures affect the contact area or 'footprint' of the tyre on the

pavement surface and influence the stresses and strains within the surface layers. Increasing axle loadings also require higher values of pavement stiffness in order to limit the stresses transmitted to the subgrade.

The more detailed maintenance assessment method described in Section 4.5.3 will often be suitable for industrial areas used by normal highway vehicles, although growth factors may not be necessary if the pavement is serving a facility with a fixed capacity.

On haul roads used by off-highway vehicles, axle loads may be very high and computer-based analytical design techniques will often be appropriate (these are discussed later in Section 4.8). The highest tyre pressures are usually associated with airfields and dock pavements. More details can be found in the publications *Guide to Airfield Pavement Design and Evaluation*[23] and the *Structural Design of Heavy Duty Pavements for Ports and Other Industries.*[24]

4.5.5.   The influence of vehicle speed on pavement design
The speed of heavy vehicles has an influence on three aspects of flexible pavement design. Analytical pavement design, discussed later in Section 4.8, permits the adjustment of the dynamic stiffness of the pavement to reflect traffic speed but the temperature of the layer (influenced, of course, by environmental conditions) is a much more dominant factor than speed.

Speed is much more relevant to the performance of the wearing course. The resilient modulus of the wearing course depends upon the rate at which the load is applied. Low speeds and high temperatures increase the risk of wheel track rutting, as illustrated in Fig. 4.4. Lateral forces due to braking and tight turning movements also need to be considered in selecting the characteristics of the wearing course (see Section 4.13).

4.6.   Design of pavement foundations
The aim of foundation design is to select a sub-base material and thickness that is appropriate to the strength of the underlying subgrade and the anticipated loads which will be imposed by construction traffic. The result should be a foundation which performs satisfactorily without excessive deterioration.

A sub-base layer must have sufficiently high stiffness to limit the stresses transmitted through it to a level which can be sustained by the underlying weaker material. This is especially important during construction when the stresses imposed directly on the sub-base are higher than those subsequently applied through overlying asphalt layers.

The sub-base layer must also contribute to the overall stiffness of the completed foundation. Adequate stiffness means that the roadbase can be sufficiently compacted and fatigue cracking of the asphalt layers under traffic loads will be resisted. In addition, the layer must have sufficient strength and deformation resistance to perform without itself suffering excessive deterioration. The approach set out in HD 25[8] is to define a 'standard' foundation which gives the same level of support to the asphalt layers, whatever the characteristics of the subgrade.

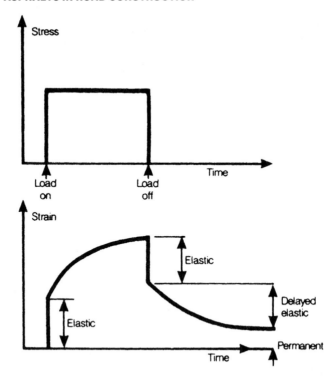

Fig. 4.4   Wheel track rutting

4.6.1.   Assessment of the subgrade

Unless unweathered rock is present at or above formation level, subgrade soils are not normally strong enough to carry construction traffic without distress. There is no single test method which allows the engineer to predict the susceptibility of soils to permanent deformation and distress under traffic. Index tests have been developed which provide guidance, but for major works there is no substitute for the experience of an engineer with a sound knowledge of the local soil types.

Ideally, a knowledge of the stiffness modulus and shear strength of the subgrade is required to design the thickness of the pavement layers. These parameters depend upon the type of soil, the degree of remoulding resulting from the earthworks phase and the effective stress levels. Effective stress is, itself, dependent upon the stresses imposed by the overlying layers (which may have changed over time), pore water pressure or suction and the position of the water table. The large number of factors involved means that simplification is necessary for most pavements with relatively straightforward ground conditions. In most cases, the California Bearing Ratio (CBR) is used as a simplified index for subgrade assessment.

Chapter 2 of HD 25[8] outlines a number of procedures for estimating the appropriate CBR value to be used for foundation design. The in situ

Table 4.7. Estimates of equilibrium CBR at subgrade level[8]

Type of soil	Plasticity Index	Estimated CBR (%)
Heavy clay	50+ 40	2 2–3
Silty clay	30 20	3–4 4–5
Sandy clay	10	4–5
Sand — poorly graded — well graded	–	20 40
Well graded sandy gravel	–	60

test methods in BS 1377[26] can also be used at the design stage, if the proposed formation level can be accessed on site. In the absence of detailed site investigation data, the estimated values in Table 4.7 may be used as guidance.

If the estimate of CBR is too low, it may be necessary to deal with areas of sub-base distress during the construction of the pavement.

4.6.2. Sub-base and capping

The grading of granular sub-base in Series 800 of the SHW[20] is intended to produce a dense and relatively stiff layer which is reasonably impermeable and sheds water during construction. A properly compacted sub-base can be expected to have a CBR of at least 30% and a stiffness modulus of 50 MPa.

A capping layer is used to improve weak subgrades by using lower cost material below the sub-base layer. A properly compacted capping layer material can be expected to have a CBR of at least 15% and to provide a good platform for the placing and compaction of the sub-base layer. The HD 25[8] approach to the design of a 'standard' foundation is summarized in Table 4.8.

The thickness of the sub-base layer is increased at values of CBR below 15%. Over most clays, the engineer may choose a thicker layer made up of a capping layer below a minimum depth sub-base layer instead of a single sub-base layer. The choice is dictated by local availability and cost of materials and the earthworks balance of the contract.

At values of CBR well below 2%, the soft subgrade may not be able to support construction of the foundation layers. One or a combination of the following techniques may be used to improve the subgrade, with the aim of creating a layer with a CBR of 2%

- remove and replace the subgrade material with granular fill to a depth of at least 0.5 m
- soil stabilization using lime or cement
- a geosynthetic or geogrid membrane and/or

Table 4.8. Capping and sub-base thickness design[8]

CBR value (%)	Design options and (total) thickness: mm	
<2	Subgrade improvement required (see text)	
2	150 sub-base over 600 capping (= 750)	
	sub-base only	capping + sub-base
2.5	350	400 + 150 (= 550)
3	300	350 + 150 (= 500)
4	270	300 + 150 (= 450)
5	220	250 + 150 (= 400)
8	190	210 + 150 (= 360)
10	170	190 + 150 (= 340)
≥15	150	—

- improve drainage but recognize that it takes time to establish equilibrium moisture conditions.

Series 600 of the SHW[20] permits a wide range of capping layer materials to be used. The principle of using low cost locally available materials which give a cost effective design solution is only feasible if there is a wide choice of acceptable materials. Since the performance criteria are not onerous, capping layers provide ideal 'low risk' locations for the use of recycled and low grade aggregates. The two key criteria are that the layer can be compacted properly and is able to support site traffic with the minimum of distress.

A layer thickness of 150 mm is the minimum which can be properly spread, levelled and compacted. This minimum thickness may need to be increased for coarse capping layers with large aggregate particles. Over low CBR clays, the first layer should be at least 200 mm thick because aggregate will be punched into the subgrade layer. This is allowed for in the HD 25[8] design charts.

Chapter 3 of HD 25 also mentions the use of asphalt in place of granular sub-base. This is discussed more fully in Section 4.9.

4.6.3. Drainage

The stiffness and shear strength of both the subgrade and foundation layers are adversely influenced by increases in moisture content. Incoming water must be excluded and an escape route provided for water already in the foundation. Unless the subgrade is permeable, a foundation drainage system is needed to ensure that the water table is below the top of the subgrade — ideally at least 300 mm below formation level.

Drainage is equally important for small works and local widening. If the ground water regime is not properly considered, it is easy to create a foundation layer that acts as a sump for local ground water.

4.6.4. Frost resistance

Series 700 of the SHW[20] specifies that granular material within 450 mm of the finished pavement surface must not be frost susceptible. The Frost Heave Test is defined in BS 812: Part 124.[27] Chapter 3 of HD 25[8] permits the 450 mm limit to be adjusted in coastal areas where meteorological data indicates a low incidence of frost.

One consequence of the 450 mm rule is that the capping layer with sub-base option illustrated in Table 4.8 may not be practical. It may be simpler to use a thicker layer of granular sub-base material which is known not to be frost susceptible. Any doubts about the frost susceptibility of the subgrade may also effectively lead to a minimum pavement construction depth of 450 mm even if the design thickness is less than that value.

The details of the specimen preparation protocol in the Frost Heave Test mean that it can only strictly be applied to granular sub-base but the test has been used to study other materials. A useful example of these studies is TRL Report 213[28] which shows that the addition of 2% of cement or bentonite to a granular gravel sub-base can virtually eliminate frost heave.

It should be noted that the UK is virtually alone in Europe in having a testing and material selection regime to control the potential for frost heave. Scandinavian countries with severe frost problems use thick and free-draining capping and sub-base layers to ensure that water is not trapped in the foundation in the first place. A European Standard Freeze/Thaw Test[29] is currently being finalised. This Test assesses the resistance of individual aggregate particles to freeze/thaw action but it does not consider frost action in a compacted layer.

4.6.5. Stabilized sub-base and design for performance

The HD 25[8] approach recognizes the use of cement stabilized sub-base below rigid concrete pavements. This ensures a stable platform for the paving train and minimizes the risk of water penetration but takes no account of the additional strength of the stabilized material.

There is a growing interest in the use of other ways of stabilizing both subgrade and sub-base materials, particularly as these techniques are well established in parts of France and adjacent European countries which do not have readily available supplies of natural aggregate. A recent report in this subject area is TRL Report 248.[30] The work concluded that granular sub-base could be replaced by thinner layers of stabilized mixtures. The work also indicates that modest adjustments to the design thickness of a stabilized sub-base can give cost effective layers with an increased structural contribution. The results are summarised in Table 4.9.

This work is leading towards a 'performance' approach to foundation layer design. In situ test methods are also being evaluated to complement a semi-analytical approach to foundation design but progress is slow. The

Table 4.9.   Use of stabilized sub-base[30]

Sub-base mixture	Thickness: mm
Granular sub-base Type 1	405
Cement/lime soil Type L	345
Cement/lime soil Type H	225
Cement bound mixture Type 1	205
Cement bound mixture Type 2A	150

search for cost effective, rapid and contractually reliable test equipment and testing procedures is proving difficult but must be successful if the use of recycled and low grade aggregates in low risk layers is to be encouraged.

4.7.   Design of asphalt thickness — the empirical approach

The design philosophy adopted in HD 26[9] is based upon the pavement research work of TRL. The approach is continually developing as knowledge of actual pavement performance is blended with a better understanding of asphalt properties in response to ever increasing traffic levels.

A good design should consider not only the initial cost but also the expected maintenance costs and the costs incurred by the road user over time, as illustrated in Table 4.10. An analysis of these total direct and indirect costs is known as whole life costing.

A minimum whole life cost is generally achieved when a design life of 40 years, with appropriate maintenance, is assumed. There are now two accepted approaches to achieving this design life

- traditional construction, with major maintenance at mid-life or
- a thick pavement of indeterminate life, requiring only periodic surface treatment.

4.7.1.   Staged construction — design for 20 years initial life

The traditional approach to major pavement design derived from LR 1132[4] assumes that the first stage of construction is designed for the traffic loading predicted over 20 years. At this point, the pavement is expected to reach the onset of the investigatory phase. A detailed assessment of maintenance needs (discussed in Section 4.11) and appropriate structural treatment (discussed in Section 4.12) is then used to strengthen the pavement for the remainder of the 40 year design life.

The term 'investigatory phase' is to be preferred to 'critical condition'. The critical condition is defined by measurements of pavement deflection

Table 4.10.   Elements of whole life costing[9]

Works costs	User costs
New construction	Traffic delays
Maintenance works	Accident costs
Residual value at design life	Fuel use/tyre and vehicle wear

(see Section 4.11.4). It marks the point in time when structural strength becomes unpredictable. The preferred term indicates that monitoring of the pavement should be intensified. There is a risk that the pace of deterioration will quicken but there must be actual evidence of deterioration for the design of the mid-life maintenance treatment to be cost effective. Additional works to correct deficiencies in skid resistance (see Section 4.10) may also be required at other times during the life of the pavement.

4.7.2.   Pavements of indeterminate life
Traditionally, the term 'indeterminate life' has been used in connection with flexible composite pavements with a cement bound material (CBM) roadbase (still often known as lean mix roadbase). CBM roadbase often has transverse shrinkage cracks but these do not generally have a significant effect on the structural performance of the pavement. Provided that the cracks are protected against the ingress of water, the rate of deterioration can be expected to be very low giving a long but 'indeterminate' design life.

Recent work by TRL into the design of long life flexible pavements has been published as TRL Report 250.[31] This study shows that the concept of indeterminate design life is also applicable to thick asphalt pavements with heavy traffic. Many thick pavements designed to have mid-life strengthening have been shown to have very stable structural characteristics and have not needed the expected treatments. It is also significant that road users have not suffered the delays associated with major maintenance works.

Thin pavements progressively weaken because of a combination of fatigue cracking in the asphalt layers and structural deformation originating in the subgrade. The research evaluated four thick pavements between 11 years old and 23 years old (equivalent to 22 msa to 66 msa) and found that fatigue and structural deformation were not present. It concluded that provided the pavement is well constructed and has sufficient strength to withstand structural damage in early life (a minimum threshold strength), it will have a structural life of more than 40 years.

Deterioration does take place but only by rutting and cracking of the surface layer. Provided this surface distress is treated before the roadbase layers are affected, the lower layers will cure slowly and become stiffer with time.

4.7.3.   Roadbase and basecourse materials
HD 26[9] recognizes five different asphaltic roadbase mixtures

- dense bitumen macadam roadbase with 100 pen grade bitumen (DBM);
- hot-rolled asphalt roadbase with 50 pen grade bitumen (HRA);
- dense bitumen macadam roadbase with 50 pen grade bitumen (DBM50);
- heavy duty macadam roadbase with 50 pen grade bitumen (HDM); and

- high modulus base macadam with 15 pen or 35 pen grade bitumen (HMB15 or HMB35).

Whichever mixture is chosen, the key to good performance is the relationship between the layer thickness and the size of the aggregate. The relationships given in BS 594: Part 1[32] and BS 4987: Part 2[33] are summarized in Table 4.11.

Modern laying and compaction plant allows macadams to be used without problem at a thickness greater than that recommended by BS 4987.[33] In France, it is not unusual to work at a layer thickness of five times the aggregate size. However, working at or below the BS 4987 thickness range will always prejudice performance. Where there is any doubt users are advised to opt for a smaller size of aggregate.

Traditionally, a DBM roadbase material would be specified with a 40 mm size aggregate but this has been displaced by 28 mm aggregate in many specifications. Although material made with a 40 mm aggregate is stiffer, it is more prone to segregation and variation across the laid width than a mixture manufactured with a 28 mm aggregate. The smaller size macadam is also easier to work with in areas where hand laying is necessary and compacts to a tighter surface finish. The proposed European Standards[34] will specify a 32 mm size aggregate to replace both the current 40 mm and 28 mm materials.

HRA roadbase mixtures have fallen out of favour, initially because of cost. They are, however, often used in resurfacing works to regulate levels below a new surfacing because their higher bitumen contents are much more tolerant of variations in layer thickness than bitumen macadams which contain less bitumen. This property is particularly relevant for thin layers. Methods for controlling the deformation resistance of HRA roadbase mixtures are not well developed and its use in heavily trafficked pavements requires caution — particularly if traffic is slow. There is an example of an HRA roadbase used in a supermarket distribution centre which deformed severely after only five years' service under slow moving delivery vehicles.

The use of stiffer versions of bitumen macadam with 50 pen grade bitumen followed research at TRL published in 1982[35] and subsequent road trials which are described in TRL Research Report 132.[36] The basis for the work was a French roadbase mixture, known as Grâve-Bitume,

Table 4.11. Asphalt layer thickness and aggregate size[32,33]

Material	Aggregate size: mm	Layer thickness: mm
Macadam roadbase and basecourse to BS 4987[33]	40	90–150
	28	70–100
	20	50–80
Rolled asphalt roadbase to BS 594[32]	40	75–150
	28	60–120
	20	45–80
	14	35–65

which contains 40/50 pen bitumen. The road trials established that, for a typical trunk road pavement, the combined thickness of the roadbase and basecourse could be reduced by 16% if DBM50 was substituted for DBM and reduced by 22% if HDM was substituted.

The difference between the two 50 pen grade mixtures relates to the filler contents of the mixtures and its influence on the amount of air voids. DBM50 has a filler content of 2% to 9%; HDM has 7% to 11%. DBM50 is easier to make with hard stone aggregates when additional filler is often not available at the coating plant without additional cost.

HDM is generally easy to make with limestone because the aggregate contains significant amounts of filler anyway. However, bitumen tends to coat the filler and fine aggregate particles in preference to the coarse particles. This can lead to a partial coating which may not look good but has no detrimental effect on the structural performance of the compacted layer. HDM mixtures at the upper end of the bitumen content and filler content specifications will often have a low air voids content and can be prone to deformation. This is one of the factors that led to the introduction of Clause 929 (Design, Compaction Assessment and Compliance of Roadbase and Basecourse Macadams) in the SHW[20] (as discussed in Section 4.7.4).

Both HDM and DBM50 are difficult materials with which to work and require good compaction plant. They should not be specified for pavements which have significant areas which are not accessible to a paver.

High modulus base (HMB) was developed by TRL as a result of the evaluation of a high modulus roadbase material known as Enrobé À Module Élevé (EME) which was introduced in France as an economic measure to reduce the use of oil-derived products. The performance of a trial pavement[37] confirmed that a bitumen macadam roadbase with a grading complying with BS 4987[38] but made with a 15 pen grade bitumen had very similar properties to EME.

A series of trials on motorway reconstruction sites in 1995 and early 1996 confirmed that HMB15 can be laid and compacted in a wide range of weather conditions with little adjustment to the mixing temperatures and compaction techniques employed with materials using a 50 pen grade bitumen. The results of the trial study are detailed by TRL in Report 231[39] which concludes that HMB offers significant improvements in deformation resistance without having a detrimental effect on crack resistance. Workability is compromised to some degree with some contractors opting to compact HMB with heavy pneumatic tyred rollers. HMB35, a mixture made with 35 pen grade bitumen, is also permitted as a compromise which allows more reliable compaction in marginal weather conditions.

The use of HMB should generally be restricted to very heavily trafficked pavements until more experience is gained. There will be some maintenance uses where HMB will offer significant design advantages but the surfacing contractor must be prepared to control the compaction stage closely to gain the full advantages of the enhanced stiffness. The sub-base layer must also be stable.

A good example of the effective use of HMB is in the reconstruction of lane 1 of a dual carriageway. If the HMB design thickness is less than the depth of the existing pavement, the work can be done without exposing the sub-base layer. This will provide a firm platform for the compaction of the new roadbase layers. One disadvantage may be that surface level tolerances will be harder to achieve because the time available for compaction will be less than that available for conventional mixtures. This means that if HMB is used directly under a thin surfacing then the surface finish may be poor. The surface of HMB is also more susceptible to damage if it is trafficked.

The wider use high modulus base was first permitted by the Highways Agency in 1998. Unfortunately, monitoring of the longer term performance of initial trials during 1999 found unexpected reductions in stiffness at some sites. As a consequence, the use of HMB made with 15 pen grade and 25 pen grade bitumen was suspended in March 2000 until the Transport Research Laboratory completed further studies. HMB made with 35 pen grade bitumen remains a permitted option.

4.7.4. Clause 929: Design, Compaction Assessment and Compliance of Roadbase and Basecourse Macadams
Clause 929[20] was introduced into the SHW[20] by amendment in 1993. The aim of the clause is to allow suppliers to minimize the risk of poor performance of roadbase and basecourse macadams by adjusting the mixture composition parameters. This adjustment to composition may take the mixture outside the recipe envelope which is defined in BS 4987.[38] Clause 929[20] also recognizes the use of the Nuclear Density Gauge (see also Sections 7.10.2 and 8.4.3 of Chapters 7 and 8 of this book) to control compaction.

The March 1998 version of Clause 929[20] also formalizes the collection of data about the stiffness modulus and deformation resistance (for more on these topics see Sections 3.10.1 of Chapter 3 and Sections 9.4.7 and 9.4.9 of Chapter 9 of this book). There are currently no specified requirements for these parameters but the collection of data on a consistent basis will assist further developments in pavement design procedures.

The methodology adopted in Clause 929[20] builds upon the established compaction control procedures using the Percentage Refusal Density (PRD) Test.[40] Although the PRD test is a very effective tool for controlling the majority of asphalts, it does not prevent the use of mixtures which comply with the specification yet have no voids and are therefore prone to secondary compaction or have excess voids and are, therefore, not as stiff as the majority of mixtures.

The procedure uses a 'Job Mixture Approval Trial' to assess probable compliance and to set a target recipe. The compliance criteria are summarized in Table 4.12. Data are collected at three evenly spaced locations along the trial section.

Clause 929[20] is not true mixture design. BS 4987[38] is a series of compromise recipes which cover a wide range of aggregate types. The use

Table 4.12.   Clause 929: Job Mixture Approval Trial[20]

Sample type	Test type
Bulk sample of loose mixture (6 No.)	Composition (3 No.) Maximum density (TMSG) (3 No.)
150 mm diameter cores (12 No.)	In situ bulk density and refusal density (6 No.) Stiffness modulus and deformation resistance (6 No.)
Nuclear density measurement (3 No.)	Correlate against initial bulk density
Calculate	In situ air voids (6 No.) Refusal air voids (6 No.) Percentage binder volume (3 No.)
Compliance criteria	• Compliance with nominated target composition • Average in situ voids $\leq 8.0\%$ • Individual in situ void content $\leq 9.0\%$ • Average void content at refusal $\geq 1.0\%$ • Binder volume at each location $\geq 7.0\%$

of a target recipe frees the supplier of the constraints imposed by this compromise whilst still ensuring that the mixture retains the essential characteristics of a dense macadam.

A consequence of the use of Clause 929[20] is that it erodes the differences between traditional mixtures. Most conventional recipe basecourse mixtures have too much binder to reliably meet the minimum voids requirement of 1% at refusal. The target filler content of mixtures made with limestone is, for the same reason, often in the overlapping range of 7% to 9% which applies to both HDM and DBM50.

4.7.5.   Design curves — normal pavements

The Department of Transport first published guidance on the structural design of asphalt pavements in 1976, as Part 1 of the *Notes for Guidance on the Specification for Road and Bridge Works.*[41] The design tables were developed from recommendations in the third edition of TRL Road Note 29.[3] Road Note 29 continued to be used for pavements with a design life of less than 2.5 msa for a number of years.

Figure 2.1 of HD 26[9] represents the latest version of the design recommendations for pavements of conventional staged construction (see Section 4.7.1.). It relates design traffic in the left hand lane (the lane with the heaviest traffic loading) to the total asphalt thickness depending upon the material used for the roadbase. The design choices are summarized in Table 4.13. The design is based on total asphalt thickness to allow choices to be made about the type of wearing course and the use of a basecourse.

In the design charts, a PA wearing course and similar permeable surfacings make a lower structural contribution than a HRA wearing

Table 4.13. Summary of pavement designs — staged construction[9] (a minimum thickness of 200 mm is now recommended (see Table 4.1))

Design traffic: msa	Roadbase material			Total asphalt thickness: mm		
	DBM			190		
1		DBM50			180	
			HDM			170
	DBM			250		
5		DBM50			230	
			HDM			210
	DBM			290		
10		DBM50			260	
			HDM			240
	DBM			320		
20		DBM50			280	
			HDM			260
	DBM			350		
40		DBM50			310	
			HDM			290

course. A thickness of 30 mm should be added to the design thickness if 50 mm of a PA is used.

A basecourse is optional below impermeable surfacing such as HRA. Its use is dictated by the balance between surface regularity and the speed of placement both of which are influenced by the number of construction layers in the pavement. If a basecourse is used, it must be of a similar material type as the roadbase material in order to ensure equivalent stiffness characteristics.

Work on long life flexible pavements (discussed in Section 4.7.2 and Section 4.7.6) has indicated that a minimum threshold thickness of 200 mm will ensure that lightly trafficked pavements have a very long indeterminate life in structural terms. This has important implications. A marginal increase in initial cost can have significant consequences for long-term structural maintenance. Table 4.14, derived from Fig. C3 of TRL Report 250,[31] shows the indeterminate life of asphalt pavements. If a minimum thickness of 200 mm is used, it also means that DBM with 100 pen grade bitumen can be chosen as the preferred mixture for most housing estate roads where workability and compactability are important material requirements.

4.7.6. Design curves — long life pavements
Work on the performance of heavily trafficked pavements has confirmed that it is not necessary to use a design pavement thickness greater than that required for an 80 msa design. The HD 26[9] design chart for heavily trafficked roads is summarized in Table 4.15 (DBM has been omitted for

Table 4.14.   Indeterminate life for thin (200 mm) asphalt pavements[31]

Roadbase type	Design life: msa
DBM	1.5
DBM50	2.5
HDM	3.6

Table 4.15.   Summary of pavement designs — long life pavements[9]

Design traffic: msa	Roadbase material	Total asphalt thickness: mm
50	DBM50	320
	HDM	300
	HMB35	280
	HMB15	250
≥ 80	DBM50	350
	HDM	330
	HMB35	290
	HMB15	260

clarity). Control of the actual stiffness of the compacted roadbase material is important for heavily trafficked pavements and will be incorporated into future designs.

4.7.7.   Asphalt in rigid composite pavements

The design of rigid composite pavements set out in HD 26[9] requires the use of a 100 mm thick asphalt layer over continuously reinforced concrete roadbase (CRCR). The asphalt layers reduce water penetration into the concrete slab and the potential for corrosion of the reinforcement. The asphalt also provides protection against rapid temperature changes in the concrete. CRCR has proved to be a successful form of construction in areas where large settlements are expected, e.g. mining areas and very weak ground conditions. The continuous reinforcement can accept significant strains whilst remaining substantially intact.

Although HD 26[9] does not permit the use of an asphalt wearing course directly onto the concrete slab, recent experience using modern asphalt surfacings and polymer modified tack coats has been very successful. The use of a negatively textured asphalt surface over a concrete roadbase gives significantly less traffic noise than a conventional concrete pavement surface.

4.8.   Design of asphalt thickness — the analytical approach

Chapter 6 of HD 26[9] provides a rudimentary introduction to the analytical (often called mechanistic) approach to pavement design. Treating the road pavement as a civil engineering structure is fundamental to design, but Volume 7 is cautious about the use of the analytical approach, principally because the failure criteria used by the computer algorithms may not match those assumed in the empirical approach detailed in the rest of HD 26.[9] However, the analytical approach is a very important tool for assessing alternative designs, designing remedial works and for testing the consequences of using new materials by comparison with an established design.

The designs discussed in Section 4.7 above are based on the work which led to the publication of LR 1132.[4] This report interprets the performance of experimental roads in the light of structural theory and uses a semi-theoretical extension of performance to anticipate behaviour under elevated traffic loadings.

Systematic work on the development of methods for the design of asphalt pavements gained real momentum during the 1950s. At that time, however, little was known about the behaviour of pavements under traffic and the physical properties of pavement materials and subgrades. Given such circumstances, a theoretical approach was always going to be difficult and the empirical approach gave a better hope for usable answers in the short term. In the last fifty years, the exponential rise in computational power has provided engineers with analytical programs on CD-ROM which can run on laptop computers. Discussed below are two of the most common methods used in the UK, some information on the stiffness values and a description of two initiatives which are supported by the European Commission and which examine analytical methods on a worldwide basis.

4.8.1.   The Shell Pavement Design Method

The Shell Pavement Design Method (SPDM[©]) is a readily available and well developed example of the analytical design approach and is applicable to pavements in all countries of the world. In 1963, Shell published a set of design charts which gave layer thicknesses for flexible pavements. The charts were based on a combination of laboratory test data and the results of the AASHO Road Test.[21] In 1978, the charts were expanded to incorporate an additional range of properties giving 296 charts and the BISTRO[©] suite of computer programs.

Shell's software is modular in format. It currently consists of four modules — BANDS[©], BISAR[©], CALT[©] and SPDM[©].

- BANDS[©] predicts the stiffnesses of both bitumens and mixtures using Van der Poel's Nomograph[42] as the basis for calculating the bitumen stiffness and Uge's Nomograph[43] for calculating the stiffness of the mixture. In addition, it estimates the fatigue relationship between critical strain and the number of cycles to failure.
- BISAR[©] (Bitumen Stress Analysis in Roads) calculates stresses, strains and displacements in an elastic multi-layer system (i.e. a road) providing the

  ○ system consists of horizontal layers of uniform thickness resting on a semi-infinite base or half space
  ○ layers extend infinitely in the horizontal plane

○ each layer is homogeneous and isotropic, and
○ materials are elastic and have a linear stress/strain relationship.

- CALT$^©$ (Calculation of Asphalt Layer Temperatures) computes temperatures in asphalt layers which are either cooling having been hot laid or are subject to climatic temperature changes. It takes account of over 20 parameters including air, material and substrate temperatures, thermal conductivities, specific heats and densities of all materials, wind speed and incident solar radiation.

    These modules are used to provide input to the SPDM$^©$. It includes many preset and default values for the key parameters which are used for an outline design and subsequently fine tuned to give a detailed design. The multi-layer linear elastic algorithm uses data for

    ○ *climate* — in terms of weighted Mean Annual Air Temperature (wMAAT)
    ○ *traffic levels* — in terms of equivalent standard axles
    ○ *subgrade stiffness* — which can be related to CBR
    ○ *subgrade strain criterion* — relationship between the permissible number of axle loads and the number of standard axles (design life)
    ○ *sub-base stiffness and thickness* — allowing the use of stabilized sub-bases to be evaluated
    ○ *asphalt stiffness* — characterizing load spreading ability and induced stresses
    ○ *asphalt fatigue criterion* — relationship between the magnitude of the induced strain and the number of axle loads to failure.

- SPDM$^©$ calculates

    ○ structural thickness design for new asphalt pavements
    ○ estimated permanent deformation, i.e. rutting in asphalt layers
    ○ structural thickness design for asphalt overlays on existing asphalt pavements.

The enhanced properties of modified asphalts can also be incorporated into the design input. Additionally, SPDM$^©$ in combination with BISAR$^©$ can be used to evaluate existing structures to assess their residual life allowing effective and timely maintenance treatments to be developed.

4.8.2. NOAH$^©$

NOAH$^©$ has been developed by Nynas, a major bitumen producer. It is used as a tool for supporting rational pavement design methods that employ mathematical models to predict pavement performance. It relies on the assumptions most commonly made in this field, starting by modelling the pavement as a multi-layered structure. Materials are assumed to behave as homogeneous elastic bodies characterized by their elastic stiffness, which may depend on environmental factors such as temperature. Traffic loads are represented as uniform pressures over circular areas, applied perpendicularly to the surface of the model pavement, simulating the imprints of vehicle tyres.

Defined stiffness values enable the calculation of the stresses and strains developed at different levels within a pavement structure under an applied traffic load. The life of a pavement is directly related to the ability of its constituent materials to withstand the magnitude and repeated application of these stresses and strains.

Structural factors such as layer thickness and material properties, environmental factors and loading factors can be treated as design variables. NOAH© automatically calculates, for each set of assigned values, the response of the pavement in terms of the geerated stresses and strains. Knowing theses values, it is possible to assess the effect of changing particular variables and thus optimize a design.

Recently, a finite element module has been incorporated within NOAH©. This allows the engineer to assess the effect of unbound layers on the performance of the pavement.

4.8.3. Use of stiffness data in analytical design

The stiffness values used in analytical pavement design are not those derived by measurement of cores taken from freshly laid asphalt. The differences are related to the frequency of the load application and the ageing effects of bitumen in service both of which lead to an increase in actual stiffness. Revised pavement design curves reflecting four grades of asphalt roadbase stiffness have been developed by TRL. Clause 944 of the SHW[20] links the design curves to the measured stiffness of cores taken to control compaction of the roadbase layers. Stiffness and its measurement were discussed further in Section 3.10.1 of Chapter 3.

4.8.4.   COST333/AMADEUS

The need for a more comprehensive analytical design method for pavements has been recognized by the Directorate for Transport of the European Commisssion.[44] In order to meet these objectives, the Commission is supporting two initiatives, COST333 and AMADEUS, which involves up to 20 countries in an effort to establish an integrated pavement design method.

COST333, Development of a new Asphalt Pavement design method, deals with information gathering and the selection of key parameters — performance requirements, traffic, climatic conditions, materials and models. AMADEUS, Advanced Models for Analytical Design of EUropean pavement Structures, examines existing methods. It contains an inventory of these, a practical evaluation and validation of each against actual situations. The methods currently being examined and the factors attributable to each are listed in Table 4.16.

4.9. Widening of pavements and haunch strengthening

HD 27[10] details a number of construction methods and techniques which support other parts of Volume 7,[2] including a chapter about widening of pavements. The widening of existing motorways and other trunk roads is an important technique for increasing the capacity of the network with minimal additional land use. Widening is generally undertaken to add

Table 4.16.   Analytical pavement methods being considered under AMADEUS[44]

Software name	Method used as response model	Type[b]	Non-linearity	Rheology	Anisotropy	Interface	Climatic effects	Dynamic loading	Axle spectrum	Tyre characteristics	Stochastic	Crack propagation	Thermal effects	Cumulative damage	Fatigue	Permanent deformation
APAS-WIN	Multi-layer	3					Y	Y	Y	Y			Y		Y	
AXYDIN	Axisymmetric FEM[a]	1					Y		Y	Y				Y	Y	Y
BISAR/SPDM	Multi-layer	3				Y	Y		Y	Y					Y	
CIRCLY	Multi-layer	3				Y				Y						
CAPA-3D	3D-FEM	3	Y	Y	Y	Y		Y		Y				Y	Y	Y
CESAR	3D-FEM	3	Y	Y	Y	Y		Y		Y	Y	Y	Y	Y		
ECOROUTE	Multi-layer	1		Y	Y	Y						Y				
ELSYM 5	Multi-layer	1														
KENLAYER	Multi-layer	2	Y													
MICHPAVE	3D-FEM	1	Y													
MMOP	Multi-layer	2				Y	Y	Y	Y	Y	Y			Y	Y	Y
NOAH	Multi-layer	3			Y	Y	Y		Y	Y	Y	Y			Y	
ROADENT	Multi-layer	2	Y		Y	Y	Y	Y	Y	Y					Y	Y
SYSTUS	3D-FEM	2				Y	Y					Y	Y	Y		
VAGDIM 95	Multi-layer	1		Y				Y	Y	Y					Y	Y
VEROAD	Multi-layer	3		Y					Y	Y	Y					
VESYS	Multi-layer	3					Y				Y			Y	Y	Y

[a] FEM = Finite Element Method

[b] Type 1, response models (i.e. those which only provide results in terms of stresses and strains); Type 2, response + partial performance (i.e. those which consider the effects of loading, climate, etc. on rutting, crack initiation, etc. but do not provide a full design procedure); Type 3, full design procedure (i.e. those which provide a recommended structure (thicknesses, materials) and long-term performance predictions)

lanes on major roads. Widening and haunch strengthening of sub-standard roads is a key tool in cost effective maintenance of the local network, particularly where vehicles run over the unrestrained edges of narrow roads. The same techniques may be used to add a marginal strip or safety area to an unkerbed major road.

4.9.1. Widening with additional lanes

As with the design of structural maintenance treatments (discussed in Section 4.12), major projects to widen and upgrade the existing network provide a design challenge which often has many more constraints than those which apply to works of new construction. The greatest challenges are keeping traffic disruption to a minimum and ensuring the safety of all involved in the works.

Section 2 of HD 27[10] discusses the design concepts of symmetrical and asymmetrical widening linked to improvements to line, crossfall and gradient. Widening is also an opportune time to consider the structural assessment of those parts of the existing road which need to be incorporated into the new pavement. The ideal solution is a 'combined' pavement with broadly equal design life throughout its length.

The state of the foundation under an existing pavement often provides a good indication of the likely equilibrium foundation condition for the widening element of the new pavement as long as the drainage system is still effective. The plane of the new foundation must also reflect probable drainage paths in the combined construction. Ponding in unbound layers resting against a bound layer must be avoided by design in the long term and by good planning during construction.

The need for effective drainage may lead to the new portion of the pavement being thicker than that which is strictly required by structural design procedures. It may also be appropriate to use stiffer materials and analytical design techniques to balance the demands of higher traffic levels on extra lanes with the depth occupied by the existing pavement.

Longitudinal joints in the asphalt layers must be made against an undisturbed vertical edge with roadbase stepped into the sub-base by at least 150 mm and surfacing stepped into the roadbase by at least a further 150 mm. Joints are a line of potential weakness in any structure. Joints should avoid wheel tracks except where one crosses a lane diagonally as a consequence of a change in carriageway width.

4.9.2. Haunch strengthening and local widening

The overrun of the edge of a narrow rural carriageway leads to structural damage which can be treated more effectively by a haunch strengthening than by overlay to the whole width. The maintenance history of many rural roads consists of incremental widening by patching local overrun areas and hiding the weak areas under surface dressing. This widening by stealth eventually leads to a nearside wheel track with little construction depth and often without a proper foundation compared with the centre of the pavement. This eventually results in structural failure under increasingly heavy local traffic loads. The situation is often exacerbated

Table 4.17. Use of reused materials in local widening

Road type	Type 3	Type 4
Traffic level: msa	0.5 to 2.5	Up to 0.5
Surfacing, conventional Basecourse, conventional	40 60	40 60
Roadbase, conventional Recycled planing with foamed bitumen Cold rejuvenated planings Granulated asphalt and planings	90 150 (Basecourse not needed) 300 320	50 110 (Basecourse not needed) 170 180

by the failure or absence of a subgrade drainage system and poor maintenance of adjacent ditches and watercourses.

TRL Report 216[45] summarizes the findings of a series of research projects funded by the County Surveyors' Society (now CSS). The research was carried out on a very practical basis, guided by engineers closely involved with maintenance of the local road network. The report also investigates the reuse of highway materials in repairing haunches.

The design method outlined in Report 216[45] emphasizes the need to consider control of the moisture content of the subgrade at formation level and reduce the ingress of water into the construction. The provision of a filter drain may be appropriate if a suitable outfall is available but it is more important to avoid the haunch becoming a drainage sump between the verge and existing carriageway. Asphalt substitution (see Section 4.9.3) is a technique which may minimize the need to excavate below the water table and allow better compaction in a restricted width. The use of a geosynthetic separator at foundation level is also encouraged when the CBR at formation level is less than 5%. Geosynthetic separators are plastic blankets permeable enough to permit the passage of water but impermeable enough to prevent the migration of soil particles.

Report 216[45] also makes a significant contribution to the understanding of the equivalence between traditional and reused materials. Plate bearing tests can be used to monitor equivalent performance for the sub-base layer. The equivalence with conventional hot asphalt mixtures derived by the research is illustrated in Table 4.17.

4.9.3. Asphalt substitution

The technique of substituting asphalt roadbase for granular sub-base has been evaluated in TRRL Technical Report 303.[46] The report looked at substitution in new construction and also identified the benefits of using the technique in reconstruction.

Bus routes in urban housing estates have derived particular benefit from the technique. Surface levels are constrained by kerb and drainage details and formation level is often fixed by a complex network of public utility apparatus. In such situations, the subgrade will usually be at a

Table 4.18.   Granular sub-base thickness and equivalent asphalt roadbase thickness[45]

Nominal CBR value %	Thickness of granular sub-base: mm	Equivalent thickness of 100 pen asphalt roadbase: mm
<2	500	140 (with geosynthetic separator)
2–4	500	140 (with geosynthetic separator)
5–7	250	70
8–14	200	60
≥15	150	40

higher equilibrium CBR value than at initial construction. Some or all of the existing sub-base can be removed to provide sufficient depth for a new pavement with additional roadbase placed in order to balance the effect of removing the sub-base.

Technical Report 303 suggests that a nominal substitution value of 30 mm of asphalt for each 100 mm of granular sub-base is necessary to give a structurally equivalent design, but this allows for poor compaction at formation level. TRL Report 216,[45] discussed above, gives more details about equivalence and this is summarized in Table 4.18.

The key to effective use of the technique is having a new formation level that can support the reconstruction operation, and choosing plant which will not damage any of the public utility apparatus. It may be appropriate to consider using a tracked paver or even hand laying critical areas. The design thickness may also need to be increased slightly if small rollers have to be used without vibration to minimize the risk of damage.

4.10.   Assessment — skid resistance

Skid resistance is a measure of the friction generated between a pavement surface and a vehicle tyre. A skid occurs when the available friction is insufficient to counter the forces imposed by a moving vehicle. At a particular location, the available friction depends upon the texture of the road surface, the properties of the tyre, vehicle speed and weather conditions.

In the UK, many areas produce aggregates with good resistance to polishing. The report published by Travers Morgan[47] in 1993 listed 98 active quarries and slag sources in mainland Great Britain producing aggregates with a Polished Stone Value (PSV) of 55 or more. The number of sources in each PSV class is shown in Table 4.19.

Table 4.19.   Sources of high PSV aggregate in the UK 1993[47]

Class of PSV	Number of sources
55–60	54
60–65	35
65–68	3
68+	6

Unfortunately, the best polish-resistant aggregates are less readily available and also the most expensive. The aim of HD 28[11] is to achieve a balance between satisfactory skid resistance performance over the service life of the surface layer and the economic use of available aggregates.

The detailed and fascinating history of the experiments, road trials and engineering ingenuity leading to the introduction of skid resistance standards is explained in an excellent book by Hosking.[48] The current specification for skid-resistant surfaces is the latest output from a long and detailed research programme which continues to keep pace with growing traffic levels. Recent research findings have been published as TRL Report 322.[49] TRL Report 367[50] details parallel work on the influence of surface texture depth on skid resistance at high and low speeds.

4.10.1.   Surface texture

When a pavement surface is dry the friction between tyre and surfacing is high. The available friction permits full use to be made of the braking and cornering performance of the vehicle. A wet surface presents a more complex situation because the presence of water affects the design of both the surface and the tyre. HD 28[11] concentrates on wet skid performance by considering both microtexture and macrotexture, as illustrated in Fig. 4.5.

Microtexture is the fine component of surface texture formed by the small interstices on the surface of aggregate particles. It is the main contributor to sliding resistance and is particularly important at low speeds (up to 50 km/h). Macrotexture is the coarse component of texture created by the shape of the aggregate particles and the spaces between them. The combination of macrotexture and tyre tread pattern provides the drainage paths to disperse water from the contact area between tyre and pavement surface.

Figure 4.6 illustrates how the importance of texture depth increases with speed. Recent research into locked-wheel friction over a wide range of speeds, published as TRL Report 367,[50] indicates that poor texture depth can have a significant adverse impact at low speeds. This is an important finding. New equipment has invalidated the assumption that texture depth is not a significant factor for low speed roads.

Until recently, an understanding of the concepts of microtexture and macrotexture had been sufficient to explain the surface characteristics of

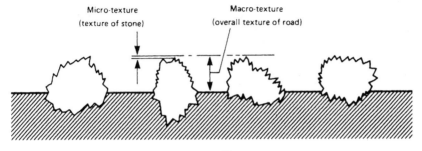

Micro-texture
(texture of stone)

Macro-texture
(overall texture of road)

Fig. 4.5.   Microtexture and macrotexture[48]

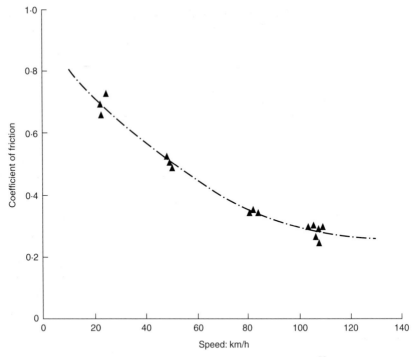

Fig. 4.6.   Typical friction–speed curve for hot rolled asphalt[50]

traditional HRAs and surface dressings. However, recent advances in surfacing technology along with consideration of the noise characteristics and spray performance have resulted in the introduction of a number of additional terms.

- Megatexture is a component of surface profile which is important in the generation of tyre noise.
- Random texture is used to describe the texture of a surface that does not follow any specific direction or pattern in the horizontal plane and is found in most asphalt surfaces.
- Transverse texture follows parallel lines perpendicular to the normal direction of traffic. An example is grooves cut in the surface of airfield runways.
- Positive texture is formed by particles which protrude above the plane of a surface. It is typically found in chipped HRA and surface dressings.
- Negative or indented texture is created by the voids between particles with upper surfaces forming a generally flat plane. Negative texture is found in many of the proprietary thin surfacings.
- Porous texture is found in negative texture materials which also allow water to pass through the layer. Porous asphalt is the best example of this.

Chapter 5 of HD 36[17] provides further illustration of the various types of surface texture and links skid resistance requirements with surface noise characteristics (noise is discussed in Section 4.13.2).

4.10.2.   Measurement of microtexture — the PSV Test

The microtexture of a surface is related to the polishing resistance of the aggregate particles under the action of tyre forces. Since microtexture exists at a microscopic level, it cannot be monitored directly on a practical scale. The measurement of low-speed-skid resistance is used as a surrogate for microtexture measurement.

The Stanley Skid Resistance Tester (SRT)[51] was designed to provide a simple means of assessing the skid resistance of a pavement surface. The results represent the skid resistance of a patterned tyre skidding at 50 km/h. When combined with the polishing machine first developed by TRL during the 1950s, the SRT forms the basis of the Polished Stone Value (PSV) Test details of which can be found in BS 812: Part 114.[51] The PSV Test was first published in 1960. It gives a ranking index for aggregate performance and uses 'control' aggregates to ensure that the ranking remains consistent over time. The polishing element of the test uses emery abrasives over six hours to produce a state of polishing similar to equilibrium conditions at average traffic levels.

The PSV Test is the most complex of the set of tests for aggregates[27,51–65] (for more on this topic see Section 1.12.6 of Chapter 1 of this book). It is important that tests are carried out by an experienced laboratory accredited by the United Kingdom Accreditation Service (UKAS) (or equivalent EC body) specifically for the PSV Test method. (UKAS is the UK national accreditation body responsible for assessing and accrediting the competence of measurement, testing, inspection and certification of systems, products and personnel.) Many other countries have tried to develop test methods with a similar intent but have departed little from the principles of the British Standard test. BS EN 1097–8[66] is based on the current procedures used in laboratories accredited by UKAS. Only samples taken from recently processed aggregates will give reliable results and the mean value of a number of test results will be more representative of the polishing resistance of an aggregate source than a single test.

Hosking[48] explains how gritstones were studied in the 1950s following failures which could have led to engineers discriminating against all gritstones, ironically the aggregates with the best resistance to polishing. Further study led to the conclusion that resistance to polishing is influenced by the degree of bonding between mineral constituents, a complex topic related to geological origins and history. The West Wycombe experiment laid in 1955[48] also highlighted the importance of adequate abrasion resistance to complement resistance to polishing. One aggregate with an Aggregate Abrasion Value (AAV) of 17 wore away rapidly.

The AAV Test is currently published as BS 812: Part 113.[62] This test is also included in BS EN 1097–8.[66] It is important to note that the AAV Test is designed specifically to model the resistance to abrasion of a single surface of an aggregate particle. The Micro-Deval Test models resistance to wear of the whole surface of a particle in the presence of water and has been published as BS EN 1079-1.[67] A reliable correlation between AAV and Micro-Deval values does not exist for all types of polish-resistant aggregates. Gritstones are particularly affected by the water in the Micro-Deval Test.

4.10.3.   Aggregate selection

Table 3.1 of HD 36[17] matches the required PSV values to the horizontal and vertical geometry and location in relation to junction proximity of the area under consideration.

4.10.4.   Measurement of macrotexture

It was shown in Section 4.10.2 how macrotexture helps to ensure rapid drainage of surface water away from the contact point between a vehicle tyre and the pavement surface. Surface texture depth is a measure of macrotexture. The simplest method of measuring surface texture depth is the Sand Patch Method detailed in BS 598: Part 105.[68] The principles of the test are illustrated in Fig. 4.7. Although the test[68] can be time consuming and needs a dry surface, it is relatively simple and readily verifiable by both contractor and client.

Work at TRRL (now TRL) in the 1970s established the validity of a statistical method of texture depth measurement involving the calculation of the standard deviation of a series of displacement measurements along the surface. This work led to the development of the Mini Texture Meter (MTM) and the TRL High-Speed Texture Meter (HSTM).

Use of the MTM is recognized in BS 598: Part 105.[68] Extensive experience on chipped HRA surfaces established the use of the MTM as a

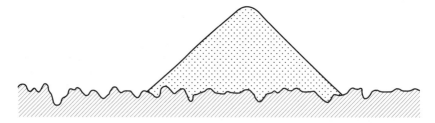

$$\text{Texture Depth} = \frac{\text{volume of sand}}{\text{area of patch}}$$

Fig. 4.7.   Sand Patch Method for measuring texture depth[68] monitoring and contract compliance tool. A reasonably reliable correlation can be established between sand patch values and the Sensor Measured Texture Depth (SMTD) using the MTM of a new HRA surface, if the requirements for the detailed calibration and the adjustment regime are met. However, there is not a general acceptance of MTM results for asphalts which have negative texture. The Sand Patch Method may be low technology but all concerned can see the results on the road.

The HSTM is used to measure the in-service texture depth of surfaces at speeds up to 100 km/h in normal traffic conditions. It is used as a network monitoring tool. HSTM results have been used as the basis of recent research which shows that high speed friction can be predicted from a combination of skid resistance and SMTD.

4.10.5. Measurement of skid resistance

The major increase in motor traffic that occurred in the 1920s resulted in frequent skidding accidents. This led to a road experiment on the Kingston Bypass in 1930, which established the principle of assessing skid resistance by measuring Sideways Force Coefficient (SFC) using a wheel mounted at an angle of 20 °C to the direction of travel.

Hosking[48] details this early work, which led to the formation of a small fleet of motor cycle and sidecar combinations to monitor the country's road network. The outbreak of war in 1939 led to the postponement of this ambitious project but the principles have been carried forward to the current fleet of machines which are used, predominantly, to measure skid resistance. Use of the Sideways Force Coefficient Routine Investigation Machine (SCRIM) is detailed in Chapter 3 of HD 28.[11] Annex 1 gives requirements for SCRIM calibration and Annex 2 sets out the operational procedures.

In the UK there are two devices in common use for measuring skid resistance: SCRIM and the GripTester. The SCRIM apparatus is mounted in a purpose built van consisting of a 12 t gross weight chassis cab with a purpose-built box containing a water tank, computational equipment and other apparatus necessary for the testing. The SCRIM measurements are taken by means of a constantly skidding fifth wheel inclined at 20° to the direction of travel and this is mounted on a frame which is separate from the ordinary driving wheels. The vehicle travels at a speed of 50 km/h for testing purposes although sharp bends and roundabouts are tested at 20 km/h. Since skidding is a problem in wet road conditions, water is jetted onto the road surface in front of the measuring wheel for simulation. One version of this machine is shown in Fig. 4.8.

The GripTester is a trailer whose measuring wheel is forced to slip at 14.5% of the distance travelled. It is relatively small (some 1000 mm long by 800 mm wide) and can be towed behind any vehicle which has a tow bar and can accommodate a water tank. Although the GripTester differs from SCRIM in both its size and its design principle, it correlates well with SCRIM. In September 1993, TRL prepared a Report[69] for the County Surveyors' Society on a correlation trial between four SCRIMs and four GripTesters which showed a strong linear correlation between

Fig. 4.8. SCRIM apparatus. Photograph courtesy of WDM Ltd

the two machines (in mathematical terms, there was a correlation coefficient of 0.977% which is very good). The GripTester is designed and manufactured by Findlay, Irvine Ltd of Penicuik near Edinburgh and is shown in Fig. 4.9.

The SCRIM apparatus is the bespoke national standard. Although skid measurements of various types have been taken since the 1920s, SCRIM became the national standard only around 1988. It is a relatively fast means of measuring the levels of skid resistance in networks. The

Fig. 4.9. GripTester apparatus. Photograph courtesy of Findlay, Irvine Ltd

principal disadvantage is financial with the equipment costing well over £100 000. Most highway authorities simply hire the equipment and operators for survey purposes. The GripTester costs around £20 000 for the trailer and data collection equipment. GripTesters tend to be used for relatively small volume testing but a few authorities use them for testing the whole network. Its range is limited only by the size of the water tank that can be carried by the towing vehicle. Although the GripTester has been shown to produce results that can be easily converted to accurate SCRIM values, it is not approved for use by the Highways Agency and so cannot be used on motorways and other trunk roads.

Traffic levels, seasonal variations and surface temperature all have an effect on the SFC of a road surface. The SFC tends to rise in winter when wet and cold weather creates a gritty detritus which roughens the surface. In drier summer conditions, surface detritus is dusty. This polishes the surface, resulting in a reduction in SFC. In order to normalize these effects, three SCRIM surveys should be carried out between May and September to establish the mean condition. This Mean Summer SCRIM Coefficient (MSSC) is used as the basis for assessment and comparison.

In time, the winter recovery may not be enough to balance the summer polishing. Skid resistance will then fall to a level at which polishing and recovery are at an equilibrium level which will be broadly maintained as long as traffic is constant.

4.10.6. Skid resistance and accident risk

The skid resistance at a location can be related to accident risk by selection of the appropriate Site Category from Table 3.1 of HD 28.[1] This risk management approach leads to the identification of a risk rating for the site. The Site Categories are the same as those used in the selection of PSV for new surfacings (see Section 4.13.1 — the right-hand column of Table 4.22 shows the investigatory level for each Site Category). If the MSSC is found to be at or below this value, detailed site investigation and risk assessment should be carried out.

Chapter 3 of HD 28[11] gives extensive guidelines about this investigation and assessment phase. If remedial action is required, complementary structural maintenance should also be considered (see Section 4.11). Knowledge of the history of the surface and the original PSV/AAV of the coarse aggregate may also influence the choice of materials for the remedial works. This knowledge is part of the feedback loop needed to ensure effective design of the wearing course and a balanced economic decision about the PSV of aggregates used in new surfaces.

4.11.   Assessment — structural conditions

The keys to cost effective design of structural pavement maintenance are

- *assessment* — a thorough knowledge of pavement condition and
- *timing* — when the best use can be made of the residual properties of existing pavement materials.

Table 4.20.   Maintenance assessment[12,13]

HD 30[13] — procedure	HD 29[12] — methods
2. Routine assessment (supported by traffic and construction data)	2. High-speed road monitor 3. Visual condition survey (CHART[70]) 4. Deflectograph
5. Detailed investigation (with coring, test pits and testing)	5. Falling Weight Deflectometer Ground radar (HA 72[14])

The aim of HD 30[13] is to describe logical routine assessment procedures which monitor trends in condition with time and detect defects at an early stage. Once defects are suspected, more detailed investigation is triggered to provide information for the detailed design of maintenance treatments, discussed in Section 4.12 below. HD 29[12] gives details of the methods which can be used to support the assessments detailed in HD 30.[13] The links between the chapters in the two documents are illustrated in Table 4.20.

4.11.1.   Routine structural assessment programme
Routine structural assessment should be carried out throughout the life of the pavement on a systematic basis, regardless of whether or not defects are found. The investigatory phase of pavement life (discussed in Section 4.4) can best be detected by monitoring trends in pavement condition over time, using data collected and presented in a consistent way.

Three methods are used in combination to form a routine structural assessment programme for trunk roads and motorways

- High Speed Road Monitor (HRM)[12];
- visual condition surveys (VCS) and
- deflectograph surveys.

The HRM[12] is relatively quick and inexpensive and is used to survey trunk road network on a two year cycle. Deflectograph surveys are slower but relatively expensive and are generally carried out on a five-year cycle. Visual condition surveys are slow and labour intensive but can provide details about specific areas of interest.

4.11.2.   The High Speed Road Monitor
The machines used to measure macrotexture make use of a contactless displacement transducer. This device was originally designed by TRRL for a machine known as the High-Speed Profilometer. Further development led to the High Speed Road Monitor (HRM),[12] which has four sensors above the nearside wheel track and a fifth sensor at its midpoint.

The HRM consists of a van and trailer fitted with laser sensors and other devices. It measures road surface condition whilst operating in normal traffic flow and can be considered as a screening tool. The processed results identify areas of poor texture, rutting and profile.

Changes in longitudinal profile or riding quality over time are monitored using a parameter known as 'Proportional Change in Variance' (PCV). An elevated PCV is an indicator of potential structural weakness. This means that HRM survey results can be used to trigger more detailed local surveys.

The latest TRL development of HRM is known as HARRIS — the Highways Agency Routine Road Investigation System. HARRIS also uses video image collection equipment to automatically identify cracks which are at least 2 mm wide.

4.11.3.   Visual condition surveys

Visual condition surveys (VCS) originally provided a systematic basis for recording information about visible physical defects such as cracking and rutting (nowadays, HRM[12] has largely replaced the VCS). The CHART[70] (Computerised Highway Assessment of Ratings and Treatments) is a very rudimentary pavement management system (PMS) which was used to formulate a maintenance programme for pavements having an asphalt surface. The output from CHART indicates sections of road that are sub-standard, suggests maintenance treatments and indicates relative priorities. CHART has now largely been abandoned in favour of a system of defect compilation. Defects are identified either by inspection from vehicles moving at the speed of prevailing traffic or from a walking survey. Details are entered into a data capture device (DCD), basically a laptop computer. The data are then downloaded into a database running on a workstation. This database (called a Routine Maintenance Management System or RMMS) allows interrogation of the defects. HERMIS, T-Road and ORACLE are databases used in the UK for this purpose. The RMMS used in Scotland is described in SH 4.[71] The defects most often related to structural deficiencies are wheel track cracking and significant rutting.

4.11.4.   Deflection surveys

A deflection survey provides a direct measurement of the structural characteristics of a flexible pavement. In simple terms, the deflectograph measures the depth of the depression in the pavement surface produced by the mass of its own rear wheels. The results are recorded electronically for further processing. Typical average deflections are between 0.2 mm and 0.6 mm, but it should be noted that the actual deflection may be greater because the measurement datum can lie within the deflection bowl.

The deflection beam was developed by Benkelman in the United States as part of a study published in 1953 as the WASHO Road Test.[72] TRRL used this work to develop a standard procedure for use in the UK, resulting in the purchase of a Lacroix Deflectograph from the Laboratoire Central des Ponts et Chaussée (LCPC, the French equivalent of TRL) in 1967. The first deflectograph was built for LCPC in France in 1956. The machines were in general use by the late 1960s. The TRRL/LCPC machine had to be modified for use in the UK and a second machine to TRRL's specification was purchased in 1970. This version of the deflectograph was built upon a Berliet chassis and is the template for the dozen or so machines used in the UK since then. Most have been built in Bristol by WDM Ltd on a Mercedes chassis and are now formally known as a Pavement Deflection Data Logging Machines (PDDLMs).

In simple terms, the deflectograph is an automated version of the Benkelman beam. The principles of measurement are illustrated in Fig. 4.10.

Two beams are mounted on a T-frame with one beam for each pair of wheels on the rear axle. The vehicle is driven at about 2 km/h and records the deflection at intervals of about 3.5 m.

All of the work came to fruition in 1978 with the publication of three complementary TRRL laboratory reports — LR 833,[22] LR 834[73] and LR 835.[74] The publication of the reports and the associated DEFLEC computer program provided a systematic basis for the design of structural overlays.

There are detailed differences between the deflection beam and deflectograph methods. In the DEFLEC computer program, deflectograph readings are converted to equivalent beam measurements. DEFLEC was superseded in 1993 by an improved analysis program called PANDEF, which uses deflectograph readings as the basis for calculations. Whichever method of measurement or calculation is used, it must be remembered that the method is empirical. The testing must be carried out in full compliance with the calibration and operating procedures in the appropriate annex of HD 29.[12]

Deflection values for a flexible pavement are influenced by temperature, so all measurements are mathematically corrected to a temperature of 20 °C. The correction factor is calculated by PANDEF from data showing the thickness, type, age and structural integrity of the asphalt layers. PANDEF calculates a parameter known as Equivalent thickness of Sound Bituminous Material (ESBM).

Tables 4.1 and 4.2 of HD 29[12] identify survey categories based on survey date, road temperature and asphalt layer thickness. An actual road temperature close to 20 °C will be more reliable than a measurement taken

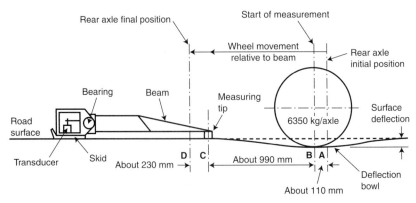

Fig. 4.10. Principles of deflection measurement[73]

at 10 °C and mathematically corrected. A dry subgrade during summer months can also lead to anomalously low deflections, so surveys done between mid-June and mid-September should not be used for detailed design of maintenance treatments.

Whatever the survey category, the results of a single deflectograph survey should not be used in isolation. Trends between surveys and complementary information from other sources should be considered.

4.11.5.   Using deflection to estimate residual life
At the time that TRRL was developing procedures for deflection measurement, work began to establish a database linking deflection to pavement performance. This database has been expanded routinely and is used to formulate the mathematical algorithms used in the PANDEF analysis program. Analysis and interpretation of deflectograph results are detailed in Annex 3 of HD 29.[12]

The principle of using a deflection reading to evaluate pavement life is illustrated in Fig. 4.11, which is taken from the Deflection Design Method in LR 833.[22] The traffic loading since construction must be known for accurate analysis. This establishes a position on the horizontal axis. The vertical axis is used to plot the measured deflection.

Consider point A on Fig. 4.11 with a cumulative traffic loading of 1.5 msa and a deflection of 35 units. The upwardly turning horizontal lines represent the change in deflection over time and are known as trend lines. A line drawn from point A roughly parallel to the adjacent trend lines cuts the 0.50 probability critical condition line at about 9 msa, which suggests a residual life of 7.5 msa for the 'average' road.

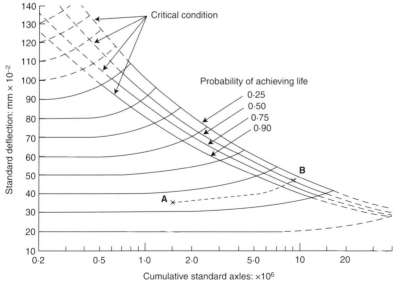

Fig. 4.11.   Relationship between deflection and pavement life — asphalt roadbase[22]

The probability lines show that there is a range of answers, there being a probability of 0.65 that the residual life lies between 5.5 msa and 8.5 msa. This is why Annex 3 of HD 29[12] replaces the term 'critical condition' with the term 'investigatory condition', thus emphasizing the need for detailed monitoring and evaluation rather than the use of the program output in isolation.

Annex 3 of HD 29[12] now recognizes three categories of pavement life, reflecting a greater understanding of the performance of thick asphalt pavements

- determinate life pavements (DLPs)
- upgradeable to long life pavements (ULLPs), with a bitumen bound roadbase and
- long life pavements (LLPs), with at least 300 mm of asphalt and low deflections.

The categories are identified by the PANDEF program.

4.11.6. *Planning an investigation — data review*
Data review marks the point when routine assessment has triggered an awareness of a potential problem or that actual pavement distress is visible. A systematic approach is needed at this stage in order to ensure that any detailed investigation establishes the root cause/s of the problem. Thorough investigation can be time consuming and expensive but without it there is a real risk that remedial action will be ineffective in either structural or economical terms. Ideally, all of the following data should be available to allow the detailed investigation to be planned

- traffic flow and composition data, including growth factors
- surface condition data, including HRM[12] (if available) and CHART[70] and
- construction and maintenance history, including pavement type and layer thicknesses.

It will also be appropriate to consider skid resistance data (as discussed in Section 4.10) if surface renewal is likely to be part of the structural maintenance treatment.

The aim of data review is to find possible associations between areas of deterioration and other characteristics so that each area of the pavement can be placed into one of three categories

- areas with visible distress which require detailed investigation
- areas with high deflection and low residual life (as discussed in Section 4.11.5.) but with little visible distress which need limited investigation to demonstrate structural soundness or
- areas with no visible distress but anomalous deflection or HRM survey results which require more monitoring in future years.

Poor correlation between deflection survey results and surface condition is often caused by adverse changes in drainage conditions. Drainage surveys can be an important part of the review because it is usually easier to correct drainage faults before the pavement is permanently damaged than to carry out structural maintenance treatments which may ultimately be rendered ineffective by the same drainage faults. There is an instance of a dual carriageway road in the north of England which had very high deflections and standing water in the sub-base. A new subgrade drainage system and a thin overlay was much cheaper than reconstruction and has subsequently given twelve years of excellent service.

4.11.7. The detailed investigation

Chapters 4, 5 and 6 of HD 30[13] provide much information about planning a detailed investigation using combinations of road cores, test pits and associated tests. The outcome of the investigation should explain the mechanisms that caused the pavement deterioration and provide information about the possible treatment options to enable the maintenance treatment to be economically designed.

It is often difficult to draw conclusions about weaknesses in pavement layers. More than one type of observation or measurement may be needed to provide supporting evidence. Areas exhibiting distress should also be compared with sound areas because differences can give valuable clues. Deterioration mechanisms may be broadly classified into those associated with traffic loading such as fatigue cracking and surface deformation and those connected with climatic factors such as thermal cracking and ageing of bitumen.

The limitations of road cores must be recognized at this stage. Cores are relatively easy to obtain and give reliable data about bound layer thickness and type. They can be used as test specimens for laboratory tests for stiffness and rutting resistance. However, they do not provide samples which are large enough to allow reliable analysis of material composition and provide only limited access to the foundation or subgrade. Test pits are labour intensive but will provide a large bulk sample of asphalt for further analysis, samples from the lower layers and access for the assessment of in situ bearing capacity. If drainage problems are suspected, test pits at the edge of the pavement may also be appropriate.

Table 6.1 and Annex A of HD 30[13] provide guidance on the assessment of flexible composite pavements with a cement bound roadbase. Composite pavements are difficult to assess because there may be load transfer across the cracks in the roadbase.

The root causes of pavement distress can be established by both planning a detailed investigation to collect as much relevant and accurate information as possible and also interpreting the data in a logical and systematic manner.

4.11.8. Falling weight deflectometer and ground radar

Chapter 5 of HD 29[12] gives details of two specialist techniques which are extremely useful tools for detailed investigation — the Falling Weight Deflectometer (FWD) and ground radar.

The FWD is fundamentally different to the deflectograph. It uses an impact load instead of a rolling wheel to create a deflection. The FWD also measures the shape of the deflection bowl, allowing an estimate of the stiffness of the pavement layers and the identification of weak foundations. The FWD can also be used to assess joints and cracks in cement bound roadbase and rigid pavements. The FWD is discussed in more detail in Section 3.10.2 of Chapter 3.

Ground penetrating radar is detailed fully in the advisory document HA 72.[14] The equipment consists of a short impulse radar transmitter/receiver which is towed behind a vehicle at about 4 km/h. A continuous profile can be derived by measuring the strength of the radar signal reflected by internal surfaces and changes in the pavement structure.

4.12. Design of maintenance treatments

It is said that a good engineer can do for a shilling what any fool can do for a pound. The design of cost effective asphalt maintenance treatments is an example of where this adage is particularly appropriate. The prescriptive empirical approach to the design of new pavements requires little imagination. On the other hand, the large number of variables which should be considered in designing maintenance measures present a real challenge — an opportunity for a designer with skill to demonstrate the science, art and practice of pavement engineering.

4.12.1. Design of structural maintenance

Table 6.1 and Chapter 7 of HD 30[13] imply a choice between

- wearing course replacement
- overlay — thick or thin, or
- reconstruction.

These techniques are generally used in combination. They must be linked with a consideration of horizontal variation — both along the length of pavement and over the cross section — and vertical constraints such as bridge clearances, adjustment to edge details, drainage features and safety fences. As explained in Section 4.11.6, it is also important that drainage defects, which adversely affect the properties of the pavement foundation, are corrected. Timely remedial work on the drainage system can remove the need for a structural overlay in subsequent years. The aim of the maintenance treatment should be to create a pavement with a uniform structural performance along its length.

4.12.2. Wearing course replacement

Replacement of the wearing course alone is unlikely to make any significant contribution to the overall structural strength of the pavement, but the timely removal of a rutted or cracked surface can check the onset of deterioration in the lower layers. In multi-lane carriageways, an 'inlay' replacement of the nearside lane is often a cost effective short-term solution.

4.12.3. Overlay

The approach to overlay design makes maximum use of the structural contribution of the existing pavement layers, providing they are intact.

The overlay thicknesses calculated by the PANDEF program are derived empirically from long-term monitoring of pavement performance. LR 833[22] published overlay design charts (such as those depicted in Fig. 4.12) which relate measured deflection and required traffic loading to overlay thickness.

When LR 833[22] was published, a 40 mm thick overlay of HRA was the thinnest practical option. The availability of proprietary thin surfacings means that this is no longer true (see Section 4.13).

The calculated overlay thickness assumes that DBM100 is used. Stiffer materials such as HDM can be used to gain the required structural performance from a thinner layer. The reduction in thickness can be estimated from the empirical design charts or calculated using analytical design techniques.

A number of factors may influence the choice of overlay. Seriously damaged surfacing must always be removed to avoid undermining the integrity of the new surface. Local reconstruction of short lengths of failed pavement may also reduce the overall thickness of overlay required. The economic balance between reconstruction of the heavily trafficked lanes and thicker overlay of all lanes needs to be evaluated for multi-lane carriageways.

Growing road traffic levels and the need to minimize disruption have encouraged engineers to employ strengthening techniques which leave the road available to the road user during the working day. Layers of basecourse mixtures made with skid resistant aggregates can be used to build up overlays across a number of lanes using overnight contraflow or variable single lane traffic systems. The basecourse is used as an intermediate surfacing which is available for use as a daytime running lane. Reconstruction can only be done under such conditions if its extent

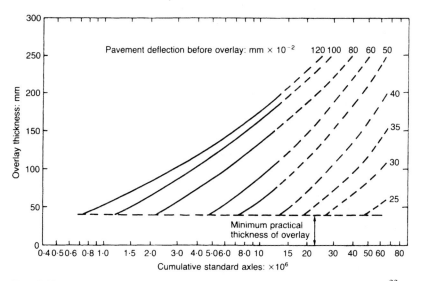

Fig. 4.12. Overlay design chart for pavements with bituminous roadbases[22]

is limited. It is very difficult to build up more than two layers of asphalt overnight because each layer needs time to cool.

Similar multi-layer overlay techniques have become standard practice for the strengthening of operational airport runways during 'no-flying' hours. Work on operational runways requires careful attention to temporary ramp details and cleanliness of the intermediate runway surface. Although the cost of a thick overlay is greater than local reconstruction with a thinner overlay, this cost is offset by the operational benefits to the pavement operator.

4.12.4.   Reconstruction

Reconstruction should ideally only remove defective material, leaving as much sound material in place as is possible. The widespread use of cold milling techniques has given much greater control of the excavation element of reconstruction.

The thickness of the new layers should be based on the design criteria in HD 26[9] as discussed in Section 4.7. It is often possible to select a design for the new pavement that is thinner than the existing asphalt layers by choosing materials carefully. This allows some of the existing asphalt (even if it is cracked) to be left in place over the sub-base, protecting it from adverse weather and providing a firm platform for compaction of the new asphalt layers. If the residual foundation is likely to make a significant structural contribution, the design of the new asphalt layers can be fine tuned using analytical techniques.

It is not usually appropriate to disturb the unbound sub-base and capping unless there is evidence that the foundation has been weakened by degradation, moisture or contamination.

4.12.5.   Minor repairs and surface treatments

HD 31[15] discusses the maintenance of asphalt roads by concentrating on methods used for pavement repairs, including

- surface treatments
  - crack sealing
  - surface dressing
  - thin surfacings
  - slurry and microsurfacings
  - retexturing.

- minor maintenance
  - patching
  - repairs to porous asphalt
  - trench reinstatements.

- major maintenance
  - cold planing and regulating levels
  - overlays and resurfacing
  - reflective crack treatment
  - haunching and widening

- ⊃ carriageway reconstruction
- ⊃ asphalt substitution.

- recycling
  - ⊃ in situ and central plant
  - ⊃ hot and cold techniques
  - ⊃ the Repave and Remix processes.

It is intended that the chapter on the Repave and Remix processes will be moved, without modification, to form Chapter 12 of HD 37.[18] Much of HD 31[15] is repeated in or extended by other parts of Volume 7. Most of the detail is covered in other parts of this Book, including Section 4.9 which considers haunch widening and asphalt substitution, Section 4.13 which discusses asphalt surfacings and Chapter 9 which details surface treatments and surface dressings.

4.13. Asphalt surfacing and surfacing materials

Safe roads which represent good value for money require careful choice of a suitable wearing course. The choice has been HRA for many years, with bitumen macadam and surface dressing for secondary uses. Much has changed in recent years, as suppliers of the new generation of asphalts have responded to the demands for reduced surface noise, minimum delays at road works, improved riding quality and enhanced deformation resistance. These initiatives have been supported by the development of performance-related specification clauses and an internationally recognized type approval and assessment system.

In Volume 7, HD 36[17] and HD 37[18] are complementary documents which give guidance on the choice of an appropriate asphalt surfacing system to suit the location and the required characteristics. Both documents include an impressive list of references. Table 4.21 is based on the asphalt parts of Table 2.1 in HD 36.[17]

Many asphalt suppliers have developed additional controls for their SMA mixtures and market them as thin surfacings. Most thin surfacings consist of a mixture in which the coarse aggregates form a skeletal structure in the same way as SMA and PA.

4.13.1. Aggregate selection

Research which is detailed in TRL Report 322[49] indicates the following.

- Aggregates do not necessarily continue to polish as traffic level increases. Lower PSV values than were once specified are now known to give adequate skid resistance at high traffic levels because the balance between the polishing and wearing actions changes.
- The relationship between PSV value and skid resistance depends on the type of site and the required level of skid resistance. For example, the additional braking forces on approaches to slip roads result in the pavement having a lower level of skid resistance than on adjacent sections of carriageway.
- Skid resistance increases with PSV and reduces with heavier traffic,

Table 4.21. Summary description of asphalt surfacing materials[17]

Material	Summary description
Hot-rolled asphalt (HRA)	A traditional surfacing material in the UK, formed by rolling bitumen coated chippings into a sand asphalt mat to provide texture and skid resistance
Porous asphalt (PA)	A mixture in which the coarse aggregate particles form a skelatal structure with interconnecting voids which allow water to drain within the compacted layer. Usually employed in the UK for its low noise characteristics, PA also reduces spray in wet weather
Stone mastic asphalt (SMA)	A mixture in which coarse aggregate particles form a skelatal structure filled with a binder-rich matrix of fine aggregate and filler. SMA mixtures need careful control to avoid loss of surface texture
Thin surfacing	Proprietary surfacings laid at a range of thicknesses to restore skid resistance and ride quality. Spray may be reduced and surfaces are usually quieter than HRA
Bitumen macadam	A traditional mixture similar to asphaltic concrete with a continous aggregate, moderate deformation resistance and low texture depth

but a constant term related to the properties of the surfacing material and road conditions dominates.

These principles are illustrated in Table 4.22.

Reference must be made to the complete table in HD 36[17] for other sites (a more detailed version of this table is shown as Table 1.15 in Chapter 1). More difficult sites will require natural aggregates with a PSV

Table 4.22. Summary of Polished Stone Value requirements[49]

Site Category	Site characteristics	Commercial vehicles lane/day at design life			Investi-gatory level
		55 PSV	60 PSV	65 PSV	
A, B	Motorway mainline Dual carriageway (non-event)	2000–3000	3000–5000	Over 5000	0.35
A1	Motorway approach to slip roads	750–2000	2000-4000	Over 4000	0.35
C, D	Single carriageway (non event) Dual carriageway (minor junction)	750–1000	1000–2000	2000–6000	0.40
E, F, G1, H1	Most single carriageway event sites	Up to 250	250 to 750	750 to 2000	0.45

Table 4.23. Aggregate Abrasion Value requirements[17]

Surfacing type	Commercial vehicles/lane/day at design life			
	AAV = 10	AAV = 12	AAV = 14	AAV = 16
Positive texture	No limit	1750 max.	250 max.	Not permitted
Negative texture	No limit	No limit	2500 max.	1000 max.

of at least 68. The use of high skid resistance surface treatments (see Clause 924 of SHW[20]) using artificial aggregates such as calcined bauxite is specified for very difficult sites, using the terminology 70+.

An economic design will restrict the higher PSV levels to the parts of the site where they are definitely needed, such as near junctions and approaches to pedestrian crossings. The design must also reflect the practical difficulties of producing and laying variable quantities of surfacing material with different aggregate PSV levels.

The maximum Aggregate Abrasion Value (AAV) required to complement the chosen minimum PSV depends upon the type of asphalt surfacing. Chippings in a material with positive texture are more exposed to wear than particles held in the interlocking stone matrix of a wearing course with a negative texture. The principles of Table 3.2 of HD 36[17] are illustrated in Table 4.23.

Choosing the appropriate minimum PSV and maximum AAV values which best match site conditions and traffic loadings should result in satisfactory performance before the investigatory levels of skid resistance are reached (see Section 4.10.6 above). PSV and AAV are discussed in further detail in Section 1.12.6 of Chapter 1.

4.13.2. Tyre-generated noise

Rapid traffic growth in recent decades means that the type of noise which most affects everyday lives is traffic noise. The abatement of traffic noise is increasingly an environmental concern which must be considered when choosing a surfacing. Chapter 5 of HD 36[17] explains the various interacting factors which contribute to tyre-generated noise.

Traffic noise generated by a road vehicle originates from three sources — the engine and transmission system, the interaction between tyre and pavement surface and the air circulation around the vehicle. At speeds over 120 km/h, air circulation is the dominant factor; below 50 km/h, mechanical noise predominates.

The motor industry has achieved significant reductions in engine and transmission noise in recent years particularly for heavy commercial vehicles. This has increased the importance of reducing the noise contributed by the characteristics of the pavement surface of inter-urban roads. In the UK, this has been done by changing from HRA to PA or to the new thin surfacings. This change reflects a move from surfaces with a positive macrotexture to those with a negative texture.

The shock waves produced by the impact between the rolling tyre and the road are reduced because the upper plane of an indented textured surface is smooth compared with a traditional chipped surface. The relative porosity of the surface also influences the amount of noise caused when air is expelled from the tread pattern on contact with the surface and sucked in behind the tyre contact point. Very porous asphalts not only minimize this noise but also absorb some of the noise energy within the asphalt layer.

4.13.3.   British Board of Agrément/HAPAS

Proprietary thin surfacings have been recognized in contract documents for some years using a system known as Type Approval/Registration (see Appendix E of SHW[20]). This system was administered by TRL and provided a systematic basis for the approval of competing branded systems and monitoring of the first two years of service. However, the Type Approval/Registration system does not provide independent assurance that the characteristics of these asphalts will be consistent with time.

The latest version of Clause 942 in SHW[20] requires proprietary asphalts to be approved by the British Board of Agrément (BBA) using the Highway Authorities Product Approval Scheme (HAPAS)[75] which was fully established in 2000.

HAPAS approval for proprietary asphalts requires their characteristics to be assessed using a series of defined tests, to create a matrix of performance parameters.

Once the characteristics have been established and confirmed by two years' service, consistency is monitored by a quality plan approach linked to routine BS EN ISO 9002 surveillance undertaken by an independent certification body as part of the Sector Scheme[76] covering the quality assurance of asphalt production.

4.13.4.   Hot rolled asphalt wearing course

Chapter 2 of HD 37,[18] entitled 'Laying Bituminous Surface Courses', contains advice on laying chipped HRA. The constitution of HRA with precoated chippings is specified in BS 594[32,77] and this material has been the traditional choice for the surfacing of high speed roads. It has performed very well in most situations and has proved to be durable in service often giving in excess of twenty years service on the trunk and principal road network.

HRA has fallen out of favour in recent years for a number of reasons, some technical and some economic. The principle technical reason is that HRA has proved susceptible to wheel track rutting on heavily trafficked pavements. In order to minimize the risks, Clause 943 of SHW[20] was introduced. This Clause controls the volumetric design of the mixture more closely and encourages the use of modified binders. In practice, its implementation has been overtaken by the SMA mixtures and thin surfacings, which give greater and more reliable resistance to wheel track rutting. The economics of site operations and traffic management are also much simpler without the chipping machine. (Chipping machines exceed 4 m in width. Given the need for working space plus a safety zone at the side of the chipping machine, it is often impossible to work on one lane and keep the other lane open on a two-lane 7.3 m wide carriageway. In this respect, the use of an unchipped surfacing becomes mandatory. Chipping machines are covered in more detail in Section 6.6.2 of Chapter 6 of this Book.)

In the longer term, HRA remains the most resource effective way of using aggregates of high PSV. Its technical shortcomings can be overcome by close control and choice of materials, but HRA is being neglected in the quest for minimum disruption and lower noise for the road user. The price of that neglect could be a longer term shortage of high PSV aggregate.

4.13.5. Porous asphalt
Chapter 5 of HD 37[18] is entitled 'Porous Asphalt'. Much has been written about porous asphalt (PA) and there are some excellent examples of its use. There have also been some failures and there are significant areas of Britain where PA has not been used.

The open structure of PA places great stress on the binder. In order to minimize risk for the client, it seems possible that the specification for PA (Clause 938 of SHW[20]) will be overtaken by porous variants of thin surfacing systems controlled by the HAPAS[75] Scheme. A number of European countries routinely use PA with smaller aggregate sizes than the 20 mm size specified in Clause 938 of SHW,[20] some as small as 10 mm.

4.13.6. Thin surfacings
Chapter 6 of HD 37[18] deals with 'Thin Wearing Course Systems'. It recognizes three thickness ranges for proprietary thin surfacings and explains the control imposed by the HAPAS Scheme. Thin surfacings have their origins either in France — particularly systems less than 25 mm thick — or in the UK using developments of SMA from Germany.

There are a large number of competing branded asphalts on the market, all offering different variations and combinations of aggregate size, air voids content, binder type and bond coat type. Polymer modified bitumens are used in many of the systems, either in the mixture, in the tack coat or in both. The HAPAS Scheme can control extravagant claims by suppliers and give assurance of short-term performance. Unfortunately, judgements about longer-term performance rely on the experience and reputation of the supplier and an assessment of the asphalt technology involved.

Good polymer modified bitumens will give better performance than conventional bitumens provided that they have been selected correctly. Smaller aggregates with good shape will give better particle interlock but will not achieve texture depth requirements if the mixture grading is not closely controlled at the asphalt plant. Smaller aggregates will be quieter than larger aggregates but may have less hydraulic conductivity when used in a porous mixture. More porous mixtures will give lower tyre noise and better spray reduction characteristics but denser mixtures will generally be less susceptible to ageing and degradation caused by water.

There are thus many conflicting aspects and the choice of surfacing is not made easier by certification. One feature of Clause 942 of SHW[20] is that it includes an extended guarantee period for the surfacing. This moves the balance of risk closer to the system producer and the surfacing contractor. Good contractors have refined their site practices to minimize their risks — close attention to detail is the key to success.

4.13.7.   Stone mastic asphalt

SMA was introduced into the UK following publication of TRL Report PR 65[78] in 1994. Its wide use across continental Europe has been reviewed[79] by the European Asphalt Pavement Association (EAPA).

Initially, it was assumed by some that the draft specification suggested by TRL could be used as the basis of a conventional recipe specification. However, it was quickly realized that surface texture was vulnerable to small variations in binder content and aggregate grading. This led to the development of proprietary SMA mixtures which are now classified as thin surfacings.

Most asphalt plants can supply generic SMA made with penetration grade bitumens and cellulose fibres for low speed uses. There are many examples of very good dense SMA surfacings in private drives, housing estates, delivery areas and haulage yards.

4.13.8.   Bitumen macadam wearing course

The bitumen macadam wearing courses defined in BS 4987[38] are relegated to the miscellaneous Chapter 13 of HD 37.[18] Their generic recipe specification has stood the test of time but performance under modern traffic conditions is often variable. The use of stiffer bitumens to overcome marking by vehicle tyres can lead to premature fretting in hand-laid areas. Surface texture is often poor after a few years, resulting in an early maintenance need.

The general market for bitumen macadam wearing courses is steadily being eroded by more durable, but more expensive, SMAs.

4.13.9.   Retexturing of asphalt surfacing

Asphalt technology has done much to improve the durability of asphalt surfacings to the point where surface failure is often caused by lack of skid resistance rather than defects such as cracking or fretting.

Chapter 11 of HD 37[18] discusses the various techniques available to restore the skid resistance and texture depth of the wearing course. These can be categorized into impact methods, cutting or flailing methods and fluid action methods. The most effective and controlled method uses self-contained specialist vehicles and a bush hammering technique. TRL work to review the long-term effectiveness of the technique is published in TRL Report 299.[80]

4.14. Summary

This chapter has considered the design of road pavements by reviewing the contents of Chapter 7[2] of the DMRB.[1] It considers both the design of new pavements and the maintenance of the established highway network. The Chapter refers to a wide range of research work which illustrates how highway engineers have sought to respond to the ever increasing demands on the highway network.

Recent years have seen a growing emphasis on the efficient maintenance of the highway network. A sound knowledge of pavement technology will help to ensure that those responsible for that work can do so in a way that displays their engineering skills in the same way that new construction did for previous generations.

Recent years have also seen many changes to the asphalt mixtures used in pavements. The growing interest in more sustainable pavement construction and maintenance techniques will continue the pressure for change. Adopting design procedures to keep pace will maintain the engineering challenge for future years.

References

1. HIGHWAYS AGENCY *et al. Design Manual for Roads and Bridges.* TSO, London, various dates, 1–15.
2. HIGHWAYS AGENCY *et al. Design Manual for Roads and Bridges, Pavement Design and Maintenance.* TSO, London, various dates, 7.
3. DEPARTMENT OF THE ENVIRONMENT. *A Guide to the Structural Design of Pavements for New Roads.* HMSO, London, 1970, Road Note 29.
4. POWELL W. D., J. F. POTTER, H. C. MAYHEW and M. E. NUNN. *The Structural Design of Bituminous Roads.* TRL, Crowthorne, 1984, LR 1132.
5. HIGHWAYS AGENCY *et al. Design Manual for Roads and Bridges, Pavement Design and Maintenance, General Information.* TSO, London, 1999, 7.1.1, HD 23/99.
6. HIGHWAYS AGENCY *et al. Design Manual for Roads and Bridges, Pavement Design and Maintenance, Technical Information.* HMSO, London, 1995, 7.1.2, HD 35/95.
7. HIGHWAYS AGENCY *et al. Design Manual for Roads and Bridges, Pavement Design and Maintenance, Traffic Assessment.* HMSO, London, 1996, 7.2.1, HD 24/96.
8. HIGHWAYS AGENCY *et al. Design Manual for Roads and Bridges, Pavement Design and Maintenance, Foundations.* HMSO, London, 1994, 7.2.2, HD 25/94.
9. HIGHWAYS AGENCY *et al. Design Manual for Roads and Bridges, Pavement Design and Maintenance, Pavement Design.* HMSO, London, 1994, 7.2.3, HD 26/94.
10. HIGHWAYS AGENCY *et al. Design Manual for Roads and Bridges, Pavement Design and Maintenance, Pavement Construction Methods.* HMSO, London, 1994, 7.2.4, HD 27/94.
11. HIGHWAYS AGENCY *et al. Design Manual for Roads and Bridges, Pavement Design and Maintenance, Skidding Resistance.* HMSO, London, 1994, 7.3.1, HD 28/94.
12. HIGHWAYS AGENCY *et al. Design Manual for Roads and Bridges, Pavement Design and Maintenance, Structural Assessment Methods.* HMSO, London, 1994, 7.3.2, HD 29/94.
13. HIGHWAYS AGENCY *et al. Design Manual for Roads and Bridges, Pavement Design and Maintenance, Structural Assessment of Road Pavements.* TSO, London, 1999, 7.3.3, HD 30/99.
14. HIGHWAYS AGENCY *et al. Design Manual for Roads and Bridges, Pavement Design and Maintenance, Use and Limitations of Ground*

Penetrating Radar for Pavement Assessment. HMSO, London, 1994, 7.3.4, HA 72/94.

15. HIGHWAYS AGENCY *et al. Design Manual for Roads and Bridges, Pavement Design and Maintenance, Maintenance of Bituminous Roads.* HMSO, London, 1994, 7.4.1, HD 31/94.

16. HIGHWAYS AGENCY *et al. Design Manual for Roads and Bridges, Pavement Design and Maintenance, Maintenance of Concrete Roads.* HMSO, London, 1994, 7.4.2, HD 32/94.

17. HIGHWAYS AGENCY *et al. Design Manual for Roads and Bridges, Pavement Design and Maintenance, Surfacing Materials for New and Maintenance Construction.* TSO, London, 1999, 7.5.1, HD 36/99.

18. HIGHWAYS AGENCY *et al. Design Manual for Roads and Bridges, Pavement Design and Maintenance, Bituminous Surfacing Materials and Techniques.* TSO, London, 1999, 7.5.2, HD 37/99.

19. HIGHWAYS AGENCY *et al. Design Manual for Roads and Bridges, Pavement Design and Maintenance, Concrete Surfacing and Materials.* TSO, London, 1997, 7.5.3, HD 38/97.

20. HIGHWAYS AGENCY *et al. Manual of Contract Documents for Highway Works, Specification for Highway Works,* 1. TSO, London, 1998.

21. HIGHWAY RESEARCH BOARD. *The AASHO Road Test Report 5.* National Academy of Science, Washington DC, 1962.

22. KENNEDY C. K. and N. W. LISTER. *Prediction of Pavement Performance and Design of Overlays.* TRL, Crowthorne, 1978, LR 833.

23. PROPERTY SERVICES AGENCY. *A Guide to Airfield Pavement Design and Evaluation.* BRE, Watford, 1989.

24. KNAPTON J. and M. MELETIOU. *The Structural Design of Heavy Duty Pavements for Ports and Other Industries.* The British Precast Concrete Federation Ltd, Leicester, 1996.

25. WHITEOAK C. D. *The Shell Bitumen Handbook.* Shell Bitumen, Chertsey, 1990.

26. BRITISH STANDARDS INSTITUTION. *Methods of Test for Soils for Civil Engineering Purposes, Compaction-Related Tests.* BSI, London, 1990, BS 1377: Part 4.

27. BRITISH STANDARDS INSTITUTION. *Testing Aggregates, Method for Determination of Frost Heave.* BSI, London, 1989, BS 812: Part 124.

28. WEBSTER D. C. and G. WEST. *The Effect of Additives on the Frost Heave of a Sub-Base Gravel.* TRL, Crowthorne, 1989, RR 213.

29. BRITISH STANDARDS INSTITUTION. *Tests for Thermal and Weathering Properties of Aggregates, Determination of Resistance to Freezing and Thawing.* BSI, London, 1999, BS EN 1367-1.

30. SHADDOCK B. C. J. and V. M. ATKINSON. *Stabilised Sub-Bases in Road Foundations. Structural Assessment and Benefits.* TRL, Crowthorne, 1997, LR 248.

31. NUNN M. E., A. BROWN, D. WESTON and J. C. NICHOLLS. *Design of Long-Life Flexible Pavements for Heavy Traffic.* TRL, Crowthorne, 1997, LR 250.

32. BRITISH STANDARDS INSTITUTION. *Hot Rolled Asphalt for Roads and other Paved Areas, Specification for Constituent Materials and Asphalt Mixtures.* BSI, London, 1992, BS 594: Part 1.

33. BRITISH STANDARDS INSTITUTION. *Coated Macadam for Roads and other Paved Areas, Specification for Transport Laying and Compaction.* BSI, London, 1993, BS 4987: Part 2.

34. BRITISH STANDARDS INSTITUTION. *Aggregates for Bituminous Mixtures and Surface Dressings for Roads and other Trafficked Areas.* BSI, London, 1999, pr EN 13043.

35. LEECH D. *A Dense Coated Roadbase Macadam of Improved Performance.* TRL, Crowthorne, 1982, LR 1060.

36. NUNN M. E., C. J. RANT and B. SCHOEPE. *Improved Roadbase Macadams, Road Trials and Design Considerations.* TRL, Crowthorne, 1987, RR 132.

37. NUNN M. E. and T. SMITH. *Evaluation of Enrobé À Module Élevé (EME), A French High Modulus Roadbase Material.* TRL, Crowthorne, 1994, PR 66.

38. BRITISH STANDARDS INSTITUTION. *Coated Macadam for Roads and other Paved Areas, Specification for Constituent Materials and for Mixtures.* BSI, London, 1993, BS 4987: Part 1.

39. NUNN M. E. and T. SMITH. *Road Trials of High Modulus Base for Heavily Trafficked Roads.* TRL, Crowthorne, 1997, Report 231.

40. BRITISH STANDARDS INSTITUTION. *Sampling and Examination of Bituminous Mixtures for Roads and other Paved Areas, Methods of Test for the Determination of Density and Compaction.* BSI, London, 1996, BS 598: Part 104.

41. DEPARTMENT OF TRANSPORT. *Notes for Guidance on the Specification for Road and Bridge Works.* HMSO, London, 1976.

42. VAN DER POEL C. A General System Describing the Viscoelastic Properties of Bitumen and its Relation to Routine Test Data. *Journal of Applied Chemistry*, 1954, **4**, 221–36.

43. BONNAURE F., G. GEST, A. GRAVOIS and P. UGÉ. A New Method of Predicting the Stiffness of Asphalt Paving Mixtures. *Journal of the Assoc. Asphalt Paving Tech.*, 1977, **46**.

44. DE LURDES ANTUNES M. and L. FRANCKEN. Development of a Bituminous Pavement Design Method for Europe: COST333 Action and AMADEUS Project. *2nd European Road Research Conference.* Directorate-General VII Transport, European Commission, Brussels, June 1999.

45. POTTER J. *Road Haunches, A Guide to Re-Usable Materials.* TRL, Crowthorne, 1996, Report 216.

46. LEECH D. and M. E. NUNN. *Asphalt Substitution, Road Trials and Design Considerations.* TRL, Crowthorne, 1991, TR 303.

47. THOMPSON A., J. R. GREIG and J. SHAW. *High Specification Aggregates for Road Surfacing Materials.* Travers Morgan Ltd, East Grinstead, 1993, Technical Report.

48. HOSKING R. *State of the Art Review 4, Road Aggregates and Skidding.* HMSO, London, 1992.

49. ROE P. G. and S. A. HARTSHORNE. *The Polished Stone Value of Aggregates and In-Service Skidding Resistance.* TRL, Crowthorne, 1998, Report 322.

50. ROE P. G., A. R. PARRY and H. E. VINER. *High and Low Speed Skidding Resistance, The Influence of Texture Depth.* TRL, Crowthorne, 1998, Report 367.

51. BRITISH STANDARDS INSTITUTION. *Testing Aggregates, Method for Determination of the Polished Stone Value.* BSI, London, 1989, BS 812: Part 114.

52. BRITISH STANDARDS INSTITUTION. *Testing Aggregates, Methods for Determination of Physical Properties.* BSI, London, 1995, BS 812: Part 2.

53. BRITISH STANDARDS INSTITUTION. *Testing Aggregates, General Requirements for Apparatus and Calibration.* BSI, London, 1990, BS 812: Part 100.
54. BRITISH STANDARDS INSTITUTION. *Testing Aggregates, Guide to Sampling and Testing Aggregates.* BSI, London, 1984, BS 812: Part 101.
55. BRITISH STANDARDS INSTITUTION. *Testing Aggregates, Methods for Sampling.* BSI, London, 1989, BS 812: Part 102.
56. BRITISH STANDARDS INSTITUTION. *Testing Aggregates, Method for Determination of Particle Size Distribution, Sieve Tests.* BSI, London, 1985, BS 812: Part 103, Section 103.1.
57. BRITISH STANDARDS INSTITUTION. *Testing Aggregates, Method for Determination of Particle Shape, Flakiness Index.* BSI, London, 1989, BS 812: Part 105, Section 105.1.
58. BRITISH STANDARDS INSTITUTION. *Testing Aggregates, Method for Determination of Moisture Content.* BSI, London, 1990, BS 812: Part 109.
59. BRITISH STANDARDS INSTITUTION. *Testing Aggregates, Method for Determination of Aggregate Crushing Value (ACV).* BSI, London, 1990, BS 812: Part 110.
60. BRITISH STANDARDS INSTITUTION. *Testing Aggregates, Methods for Determination of Ten Per Cent Fines Value (TFV).* BSI, London, 1990, BS 812: Part 111.
61. BRITISH STANDARDS INSTITUTION. *Testing Aggregates, Method for Determination of Aggregate Impact Value (AIV).* BSI, London, 1990, BS 812: Part 112.
62. BRITISH STANDARDS INSTITUTION. *Testing Aggregates, Method for Determination of Aggregate Abrasion Value (AAV).* BSI, London, 1990, BS 812: Part 113.
63. BRITISH STANDARDS INSTITUTION. *Testing Aggregates, Method for Determination of Water-Soluble Chloride Salts. BSI, London, 1988, BS 812: Part 117.*
64. BRITISH STANDARDS INSTITUTION. *Testing Aggregates, Methods for Determination of Sulphate Content.* BSI, London, 1988, BS 812: Part 118.
65. BRITISH STANDARDS INSTITUTION. *Testing Aggregates, Method for Determination of Soundness.* BSI, London, 1989, BS 812: Part 121.
66. BRITISH STANDARDS INSTITUTION. *Tests for Mechanical and Physical Properties of Aggregates, Determination of the Polished Stone Value.* BSI, London, 1999, BS EN 1097-8.
67. BRITISH STANDARDS INSTITUTION. *Tests for Mechanical and Physical Properties of Aggregates, Determination of the Resistance to Wear (Micro-Deval), Part 1.* BSI, London, 1996, BS EN 1097-1.
68. BRITISH STANDARDS INSTITUTION. *Sampling and Examination of Bituminous Mixtures for Roads and other Paved Areas, Methods of Test for the Determination of Texture Depth.* BSI, London, 2000, BS 598: Part 105.
69. ROE P. G. *A Comparison of SCRIM and GripTester—Report of a Further Trial in September 1993,* unpublished Report. TRL, Crowthorne, 1993, PR/H/58/93.
70. HIGHWAYS AGENCY *et al. Design Manual for Roads and Bridges, Pavement Design and Maintenance, Suite of Highway Maintenance Analysis Programs, Program Suite HECB/R/16 (CHART).* HMSO, London, 1977, H 77.
71. HIGHWAYS AGENCY *et al. Design Manual for Roads and Bridges, Pavement Design and Maintenance, Scottish Routine Maintenance*

Management System. HMSO, London, 1986, SH 4.

72. HIGHWAY RESEARCH BOARD. *The WASHO Road Test*, Board Special Report No. 18. Highway Research Board, Washington DC, 1953.

73. KENNEDY C. K., P. FEVRE and C. S. CLARKE. *Pavement Deflection, Equipment for Measurement in the United Kingdom.* TRL, Crowthorne, 1978, LR 834.

74. KENNEDY C. K. *Pavement Deflection, Operating Procedures for Use in the United Kingdom.* TRL, Crowthorne, 1978, LR 835.

75. BRITISH BOARD OF AGRÉMENT. *Guidelines Document for the Assessment and Certification of Thin Surfacing Systems for Highways.* BBA, Watford, 1999.

76. SECTOR SCHEME ADVISORY COMMITTEE FOR THE QUALITY ASSURANCE OF THE PRODUCTION OF ASPHALT MIXES. *Sector Scheme Document for the Quality Assurance of the Production of Asphalt Mixes*, Sector Scheme No 14. Quarry Products Association, London, 1998.

77. BRITISH STANDARDS INSTITUTION. *Hot Rolled Asphalt for Roads and other Paved Areas, Specification for the Transport, Laying and Compaction of Rolled Asphalt.* BSI, London, 1992, BS 594: Part 2.

78. NUNN M. E. *Evaluation of Stone Mastic Asphalt (SMA), A High Stability Wearing Course Material.* TRL, Crowthorne, 1994, PR 65.

79. EUROPEAN ASPHALT PAVEMENT ASSOCIATION. *Heavy Duty Surfaces — The Arguments for SMA.* EAPA, Bleukelen, The Netherlands, 1998.

80. ROE P. G. and S. A. HARTSHORNE. *Mechanical Retexturing of Roads, An Experiment to Assess Durability.* TRL, Crowthorne, 1998, Report 299.

5. Production: processing raw materials to mixed materials

5.1. Preamble

This Chapter considers the types of rock used in manufacturing asphalts in the United Kingdom and examines the methods used to quarry these materials. Different asphaltic production methods are then discussed and their advantages and disadvantages summarized.

5.2. Raw materials

The raw materials, or coarse aggregates, used in roadstone production must be hard, clean and durable. The term aggregate is defined in BS 6100: Part 6[1] as 'granular material, either processed from natural materials such as rock, gravel or sand, or manufactured such as slag'. The UK is fortunate since there is an abundant supply of different aggregates. Its availability and relatively low cost contribute to the supply being somewhat taken for granted, although it is obviously not a limitless resource.

The roadstone industry uses crushed rocks, sand and gravel and artificial materials such as blast furnace slag, or steel slag as the main ingredients for the construction of flexible pavements. The decline in the steel industry has limited the availability of slag.

5.2.1.   Crushed rocks

Crushed rocks are classified relative to their geological origins and these are termed igneous, sedimentary and metamorphic.

Igneous

Rocks formed from a hot liquid, or magma, cooling either above, or beneath, the Earth's surface. These rocks have a crystalline structure. Examples are basalt, diorite, dolerite, gabbro, granite.

Sedimentary

The action of ice, wind or water on a rock's surface, or organic remains, results in the formation of smaller particles. These particles are deposited on lake or river beds, the sea floor or depressions in the Earth's surface. This sediment accumulates, becomes buried and compaction together with cementation over the years removes the air and water to form sedimentary rocks. Examples are gritstone, limestone, sandstone.

Metamorphic
Heat, pressure and/or chemical substances deep beneath the Earth's surface act on existing igneous and sedimentary rocks to form a new type called metamorphic rock. Examples are gneiss, hornfels, marble, quartzite, slate.

5.2.2. Sand and gravel
Sand and gravel are the products of mechanical and chemical weathering of rocks such as granite and sandstone and are, therefore, a type of sedimentary deposit. Weathering by heat, ice, wind and moving water transports the rock fragments, this movement results in further mechanical working and transforms the sand and gravel into rounded and smooth irregular shapes. The deposits are found along the sea shore, along river valleys and on the sea floor.

Gravels are usually classified by their major rock type, e.g. flint gravel. Gravels can also contain a variety of several different rocks.

5.2.3. Blast furnace slag and steel slag
Although limited in availability, BS 594: Part 1[2] and BS 4987: Part 1[3] allow the use of a process manufactured material as the coarse aggregate in a flexible pavement. The British Standards allow the use of '(c) blast furnace slag conforming with BS 1047[4] and (d) steel slag, either electric-arc furnace or basic oxygen slag, which shall be weathered until it is no longer susceptible to falling. The compacted bulk density shall be between $1700\,kg/m^3$ and $1900\,kg/m^3$ when tested in accordance with BS 1047: 1983.'

5.3. Resources
The UK has a wide variety of quality aggregates with specific and different characteristics. A simplified geological survey is shown in Fig. 5.1. This shows the distribution of the major rock formations throughout the country. The map indicates that the south east of England is predominantly sand and gravel, clay and chalk with very little hardrock.

The South East is a highly populated region creating considerable development and therefore has a need for quality materials. Although some roadstone production can be met from local resources, the largest percentage is transported from the South West and to a lesser degree from the Midlands and by sea from Ireland and Scotland.

The major suppliers of aggregates have developed sites containing large asphalt mixing plants with capacities from 400 t/h up to 800 t/h in and around the London area. These plants have rail off-loading and storage facilities and they can have aggregates delivered in loads of up to 5000 t capacity by rail as well as deliveries by truck.

The majority of aggregate movement is by road transport since most mixing plants are local to the rock source. With the advent of fewer but larger capacity asphalt mixing plants the use of both rail and sea transport is likely to increase.

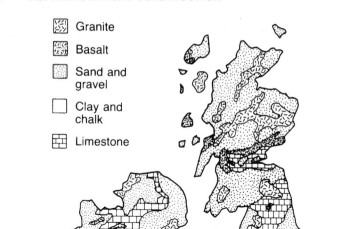

Fig. 5.1.   Simplified geological map of the UK showing major distribution of rock types used in the roadstone industry

5.4.   Methods of producing aggregates

There are numerous types of rock making up the igneous, sedimentary and metamorphic classifications. Each rock has its own characteristics, e.g. Aggregate Abrasion Value, Aggregate Crushing Value, Aggregate Impact Value, Polished Stone Value. These properties are discussed in more detail in Chapter 1. These inherent qualities together with local geological experience and engineers' expertise within the quarrying industry govern the type and size of equipment selected for an aggregate processing plant.

Plants processing materials for road-making have to produce a wide range of sizes with restrictions on shape, the preference being cubical. They should have the ability to operate at varying capacities and be flexible enough to alter the product sizes when the need arises. Therefore, the aim is to produce aggregates to the required sizes, with an optimum shape, at the capacity needed and in an economical manner. Crushed rocks used in roadstone production have a maximum Flakiness Index of 45,[2] whereas rolled in precoated chippings have a maximum allowed

Flakiness Index of 25.[2] The Flakiness Index is an indication of the aggregates shape, the lower the figure the more cubical the product.

The stone is reduced in size by a crusher. There are four types of crusher: jaw, gyratory, impact and cone. To produce the required size and optimum shape the rock is reduced in stages and plants can have one, two, three or more stages depending on the application. The shape is controlled by the type of crusher employed, the reduction ratio, the crusher setting, the rate of feed and the physical characteristics of the stone itself. The best cubical shape is achieved at, or around, the crusher setting and with a choke feed on the cone, jaw and gyratory machines.

Plants are usually rated at an operational efficiency of 80% to allow for blockages within the crushers, or breakdowns. Therefore, if 240 t/h of material are required the plant would have to be rated at 300 t/h minimum in order to meet this demand.

Aggregate processing plants can be roughly divided into three categories: hardrock, limestone and sand and gravel plants. The industry loosely uses the term limestone crushing and screening plants to describe plants for soft and medium hard limestones where the silica content is less than 5%. Limestones can be hard with much higher silica contents.

5.4.1. Hardrock crushing and screening plant

An example of the flow diagram for a hardrock crushing and screening operation is shown in Fig. 5.2.

In preparation to process rock, the overburden, consisting of topsoil, subsoil, clay and small loose rocks, is removed and holes are drilled at predetermined distances for the blast charge. Blasting generally occurs once per day; some of the larger quarries can blast up to 30 000 t or more at a time. Great care and skill are needed when blasting, not only to ensure that the blast itself is safe and contained but to produce a stable quarry face afterwards. Quarries tend to be deep and therefore, in the interests of safety, the face is worked in levels, or benches. Typically these benches can each be 15 m deep but this depends on the rock characteristics and the depth of the rock seam. Many workings go below the water table and in these instances a pumping system is used.

The maximum size feed material to the plant illustrated is an 800 mm cube. Cost effective primary blasting means there will be a small percentage of rock which is too large for the primary crusher. Secondary breaking is the term used for reducing these oversize rocks. Examples of this process are further drilling and blasting, the drop ball, the hydraulic hammer, the impact breaker and the rock breaker. The blasted rock is loaded and hauled from the face to the plant by large dump trucks which can have a capacity up to 100 t.

The rock is fed into a lump stone chute equipped with a chain curtain to control the flow. The material discharges onto a reciprocating tray feeder with a built-in grizzly section. A 'grizzly' is a term used by crushing plant manufacturers for a vibrating grid with bars in one direction only. The grizzly bars are set at 50 mm spacing, allowing the natural fines to bypass

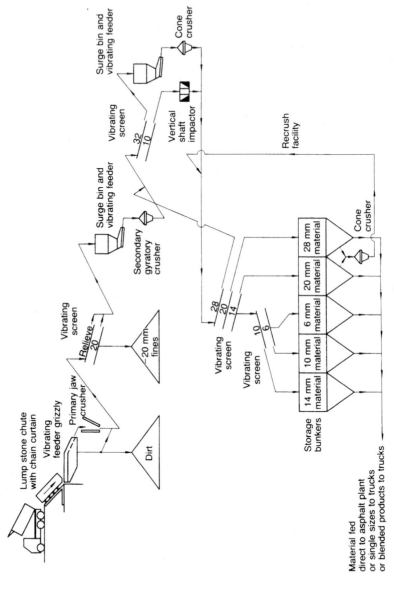

Fig. 5.2. Flow diagram of a hardrock crushing and screening plant[5]

Fig. 5.3. Direct feed of stone into a gyratory crusher. Photograph courtesy of Quarry Management

the crusher, or go to a dirt stockpile if the quality is poor. The 1200 mm × 1000 mm primary jaw crusher operates at 150 mm closed side setting and therefore produces a maximum size of approximately 230 mm. The crushed rock and the 'good' natural fines are passed over a scalping screen where the minus 20 mm material is removed.

Plant capacities in excess of 600 t/h would use a primary gyratory crusher instead of the jaw. In such applications, the gyratory crusher does not need a feed station since the dump trucks can tip directly into the machine. Although a good cubical shape is produced the primary gyratory is relatively expensive and requires a large headroom for installation. However, at high capacities the cost per tonne of material crushed is lower than with other crushers. Figure 5.3 shows material being fed into a primary gyratory crusher.

The plus 20 mm material from the scalping screen is transferred to a surge hopper. This facility enables the secondary crusher to have a controlled choke feed. An electric vibrating feeder delivers the material to a secondary gyratory which operates at 40 mm closed side setting. A gyratory crusher is used here as a secondary machine to take advantage of the high reduction ratio available and the good product shape. The crushed material is conveyed to a two-deck vibrating screen.

The plus 32 mm material is fed to a surge bin, electric vibrating feeder and a tertiary cone crusher operating with a 19 mm closed side setting. A cone crusher with this type of non-segregated feed grading has a good reduction ratio and performs effectively with hard abrasive materials. It is economical and crushing results in a cubical product near to the closed side setting providing a choke feed is maintained.

The minus 32 mm plus 10 mm aggregate is directed into a vertical shaft impactor; which operates on the crushing principle of throwing stone by centrifugal force against captive stone. Vertical shaft impactors are useful third- or fourth-stage crushers and produce a good cubical shape. On occasion these machines are used to perform a shaping duty on poor quality materials.

The crushed materials, together with the minus 10 mm from the two-deck screen, are conveyed to the final sizing screens and the single sized materials are stored in bunkers. The plus 28 mm material is recirculated to remove the oversize stone.

Most overseas crushing plants store the materials in ground stockpiles instead of bunkers and many transportable and mobile plants in this country also adopt this approach. This layout reduces the cost considerably but can lead to contamination and it exposes the crushed aggregates to adverse weather conditions. A material with a higher moisture content will be more expensive to dry at a later stage in the roadstone production process. Storage of the crushed products also enables a recrush facility to be more easily incorporated. A cone crusher fed from the 28 mm and 20 mm aggregate bins provides a more versatile arrangement should market requirements change.

It is not unusual for hardrock plants to have extensive storage facilities including larger and smaller sized products. The materials from the bunkers can be fed directly to an asphalt plant, single sizes or blended products to trucks.

5.4.2. *Limestone crushing and screening plant*

The extraction from the ground of a softer, less abrasive material such as limestone, with a low silica content, is almost identical to the hardrock process. The overburden is removed, the rock is blasted, secondary breaking takes place where necessary and the stones are transported from the quarry face to the plant in large dump trucks. A limestone crushing and screening plant is shown in Fig. 5.4.

In the flow diagram shown in Fig. 5.5 the trucks discharge into a lump stone chute. The largest feed size to the plant is a 900 mm cube and a chain curtain is used to control the flow. A vibrating feeder grizzly removes the minus 50 mm material to either bypass the crusher or feed to a dirt stockpile. The plus 50 mm material is discharged into a primary horizontal impact breaker.

The impact breaker works on the principle of reducing the material size by the impact of the rock against hammers which are attached to a high-speed rotor. The broken rock then strikes breaker bars located on the inside of the machine resulting in more reduction and the stone falls again

Fig. 5.4.   Limestone crushing and screening plant. Photograph courtesy of Hanson Aggregates (Southern)

onto the revolving hammers for further reduction. The product size is determined by the position of the breaker bars relative to the rotor hammers and the rotor speed can be adjusted to cater for different feed materials.

A high reduction ratio can be achieved with this machine. A maximum feed size of 900 mm results in a product size of minus 100 mm and a good cubical shape. This machine can achieve the equivalent of two stages of crushing in a hardrock plant. The feed material must be relatively soft, not too abrasive and have a silica content of less than 5%. The crushed aggregates are conveyed to a surge bin which provides the remainder of the process with a controlled feed by eliminating the inconsistent dump truck discharges experienced at the primary crushing stage. An electrical vibrating feeder discharges the material from the surge bin to a conveyor and two-deck vibrating screen.

The plus 20 mm material is fed to a secondary horizontal impact breaker, the minus 20 mm bypasses the crusher. The secondary impactor operates on a similar principle to the primary impactor. However, the product size is better controlled and some machines incorporate grids at the discharge outlet. Although the reduction ratio is high compared with a

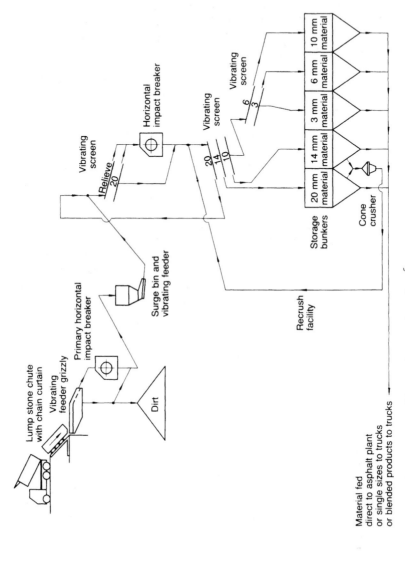

Fig. 5.5. Flow diagram of a limestone crushing and screening plant.[5]

jaw crusher, it is less than with the primary impactor and again, a cubical shape is obtained throughout the product sizes. The crusher discharge and the bypass material are conveyed to the final sizing screens. Material sizes of 20 mm, 14 mm, 10 mm, 6 mm and 3 mm are stored in bunkers. The plus 20 mm is returned for further reduction.

As with the hardrock crushing and screening plants, a low-cost option is for the material to be stored in ground stockpiles but adverse weather can affect the finished products. Contamination can also result and there is the cost of additional drying at the asphalt plant.

A cone crusher is situated between the 20 mm and 14 mm bins to produce a recrush option should market needs change. This is not necessarily considered to be a further crushing stage but a design feature to make the plant's production more flexible.

As the setting of an impact breaker and the reduction ratio are not as significant in influencing the product shape with these low silica content materials, the process described only needs two crushing stages. If the feed material could be supplied cost effectively at 100 mm down then only a single-stage crushing operation would be required.

5.4.3.   Sand and gravel processing plants

Sand and gravel can be dredged from the sea or dug out of the ground. Marine dredged gravels are only used for roadstone production when hardrock and limestone are in short supply. Care must be taken with marine gravels to ensure that the chlorides are removed before the application of bitumen otherwise stripping of the bitumen will occur.

Land-based sand and gravel is found in two forms, a wet pit or a dry pit. A wet pit is where the majority of the deposit is under water; this material can be won by a suction dredger or a drag line excavator. In a dry pit the sand and gravel can be worked using a loading shovel, a drag line excavator or a scraper together with a tractor. The water can be pumped out of a wet pit and the deposit worked as a dry pit. Deposits can be up to, say, 8 m thick but they usually average about 5 m. The overburden is generally shallow, say from 0–2 m thick.

Many boreholes are taken when excavating for sand and gravel and the samples are then tested in the laboratory. The ratio of sand to gravel is of prime interest along with the geological composition of the gravel since these factors govern the design of the processing plant. Dump trucks can be used to transport the sand and gravel from the workings to the plant or a field conveyor system can be installed. Such a plant is shown in Fig. 5.6.

The sand and gravel processing plant flow diagram shown in Fig. 5.7 includes a field conveyor from the deposit to the plant. The as-dug material is conveyed to a vibrating grizzly with the bars set at 50 mm spacing. The plus 50 mm is stockpiled and the minus 50 mm material passes to a large surge hopper.

A fixed-speed belt feeder extracts the material from the hopper by way of a conveyor to a de-sanding screen equipped with water sprays. The aggregate is discharged into a rotating washer barrel where water is

Fig. 5.6.   Sand and gravel plant. Photograph courtesy of Allis Mineral Systems (UK) Ltd

added. The washed aggregate is then passed over a reject vibrating screen, again fitted with water sprays.

The sand and water are flumed to a sump and separated into coarse and fine fractions. The dirty water is pumped to lagoons where the silt settles out prior to the water being recirculated to the plant. The plus 20 mm material from the reject screen is transferred to a surge bin. An electric vibrating feeder discharges the material to a cone crusher and the crushed gravel is returned to the reject screen. The minus 20 mm plus 4 mm material is conveyed to a grading screen separating out 20 mm, 14 mm, 10 mm and 6 mm products to ground stockpiles. These single sizes can be stored in bins and discharged, individually or blended, to trucks. A stockpile is a good means of allowing the gravel to de-water before further processing.

5.5. Methods of producing asphalts[6]

The selection of the type of plant for producing asphalts is never easy since the circumstances surrounding each purchase are usually different. The plant could be required for a quarry where the aggregates can be controlled and are know to be in plentiful supply, or a depot site where the aggregate source is likely to change. The type of plant chosen is influenced by the markets in which the materials are to be sold since they determine the product requirements. The duty required will not only be directly related to the market potential but also to the number and type of plants in the surrounding area. The budget will play a major role in determining the equipment selected, as will the running costs. Finally, the type of plant will be influenced by local site restrictions such as noise, availability of space, height and even emission regulations should the plant be destined for overseas.

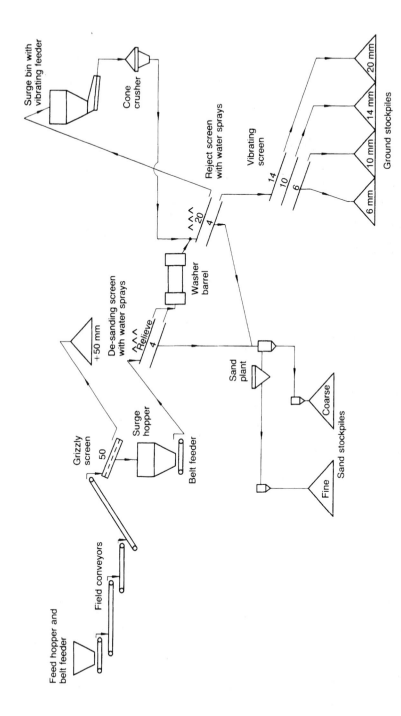

Fig. 5.7. Flow diagram of a sand and gravel plant

There are two basic categories of production facilities, batch plants and continuous plants. Batch plants are conventional asphalt batch plants and batch heaters while continuous plants can be categorized into drum mixers and counterflow drum mixers.

5.5.1. Conventional asphalt batch plants

A conventional asphalt batch plant is shown in Fig. 5.8.

The aggregates specified in the final mix are loaded into the feed hoppers, a different hopper is used for each ingredient. The feeders control the flow of material from the hoppers and provide a proportional feed. Each feeder is set to give the required percentage in the finished mix. A controlled and accurate feed rate is needed otherwise energy will be wasted on unused materials. The feed materials are conveyed to the dryer, consisting of a rotating cylinder with a burner mounted at the discharge end. The cold and wet aggregates are dried and then heated within the cylinder. Inside the dryer is a series of lifters which are designed to cascade material across the diameter of the drum. The lifters at the feed end produce a dense material curtain to expose the aggregate and sand to heat from the burner. At the discharge end, the lifters hold the material around the circumference of the dryer enabling the flame to develop and complete combustion to occur. When combustion takes place within the dryer the gases expand. The products of combustion and the water vapour produced in the drying process are removed by an exhaust system. Figure 5.9 shows a Titan 2000 conventional asphalt batch plant near Dublin.

A contraflow air system operates on the majority of stone dryers where the hot exhaust gases travel in the opposite direction to the material flow. This arrangement is an efficient process since the exhaust gases pre-heat the incoming material. The exhaust system creates an air speed within the dryer so that fine particles of dust are also picked up. Figure 5.8 illustrates a dry collection system, the coarse particles, approximately $+75$ μm, are removed by the skimmer and returned to the plant via the hot stone elevator. The fine particles of dust are removed from the exhaust air by a bag filter. Emission levels of $20\,\text{mg/m}^3$ are achieved using this method and all of the dust is reclaimed by the process. The material discharges from the dryer at temperatures ranging from 120–200 °C and is fed into the hot stone elevator. The elevator transfers the material to the screen on the mixing section. Asphalt plant screens used in the British market grade four sizes plus rejects, or alternatively six sizes plus rejects. The material discharges from the screen into the hot stone bins, internal chutes enable the bin compartments to overflow to a single point.

The mix recipe selects the required aggregates to be weighed in the batch hopper and simultaneously the bitumen and filler are weighed off. The ingredients are then emptied into the paddle mixer until the contents are fully coated. Typically, mixing cycle times of 45 seconds and 60 seconds are possible, depending on the specification. The mixer can load directly to trucks or to a skip and mixed material storage.

Ancillary equipment includes heated bitumen storage with a ring main to the bitumen weigh hopper, imported and reclaimed filler systems

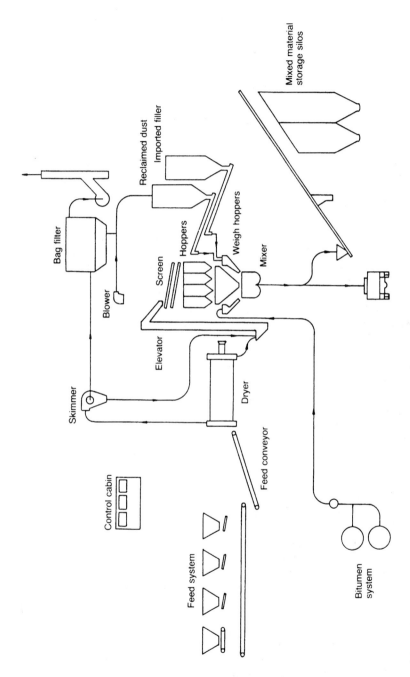

Fig. 5.8. Flow diagram of a conventional asphalt batch plant

Fig. 5.9.   Conventional asphalt batch plant. Photograph courtesy of Tracey Enterprises Ltd

consisting of a silo, rotary valve and screw conveyor to the filler weigh hopper.

Between 10 and 15% of reclaimed mixed material can be added to a conventional asphalt batch plant by a belt feeder and conveyor directly into the batch weigh hopper. In this arrangement, the reclaimed material is considered to be an additional ingredient in the mix. The reclaimed material can also be added directly into the mixer. The equipment consists of a belt feeder and a conveyor fitted with a belt weigher.

Up to 50% of reclaimed material can be added to the mix using the parallel drum method. The material is heated to between 80 and 120°C, weighed in a separate vessel and then screwed into the paddle mixer, a flow diagram of the process is shown in Fig. 5.10. The exhaust from the reclaimed, or black, dryer is directed into the discharge end box of the virgin drum where the fumes are incinerated. Table 5.1 lists the advantages and disadvantages of conventional asphalt batch plants.

5.5.2.   Batch heater plant
A flow diagram for a batch heater plant is shown in Fig. 5.11. This process is different from the conventional asphalt plant since each batch is manufactured individually. Pre-graded aggregates are fed to the plant in a single batch by time controlled feeders; the time is directly related to the

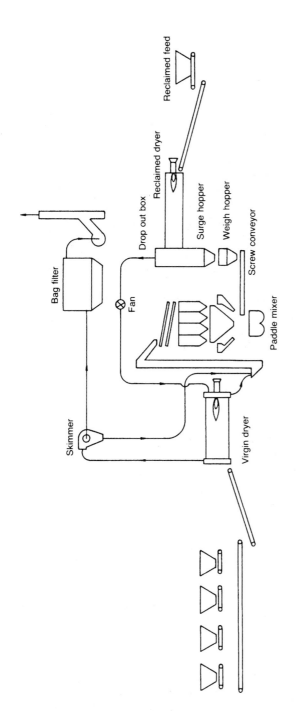

Fig. 5.10.   Flow diagram of a batch plant with dryer for recycled material

Table 5.1.   Advantages and disadvantages of conventional asphalt batch plants

Advantages	Disadvantages
Ability to manufacture all materials to BS 594 and BS 4987	Heat wasted on rejected and overflow material
Emission to atmosphere within acceptable limits	Relatively high maintenance costs
Inconsistent feed materials tolerated	High production costs compared with drum mixer process
Small tonnages possible	Relatively high capital cost
Production of materials at all temperatures	Capacity restricted to mixer size and mixing cycle
Mixed material storage is not essential	
High percentage of reclaimed material can be added	

proportion set in the mix recipe. The materials are conveyed to a check weigh hopper where the wet feed weight is determined. An allowance is made for the moisture content in the feed, the control system then calculates the bitumen and filler contents and proceeds to weigh these ingredients.

Drying and heating of the materials takes place in a short rotary drum and the discharge temperature is controlled by the time the materials are kept in the heater. All materials are discharged into a paddle mixer where a homogeneous mix is produced. A typical mixing cycle time is two minutes but this depends upon the moisture content in the feed and the material specification. The mixer can discharge directly to trucks or, alternatively, to mixed material storage. Some plants have an additional weigh vessel between the batch heater and the paddle mixer to determine the dry aggregate weight. The bitumen content is then calculated from this dry batch weight. The skimmer can be arranged to discharge on a discrete basis into the paddle mixer.

The bitumen storage and ring main, the filler storage and the bag filter exhaust system are generally as described previously. If moisture contents greater than 2% overall are experienced then a pre-dryer is used for the sand, as shown in Fig. 5.11. The sand is then called up from the dry storage. Between 10 and 15% of reclaimed material can be added to the mixer by means of a belt feeder and weigh conveyor.

Figure 5.12 shows twin batch heater plants fed directly from a quarry plant. Table 5.2 lists the advantages and disadvantages of batch heater plants.

5.5.3.   Drum mixer
A flow diagram of a drum mixer is shown in Fig. 5.13.

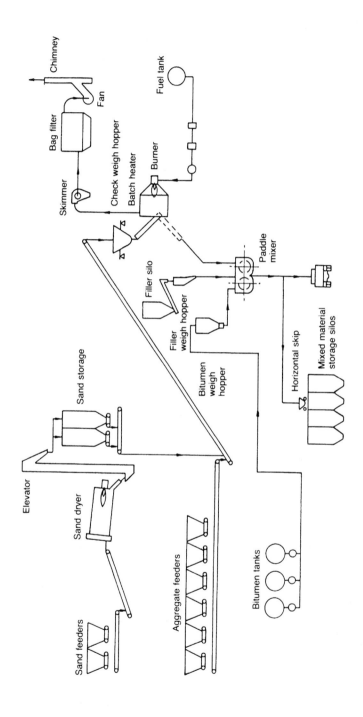

5.11. Flow diagram of a batch heater

Fig. 5.12. Twin batch heater plants fed directly from a quarry plant. Photograph courtesy of Hanson Aggregates (S Wales)

The feed section is one of the most important aspects of the drum mixer process. It is essential for the success of the finished product that pre-graded aggregates are used. The feeders control the flow of material to the plant at a rate proportional to the quantities called for in the mix. Volumetric feeders are generally employed although an option sometimes preferred is the use of weigh feeders especially for sand and dust. Wet sand and dust can be difficult to handle and they do not always flow evenly or readily. The use of a volumetric feeder

Table 5.2. Advantages and disadvantages of batch heater plants

Advantages	Disadvantages
Small batches economically produced	Single-sized feed materials
No material wastage	Maximum 2% moisture in feed materials
Ability to manufacture all materials to BS 594 and BS 4987	Pre-drying required if moisture content greater than 2%
Quick recipe change	Relatively high capital cost
Mixed material storage is not essential	Relatively high maintenance costs
Number of feed materials is only limited by quantity of feeders	
Reclaimed material can be added	

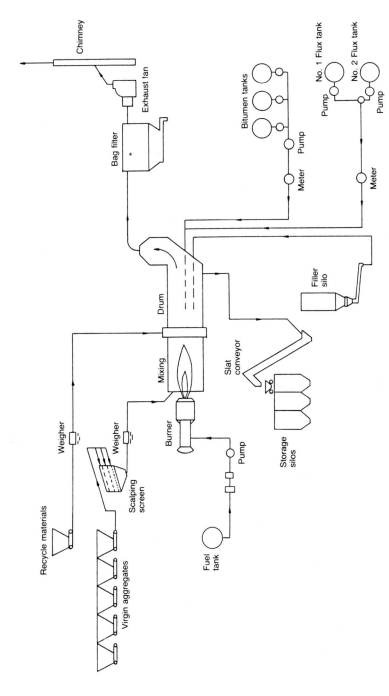

Fig. 5.13. Flow diagram of a drum mixer with recycling facility

cannot therefore be relied on and in these instances a weigh feeder is used.

A scalping screen can be incorporated into the feed system to remove unwanted oversize material when the quality of the aggregates cannot be guaranteed. For example, this would be appropriate if the plant is fed from ground stockpiles.

The materials pass over a belt weigher before entering the drum. The mass passing over the weigher supplies the master signal to the ratio controller which determines the flow rates for the other mix ingredients making allowance for the moisture content in the feed.

The drum mixer consists of a rotating cylinder with the burner mounted at the feed end. Lifters around the circumference of the drum keep the material away from the flame, followed by a lifter pattern which forms a dense cascading material curtain. After drying and heating, the lifters are designed to mix the aggregates with bitumen inside the drum. The binder is usually added from the discharge end to prevent degradation which can occur if it is too close to the burner flame. Imported or reclaimed filler can also be added to the mixing zone.

One of the most accurate methods of providing a continuous flow of filler is to employ a 'loss-in-weight' system. Special attention is given to conditioning the filler and it is continuously reverse weighed against a known calibration rate. A pneumatic conveyor transfers the filler to the drum mixer. An exhaust system removes the products of combustion and

Fig. 5.14.   Drum mixer plant. Photograph courtesy of Cedarapids USA

Table 5.3. Advantages and disadvantages of drum mixers

Advantages	Disadvantages
Economical plant for long production runs of one material specification	Unable to manufacture all BS 594 and BS 4987 materials
Number of feed materials is only limited by quantity of feeders	Single-sized feed materials
Dust removed from process is minimal	Mixed material system essential
High capacities readily achieved	Small batches are uneconomical
Easily adapted to mobile design	Possibility of fume emission with high temperature mixes
Relatively low maintenance costs	Wastage at beginning and end of production
Reclaimed material can be added	

a filter prevents the inherent dust particles being emitted into the atmosphere. Due to the nature of the process the dust collected is considerably less than that which is recovered from a conventional dryer. Up to 15% of reclaimed mixed material can be added to the drum mixer and successfully mixed before fuming becomes a problem. As the process is continuous, large tonnages are possible in a relatively short timescale, a mixed material system is therefore needed. Belt conveyors, travelling skips and slat conveyors have all been used for these storage schemes.

Figure 5.14 shows a Cedarapids Standard Havens drum mixer plant operating in San Francisco. Table 5.3 lists the advantages and disadvantages of drum mixers.

5.5.4. Counterflow drum mixer
Figure 5.15 shows the layout of a counterflow drum mixer.

There are two major differences between the counterflow drum mixer and the drum mixer. Firstly, an extended burner is used and this is mounted at the discharge end of the drum and secondly, the mixing chamber is completely separated from the drying and heating zones.

The drying section of the counterflow drum mixer is identical to a conventional dryer, pre-heating of the feed materials results giving a more efficient drying method. The burner flame cannot be in direct contact with the bitumen since the mixing zone is isolated. As a result, the possibility of fumes is reduced considerably. A scavenger fan removes the hot gases and any fumes present from the mixing chamber are incinerated by the burner flame.

Bitumen, filler and reclaimed material are added to the mixing zone as called for by the specification. The drying and heating section can be used purely as an aggregate dryer if required. Up to 50% reclaimed material can be added to the counterflow drum mixer.

The principle of operation and associated equipment such as feeders, scalping screen, weigh conveyor, exhaust system, bitumen system, filler

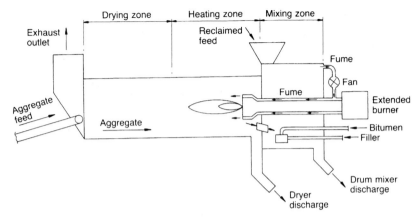

Fig. 5.15.   Flow diagram of a counterflow drum mixer

storage and mixed material are as described for the drum mixer. This process evolved from the drum mixer to meet the need for a continuous plant with improved emission levels and a reduced fume content when mixing at higher temperatures or using different bitumens.

Table 5.4 lists the advantages and disadvantages of counterflow drum mixers.

5.5.5.   Summary of main features of asphalt production plants
Table 5.5 summarizes the main features of asphalt production plants.

5.6.   Controlling manufacture
5.6.1.   Aggregate processing plants
Control of the feed material in an aggregate processing plant is important

Table 5.4.   Advantages and disadvantages of counterflow drum mixers

Advantages	Disadvantages
Most economical plant for long production runs of one material specification	Wastage at beginning and end of production
Possibility of fume emission is reduced when producing high temperature mixes	Small batches are uneconomical
Ability to manufacture all materials to BS 594 and BS 4987	Single-sized feed materials
Number of feed materials is only limited by quantity of feeders	Mixed material system essential
High capacities readily achieved	
Relative low maintenance costs	
Reclaimed material can be added	

Table 5.5.   Summary of asphalt production plants

Type of plant	Uses	Remarks
Conventional asphalt batch plant	The most common type of mixing plant in the UK used for small batch production and up to the medium capacity range of 240–300 t/hr	Highly flexible production and capable of manufacturing all British Standard specifications
Batch heater	Ideal for small production requirements with the facility to change specifications quickly	Maximum allowable moisture content of 2%; a pre-dryer is needed if moisture is higher
Drum mixer	Economical plant for the production of large quantities of one material specification	
Counterflow drum mixer	Selected for the continuous production of large quantities of one material specification when the control of fume emission is critical	

for several reasons. Jaw, gyratory and cone crushers should be choke fed to optimize plant performance and obtain good product shape. Maximum feed sizes allowed to enter the crushers should be observed in order to provide continuity of production as a blockage within a crusher, or chute, can cause considerable delays. Unwanted materials like soil, clay and very fine rocks should be prevented from entering the plant since these reduce overall throughputs.

Tramp iron (unwanted iron or steel in the aggregates, e.g. an excavator tooth) should be avoided in the feed since this can permanently damage the crusher wear parts. A metal detector is therefore usually fitted before the secondary crushing stage. An audible warning can be given to detect the metal and/or a magnet used to remove it. The majority of crusher wear parts are made from manganese steel and as this material is non-magnetic, electronic detectors have to be used to identify any of these loose objects within the process.

A limestone quarry is shown in Fig. 5.16.

Larger aggregate plants are controlled by microprocessors and these eliminate the need for costly, extensive electrical relays and their associated wiring. This method of control also enables the status of the plant such as material flow, motors on/off, doors open/closed or water supply on/off to be monitored and alarmed if a fault condition occurs. Individual crushers can be microprocessor-controlled to automatically adjust, or maintain, the discharge setting, e.g. in a cone crusher. This is a useful feature since a granite with a high silica content can noticeably wear the liners of a cone crusher in a very short time, say four or five hours, and such crushers require regular re-setting to maintain a quality product.

Automatic overload protection is available on jaw, cone and gyratory crushers by means of a hydraulic system. Tramp iron and other

Fig. 5.16.   Aerial view of a limestone quarry. Photograph courtesy of Hanson Aggregates (Southern)

uncrushable materials are allowed to pass through the crushers without costly damage to the liners and the original setting is then adopted. This would be in addition to the metal detector discussed earlier. The correct screen mesh sizes need to be fitted to gain full control of the product sizes. The meshes should be inspected regularly for proper tensioning and holes caused by wear. The individual units of new aggregate plants are usually sheeted which makes the plant more aesthetically acceptable as well as reducing noise and dust emissions. The nuisance dust from the conveyor transfer points and housings is collected in a dry bag filter and the 'cleaned air' emitted to atmosphere.

A recrush facility enables the plant to be more versatile with the sizes being manufactured. This feature incorporated in the design is useful when the market trends change. The larger sized materials 40 mm, 28 mm and 20 mm are reduced to the more popular smaller sizes 14 mm, 10 mm and 6 mm.

The final products are either stored in stockpiles on the ground, or in storage bunkers. In either case, contamination of the aggregates should be prevented. Stockpiles should be on hardstanding such as the quarry floor, or concrete to prevent contamination from below and separated to avoid overspill. Storage bunkers should be fitted with high level indicators or overflow chutes. Having invested substantial amounts of time and money to ensure that the product size and shape is correct, degradation of the material should also be prevented. Rock ladders are used within stockpiles and storage bunkers to control the flow of large aggregates from a height and overcome unwanted breakages.

Laboratory inspection of the materials, both during and after production, is an important aspect of controlling manufacture and quality. The results of the samples from within the process indicate whether the plant is functioning correctly and enable adjustments to be made where necessary. The results from the final material samples determine whether the product is acceptable for sale. Single sized products will be checked against BS 63: Part 1.[7] The sample must be representative of the bulk material and the personnel undertaking the testing must be skilled operatives. Methods of sampling and testing aggregates are given in BS 812: Part 102[8] and Section 103.1.[9] More details on this subject are contained in Chapter 9.

5.6.2. Hot rolled asphalt and coated macadam plants
Storage of raw materials
HRA and coated macadam plants can either be located adjacent to the aggregate plant in the quarry, or remote at a depot site. If the plant is remote then the aggregates will be transported by truck or rail for larger operations. The quantity of materials delivered by rail necessitates an automatic off-loading scheme incorporating underground discharge hoppers, tripper conveyor and covered storage bays. The material is recovered from storage by an underground feed system to the asphalt plant. Both truck and rail delivered aggregates should be stored on hardstanding and separated to avoid contamination. This method of storage along with a quality feed material provides a consistent feed to the asphalt plant reducing overflow, wastage and fuel costs.

The ideal situation is for an asphalt plant to be fed directly from the aggregate plant bunkers. The product sizes can be more easily controlled by matching the screen mesh sizes on both plants and, since the material has not been exposed to the weather since entering the quarry plant, the moisture content will be less. This layout can reduce the overall feed moisture content by up to 2% and is therefore more economical.

The filler silos containing imported and reclaimed materials can be subject to the formation of condensation on the inside walls under certain weather conditions. As filler is a hygroscopic material, it has a tendency to absorb moisture and the contents of the silos can become solid if left for prolonged periods. It is good practice to order only the quantity of imported filler needed for planned production runs and leave the silos empty when the plants are not operational for extended periods.

Bitumen needed for production is stored within the temperature ranges laid down by the specification. These temperatures are given in BS 594: Part 1[2] and BS 4987: Part 1.[3] The bitumen is usually circulated in a ring main system up to the mixing plant before usage to reduce heat loss in the pipework even though the pipes are heated and insulated. To prevent the bitumen hardening through oxidation the material should not be stored at high temperatures for long periods especially if no new bitumen is added. Prolonged storage with no usage should be approximately 25 °C above the bitumen softening point and recirculation stopped. Bulk reheating of the bitumen should be carried out intermittently to prevent local overheating around the heating member. Bitumen can ignite instantaneously if temperatures above 230 °C are reached so such temperatures should be avoided.

Plant capacities
There are two basic constraints on the output of an asphalt and bituminous macadam plant, the drying capacity and the mixing capacity. The drying process is complex. Some of the factors which directly affect the capacity are

- altitude
- air temperature
- type of feed material and grading
- feed material temperature
- moisture content in feed material
- moisture content in discharge material (usually less than 0.5%)
- dryer discharge temperature
- exhaust temperature
- dryer angle
- calorific value of the fuel
- exhaust air volume available
- lifter pattern in dryer
- efficiency of burner.

The mixing process is also complicated, and some of the factors affecting the capacity and whether the mix is homogeneous are

- mix recipe
- mixer live zone
- paddle tip speed
- size of paddle tips
- mixer arm configuration
- method of feeding ingredients
- temperature of materials
- mixing time (some mixes have a minimum mixing time).

One of these processes will determine the overall plant capacity. In a conventional asphalt batch plant and a batch heater the two processes are separate and can therefore be easily identified and considered. In a drum mixer and counterflow drum mixer the drying and mixing are combined and require more detailed consideration. Basically, the feed material must be completely dried and heated before the bitumen is added. The drying constraints are therefore similar to the batch plant dryers. The mixing time in a continuous plant is controlled by adjusting the throughput, altering the drum angle, adjusting the position of the bitumen injection pipe within the drum, altering the design of the mixer lifters and altering the speed of the drum. In practice, once these variables are set during commissioning they are not usually altered.

To determine the equipment sizes for an asphalt and bituminous macadam plant the manufacturer requires the following information

- maximum plant capacity
- type of feed material
- average moisture content of feed material
- type of mixes to be produced
- temperature of mixes
- special constraints, e.g. minimum mixing time.

Plant control

With the exception of the batch heater, which has fixed-speed time-controlled feeders, most plants have variable speed feeders with a 20:1 turn down ratio (the feeders operate from 5 to 100% of the belt speed or capacity). Volumetric belt feeders generally have a repeatability accuracy of ±0.5% and vibrating feeders are around ±2%. This repeatability and turn down ratio enables an accurate plant feed to be obtained and, providing the feed materials are within the specified size limits, this also ensures an economic control of the materials, since heat will not be wasted on unused aggregates. Low level indicators can be fitted to the feed hoppers with elevated lights above the hoppers to warn the shovel driver of an impending shortage. The feeders can be equipped with no-flow indicators and after an initial warning and a pre-set time delay, the plant will shut down if prescribed.

The burner mounted on the dryer, or drum mixer can have turn down ratios of up to 10:1. Different plant throughputs and different dryer discharge temperatures can therefore be easily accommodated with an economic use of the fuel. The burner is automatically controlled by a temperature sensing device in the dryer discharge chute. A non-contact infra red pyrometer is normally used to record the temperature as the response times are a few seconds compared with a thermocouple probe which can take up to 15 seconds.

To sustain combustion an exhaust system is connected to the dryer or drum mixer. As air is drawn through the drum, small particles of dust are picked up. This must be removed before discharge to atmosphere. A dry bag filter dust collector is employed to remove the dust and plants are currently being guaranteed to meet particulate emission levels of 20 mg/m³. The dust collector also pulls from nuisance points on the plant and controls the overall emission level. The air volume is controlled by a variable damper in the system, usually fitted prior to the fan on the clean side of the filter. The damper is controlled by a pressure reading taken at the burner end of the dryer.

Continuous level indicators can be installed in the hot stone bins of conventional asphalt batch plants. These give an indication of the material level within each compartment and provide a means of trending. From this information the feeders can be set up during commissioning and adjusted when, and if, the feed aggregates become inconsistent. Again this enables the plant to operate more economically with reduced overflow.

Temperatures of the sand in the hot stone bins and the bitumen in the ring main are taken by thermocouple probes and recorded. These readings confirm that the temperatures are correct before mixing commences.

Although the British Standards allow tolerances of $\pm0.6\%$ on the accuracy of the bitumen mass in the mix, plants are expected to perform with an accuracy of $\pm0.1\%$. Bitumen is the most expensive ingredient and therefore economics dictate the importance of this requirement. Batch plants are equipped with load cell weigh systems with repeatable accuracies of $\pm0.3\%$ for the bitumen and filler hoppers. Continuous plants use a load cell in the form of a belt weigher and accuracies of $\pm1\%$ are guaranteed.

Fully automatic microprocessor control systems are used on all types of asphalt plants. On batch plants the microprocessor provides accuracy by calculating the bitumen content as an actual percentage of the aggregate and filler content achieved. Outputs are maintained at an optimum by weighing off a large percentage of the bitumen at the same time as the aggregate. The required bitumen is then calculated relative to the actual aggregate and filler content. Self-learning in-flights (to provide accurate weighing) for the hot stone bin discharges and alternative bin selection are also features incorporated in conventional asphalt plants which help throughput whilst maintaining accuracy.

The use of microprocessor control allows the operator to adopt a passive role during production since the plant is operated from the input data. Providing these data are correct and no faults are recorded, the finished mix will be correct. The operator is relieved to a certain extent of the responsibility and therefore the quality of the product is not operator dependent. The monitoring of process faults like sticking doors, tare (weight) faults, stopped motors, broken chains, or sticking valves reduces down time and improves the quality of the products. In addition, the microprocessor can stop the plant from mixing if acceptable pre-set limits are not met. These faults are logged and recorded on the system again providing quality control.

An infra red pyrometer can be positioned, at the paddle mixer discharge and the mixed material bin outlets, to read and record the final discharge temperature. Microprocessor control can prevent contamination of different mixes within the mixed material system by remembering the type and quantity of mix in each bin. Storage silos mounted on load cells can accurately weigh out mixed material to trucks, avoiding wastage and saving time.

Full microprocessor control systems offer VDU screens with plant mimic diagrams, the ability to store over two hundred recipes, production data, storage data, a daily job queue, print-out of mix results, data, time, etc. The print-out and storage data are important elements of quality assurance since each batch and truck load can be traced, confirmation of a manual or automatic mix is also recorded. The controls information can be limited to the laboratory, the weighbridge and the main office accounting system, allowing information to pass in both directions.

Production and operating conditions
To obtain optimum fuel usage a constant and uniform feed to the plant is needed. The aggregates should preferably be taken from the same place on the stockpile each time, avoiding the lower stock near to the ground and surface water from entering the loading shovel. This procedure will ensure that the minimum amount of water enters with the feed aggregates and provide uniform drying conditions. The dryer, burner and exhaust system operate more efficiently under stable conditions.

The dryer has a minimum capacity directly related to the effectiveness of the lifter pattern, the turn down ratio of the burner and the maximum inlet temperature permitted to the bag filter. The majority of filters in the UK are equipped with Nomex material for the bags. This material has a maximum operating temperature of approximately 200 °C although higher peak temperatures can be accommodated. With a small aggregate feed to the dryer, the burner on minimum flame and a discharge temperature of say, 190 °C, there will be occasions when too much heat is available and the inlet temperature probes to the filter will automatically switch off the burner fuel supply. The dryer angle can be lowered to try to overcome this by increasing the drum loading, but this will not resolve the problem when a small production rate is required. As a general rule, the minimum drying capacity for a plant equipped with a bag filter is approximately 50% of the capacities quoted for hot rolled asphalts. This constraint does not apply to plants fitted with wet collection exhaust systems.

As the heated aggregates pass from the dryer to the mixer, there is an initial temperature loss of between 10 °C and 20 °C depending upon the required mix temperature. During long production runs, this temperature difference can be reduced to between 0 °C and 5 °C.

When drying and heating gravels, or marine gravels, for use in roadstone production it is important not to overheat the aggregate. Overheating can cause 'sweating' of the aggregate later in the process and this will eventually lead to stripping of the binder. For this reason, and as an aid to adhesion since crushed gravels still tend to be slightly rounded, the addition of 2% Portland cement or hydrated lime at the mixing stage is recommended.[2,3]

The efficiency of the vibrating screen on a conventional batch plant changes with different flow rates. Consistent feed gradings and constant flow rates are, therefore, preferred so that the aggregate separation into each hot stone bin compartment can be predicted. The continuous level indicators in each compartment are a useful check since, in practice, a consistent feed grading is difficult to achieve.

During production, care is taken not to contaminate hardrock mixed with limestone. However, the reverse situation can usually be accommodated. In a conventional asphalt plant, the hot stone bins are completely purged when the type of feed material changes. Purging can lead to considerable delays when changing specification. As each batch is manufactured individually with a batch heater, this situation does not arise. Continuous mixing plants can change specifications during production providing the same type of feed material is used, otherwise the plants are purged.

The ideal production schedule is to manufacture the lower temperature materials first and progress to the higher temperature specifications. In reality, this is not always possible and care has to be taken not to overheat a low-temperature material that is required in the middle of a high-temperature production run. Overheating can lead to oxidation and hardening of the binder and this may well change the mix characteristics which can have an adverse effect on the life of the pavement.

The paddle mixer has a recommended minimum capacity of 50% of the nominal capacity. Mixers do not function satisfactorily below this level as a homogeneous mix does not result and oxidation of the bitumen can occur. Prolonged mixing of a full or partial batch does not necessarily mean a more homogeneous mix. Extended mixing times can lead to binder oxidation and subsequent hardening.

To maintain the quality of the product, the equipment should be recalibrated on a regular basis. The feeders, the burner, the temperature measuring equipment and the load cell weighgear should all be inspected within a planned maintenance schedule to attain the Quality Assurance standards needed.

Mixed material storage

Mixed material silos are manufactured in both square and circular forms with bin capacities ranging from 30–200 t each. The top, the top doors, the sides, the conical section, and the discharge doors are all insulated with material which can be up to 150 mm thick. Heaters are usually attached to the discharge doors and the conical section and although the bin sides can be heated it is not common practice. The heating can be electric or indirect by a hot oil heater and the heating system usually allows the temperature of each individual silo to be controlled separately. The heat is applied purely to maintain the temperature by replacing lost heat and not as additional heat for a mix with a low temperature. Overheating of the material in the silo can cause the bitumen to strip from the aggregate.

Mixed material silos are available with grease- or oil-sealed doors at the inlet and discharge, with the facility to add an inert gas. This design feature provides extended storage times by preventing the mix coming into contact with the atmosphere and oxidizing.

Even without the sealing and the addition of an inert gas extended storage is possible. To achieve the best results the silo is completely filled. Dense materials store better over a longer period. For a 12 hour period the heat loss is minimal, say a maximum of 3 °C and providing the outlets are large enough, the silos will discharge freely. If the material is to be stored for two to three days then a crust forms on the top and around the discharge outlet. This can be overcome by discharging one tonne of material at regular intervals, say every nine hours. Again, the temperature loss over these few days will be minimal, probably a maximum of 20 °C due to the effectiveness of the heating and insulation. Hot rolled asphalts and open-textured materials are not recommended for extended storage because of the difficulty in discharging from the silo outlets.

Sampling
Production must be monitored to ensure that the aggregate grading and bitumen content comply with the specification but not every batch can be tested. Samples are therefore taken at pre-determined intervals and these samples must be representative of the bulk material. Methods of sampling and achieving representative samples are given in BS 598: Part 100[10] for bituminous mixtures and BS 812: Part 102[8] for aggregates, sands and fillers. Sampling should only be performed by skilled personnel using well maintained and calibrated equipment.

Samples should be taken of all materials delivered to the plant, i.e. aggregate, bitumen and filler. If the asphalt plant is located in the quarry with the aggregate plant then part of this requirement will have been fulfilled.

The aggregate samples from the hot stone bins are taken either from special sampling devices on the bins or directly through the paddle mixer. This tests the gradings within each bin and gives an indication of the efficiency of the screen. The results from these samples determine the amount to be weighed off from each bin during the weighing cycle. This exercise should be carried out at regular intervals to ensure aggregate grading control.

Samples of the mixed materials can be taken from the paddle mixer or drum mixer discharge and the mixed material silo discharge. The mixer and drum mixer discharge sample determines whether a homogeneous mix is produced and the silo discharge sample confirms that the material has stored favourably or indicates whether segregation or other deterioration has occurred. During production, samples are usually taken from the truck load of mixed material. The paddle mixer, drum mixer and silo discharge samples are taken during commissioning and calibration and subsequently only if problems occur.

Tests on the samples take time, as discussed in Chapter 9 and a large quantity of material may have been manufactured and laid before the results are available. However, sampling is essential for calibration purposes and good quality control during production. It also gives confidence in the performance of the plant and, therefore, has to be undertaken on a regular basis.

5.7. Summary
This Chapter has dealt with the production of asphalts. It has considered the availability of aggregate sources in the UK and how these basic materials are processed using different types of crushing and screening plants. The various categories of asphalt production plant were examined. Finally, different factors affecting the control of manufacture were discussed.

References

1. BRITISH STANDARDS INSTITUTION. *Building and Civil Engineering Terms, Concrete and Plaster, Aggregates*. BSI, London, 1995, BS 6100: Part 6, Section 6.3.
2. BRITISH STANDARDS INSTITUTION. *Hot Rolled Asphalt for Roads and other Paved Areas. Specification for Constituent Materials and Asphalt Mixtures*. BSI, London, 1992, BS 594: Part 1.
3. BRITISH STANDARDS INSTITUTION. *Coated Macadam for Roads and other Paved Areas. Specification for Constituent Materials and for Mixtures*. BSI, London, 1993, BS 4987: Part 1.
4. BRITISH STANDARDS INSTITUTION. *Specification for Air-Cooled Blast Furnace Slag, Aggregates for Use in Construction*. BSI, London, 1983, BS 1047.
5. ALLIS MINERAL SYSTEMS 'UK' LTD. Based on flow diagrams courtesy of Goodwin Barsby Division, Leicester.
6. MOORE J P. Coating Plant Section. *Quarry Management*, May 1992.
7. BRITISH STANDARDS INSTITUTION. *Road Aggregates, Specification for Single-Sized Aggregate for General Purposes*. BSI, London, 1987, BS 63: Part 1.
8. BRITISH STANDARDS INSTITUTION. *Testing Aggregates, Methods for Sampling*. BSI, London, 1989, BS 812: Part 102.
9. BRITISH STANDARDS INSTITUTION. *Testing Aggregates, Method for Determination of Particle Size Distribution, Sieve Tests*. BSI, London, 1985, BS 812: Part 103, Section 103.1.
10. BRITISH STANDARDS INSTITUTION. *Sampling and Examination of Bituminous Mixtures for Roads and other Paved Areas, Methods for Sampling for Analysis*. BSI, London, 1987, BS 598: Part 100.

6. Surfacing plant

6.1. Preamble

This Chapter describes the main items of plant which are used to lay asphalts. As well as describing the elements which combine to make what is often called a 'paving train', it considers delivery systems and specialist plant associated with surfacing contracting.

Whilst significant strides have been taken in the introduction and use of new materials, it is still generally true to say that the equipment used to provide the surface itself is still at the discretion of the individual paving contractor, although the introduction of thin surfacings has enhanced the reputation of certain types of paver. The choice of equipment is always based on price, tempered by the loyalty of the surfacing contractor to a particular make of equipment based on previous experience with that manufacturer.

Whilst there is no substitute for skill and experience, the asphalt industry is suffering from a drain in such personnel caused by a combination of early retirement packages and a lack of recruitment. The current lack of continuity of surfacing work tends to support the view that this situation may not change in the foreseeable future. At the same time, clients are demanding high levels of quality in the finished product and, in many instances, that surfacing and reconstruction works be carried out under demanding lane rental conditions. This general situation has given rise to the use, in some instances, of specialist laying gangs which travel from contractor to contractor on a national basis, making use of the contractor's own plant. This can result in a lack of continuity of care for individual items of surfacing equipment.

Paving plant manufacturers, recognizing these problems and the demands from the contractors for equipment to be cost effective, have reacted by supplying machines which have increasing degrees of sophistication and more automation making them less susceptible to the effects of neglect.

Notwithstanding, it is essential that the paving train operates efficiently if surfacing contractors are to be able to lay a wide range of materials to a variety of increasingly more demanding specifications. Plant is clearly an important element in all surfacing operations.

6.2. Lorries

Asphalts are manufactured to close tolerances at elevated temperatures in modern plants that are controlled by computer. It is important that the lorries which transport the manufactured materials to site do so in a manner which maintains the temperature of the mixture, minimizes segregation and permits efficient transfer to the paver. Lorries used for this purpose are often described as 'asphalt wagons' and their importance should not be underrated.

The speedy and efficient delivery of material to the paver is a vital and integral element in the success of a surfacing operation. Interrupted or discontinuous flow of mixed materials will give rise to additional costs and may result in problems with the finished mat.

British Standards[1,2] require delivery vehicles to be 'insulated and sheeted to prevent an excessive drop in temperature and to protect against adverse weather conditions'.[1]

A useful rule of thumb regarding lorry capacity is 5 t per axle, so a four-wheeled lorry with two axles will carry 10 t. Similarly, an eight-wheeled lorry with four axles will carry 20 t. Overloading of delivery lorries using public highways is a prosecutable offence and hauliers and suppliers must be very careful to ensure compliance with the appropriate regulations (Road Traffic Act 1988).

Responsible surfacing contractors will only use lorries that are equipped with reversing alarms whether they are owned or hired. (A legal requirement for lorries to have reversing alarms is currently being considered.[3]) Similarly, each driver (along with all other staff) should be inducted as part of the Health and Safety Plan under the Construction (Design and Management) Regulations 1994 for each different site. Liaison between the surfacing operation and the manufacturing plant has always been extremely important and the majority of surfacing supervisors will correlate the required delivery rates with the speed of laying (for more on this topic see Section 7.6.2 of Chapter 7).

In the continual search for cost effectiveness and increased efficiency, the round trip times of each lorry become more important. Delays at the paver or the asphalt plant will cause discontinuities at the surfacing operation with associated problems in mat finish and profitability of the operation. Indeed, local traffic congestion caused by the surfacing operation itself is a potential delay which is often forgotten.

Paving trains only have the opportunity to achieve maximum efficiency when laying large tonnages, for example when surfacing an airfield. In such circumstances, a few minutes saved in unsheeting, releasing the lorry tailgate and raising the tipper body can make a significant difference to the productivity. Recognizing all these factors, modern asphalt delivery lorries are sophisticated and powerful units which frequently incorporate an 'easy sheeting' system, automatic tailgate opening together with relatively fast tipping speed. A typical example is shown in Fig. 6.1.

It is interesting to note that the effect on productivity of the time taken to raise the body of an asphalt tipper has been recognized for some time in

Fig. 6.1. A modern asphalt lorry. Photograph courtesy of E and JW Glendinning Ltd

the USA. As a consequence, vehicle manufacturers reacted by producing units which raise to full height in a total of 4 s. A machine used in the USA, called the 'Contrail', is a non-tipping articulated trailer with a conveyor bed which transports the mixed material directly into the hopper of the paver. In a country such as the USA, which has vast road networks, there exists the opportunity to develop equipment which will be used to capacity on a regular basis. This is not always true of the UK. (Each year in the USA, some 700 M t of asphalt are laid, whereas in the UK the figure falls short of 30 M t — the ratio of populations is approximately 4.5:1.)

The current generation of delivery lorry is an expensive and sophisticated piece of equipment which operates within a framework of strict legislation and, together with the driver, forms a vital piece of the paving train. Unless the asphalt is delivered at the required time in good condition then even the most sophisticated laying equipment and vastly experienced personnel will be singularly ineffective.

6.3. Pavers

The UK market for pavers was around 90 machines in 1999 with peak sales, earlier in the 1990s, of around 150 models of all types. Pavers are available in a wide variety of sizes. Some can lay asphalts as narrow as 1 m whilst other machines can lay as wide as 16 m with special widths being available from manufacturers upon request. In the UK, the potential for laying mats as wide as the latter figure is extremely limited. The standard paver is around 2.5 m. The standard width of a lane on a major carriageway in the UK is 3.65 m. Figure 6.2 shows a modern asphalt paver.

Fig. 6.2. A modern asphalt paver, a Dynapac F 16-6W. Photograph courtesy of Dynapac

The basic function of the paving machine is to place materials onto the existing surface, at predetermined widths, to a designated thickness or level, to a required smoothness, all ready for final compaction. Despite the fact that modern pavers are sophisticated pieces of plant and becoming more so, this fundamental philosophy remains constant.

All asphalt pavers consist of a 'tractor unit' and a 'screed'. The tractor unit is the source of forward propulsion and provides motive power to all the paver systems. In addition, it acts as a delivery conveyor to the screed and provides the platform for the screed and what is called the 'tow point'. Asphalt lorries reverse into the paver in order to discharge their loads. Since the hopper of the paver has insufficient capacity to hold the entire load, the paver must have the power to push the truck until the body of the truck has been emptied. Thus, the combination of tractor unit and screed, i.e. the paver, must be designed to provide sufficient grip and power to propel the delivery lorry during discharge of the mixed material.

Traction is provided either via drive wheels with front steerable bogey wheels or tracks. Most operators are familiar with wheeled pavers which have the advantage that they can be driven over short distances within a site or between sites. The drive wheels are independently braked which enables the operator to slew the paver where demanded by tight radii. Tracked machines provide superior grip in poor ground conditions, better manoeuverability by an experienced operator and a stable platform which assists when laying thin surfacings which are becoming the preferred option for many clients. Wheeled pavers continue to dominate the market in the UK with tracked machines taking some 8–10% of sales in 1999. Figure 6.3 shows a schematic diagram of a wheeled paver and illustrates the basic operation of an asphalt paver.

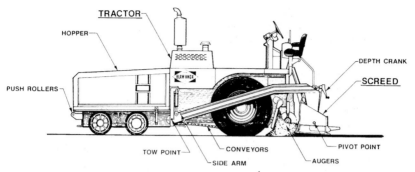

Fig. 6.3.   Schematic diagram of a wheeled paver[4]

6.3.1.   Elements of a paver
Push rollers
Rotating rollers which, during laying, are in contact with the lorry tyres whilst discharge takes place, adjustable for height and used to push the delivery lorry.

Hopper
Acceptance area for mixed material, paver conveyor bed extends to the front of hopper area. Hopper sides, or wings, fold hydraulically towards the centre, following departure of lorry, to ensure that all material is channelled into the conveyor area.

Conveyors
Transport material from the hopper through the paver to the augers in front of the screed. Conveyor speed is hydrostatically variable to ensure adequate head of material at the screed.

Augers
Distribute the material from the conveyor to the full width of the front of the screed. Variable speed and height adjustable.

Screed
Spreads and partially compacts material to appropriate width and depth criteria. May be 'tamping' or 'vibrating' or a combination of both.

Side arms
Forming the 'tow' system for the screed; the tow points on each side of the tractor unit are height adjustable and dictate the thickness of the mat. Note that the further the tow points are from the screed, the better the rideability of the finished mat. Hence, long wheelbase pavers have a greater 'smoothing' capability than single axle or short wheelbase machines.

6.3.2.   How asphalt flows through a paver
Material is discharged from the delivery lorry, in a controlled tipping action, whilst the paver moves forward. Material is drawn along the

conveyor (or flight bed) through adjustable flow control gates at the rear of the hopper to be deposited at the augers which, in turn, distribute the mixture to the full width of the screed. The mixed material is limited to the width of the screed by side gates attached to the screed. It is very important that the height of the asphalt (usually described as the 'head of material' and discussed in more detail later), which is in front of the screed, is maintained at a constant level. Failure to do so may well cause poor rideability characteristics in the finished mat. The paving width, thickness and speed of travel will dictate the settings of the individual parts of the flow system.

A simple volumetric calculation links forward speed to targeted tonnages per shift or hour. Since material deliveries form an integral part of the paving operation, it is important that the supplier is fully informed of hourly needs.

6.3.3. Action of the screed

The paving machine as a unit is basically a mechanism for accepting and delivering material to the screed. It is important, therefore, to understand the basics of how the screed functions and this is illustrated in Fig. 6.4.

When paving, the screed is pulled by the tractor unit via the tow point and is thus free to float upon the material distributed in its path by the auger system. This mechanism explains why the screed fills hollows and rides over high spots in the underlying layer. The screed will 'float' upwards or downwards until the forces acting on the screed plate are in equilibrium.

Initial mat thickness is determined by the angle of attack, i.e. the angle of the screed plate in relation to the layer which is being covered. If the angle of attack remains unaltered, the paver will continue to lay the initial thickness of the material. Alterations to mat thickness are effected by changing the height of the tow point, which causes the screed to float upwards or downwards until equilibrium is reached.

A useful rule of thumb is that a ratio of 8:1 applies between vertical changes at the tow point and alterations to the mat thickness. For example,

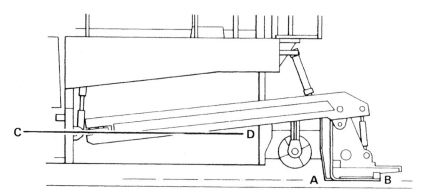

Fig. 6.4. Action of a floating screed[4]

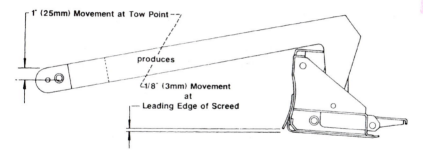

Fig. 6.5. Effect of changing the height at the tow point[4]

a 25 mm change in the level of the tow point will result in a 3 mm change in mat depth, as illustrated in Fig. 6.5. However, the change is not instantaneous and the paver will move forward a distance of approximately five times the length of the side arms. This is termed the 'natural paving length' and underlines the point made previously that the longer the paver the better the rideability of the mat. This value, like the depth ratio above, will vary for each brand and model of paver. A good foreman and operator are completely familiar with the characteristics of their own machines.

As the paver rises and falls over bumps and hollows in the layer being overlaid, the corresponding changes to the tow point height will alter the angle of attack and hence increase or decrease the mat thickness, and this is also illustrated in Fig. 6.5. However, if the frequency of irregularities in the underlying layer is less than the natural paving length of the machine, the screed will not fully reflect such discontinuities. It is the case, however, that irregularities are significantly reduced with each successive layer.

6.3.4. The screed

A full size paver has a screed around 2.5 m wide. Smaller pavers are available for surfacing footways and similar smaller areas and some are capable of laying at a width of 1 m.

Screeds can be 'fixed width' or 'hydraulically extending'. The vast majority of modern pavers are equipped with hydraulically extending screeds and, indeed, many manufacturers no longer offer pavers with fixed-width screeds. On a paver equipped with a fixed screed, changes in width are accomplished by bolting on extension boxes which increase the basic 2.5 m width to the required value. Minor reductions are effected by slotting 'cut off shoes' onto the inside of the side gates of the screed.

Hydraulically extending screeds were frequently unreliable when they were first introduced. However, they have evolved substantially over recent years and they now match the high performance of fixed screeds whilst offering significant savings in time and, consequently, cost. Indeed, orders for new pavers with fixed screeds are now comparatively rare.

Some pavers, whether incorporating fixed or extending screeds, make use of the mechanisms shown in Figs 6.6, 6.7 and 6.8 to impart compaction. Each of these systems assists the flow of material under the screed during paving operations and provides differing levels of initial compaction.

Tamping screeds
The tamping system consists of, effectively, a blade edge oscillating in a vertical or slightly inclined plane ahead of the screed plate. The tamping action 'tucks' the material under the leading edge of the screed resulting in a degree of initial compaction during forward paving which, when properly set, will result in a mat which is uniform and free of any tearing. The absence of any tamping or vibrating action would produce a mat torn and full of blemishes and holes as a result of the screed being towed over

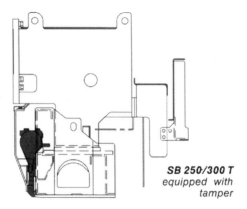

SB 250/300 T
equipped with
tamper

Fig. 6.6.   *Screed equipped with tamper. Diagram courtesy of Vögele*

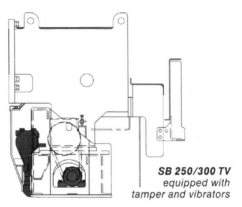

SB 250/300 TV
equipped with
tamper and vibrators

Fig. 6.7.   *Screed equipped with tamper and vibrators. Diagram courtesy of Vögele*

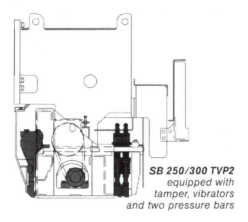

SB 250/300 TVP2
equipped with
tamper, vibrators
and two pressure bars

Fig. 6.8. Screed equipped with tamper, vibrators and two pressures bars.
Diagram courtesy of Vögele

the asphalt. The tamping speed is varied by the hydraulic drive, and settings for differing materials and paving speeds are a matter of experience. The height of the tamper blade must be set correctly and the general rule for telescopic screeds is 2 mm above the soleplate of the screed. The leading edge of the screed is angled which requires settings to be checked by means of a straight edge across the soleplate. The screed should be heated to paving temperature before the tampers are switched on since cold bitumen may have built up, jamming the tamper system. The tampers should be checked regularly for wear, and renewed as required as part of the maintenance programme for the paver.

Tamping/vibrating screeds
Vibration systems are added to the screed and these increase the density of the laid mat. They also provide a better surface texture/finish with some types of asphalt. It is common, for example, to make use of a combination screed when laying Marshall asphalt on airfield runways and taxiways where mat density and texture are of paramount importance.

Vibration is transmitted through the screed frame to the material via the surface of the screed soleplate. It is produced by an eccentrically weighted rotating shaft or similar system, depending on paver manufacturer. The amount of vibration is determined by two controllable factors, frequency and amplitude. Frequency is a function of the speed of the shaft; amplitude is varied by altering the position of eccentric weights on the shaft.

As a general rule, it is advisable to maintain a ratio of 2:1 vibrator to tamper speed. Too much vibration may cause 'fatting up' of the mat (the bitumen rises to the surface of the layer).

High compaction screeds
Some manufacturers offer what are termed 'high compaction screeds'. These screeds have been developed to offer the following advantages

- decreased reliance on subsequent compaction equipment to achieve required densities
- possible elimination of some regulating courses, i.e. single-layer construction
- environmental factors have less effect on the cooling of mats, even for thin surfacings — the mat is so well compacted by the paver that rapid cooling in adverse conditions is much less likely.

These high compaction screeds achieve the above by a number of innovations such as doubling the tampers, increasing the amplitude of vibrators and the introduction of a system of 'pressure bars'. The pressure bar, in contrast to the tamper, is in constant contact with the material and pressed down (in equilibrium) by high-frequency hydraulic impulses. The impulse frequency is variable. The advance of the pressure bar between two impulses, which results from a combination of the frequency and paving speed, is extremely short. At a paving speed of 5 m/min and a frequency of 70 Hz, for example, the advance is around 0.2 mm. This short distance is covered in only 0.007 s. The combination of rapid acceleration and high-frequency impulses results in high values of force acting on the asphalt, hence the greater degree of compaction compared to that obtained with traditional pavers.

Heating of screeds
All screeds, regardless of type or configuration, must be heated prior to the commencement of laying. The asphalt which is being laid will stick to a cold or unevenly heated screed plate causing dragging and leaving blemishes in the mat. Screed heating may be provided by a number of different sources, i.e. diesel, LPG or electricity. Most manufacturers offer a choice to the purchaser. Heating systems are designed so that the screed heats evenly along its entire length.

Traditionally, heating systems were fuelled by gas or diesel. These had to be ignited by hand by the operator. Modern machines are equipped with automatic burner ignition for diesel or gas systems and thermostatic control in order to maintain a uniform soleplate temperature. The burner controls, normally housed in a console at the back of the tractor unit, may be operated manually or automatically, as in the paver shown in Fig. 6.9.

Commonly, fixed screeds can be heated to one of two temperatures, 90 °C or 110 °C but are available with full thermostatic control. Hydro-static screeds are heated to higher temperatures than fixed screeds with the fixed portion being hotter than the extending sections. Typically the screed will be heated to around 150 °C in the fixed section and around 125 °C on the extending sections. These temperatures can usually be adjusted as high as 200 °C for the fixed centre section of an extending screed with a corresponding increase for the extremities. The temperatures are controlled by thermostats which switch the heat source on/off to maintain the desired temperature.

The following points should be remembered.

Fig. 6.9. A modern paver showing control console. Photograph courtesy of Vögele

- On pavers where the heaters are manually controlled, it should not be necessary to leave the burners switched on as the hot material will maintain the temperature of the screed.
- The screed may buckle if it is overheated. This would result in deformities in the mat.
- Operators can input the required screed temperature on modern heating systems. Thermostats ensure that the desired temperature is maintained by switching the heating system on and off automatically.

Heating of the screed can take anything from 10 to 20 minutes before the paving operation can begin.

6.3.5. Automatic control of devices
The present generation pavers incorporate a number of automatic control devices to assist in reducing variations in the finished mat and in eliminating the need to constantly adjust the paver controls.

Head of material

It is important to understand how the screed acts during paving. If the volume of material at the augers varies significantly then this will cause the screed to change height. The screed is forced upwards as the volume of material in the 'auger box' increases and the opposite effect occurs as this volume falls. It is therefore vital that the head of material is maintained at the correct level. This principle is illustrated in Fig. 6.10. The auger speeds, left and right, and the conveyor speeds, left and right, can each be set independently of each other as shown in Fig. 6.11.

Traditionally, operatives had to control the conveyors, augers and flow control gates in order to ensure that a constant and appropriate head of material was maintained in the auger box. Sensor systems are now available and these are fitted in strategic locations in the flow path of the asphalt. They automatically control the system and ensure that the correct head of material is maintained. These systems may be based on mechanical paddles, electronic switches or ultrasound sensors which operate on the same principles as marine depth sonar. Figure 6.12 shows a system which uses paddles at the ends of the conveyors to control the flow of material into the auger box. Figure 6.13 shows a system of sonic

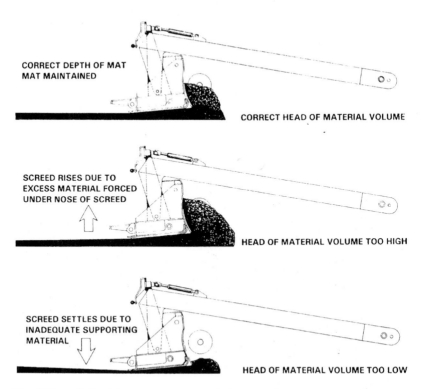

Fig. 6.10.   *Maintaining a constant head of material in the auger box*[4]

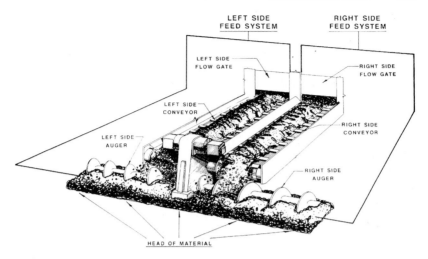

Fig 6.11.   Controlling the head of material in the auger box[4]

control. The advantage of this system is the elimination of moving parts which may be damaged or jammed.

A machine which is paving at a constant speed with the material being fed at the optimum rate will produce a mat with far superior rideability than one which is constantly stopping and starting due to an incorrectly controlled feed system.

Auger height
Pavers are capable of laying mats at thicknesses which vary from a few millimetres to approaching 400 mm. In some pavers, the height of the

Fig. 6.12.   Control of conveyors by mechanical paddles. Photograph courtesy of Vögele

Fig. 6.13.   Control of conveyors by sonic sensors. Photograph courtesy of Vögele

augers can be adjusted to accord with the thickness of material which is being laid. Auger height is an important element in achieving the correct head of material.

Automatic level control
As previously described, changes to the layer thickness can be effected by altering the height of the tow points. The tow points, located at each side of the tractor unit, are controlled by hydraulic rams which are varied by switches, one per side. These are normally fitted to the rear of the screed, again one per side. They are often duplicated on the control console. These switches are located such that the screwman can effect alterations easily.

On older pavers, changes to the mat thickness or finished height were made via a mechanical linkage from the screed to a screw thread at the tow point. Operatives would physically turn a wheel or handle and screw the tow point up or down. Hence the term 'screwman', still in use today, which refers to the operative who has the responsibility of ensuring that material is laid at the correct thickness.

Despite the fact that automatic level control devices are becoming very sophisticated, the basic concept remains quite simple. They maintain the tow points at a constant level in relation to a datum. A range of level control accessories are available with all pavers and they all operate on the basis of this simple principle.

The datum may be one of several forms including kerb, adjacent mat and tensioned reference wire. Different attachments are available to suit each form. Automatic level control devices include skis (or shoes),

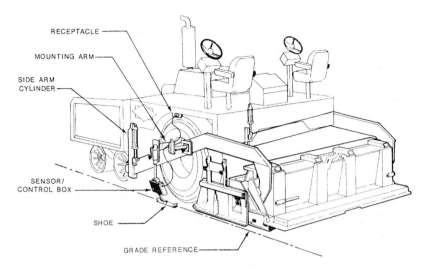

Fig. 6.14.   Paver using a ski (or shoe) as an automatic level control device[4]

levelling beams and lasers. Figures 6.14–6.17 illustrate different datum types and a variety of automatic level control devices.

Slope control
Most paver manufacturers offer accessory equipment to allow the operator to input the desired slope of the mat into the levelling system. This is useful on thicker, lower layers where a horizontal datum to one side of the paver can be supplied by kerbs or wire and the transverse

Fig. 6.15.   Paver using a ski (or shoe) as an automatic level control device and a footway as a reference. Photograph courtesy of Vögele

Fig. 6.16.   Paver using a bow and a reference wire to provide automatic level control. Photograph courtesy of Vögele

gradient of the new layer dictated by the slope control equipment. Figure 6.17 depicts such a system. The desired slope value and the actual value determined by the sensors are constantly compared through the control unit and any deviation results in an adjusting signal being transmitted to the appropriate tow point rams.

Averaging beams
It is possible to provide the paver with a mobile reference datum levelling system. Kerbs can become displaced and wires can snap. In the surfacing layers, especially in wearing courses, it is not unusual to make use of an averaging beam. These beams, towed by the tractor unit itself alongside the paver, provide each sensor pick-up arm with a level datum. They are particularly useful for providing good rideability in the finished surface since a 5 m to 8 m beam will straddle any shorter wave imperfections in the underlying basecourse without the screed reacting to them. Two types are shown diagramatically in Figs 6.18 and 6.19. Photographs of such a system in action are shown in Figs 6.20 and 6.21.

6.3.6.   Integral sprayer
Perhaps the most significant recent change to paver manufacture has been the emergence of pavers which are equipped with a spraying mechanism. These 'integral pavers' spray the surface immediately in front of the mat.

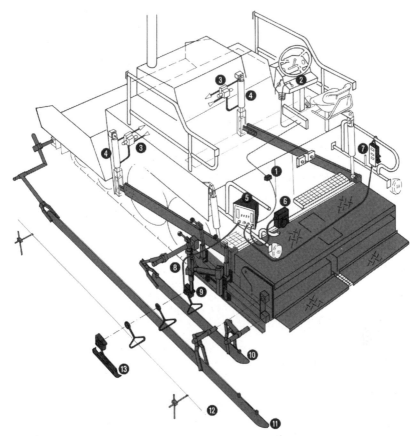

Fig. 6.17.   Paver using a bow and a levelling beam to provide automatic slope control. Photograph courtesy of Vögele

Emulsion is sprayed at a rate which is much more uniform than that achieved manually. If emulsions are sprayed manually, the results in terms of both rate and uniformity are often poor. Integral pavers are used not only for laying the traditional range of asphalts but also particularly for thin surfacings. These proprietary materials, discussed in some detail

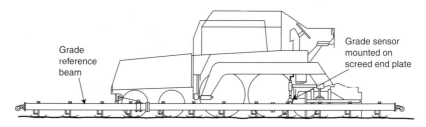

Fig. 6.18.   Paver equipped with a grade reference beam[4]

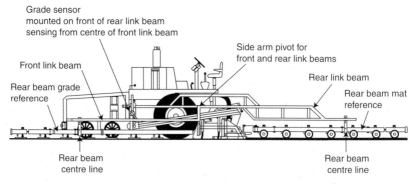

Fig. 6.19. Paver equipped with a mobile mat reference system[4]

Fig. 6.20. Laying asphalt using a mobile mat reference system. Photograph courtesy of Colas Ltd

in Sections 3.7.5, 4.13.6 and 10.7 of Chapters 3, 4 and 10 respectively, often require a tack coat which is evenly sprayed over the basecourse. The tack coat may be the normal cationic K1-40 or a polymer modified bitumen depending on the nature of the thin surfacing. However, integral sprayers suffer from problems of reliability. In 1999, these machines accounted for around 3% of total sales. This is likely to increase as such pavers become more reliable and the market for thin surfacings increases as expected. An integral paver is shown in Fig. 6.22.

Fig. 6.21. Laying asphalt using a mobile mat reference system. Photograph courtesy of Colas Ltd

Fig. 6.22. Paver with integral sprayer. Photograph courtesy of Vögele

Fig. 6.23. Blaw-Knox MC-30 asphalt feed system. Photograph courtesy of Blaw-Knox/Ingersoll-Rand

6.4. Paver delivery systems

In the USA, in an average day, the equivalent of one lane of 50 mm thick asphalt from Los Angeles to New York is laid. When a contract merits or requires high outputs, additional equipment is needed to deliver the mixed material to the hopper of the paver. In Europe and the USA, it is not uncommon for pavers to operate at large screed widths making continuous laying impossible if the paver is fed using traditional asphalt lorries. Special equipment is required to feed the paver and provide a surge store to ensure continuity of supply to the paver. An example of such equipment is the Blaw-Knox MC-30 (shown in Fig. 6.23) which is a self-propelled, wheel mounted, bulk material handling/delivery system with a surge store capacity of up to 45 t.

There are several key advantages to the use of such equipment if contract circumstances permit

- the use of larger lorries
- reduced times for delivery cycle, thus requiring fewer lorries
- laying capacity increased in excess of 50%
- continuous paving produces a smoother and more uniform surface
- systems can be fitted with a swing conveyor allowing adjacent deliveries.

The last point is illustrated in Fig. 6.24.

The demands of lane rental contracts have generated a need for high output and accurate paving to be carried out to central reserves, shoulder extensions and verge areas. Machines that permit side discharge of asphalt have been available for hire in the UK for some time. These machines are capable of laying asphalts (and uncoated materials) to varying widths and depths. The machine consists of a self-propelled unit with a hopper, sideways conveyor and side mounted screed/strike off plates. Optional equipment includes belly-mounted brushes and hydraulically extending screeds. An example is shown in Fig. 6.25.

Fig. 6.24. Use of a swing conveyor on a Blaw-Knox MC-30. Photograph courtesy of Blaw-Knox/Ingersoll-Rand

Fig. 6.25. Blaw-Knox RW-100A shoulder paving machine. Photograph courtesy of Blaw-Knox/Ingersoll-Rand

6.5. Compaction plant

Compaction of the newly laid asphalt mat is extremely important. The life of a pavement is a function of the degree of compaction achieved during construction. Poor compaction may lead to compliance problems resulting in the client rejecting the material. Indeed, some specifications used in other parts of the world reduce the amount paid against a sliding scale by comparing the densities which are achieved against the required standard. It is important to realize that the skill of the roller operators and the suitability of the compaction plant will have a significant effect on the completed layer. Roller numbers, types, working patterns and the mode of operation must be taken into account when planning surfacing operations. Material specifications will include minimum rolling temperatures with the delivery temperature parameters. Laying issues are considered in detail in Chapter 7 whilst compaction features in Chapter 8. The latter deals with the choice of roller for particular tasks. This Section considers rollers under the headings of dead weight, pneumatic tyred rollers and vibrating rollers.

6.5.1. Dead weight tandem and three-point rollers

Up to the early 1990s, most asphalts were compacted using three point smooth wheeled dead weight rollers. The term 'three-point' relates to the fact that these rollers run on three barrels, one at the front and two at the rear. Modern three-point smooth wheeled rollers have evolved to their current stage of development over many years. They are now fitted with hydrostatic drive, affording them the capability of infinitely variable speeds, such variation being effected evenly to avoid damaging the material being compacted. They incorporate water tanks and, via spray bars, water is fed to the barrels in order to prevent them sticking to the hot asphalt. Drive is through the rear barrels and steering is via the front barrel. They are singularly reliable pieces of plant and it is not unusual to find examples in use which are over twenty years old.

Dead weight rollers can be fitted with additional equipment including scarifiers and cutting wheels. This wheel, with its associated plough and a guide bar, is used to cut longitudinal joints. In order to seal the joint, this face has to be painted with bitumen before the adjacent mat is placed. Many experienced operators prefer this type of roller for cutting joints. An example of a modern dead weight roller is shown in Fig. 6.26.

Fig. 6.26. A modern dead weight three-point smooth wheeled roller. Photograph courtesy of Aveling Barford Ltd

The barrels of dead weight rollers are capable of being ballasted with water or sand and, depending on the model and the amount of ballast added, can operate at a total dead weight of between 9.5 t and 15 t.

It is important to ensure that there is a source of water available on site when using these rollers. Dry barrels leave unacceptable blemishes on hot asphalts.

6.5.2.   Pneumatic tyred rollers

Acting as dead weight rollers and capable (via ballasting) of gross weights from 8 t to 35 t depending on model and size, these rollers are particularly useful on asphaltic concrete where the strength of the compacted mixture is derived from the mechanical interlock of the aggregate.

The effectiveness of the compactive effort is dependent on the weight of the roller and the pressure of its tyres. In some models, such as that shown in Fig. 6.27, the pressure can be varied and effected whilst the roller is in operation. The kneading and flexing action imparted by the tyres is effective in sealing the surface of the mat and can be used to deal with cracks left by other rollers. It is important to ensure that the tyres gain heat as quickly as possible in order to minimize pick-up and the necessity to use lubricants on the tyre. It is not unusual to see pneumatic tyred rollers (PTRs) fitted with canvas skirts which assist in maintaining a heat build up around the wheels and tyres. The rolling effect of a PTR is

Fig. 6.27.   Bomag BW 16 R pneumatic tyred roller. Photograph courtesy of Bomag

also particularly useful in closing up the surface of a freshly laid mat but retaining good texture. Modern PTRs have hydrostatic drives and can, therefore, effect a smooth change of speed. Consequently, they are frequently used on surface dressing contracts to embed the chippings without crushing the aggregate. Ancillary equipment includes a hydraulic side impact wheel to achieve high levels of mat density at, and close to, the longitudinal joint.

6.5.3. Vibrating rollers

Vibrating rollers compact asphalts through a combination of static and dynamic loading. The range of available vibrating rollers is extensive — from small single drum pedestrian machines to large tandem rollers with both drums capable of imparting compactive effort. The increasing requirement for long life asphalt pavements has dictated improved compaction in order to produce higher densities in laid mixtures. Vibrating rollers have become, in many circumstances, a necessity in order to meet the demands of current specifications and materials.

The modern tandem vibrating roller is a sophisticated machine with the capability to vary both the frequency and amplitude. Such changes are necessary to provide the required level of compactive effort necessary for a wide range of asphalts which are laid at different depths.

Vibrating rollers have water sprinklers like dead weight rollers. Some models allow the option of pumped or gravity fed sprinkling. A vibration

Fig. 6.28. Dynapac CC 222 medium tandem vibratory roller. Photograph courtesy of Dynapac

Fig. 6.29.   Bitelli DTV 100 large tandem vibratory roller. Photograph courtesy of Bitelli

cut off switch is linked to the forward and reverse mechanism. Optional extras usually include side compactors and cutting wheels.

The relatively high cost of the larger versions of these items of plant is countered by the fact that they are, in fact, multi-functional rollers. On most machines, vibration can be switched off completely and they can be used as tandem dead weight rollers for finishing off (polishing) the compacted asphalt.

Some manufacturers offer a range of combined rollers, having a vibrating steel drum on one axle and a pneumatic tyred drum on the other axle. Figures 6.28 and 6.29 show medium- and high-capacity tandem vibratory rollers respectively.

There are some excellent compaction guides available from the roller manufacturers and the data sheets for each roller contain detailed information regarding typical speeds, compactive effort and rolling capacities for each model and variant.

6.6.   Ancillary equipment

6.6.1.   Milling machines

The removal or reduction in level of existing pavements often forms part of a resurfacing contract and the current generation of milling machines have evolved to suit the majority of circumstances encountered in executing such works. Milling machines (previously called 'planers') are available with cutting widths from 0.3 m to 2.5 m with the larger machines capable of cutting to a depth of 330 mm in one pass.

Fig. 6.30.   Bitelli SF 60 medium-capacity milling machine. Photograph courtesy of Bitelli

Milled material is loaded onto lorries via a front- or rear-mounted swing conveyor and full automatic level control is available similar to the systems used by asphalt pavers. On a 2.5 m milling machine the cutting operation is carried out by a drum, having a diameter over 1 m, fitted with over 150 renewable tungsten carbide teeth (or picks) rotating at over 5 m/s. The majority of these machines cannot be driven between sites and are transported by low loaders or on purpose built multi-functional lorries or even trailers (larger machines weigh in excess of 40 t).

The output from these machines is frequently dictated by the number of lorries available to transport the milled material (or 'planings') to licensed tips. As in asphalt paving, progress can be impeded by unsuitable or inadequately planned works. Medium-capacity and heavy-capacity milling machines are shown in Figs 6.30 and 6.31 respectively.

6.6.2.   Chipping machines

Chipped HRA wearing course has been the traditional wearing course in the UK for many years. However, except in Scotland, it is rapidly being replaced by proprietary thin surfacings. This is largely a result of the standard UK design guide[5] specifying thin surfacings as the preferred wearing course. (The section of the guide[5] relating to Scotland continues to require chipped HRA and that relating to Northern Ireland permits either chipped HRA or thin surfacings as the wearing course.) Many Local Authorities use this design guide[5] for their own roads. Nevertheless, substantial areas of chipped HRA wearing course continue to be laid and, except for small areas and special locations, a chipping machine is necessary to place the precoated chippings on newly placed asphalt.

Fig. 6.31. Wirtgen 2100 DC heavy-capacity milling machine. Photograph courtesy of Wirtgen

Originally designed in conjunction with the Road Research Laboratory (now TRL), the basic concept of the machine has remained unaltered although improvements in design have taken place. A typical chipping machine is shown in Fig. 6.32 and their operation is covered in detail in Section 7.9 of Chapter 7.

The chipping machine should not dictate the speed of the paving train but may do so if the matter of supply is not properly addressed. If storage areas, for the chippings are far apart or inaccessible then it may be possible to use a 'hoist and grab' to feed the chipping machine. This does so by running alongside the chipper. This mobile 10 t to 20 t storage facility will avoid

Fig. 6.32. Bristowes Mark V 3.7 m wide chipping machine. Photograph courtesy of Bristowes Ltd

disruption to the paving train. A 20 t load will, at typical rates of spread, chip well over 400 m length of a lane of standard width.

Since chipping machines move particularly slowly, they are transported on their own purpose-built trailer not only between sites but also within jobs to minimize the delay before starting the next area of surfacing.

6.6.3. Hot boxes

When asphalts are laid in small quantities, e.g. in potholes, they often require very little material. The difficulty in such situations is ensuring that the asphalt retains enough heat to facilitate adequate compaction. Hot boxes are simply heated storage boxes mounted on a standard lorry chassis. Heat is provided by LPG. Their capacity can be 10 t or more but they need not be filled to be effective and are suitable for carrying as little as 0.5 t of asphalt.

6.6.4. Infrared heaters

These machines have a downward-facing, horizontally mounted element fixed close to the road surface. The heat is provided by LPG. Although they can be used to heat a frozen or icy surface prior to being overlaid, their main use is to heat areas of newly laid chipped HRA wearing course where the chippings have not properly embedded. Once heated, the precoated chippings can be further embedded. Usually further chippings are added. Although the depth which is heated is limited, it can make the difference between adequate embedment and chipping loss immediately after the surface is opened to traffic. It is infinitely preferable, and much less expensive, to ensure that chippings are properly embedded at the time of laying. The emergence of thin surfacings which are, of course, unchipped will probably consign these machines to the history books.

6.7. Summary

This Chapter has examined pieces of plant used for laying asphalts. Lorries which, nowadays, are sophisticated pieces of equipment have been considered. Their contribution to the success of surfacing operations is often neglected.

Aesthetically, pavers are unusual machines which are capable of laying asphalts to the close tolerances required by modern specifications. The elements of pavers from the rudimentary tractor/screed approach to a description of the myriad of automatic control devices have been described. The key elements of successful paver operation have been discussed. The emergence and use of integral sprayers was noted. It has been shown how paver delivery systems, common in the USA but rare in the UK, can be used to increase productivity. Different types of rollers are available to compact asphalts. These have also been described and their uses reported. Finally, the remaining specialist pieces of asphalt plant — milling machines, chipping machines, hot boxes and infrared heaters — were discussed.

Over the last few years, innovation within the asphalt industry has been seen as the province of those associated with thin surfacings and other proprietary materials. There is no doubt that these mixtures warrant respect and recognition. However, there have also been major strides in other areas of asphalt technology. Striking improvements in plant have taken place giving client and contractor confidence that rigourous specifications can be met.

References

1. BRITISH STANDARDS INSTITUTION. *Hot Rolled Asphalt for Roads and other Paved Areas, Specification for the Transport, Laying and Compaction of Rolled Asphalt*. BSI, London, 1992, BS 594: Part 2.
2. BRITISH STANDARDS INSTITUTION. *Coated Macadam for Roads and other Paved Areas, Specification for Transport Laying and Compaction*. BSI, London, 1993, BS 4987: Part 2.
3. FREIGHT TRANSPORT ASSOCIATION. *1999 Yearbook of Road Transport Law*. FTA, Tunbridge Wells, 1999.
4. BLAW-KNOX CONSTRUCTION EQUIPMENT CO LTD. *Paving Manual*. Blaw-Knox Construction Equipment Co Ltd, Rochester, December 1986.
5. HIGHWAYS AGENCY *et al. Design Manual for Roads and Bridges, Pavement Design and Maintenance, Surfacing Materials for New and Maintenance Construction*. TSO, London, 1999, 7.5.1, HD 36/99.

7. Good surfacing practice

7.1. Preamble

Good surfacing practice is a subject about which little has been written. In order to perform properly, asphalts must be laid in accordance with the appropriate specification. Chapter 8 of this Book explains the importance of compaction and its effect on the lifespan of a surfacing material. This Chapter provides advice on good practice associated with the laying of asphalts, including consideration of the estimate and an economic evaluation of the laying process. It does so from a practical standpoint, i.e. from the perspective of the surfacing contractor.

The most important consideration on any building or construction site is safety. Section 7.2 considers this issue whilst the rest of the Chapter considers estimating, planning and programming, pre-start inspection, ordering, transportation and laying (including a section which specifically addresses the laying of thin surfacings including SMAs and repairs thereto). The application of chippings to an HRA wearing course is considered along with compaction, quality controls, requirements of the *Specification for Highway Works*,[1] economic considerations and remedial works.

7.2. Safety

Construction sites, especially congested roadworks, are particularly dangerous places to work. It is vitally important that everyone concerned is aware of his personal responsibility under the numerous regulations which govern activities on construction sites. This affects his own safety and that of his colleagues.

7.2.1. Site safety policy

An initial assessment must be carried out before any rules can be devised for a specific site. The aim of this assessment is to identify specific hazards and possible environmental problems. Table 7.1 demonstrates a typical site assessment form.

Completion of this form permits site safety rules to be written and these should address the following issues

- Section 7 of the Health and Safety at Work, etc. Act 1974 (HASWA), which places strict legal requirements on all construction staff

Table 7.1. Example of a site assessment form

SUPERVISOR'S SITE ASSESSMENT

1st copy=book, 2nd copy=Foreman, 3rd copy=Contracts Manager

| | Supervisor | Date of Inspection | Date of Contract |
| | Name of Contract | | Contract No |

WORKING ENVIRONMENT ON THIS SITE

Potential Hazard HEALTH, SAFETY & ENVIRONMENTAL	Risk H	Risk M	Risk L	Hazard Rating H	Hazard Rating M	Hazard Rating L	Comments	Protective Measure/Control	Further Action Proposed
Live traffic									
Site traffic									
Pedestrian traffic									
Road widths									
Gradients									
Approach visibility									
Junctions									
Schools - Children									
Bridges									
Excavations									
Underground services									
Overhead electricity									
Lifting									
Hazardous substances									
Site access & egress									
Delivery Control									
Vehicle parking area									
Sub-Contractors									
Fuel & Oil									
Gas containers									
Bitumens									
Plant & Equipment									
Waste(storage, disposal)									
Lighting									
Noise									
Vibration									
Water									
Fire Risk									
Community relations									

- minimum standards of personal protective equipment (PPE), which must be worn on site, i.e. boots, hard hats, reflective jackets/vests, all to the appropriate standard[†]
- access and egress points throughout the site
- access through the site including emergency lanes, location for material storage, marshalling areas, vehicle parking and positions for standby plant
- site speed limits which must be fixed and actively enforced
- procedures for reversing, reversing must be limited and controlled by a banksman who should be easily identifiable
- locations of buried and overhead utilities and other hazards, all of which must be clearly signed
- the danger posed by and to children and pets on site, usually arriving in delivery vehicles
- accident and emergency procedures including the location of the first aid personnel and equipment
- traffic management schemes and responsibilities
- use of amber beacons and relevant signs on vehicles
- training requirements for the site
- arrangements for welfare facilities per The Construction (Health, Safety and Welfare) Regulations 1996.

7.2.2. Site safety officer

Although every person has a responsibility for health and safety, each site should have a nominated site safety officer. In reality, it is usually the supervisor or the gang foreman who undertakes this role. It is vital that the nominated person has appropriate training in the use of Chapter 8 of the Traffic Signs Manual[4] (usually described simply as 'Chapter 8'). It is this document which prescribes the layout of signs, barriers, etc. on major roadworks. Traffic management and organization are often major factors in controlling site safety.

7.2.3. Control and monitoring

The Management of Health and Safety at Work Regulations 1992 and The Construction (Design and Management) Regulations 1994 require employers to introduce and operate procedures related to health and safety. These include

- risk assessments and those resulting from application of The Control of Substances Hazardous to Health Regulations 1988 (COSHH) carried out and communicated to operatives

---

[†] In the standard UK specification, the *Specification for Highway Works*[1] (SHW) (discussed in detail later in Section 7.13), this is described as 'high visibility warning clothing complying with BS EN 471'.[2] 'Clothing shall be to Table 1, Class 2 or 3 (Class 3 on motorways or other high speed roads)'. A 'high speed road' is defined in Clause NG 921 in the Notes for Guidance on the Specification for Highway Works[3] (NGSHW) as a road where the 85th percentile speed of traffic ≥90 km/h. Further requirements for clothing are given in SHW[1] and these should be studied in detail.

- safe systems of work — permits to work for high risk activities, and
- policies and procedures to ensure compliance with current legislation relating to work activities.

There must be a monitoring system in place in order to close the loop on the safety management system. It is essential to monitor the site layout and activities and this must be carried out by site management. Regular and detailed monitoring must be carried out by the organization's health and safety advisers to ensure that policies and procedures are being properly implemented at all times.

7.2.4. *General site safety*
New employees should undergo a general induction. Thereafter, all persons involved in site work must attend a site induction dealing with issues such as

- the site manager's rules
- emergency procedures and
- site-specific hazards.

It is the responsibility of all persons on site to ensure that the site is safe and tidy at the end of each shift. Mobile plant must be parked correctly, cabs must be locked, handbrakes applied and plant immobilized wherever possible. Hand tools must be locked away. Materials and substances, especially those displaying a COSHH warning symbol, must be stored safely in a secure location. HASWA imposes a duty of care on organizations towards not only employees and others who have cause to be on site but also to trespassers.

Liquid petroleum gas (LPG) poses a potential risk on site and must be stored and handled with extreme care. An appropriate fire extinguisher must be kept near appliances that use LPG. Operatives must be trained in tackling LPG fires. All LPG cylinders, whether empty or full, should be stored in a secure cage.

Nuclear Density Gauges (NDGs) create potential hazards on sites. Each meter should have its own secure transit box which is properly marked. Its use must be governed by a local procedure which will include the use of cones or a light barrier showing the trefoil symbol (the standard radioactivity symbol) and the wording *Unauthorised personnel keep at least two metres away*. In addition, it is essential that all site personnel are trained in contingency plans which operate in the event of an incident involving an NDM.

All operatives should hold Certificates of Competence for equipment which they are expected to operate, e.g. abrasive wheels, and for the additional activities which they undertake on site such as the establishment of traffic management measures, reinstatement works etc. These Certificates can be awarded by many organizations. Specific training requirements can be identified through the organization's safety functions and initial site assessments. It is strongly recommended that training is refreshed every three years.

Using jack hammers, floor saws, etc. for cutting joints must only be undertaken by staff who wear the appropriate PPE including goggles, ear muffs, gloves and nasal masks. Consideration should also be given to the effects of noise, vibration and dust suppression or extraction and the benefits of job rotation.

The positions of overhead cables and buried services must be clearly marked with warning signs, barriers and 'goal posts'. Where appropriate, voltages and clearance heights should be displayed. Tipping of delivery vehicles must be controlled by a trained banksman and must not be permitted near overhead high-voltage cables. Banksmen should make themselves known to drivers and should be easily distinguishable from other members of the site staff. Access and egress points, holding areas and locations for cleaning out vehicles which have discharged their loads should be clearly signed with special attention being given to reversing lorries. Most serious accidents, some of which are fatal, occur whilst heavy plant and delivery vehicles are reversing. Audible reversing warning systems should be fitted to all large items of plant and all delivery vehicles.

The Construction (Health, Safety and Welfare) Regulations 1996 require the provision of welfare facilities. These Regulations specify clearly what should be provided so far as is reasonably practical. What is provided should be suitable and sufficient and includes

- sanitary facilities
- washing facilities including running hot water
- a supply of drinking water
- accommodation for clothes used whilst working (including PPE) and those which are not used whilst working
- facilities for changing and
- facilities for rest and taking meals.

The HSE Construction Information Sheet No. 46[5] is essential reading for all those who are responsible for the provision of welfare facilities on transient sites.

Alcohol, children, animals and horseplay should be prohibited on all construction sites. Speed limits must be set and strictly enforced. The provision of lateral safety zones is required by Chapter 8[4] and their function is to separate the work and traffic passing through the site. These must be clearly marked, maintained free of equipment and plant, and no person should work therein.

7.2.5. Precautions for paver work
The following points, as appropriate, should be observed by anyone involved in paver work.

- The paver operator should always walk around the paver and check for obstructions and equipment in the hopper before mounting. Deck plates must be kept free of oil, grease, obstructions, loose equipment or other hazards. Manufacturer's safety guards must always be fitted correctly.

The paver operator must ensure that all controls are in neutral before starting the engine and he should always look around and signal before moving off. Diesel should not be used when cleaning the paver, consideration being given to alternatives which are more environmentally acceptable.

- Personnel must not walk between the paver and a reversing lorry or enter the hopper of the paver while the engine is running.
- All personnel must wear protective clothing, know the locations of fire extinguishers, and be aware of site and live traffic at all times.
- Hydraulically extending screeds should not be operated without checking for personnel or obstructions.
- Machines must be braked properly when parked or during servicing and maintenance.
- Appropriate warning measures must be employed at all times.
- Servicing and maintenance must not be carried out while the engine is running and the batteries must be disconnected when the paver is undergoing electrical servicing. All hydraulic or pneumatic systems should be de-pressurized and hopper and screed lowered before servicing is undertaken.
- During transportation on low loaders, pavers must be adequately secured. They should not carry more than the number of LPG cylinders recommended by the manufacturer of the paver. Consideration should be given to the size of LPG cylinders used since smaller cylinders are much more easily handled.

7.2.6. *Environmental issues*
There is an increasing demand for organizations to address environmental issues and on surfacing sites there are matters which may benefit from such considerations.

- The cleaning of equipment especially the paver and sprayers used with tack coat. Is diesel used and what is done with the residue?
- How is refuelling undertaken on site, manually or pumped?
- Is the method of materials storage, including LPG, liable to cause pollution?
- Do emergency procedures exist to deal with an incident?
- Can atmospheric emissions be reduced? (All engines should be switched off when not in use).
- Techniques which are technically and economically feasible to minimize or eliminate nuisance such as noise, vibration and dust should be employed. Equipment designed on the basis of Best Available Techniques Not Entailing Excessive Cost (BATNEEC) may reduce environmental damage.

7.3. Estimating for work
Tendering for works involving road construction or, more likely, road maintenance is often based on a two envelope system. One envelope contains a quality submission whilst the other is the tender submission.

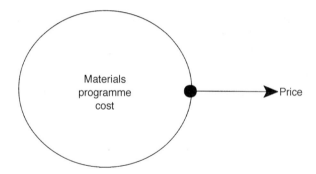

Fig. 7.1.   The estimating process

The quality submission must reach a predetermined standard before the tender submission, i.e. the price, is considered. Most surfacing firms sub-contract to the main contractor who often bases his selection purely on price considerations. This price must cover the costs associated with the production of a pavement to the specified standard. In addition, it must produce a return on capital employed, i.e. a profit. Contracts essentially consist of work which is placed by machine and/or by hand. An examination of the estimating process highlights many of the practices an efficient surfacing contractor will adopt if these aims are to be met. The estimating process is depicted in Fig. 7.1.

7.3.1.   Materials
The estimator extracts quantities and produces a schedule of materials. This serves two purposes. The first gives an initial view of the size of the contract and an experienced estimator will have an average cost per tonne laid for various materials. The second purpose is to allow the estimator to obtain quotes for materials from several suppliers. In lane rental contracts, the time constraints may be such that even if the surfacing contractor owns a coating plant, its output may not be sufficient to supply the contract. This is an important point since there may be several different sources of supply, requiring an average material price to be used in the estimate. If a problem occurs during the construction phase which results in delay and disruption to the surfacing works then this may well result in the original proportions of supply from the plants being radically different from those envisaged at the time of tender. In such circumstances, the surfacing contractor may seek additional payment.[6]

7.3.2.   Programme
Even when pricing surfacing works as a sub-contractor, the estimator must be acquainted with the programme for the works. He will have to allow for continuous working on site or for periodic possessions over a lengthy period to match the main contractor's programme. Such considerations can make a considerable difference to the surfacing contractor's prices.

It is common practice for main contractors, having won a contract, to ask potential sub-contractors for a revised (i.e. cheaper) quotation in order to reduce the price of the sub-contracted works.

7.3.3.   Cost

Assuming the estimator is aware of how the works are to be completed, he can now identify resources and outputs.

Main components of the estimate
The important elements of any estimate are

- *labour* — number, skills and time
- *plant* — types, combinations and time
- *material* — specification, laid thicknesses
- *sundry expenses* — traffic safety and control, supervision, setting out and testing
- *overheads and profit.*

Labour
The constitution of surfacing gangs varies according to the type and size of the contract and the nature of the materials being placed. However, for estimating purposes, Table 7.2 illustrates the compositions which can be used in various combinations to cover most eventualities. In order to calculate an average daily gang cost, the number of men employed is divided by the anticipated average daily tonnage laid to produce a mean cost of labour per laid tonne.

Plant
The types and uses of the various kinds of available plant are covered in detail in Chapter 6 of this book. The items of plant which may be required for a machine laying contract are shown in Table 7.3. Ancillary plant, small tools and sundry material should also be included, where appropriate. Some of the more common items and their functions are given in Table 7.4.

On larger sites, it is often the case that the entire area of surfacing is not made available at one time. Surfacing contractors should, when acting as a sub-contractor, ask for a programme which shows when particular areas will be released so that they can make an allowance within their estimate for the particular costs associated with moving on and off site.

The daily cost of plant is divided by the average daily output in tonnes to produce a cost per tonne for plant. The cost of additional site visits may be included in this calculation or presented as a separate sum.

Materials
The material content of the estimate usually accounts for around 60% of the total cost and, on a major motorway contract, may account for as much as 90% of the total cost.

While the average daily tonnage output is extremely significant, an equally important figure in any surfacing estimate is the material superage

Table 7.2. *Composition and usage of surfacing gangs**

Gang	Operative skills	Plant	Usage	Remarks
4 man hand-laying gang	Foreman, raker, roller driver, general operative	Crew bus, roller	Footways with good access	Often forms chipping gang in 12 man machine laying gang
6 man hand-laying gang	Foreman, 2 rakers, roller driver, loader driver, general operative	Crew bus, roller, shovel loader	Footways with poor access or where larger outputs are available	Often forms chipping gang and string men in 14 man machine laying gang
6 man machine-laying gang	Foreman, paver driver, screwman, raker, roller driver, loader driver/general operative	Crew bus, paver, 2 rollers, shovel loader	Small roads, housing estate roads, roadbase on larger jobs where level control is supplied	Basic machine laying gang
8 man machine-laying gang	Foreman, paver driver, screwman, 2 roller drivers, raker, loader driver, general operative	2 crew buses, paver, 2 rollers, shovel loader	Roadbase on larger jobs where level control is not supplied or where second roller is required for high output	
10 man machine-laying gang	Foreman, paver driver, 2 screwmen, 3 roller drivers, raker, loader driver, general operative	2 crew buses, paver, 3 rollers, shovel loader	Basecourse on larger jobs and motorways	
12 man machine-laying gang	2 foremen, paver driver, 2 screwmen, 2 roller drivers, chipping machine driver, 2 chippers, loader driver, general operative	2 crew buses, paver, 2 rollers, chipping machine, shovel loader	Wearing course on resurfacing jobs and housing estates	
14 man machine-laying gang	2 foremen, paver driver, 2 screwmen, 3 roller drivers, chipping machine driver, 2 hand chippers, 2 loader drivers, general oprative	2 crew buses, paver, 3 rollers, chipping machine, 2 shovel loaders	Wearing course on motorways	

* Subject to variation according to site conditions, output and material types.

Table 7.3. Plant for laying asphalt by machine

Item	Detail
Pavers	One paver plus a standby if the cc capacity. It may be sensible to ha. extending screed plus one fixed screed. screed machine be unacceptable the it woulu two fixed screeds built up to different widths.
Rollers	Two three-point deadweight rollers, both equipped with cutting wheels and scarifiers, with water sprinklers and mats. One large vibrating roller plus one smaller vibrating roller or vibrating plate for difficult, small areas.
Chipping machine	With weigh scales and trays for checking calibration.
Loading shovel	Lifting offcuts from laid mats and feeding chipping machine. Also equipped with compressor, jack hammers and hoses and asphalt cutters.
Specialist plant	Mechanical road sweeper, for use in initial sweeping of existiing surfaces before application of tack coat, and for use in picking up joint offcuts and surplus precoated chippings. Cold planing machine for use in preparing tie-ins at extremities of site, such as start and finish lines and side road entrance. Infra red heater for use in minor remedial areas, such as the removal of roller marks, re-embedment of chippings or rectifying small areas of under-compaction.

* Extending screed pavers are not normally used to lay hot rolled asphalt wearing course. However, they are very useful when laying roadbase or basecourse in varying widths to suit joint patterns or when laying in excavated areas, bellmouths, etc. where frequent changes of width of mat are necessary.

figure, i.e. the area covered per tonne of laid material. Different materials have different compositions and therefore different coverages. The properties of the individual geological aggregate used in any mixture will also influence the superage.

Superage figures for any combination of specification and aggregate source might vary by as much as 10%. The volume per unit weight is likely to be in the range 0.4–0.45 m^3/t.

$$\text{Superage in } m^2/t = \frac{1000 \times \text{volume/unit weight } m^3/t}{\text{Thickness (mm)}} \quad (7.1)$$

Examples (taking volume/unit weight as 0.425 m^3/t) now follow.

- Superage of 100 mm thick roadbase

  $1000 \times 0.425/100 = 4.25 \ m^2/t$

- Superage of 55 mm thick basecourse

  $1000 \times 0.425/55 = 7.73 \ m^2/t$

able 7.4.   Ancillary plant and small tools used in surfacing operations

Item of plant	Use in surfacing operations
Emulsion sprayer	Usually driven by an integral petrol engine (used for spraying a layer of material ready for the next coat) where there is doubt whether the existing layer will bond with the mat to be laid. A standby is often kept in reserve, particularly on larger contracts.
Bitumen boiler	Heating of bitumen to be used on joints, around street furniture, etc. Bitumen pots are also required.
Chipping boards	Placed and moved with the chipping machine to ensure that the channels, i.e. the extreme sides of the mat, are kept free of chippings for the purposes of allowing a free flow for surface water run off and for aesthetic reasons.
Shovels	Different types for shovelling materials at edges of mats and start and finnish of runs and also for hand chipping.
Tampers	Melting bitumen in joints ensuring all voids are filled, known as sweating and also hand compaction of materials in confined spaces and around street furniture.
Rakes	Hand spread (dressing) asphalt in channels, at joints and around street furniture.
Tool heaters	Heat all tools which come into contact with materials to prevent adhesion between the materials and the tools.
Mechanical sweepers	Used for sweeping carriageway before next layer, usually where roadbase or basecourse has been used for traffic.
Core drilling rig	Used for taking cores in laid materials usually to check on achieved compaction levels.
Fuel tanker	Fuel for plant.
Water tanker	Water for rollers to prevent barrels sticking to hot materials.
Tapes and steel rules	Check depths and widths to be laid and during and after laying.
Chipping trays/balance	Check that the desired rate of chipping is achieved on chipped wearing courses.
Sundries	A supply of tack coat, enough for at least two days work. Bitumen pot with burner and supply of gas. Spare bottles of LPG for the paver. Large drum of diesel, for topping up plant when fuel tanker breaks down or is late and also for cleaning the paver and the tack coat sprayer at the end of the day. Supply of spray road marking paint for delineating mat widths, chainages. Paver driving lines and spot levels. Supply of bitumen and turks head brushes for joint painting. 80 lb breaking strain cat-gut line. At least two block-up staffs or dipping stands. A 2 m rule. A 3 m straight edge for use in cutting back cross joints and checking levels across wearing course joints.

- Superage of 45 mm thick wearing course

$$1000 \times 0.425/45 = 9.44 \ \text{m}^2/\text{t}$$

The above examples are average figures. On a large contract, more accurate figures should be used taking account of the anticipated degree of compaction or air voids content, the specific gravity of the aggregate to be used and the percentage binder content of the mixture. These can then be modified, if necessary, based on the actual values of superage achieved on the site.

Experience with local materials is the most reliable basis for gauging accurate superage values. Sample calculations are given below.

Superage of DBM basecourse

(a) Data

Relative densities	2.72 — aggregate
	2.51 — sand
	2.65 — filler
	1.00 — bitumen
Specification	BS 4987: Part 1: 1987[1]
Centre of specification	61% aggregate
	33.5% sand
	5.5% filler
	100%

Binder is 4.7% of total mix

(b) Calculation

Aggregate proportion	$= 100\% - 4.7\%$
	$= 95.3\%$

So, to take account of binder content, the proportions of aggregate, filler and sand above should be reduced, proportionately, such that they total 95.3%

$$\begin{aligned}
\text{Aggregate} &= 61\% \times 0.953 &= 58.2\% \\
\text{Sand} &= 33.5\% \times 0.953 &= 31.9\% \\
\text{Filler} &= 5.5\% \times 0.953 &= 5.2\% \\
&= 95.3\%
\end{aligned}$$

$$\text{Theoretical maximum density} = \frac{100}{\dfrac{58.2}{2.72} + \dfrac{31.9}{2.51} + \dfrac{5.2}{2.65} + \dfrac{4.7}{1.0}}$$

$$= 2.453 \ \text{t/m}^3$$

So, the theoretical maximum density (voidless) of the mix is 2.453

Assume 5% air voids in laid 20 mm DBM basecourse

Compacted Density	$= 2.453 \times (1.00 - 0.05)$
	$= 2.330 \ \text{t/m}^3$

Effectively 2.33 Mg of DBM basecourse occupies $1\,m^3$

So 1 Mg of DBM basecourse occupies $\dfrac{1}{2.33}\,m^3$

$$= 0.43\,m^3$$

At 60 mm thickness, superage $= \dfrac{0.43}{0.06}$

$$= 7.2\,m^2/t$$

So, each tonne of this material at 60 mm thick will cover 7.2 m^2

Superage of chipped hot rolled asphalt wearing course.

(a) Data

Relative densities

2.72 — aggregate
2.51 — sand
2.65 — filler
1.00 — bitumen

Precoated chippings — same aggregate as bituminous mixture aggregate

Specification   BS 594: Part 1: 1992[2]
Centre of specification   34% aggregate
56% sand
10% filler
100%

Binder is 8.3% of total mix.

(b) Calculation

Aggregate proportion $= 100\% - 8.3\%$
$= 91.7\%$

So, to take account of binder content, the proportions of aggregate, filler and sand above should be reduced, proportionately, such that they total 91.7%

Aggregate $= 34\% \times 0.917 = 31.2\%$
Sand $= 56\% \times 0.917 = 51.3\%$
Filler $= 10\% \times 0.917 = \underline{9.2\%}$
91.7%

Relative density proportions in the mix

Theoretical maximum density $= \dfrac{100}{\dfrac{31.2}{2.72} + \dfrac{51.3}{2.51} + \dfrac{9.2}{2.65} + \dfrac{8.3}{1.0}}$

$$= 2.371\,t/m^3$$

So, the theoretical maximum density of the mix is $2.371\,t/m^3$
Fig. 7.2 represents a layer of chipped hot rolled asphalt wearing course

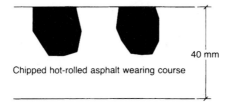

40 mm

Chipped hot-rolled asphalt wearing course

Fig. 7.2.   Physical model of chipped hot rolled asphalt wearing course

For the purposes of this calculation only, if the chippings are treated as a homogeneous layer of aggregate at the top of the wearing course then the volume occupied by the chippings can be discarded. This is shown in Fig. 7.3

Rate of spread of precoated chippings $= 12 \, \text{kg/m}^2$
Relative density of precoated chippings $= 2.72$
Density of precoated chippings $= 2.72 \, \text{Mg/m}^3$
i.e. 2.72 Mg occupies $1 \, \text{m}^3$

so 12 kg occupies $\dfrac{12}{2720} \, \text{m}^3$ $= 4.4118 \times 10^{-3} \, \text{m}^3$

So the effective thickness of chipping aggregate i.e. x in Fig. 7.3 is 4.4 mm
Effective asphalt thickness is 40 mm $-$ 4.4 mm
$$= 35.6 \, \text{mm}$$
Assume 6% air voids in laid hot rolled asphalt wearing course
Net effective Relative Density $= 2.371 \times (1.00 - 0.06)$
$$= 2.228 \, \text{t/m}^3$$
Effectively 2.228 Mg of hot rolled asphalt wearing course occupies $1 \, \text{m}^3$
So 1 Mg of hot rolled asphalt wearing course occupies $\dfrac{1}{2.228} \, \text{m}^3$

$$= 0.45 \, \text{m}^3$$

At 35.6 mm thickness superage $= \dfrac{0.45}{0.0356}$

$$= 12.7 \, \text{m}^2/\text{t}$$

So each tonne of this material at 40 mm thick will cover $12.7 \, \text{m}^2$

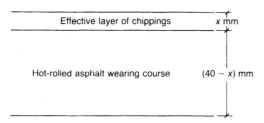

Effective layer of chippings x mm

Hot-rolled asphalt wearing course $(40 - x)$ mm

Fig. 7.3.   Mathematical model of chipped hot rolled asphalt wearing course

Obviously, the values of the parameters will have a significant effect on the calculation and the resultant superage. Since the cost of asphalts is such a large proportion of the surfacing contractor's tender, superage inaccuracies exceeding 10% are too high to be acceptable. However, this factor seldom receives the same attention as the gang size or the tonnage output despite the fact that it warrants close consideration.

At the end of this phase, the estimator has calculated the 'net cost'. It is essential that this is calculated as accurately as possible and can withstand scrutiny by others as the estimator is likely to come under pressure from the main contractor to reduce his price.

7.3.4.   Price

This phase relies substantially on the chief estimator or commercial manager knowing the market. In the case of a relatively small contract, a definition which depends on the magnitude of the individual company, it is an easy calculation to add the overhead percentage and profit as dictated by company policy. If, on the other hand, the work is a major contract then the process is rather more complicated and requires consideration of overheads, site supervision costs and profit.

Overheads

Company overheads are usually calculated on the fixed cost extracted from the annual trading results and the predicted turnover for the year. However, if the contract will produce a large proportion of the annual turnover during a short period of time then contractors will usually reduce the percentage return which is required.

Site supervision costs

These are calculated as time related costs. Accordingly, little can be done to influence them.

Profit

When dealing with major works, there are only a few surfacing contractors in the UK with the level of resources necessary to undertake the works. In determining the profit margin, the estimator must give consideration to current market conditions and the time when the works will be carried out. He must also consider the trading position of his company. Winning a large contract can establish a firm order base for the next year which allows him to increase his prices for other works in that year. There is no formula for calculating this figure, it is affected solely by the law of supply and demand and relies on the judgement of the chief estimator/commercial manager and his knowledge of the market.

Estimating programs which run on computers are now commonplace in contractors' offices. The layout of the estimating sheet will vary from company to company but a suggested layout is shown below.

Item no. 7/002 — Dense macadam basecourse 55 mm thick in carriageway, hardshoulder and hardstrip

>Material — 20 mm DBM BC
>Thickness — 55 mm
>Area — 16,750 m^2
>Superage — 7.6 m^2/t
>Tonnage — 2204 t
>Output — 551 t/day
>Duration — 4 days

(a) *Materials*

20 mm DBM BC	22.50	
Detention	0.10	
	22.60 × 2204 t	= £49,810.40

(b) *Labour*

Gang cost M8 (day)	800.00	
Travel	68.74	
Service time	90.00	
	958.74 × 4 days	= £3,834.96

(c) *Plant*

Paver	185.00	
10 t Roller	40.00	
10 t Roller	40.00	
Tractor/Compressor	32.00	
Crew bus	30.00	
Van	20.00	
Mess hut	5.64	
	352.64 × 4 days	= £1,410.56

(d) *Sundries*

Allow £50 per day		× 4 days	= £200.00
			£55,255.92

(e) *Overheads*

11%	£6,078.15

(f) *Profit*

5%	£2,762.80

	£64,096.87

(g) *Area*   16 750 m^2

(h) *Rate per m^2*   £3.83

Whilst the above is a basic item calculation which would be repeated for each item in the estimate, the system must be flexible enough to allow the information to be presented in various forms. For example, it is usual to produce an estimate summary such as that shown in Table 7.5.

Table 7.5.   Extract from the estimate

Item No	Description	Labour	Plant	Material	Misc.	Cost	O/H	Total
7/001	HMB RB 275 mm thick	1.31	0.48	21.35	0.07	23.21	3.71	26.92
7/002	DBM BC 55 mm thick	1.74	0.64	22.60	0.09	25.07	4.01	29.089
7/003	HRA WC 45 mm thick	3.72	1.54	28.05	0.13	33.44	5.35	38.79
7/004	Regulating	6.40	2.33	22.75	0.33	31.81	5.04	36.90

7.4.   Planning and programming

As is the case with all construction processes, adequate time must be spent on planning and programming to ensure that the finished product meets the requirements of the specification and that operations are carried out in a manner which is both efficient and cost effective. Adverse weather conditions and frequent interruption of the laying process coupled with a lack of attention to temperature control, compaction and site preparation are the most common reasons for poor laying and premature failure of asphalts. Programmes soon become outdated and should be reviewed regularly. Table 7.6 illustrates the points which should be considered.

The first contract programme will normally be prepared during the tender period. It will show the phasing of the works, traffic management measures, etc. and provide information which will contribute to the estimate. The programme should be reviewed on a daily, weekly or

Table 7.6.   Points to be considered in compiling a surfacing programme

Topic	Detail
Site preparation	Any preparatory work to be carried out by the surfacing contractor, usually done by the main contractor.
Labour	Numbers and skills requirement.
Plant	Numbers and types including standby plant.
Specialist plant	Cold planers, infra red heater, core drill rig, mechanical sweepers, etc.
Materials	Full specification including BS table numbers, aggregate nominal size and percentage, binder type and percentage content. Total tonnage plus daily output and hourly rate of supply which should be maximized at all times if the average used in the estimate is to be achieved. Full delivery address, together with limitations, if any, on points of entry to the site and access road. Special requirements: minimum batch tonnage in cases where there is more than one source of supply, laying special precoated chippings, early delivery to allow for testing.
Notices	Notices to Engineer's Representative, main contractor and other subcontractors.

sectional basis throughout the contract with all relevant staff providing an input. Particular events, e.g. delays, may also trigger a review of the programme.

Most programmes are produced using commercial software and include facilities which permit sophisticated monitoring and interrogation. Their use is to be encouraged.

7.5. Pre-start inspection

Surfacing operations are often the last major element on road contracts and are frequently given scant consideration by the main contractor. This often results in impossible demands being made upon the surfacing sub-contractor. No plant or materials should be ordered until the site has been thoroughly inspected by the surfacing sub-contractor (usually undertaken by the surfacing supervisor and a foreman) in conjunction with a representative of the main contractor and the clerk of works or Engineer's Representative. In preparing a surfacing programme, points which require particular attention include those listed in Table 7.7.

7.6. Ordering plant, equipment and materials

7.6.1. Ordering plant and equipment

Once an order is placed or a firm acceptance of a tender is received, the items of plant and equipment, typically as shown in Tables 7.3 and 7.4, should be ordered internally or externally, as appropriate. Provisional orders are commonplace for specialized items of plant such as chipping machines or materials such as precoated chippings where advance notice is required.

Table 7.7. Pre-start checklist

Item	Detail
Sub-base/planed area	Checks on the surface to be covered including: the absence of soft spots in the sub-base; dipping of the sub-base or existing surface; ensuring that the sub-base has been properly shaped to give the correct cambers, crossfalls, moving crowns; adequate compaction of the sub-base; adequate cleaning and sweeping of planed areas; ability to support surfacing plant and delivery vehicles.
Water supply	Suitable water supply either stop-tap box or water bowser for filling up rollers.
Access	Site entrance, road signs, ramps, etc., paving machines, rollers and delivery vehicles must have reasonable access to the area to be surfaced and access must be able to support surfacing plant and delivery vehicles.
Cleanliness	All accesses and surface to be overlaid to be free of mud, debris, etc.
Safety	Inspection of site safety rules and any hazards.

7.6.2. Ordering materials
Areas available for work

Obviously, the available area should be as large as possible. Should any isolated areas be unavailable, for example at the approach to a bridge, then a patch, say 50 m or 100 m in length, should be left unsurfaced. The actual size of the patch is not important providing it is of reasonable magnitude — the surfacing of small patches is never satisfactory since materials in small batches cool rapidly, screeds are often cold, etc. Any attempt to surface numerous small areas in such circumstances invariably leads to unsatisfactory results both in terms of variability in level control and finish.

Areas should not be considered as being available until the following matters have been addressed.

- The sub-base has been trimmed, rolled and passed by the Engineer's Representative.
- Any overlay or planed off area has been thoroughly cleaned and passed by the engineer's representative.
- All the required setting out has been undertaken. This should be clear and unambiguous. Colour codes should be used if more than one crossfall is required using pins and tapes (as discussed below). All references to redundant or spurious setting out must be removed. The surfacing gang must be wholly familiar with the system which has been used.
- All earthworks, kerb laying, pipe laying and manhole or gully construction has been completed. Street furniture (i.e. gullies, access covers, manholes, etc.) must be set at levels which are no higher than the top of the first layer of asphalt which is to be laid.
- All unnecessary site traffic has been diverted.
- Site accesses are adequate and clean. Delivery vehicles can quickly turn a site into a quagmire. Accesses must be clear, unfettered by cranes, concrete pumps or other equipment.
- All temporary sub-base or hardcore access ramps which have been formed to facilitate access into areas being excavated have been removed and replaced with timbers or hand-laid asphalt.

Supply of materials from the mixing plant

Once the work area is available the required tonnage can be calculated and an order placed. The tender rates are now largely irrelevant. The surfacing contractor should aim to maximize output given the production capability of the source of supply, the nature of the site and the size of the laying gang.

Tonnages are calculated from the area, depth and density of the material being laid. This is the easiest part of the ordering process. Timing and rates of laying are equally important if a good working relationship is to be established between the laying gang and the supplier.

Rates of laying are determined not only by paver speed but also by the width and depth of the mat to be laid and restrictions in access. Laying at a constant speed of 200 m/h and assuming a density of 2.6 t/m³ will give the following outputs

- DBM roadbase 5 m wide × 125 mm depth = 300 t/hr
- DBM roadbase 3 m wide × 100 mm depth = 150 t/hr
- DBM basecourse 3 m wide × 55 mm depth = 80 t/hr
- HRA wearing course 5 m wide × 45 mm depth = 70 t/hr

The outputs quoted above should be regarded as the maximum on all but the very largest motorway contracts. Poor access for delivery vehicles, or reversing of lorries onto the paver over long distances, will reduce these figures considerably. Increasing the speed of the paver is not a solution because if the paver lays too fast then the screed may drag and the mat will have a poor finished appearance.

In calculating the daily tonnage requirement and the timing of deliveries, allowance should always be made for the following contingencies which invariably occupy the first hour of the working period. These are

- the paver, rollers, etc. must be fully fuelled, moved onto site and positioned ready for laying to begin
- the paver screed should be built up, if appropriate
- the screed must be heated to the laying temperature
- joints must be cut, prepared and painted with bitumen
- rollers must be filled with water
- the chipping machine must be calibrated and checked
- tack coat must be sprayed
- driving lines and levels must be marked out.

Material deliveries should be specified at intervals which allow sufficient time for the previous load to be laid correctly. A discussion with the foreman is always beneficial in such circumstances, especially if some asphalt is to be laid by hand. Surfacing contractors should always seek to minimize the period when delivery vehicles are on site. Not only are excessive periods chargeable (30 minutes is normally allowed) but these vehicles are invariably needed for further deliveries to the site. Small, awkward areas should be left until later in the day when the need to return lorries to the mixing plant for another load is less urgent. Deliveries must be coordinated with meal breaks. One hour should be left at the end of each shift to allow for rolling off and preparation of the site for a prompt start on the next shift.

Finally, before orders are placed with the supplier, the specification should be written down and checked in order to minimize the possibility of errors. The required tonnage and the timing of deliveries should be made clear to the supplier. The use of a *pro forma* is useful for this purpose. The information, which is provided at the time of placing an order, should include

- the name of the surfacing contractor
- the name of the person placing the order

- contact telephone number
- contract description
- delivery locations
- lorry sizes
- tonnages of each of the materials
- full material specification/s
- start times including morning and after lunch
- delivery times or frequencies and any other special instructions.

The order is normally placed by telephone to the supplier's order clerk. He should be asked to repeat it back to the person placing the order to avoid mistakes or misunderstandings. Whenever possible, a faxed confirmation of the order should be sent. (Most computers, including laptops, now have the capability to send faxes either via a traditional telephone socket or through a mobile telephone.) On a major surfacing contract, such orders have a substantial financial value and, consequently, there can be no room for error.

Turning round delivery vehicles efficiently will assist the supplier and help foster good relations. This will lead to good service from the mixing plant. Normally, a waiting time of 30 minutes is included in the prices for materials. Where this is exceeded, a charge for the extra period (usually called 'standing time') is normally incurred. However, it may well be beneficial for the surfacing contractor to examine all delivery times to ensure that he has not incurred standing charges as a result of late delivery by the mixing plant. It may also be the case that delivery vehicles have, on average, been emptied in substantially less than the permitted period and that the mixing plant has, as a consequence, derived some benefit. It is often possible for the surfacing contractor to persuade the supplier to forego further charges where the latter has been inefficient or has incurred reduced costs because waiting times were minimized.

Gang output and motivation
Surfacing gangs are specialists in laying asphalts. They are not trained in placing sub-base, laying kerb, etc. These are operations which should have been completed before surfacing commenced. They perform at maximum efficiency when everything runs smoothly and laying is continuous.

A gang is far more likely to produce consistent work of high quality at a productive rate which is consistent with reasonable profit and the minimum of remedial works if sites are ready for asphalt when they arrive, the correct equipment is available on time and in good working order and material is supplied at a steady rate without interruption. Failure to comply with any of these basic elements may well result in reduced morale, motivation and output.

It is most important not to expect too much of a gang. If it is found that extra capacity is available, rather than order more material for the same day, it may be more prudent to discuss the workload with the foreman and increase the order for the following day. Such an approach is less likely to adversely affect the morale of the gang and should lead to better overall productivity.

If the order is changed during a working day then it is wise to confirm the total tonnage for that day with the supplier to avoid the possibility of duplication if, say, the alteration has already been communicated directly from site.

Ordering the exact tonnage of material necessary to complete surfacing up to a closed joint, such as areas of excavation and patches, calls for considerable skill and experience. Wherever possible, it is advisable to organize the day's work so that there is an adjacent area, such as a footway or a central reservation, where excess material can be laid. Where this is not possible, an extra one or two tonnes should be ordered. The superage obtained in the area of work should be monitored carefully so that the coverage can be accurately anticipated. It is wise to liaise closely with the mixing plant on the exact weight of the last load.

7.6.3. *Surfacing operations checklist*
Table 7.8 summarizes the actions and considerations necessary before surfacing work can start.

7.7. Transportation
All asphalts should be mixed and delivered within the limits specified in the contract documents. Good transportation practice is summarized in Table 7.9.

Figure 7.4 shows a delivery vehicle discharging into a paver and demonstrates a number of elements of good practice. The lorry is clean, well insulated and sheeted, and reversing under the control of a banksman. The sheeting on the lorries is removed only to the extent necessary to facilitate discharge. The dead weight roller is tight behind the paver. The vibrating roller is working to a predetermined pattern.

7.8. Laying
7.8.1. *Setting out*
All setting out must be completed before laying can commence. The foreman will usually organize this operation whilst the paver is being set up, rollers are being filled with water, sprinklers are checked, wheels are cleaned, the chipping machine is calibrated, stockpiles of chippings are checked, the pattern of delivery of chippings from the stockpile to the point of laying is agreed and tools and bitumen are heated.

A laying pattern should be worked out which maximizes the lengths of runs, thus avoiding the need for joints and hand laying. The following points should be remembered when working out a laying pattern.

- The amount of hand laying should be minimized. This includes hand spreading where the laying width is greater than the screed width (called 'bleeding out').

Table 7.8.   Plant, contractual and environmental surfacing checklist

Environmental	Weather to be expected Exposed location? Air and wind temperatures Have previous problems been encountered in the area? Schools, populated areas: inform them if possible
Contractual	Noise levels Restricted working hours Compaction specifications: method or end product Laying specifications (BS 594, BS 4987, contract specification) Has site been checked to make sure that it it is ready for surfacing?
Paver	Plan sequence of paver mat runs Size/type of paver required: fixed width/telescopic? Foreman and gang availability Source/s of material Width/length of runs Mat profile Crossfall, camber or crown? Automatic level control required?
Rollers	Number and type of rollers for each layer of construction Availability of water supply for sprinklers Cutting devices on rollers to trim joints
Chipping machine	'Dry run' to set chipping rate Preferred locations for chipping delivery and rolling temperatures Spanners for changing gauge setting Width of mat runs Obstacles, fences, width restrictions
Bituminous materials	Specifications of materials to be laid Minimum and maximum delivery and rolling temperatures Distance of site from mixing plant Contact name and telephone number of mixing plant Method of placing orders (when and by whom) Has proper amount been ordered? Are enough chippings present? Who checks material coverages (superages) on site to adjust material orders as necessary?
Other plant: Emulsion sprayer Lorries Tractors Bitumen boiler Hand tools and other Tool heaters	Quantity of emulsion needed per layer (if applicable) Are they properly insulated? Are they properly sheeted/double sheeted? Is material covered until just before use? For feeding chipping machines For cutting 'cold' joints Number of compressor guns/power of tractor Compressor hoses, etc. For painting bitumen on longitudinal/transverse joints Quantity and type of bitumen required Rakes, shovels, tampers, safety equipment (gloves, hard shoes, hard hats, high visibility jackets, etc.) For heating rakes, shovels and tampers to prevent adhesion of material
General	Procedures and responsibilities for daily maintenance of all plant Supply of fuels/oils to all items of plant Gas supply for screed heating Storage of plant overnight Forward planning of low loader movements to next job Other provisions for workplace if weather not suitable for layer Steel rulers, 30 m tape, string lines, metre rule, paint, etc. First aid equipment/fire extinguisher

Table 7.9. Good transportation practice

Topic	Detail
Delivery vehicles	Hand-laying operations invariably require smaller trucks than materials placed by pavers. Consequently, delivery vehicles should be capable of carrying the maximum weight of material consistent with site conditions and the nature of the operation. Four-wheeled lorries are only appropriate where access is difficult or hand-laying is being undertaken.
Insulation	Delivery vehicles should be insulated and sheeted (preferably double sheeted) to minimize heat loss in transit or while awaiting discharge.
Vehicle cleanliness	The inside of the body should be clean with the minimum of dust or sealing grit applied to facilitate discharge. Diesel treatment to the tailboard mechanism is worthy of particular attention providing none comes into contact with the material.
Rate of delivery	The rate of delivery should be controlled to provide continuity of laying. As indicated previously, clear instructions should be included as part of the order to the supplier. Normally bituminous mixture deliveries are specified as beginning at a particular time and continuing at a rate of 15–20 t every 5–10 min, as appropriate. When laying chipped hot rolled asphalt wearing course, two or three loads are often ordered for the start time so that continuity can be maintained.
Site address	Clear instructions should be shown on the delivery note, together with access points, turning locations, marshalling areas and limited weight or access routes. Many motorway contracts are subject to restricted access routes which must be specified.
Arrival on site	Immediately upon arrival on site the driver should report to the marshalling point or to the foreman or other authorized person. Failure to do so may delay temperature checks or sampling procedures and cause delay.
Discharge	Excessive reversing distances can be extremely dangerous. If necessary, turning points should be provided. Reversing in dangerous locations or in the vicinity of other operations while backing onto the paver must always be under the supervision of an experienced banksman.
Sheeting removal	Sheeting should only be removed as far as is necessary to facilitate discharge. Hand-laying does not normally warrant complete removal. Such removal should not take place until immediately before tipping.
Hand-laying	Hand-laying operations necessitate controlled discharge to avoid the possibility of excess material being tipped. Failure to do so involves unnecessary effort and/or material becoming cold before it is properly compacted.
Cleaning out	Once the load has been tipped, the lorry should draw fully away from the paver and cleaning out or tipping of excess material restricted to a designated cleaning out point on site.

Fig. 7.4. Lorry deliveries to a paver. Photograph courtesy of Wrekin Construction

- Paver runs and widths should be planned such that the length of runs is maximized whilst the amount of laying by hand at the beginning and end of runs and the number of joints is minimized.

Driving lines should be marked on the existing surface or lower layer. These assist the paver driver to maintain a constant line and assist roller drivers when cutting longitudinal joints in wearing courses. Figure 7.5 shows typical joint patterns and laying widths for a large car park.

In Fig. 7.5, material which must be laid by hand is limited to the start of panels 1 and 2, the start and finish of panel 3 and at the radius. Joint cutting is restricted to the ends of panels 1 and 2. The direction and order of laying panels 1, 2 and 3 maximizes the time for run 2 to cool before the

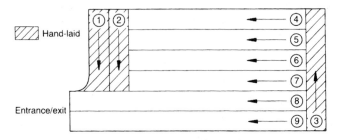

Fig. 7.5. Surfacing a car park

paver runs over the laid material in panels 1 and 2. This approach avoids the necessity of hand laying for a paver length at the end of runs 4 to 7.

An alternative method would be to lay panels 3, 4, 5, 6 and 7 before panels 1 and 2. This would avoid the paver operating on top of the fresh material on panels 1 and 2 but it requires hand laying and joint cutting at the ends of panels 4, 5, 6 and 7. On a hot day with a thick wearing course (40 mm or more), this method would have to be adopted since runs 1 and 2 would be too hot to take the weight of the paver.

All situations have to be assessed individually. Experience is an invaluable asset in deciding which approach will achieve the best results with the least effort and minimum cost. A temporary hardstanding of hardcore can be laid using an ordinary excavator around the car park entrance/exit to avoid damage to the formation. Level control is normally exercised by means of steel pins or timber stakes (or coloured tape on steel pins or timber stakes) set along both sides of the carriageway at a maximum of 10 m centres. These are set in relation to the datum levels for the contract. The tops of the pins or coloured tape are fixed at the same height above the finished surface levels, say 300 mm above roadbase finished surface levels, 245 mm above basecourse finished surface levels and 200 mm above wearing course finished surface levels.

It is essential that the pins or stakes are driven into firm ground, or preferably set in concrete, otherwise they are likely to become loose and the asphaltic layers will be laid to incorrect levels. Dipping below a gut line, held tight across the carriageway between two such pins or stakes, provides a simple but accurate means of level control. The use of chalk on kerbs to denote crown lines and stop points, crown line offsets and block heights, etc. is recommended.

Block heights should be underlined to ensure that figures are read correctly. Chainages should be marked 'CH X' to avoid the possibility of confusion between levels and chainages.

7.8.2. Use of a block-up staff
Vertical alignments of carriageways are designed to the nearest milli-metre. Accordingly, close control is required when road layers are placed.

On major carriageway repairs, where diversions or contraflows cannot be used, level control can be a problem. The requirement to maintain free-flowing traffic means that only half the carriageway is accessible. Accurate level control requires the establishment of control levels along the centre of the road. This is normally done with a painted cross or masonry nail with an associated block-up height in millimetres. Block-up staffs make it possible to hold a line at the exact height which is required and thus achieve an adequate degree of level control. Good quality staffs are made from the best materials and will withstand years of wear and tear on site. This makes them cost effective against standard rules which are initially cheaper but have a much shorter life.

The major components of a block-up staff are a 1.04 m hardwood staff with an armoured foot, a heavy duty brass rule and a stainless steel sliding cursor which locks in place and is shown in Fig. 7.6.

Fig. 7.6.   Block-up staff. Photograph courtesy of TBS Engineering

The lower edge of the cursor is set at the appropriate level on the staff. It is firmly locked in this position using a thumbscrew on the back. This screw acts on a captive stainless steel pressure plate to avoid damaging the timber. A single-filament nylon fishing line having a 40 kg breaking strain is then pulled under the locating tab below the cursor. The other end of the line is secured by looping over a setting out pin at the top of marking tape or hooked over a kerb, as required. The existing surface can now be dipped from the line to calculate the next layer thickness.

Figure 7.7 illustrates a number of errors that must be avoided when setting or checking levels. These are as follows.

- If the line breaks, the chainman may fall into the line of traffic.
- The operative cannot watch for site traffic running through the line as he must look at the rule to hold a constant level.
- Wrapping the line around fingers is dangerous and can cause injury if site traffic snags the line.
- Use of a standard rule which is too thin and will be unstable.
- The graduations on a standard rule wear off or dirty quickly when handled with soiled hands, they may thus be hard to read.
- This method relies on the operative being able to read the rule correctly and set the level accurately.
- Checking involves repeating the process.
- It is very uncomfortable to remain in this position for any length of time.

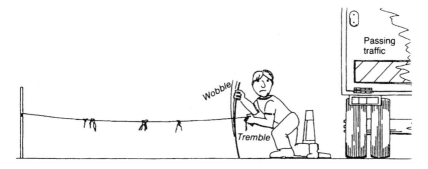

Fig. 7.7.   Setting levels in a live carriageway incorrectly

- If the line is dirty, too thick, tied in knots or has tapes on it then it may sag.
- Many operatives are not confident when measuring.

Compare this with the method illustrated in Fig. 7.8, which shows the following points.

- The line is now being held at the top of the staff. Tension is maintained by setting up 0.5 m short of the datum and the operative moving the staff sideways with his foot. This maintains constant line tension and there is no danger of injury due to line breakage.
- The operative is now free to be alert to any danger from site traffic, etc.
- The line is wrapped round a short stick which is easy to hold or let go if site traffic catches the line.
- The staff is strong and stable.
- The rule can be set by a competent person and given to the operative.
- The setting can be checked before being altered for the next chainage.
- When being set, the staff can be held at the most convenient distance to suit the operative's eyesight.
- Since the line is held round a right angle bend at the cursor and braced against the top of the staff, it is easy to maintain constant tension.
- The upright stance is comfortable for long periods of time and makes the chainman more visible to site traffic improving safety.

Fig. 7.8.   Setting levels in a live carriageway correctly

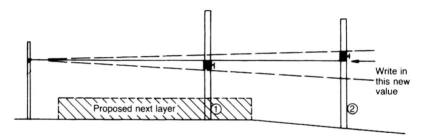

Fig. 7.9.   Offsetting control levels

- The recommended line is the best balance of strength and weight. Use of a warning board is advisable.

Figure 7.9 shows how to offset a block height at a point which will be covered by the next layer. The top of the cursor is set on the first staff at the original height to offset a block height from position 1 to position 2. The second staff is set up at the new offset location. The line is pulled in the normal manner and the cursor on the second staff slid up or down until the line coincides with the top of cursor 1. Cursor 2 is locked in position. The second staff is rocked slightly to check that the line just clears the top of cursor 1 and adjusted, if necessary. The location of the second staff is marked on the ground and the height of cursor 2 recorded.

Figure 7.10 shows how to set out levels when there is a crown line at the centre. The cursor is set at twice the value of the crown height. The staff is then used to block up on top of the right-hand kerb. The line is pulled taut between the top of the left-hand kerb and the cursor. The left half of the road can now be dipped. The procedure is reversed for the right half of the road.

Figure 7.11 shows how to set out levels where there is an obstruction such as a crash barrier. If the tapes on the pins are set at a control level, say 200 mm, and the line cannot be pulled because of site obstructions such as piles of precoated chippings, barriers or timber baulks then the staff can be inverted and the cursor set on top of the pin. The staff is now

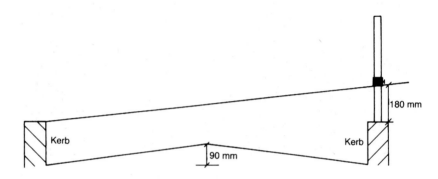

Fig. 7.10.   Setting a level where there is a crown line at the centre

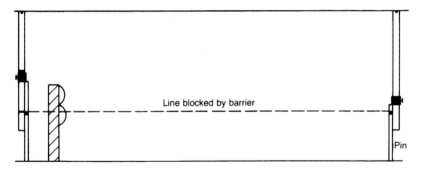

Fig. 7.11.   Setting out over an obstruction

slid through the cursor until a convenient reading is against the top of the tape, say 800 mm. This height should always be kept to a minimum to avoid errors caused by the staff not being vertical. If the height is above 0.5 m then a spirit level should be employed. The work can now be dipped using the new control of 800 mm + 200 mm = 1000 mm.

Figure 7.12 shows a laying gang dipping the top of a chipped HRA wearing course. The resultant figures are often referred to as 'dips' and they should always be recorded, usually in a dimensions (DIM) book.

7.8.3.   Laying asphalts by hand

Spreading by hand and raking of asphalts is a highly skilled and physically demanding operation and it is rare to find a good machine crew that is equally skilled at hand laying.

Fig. 7.12.   Dipping to check the finished road levels. Photograph courtesy of Wrekin Construction

Good practice when laying asphalts by hand can be summarized as follows.

- Wherever possible, material from the lorry should be tipped in small heaps, exactly where the material is to be raked. Large heaps cause unnecessary carrying and spreading of the material. They can also result in the material at the bottom of the heap being compacted by its own weight, which may well result in humps in the finished mat after rolling.
- Unnecessary raking and shovelling, which causes segregation of the material, should be avoided.
- Tipping and raking should be completed speedily to minimize heat loss and to enable rolling to begin as quickly as possible.
- Joints are the locations most likely to fail early due to the material cooling before it has been adequately compacted. At the start of rolling, joints should be treated first since all materials cool first at the joint. Alternatively, the joint may be hot ironed as raking progresses.
- Wearing course joints should be raked beyond the actual joint and onto the existing surface so that any segregated aggregate may be scattered back onto the mat with a flick of the rake. Failure to do this can result in a concentration of coarse aggregate at the joint resulting in plucking and premature failure of the mat.
- Best results are achieved by spreading the material as accurately as possible allowing extra thickness for rolling down (the amount varies for different materials but may well be of the order of 30%) followed by no more than two passes of the roller. Differential compaction resulting from laying variable thicknesses will result in small irregularities which can then be rectified by adding material to these areas before final compaction is effected.
- The provision of good access for the delivery of materials to the point of laying will markedly increase the speed and efficiency of the laying operation.

Examples of gang size calculation
Site A — a footway job. Most sites provide easy access to a standard 1.8 m wide footway via the adjacent road, although increasingly there are grass verges between the footway and the carriageway. Such a site lends itself to the use of a tractor shovel loader and a small vibrating tandem roller.

A six-man gang could reasonably lay 60 t/day with deliveries at 8 a.m., 10.30 a.m. and 1 p.m. These times, which make an allowance for tea breaks and lunch, assume 10 t of basecourse per man day laid at 40 mm thickness. One man will drive the roller, another the tractor and the remaining four men will handle the material.

Each load should be tipped at a location near to the point of laying, normally in an adjacent bellmouth on previously laid basecourse. This gives space for the tractor loader to manoeuvre. Any damage to the existing basecourse, which will be minimal, will be covered by the subsequent wearing course. Tipping hot material onto an existing wearing course marks the surface and must be avoided.

Six tonnes of material will cover approximately 366 linear metres of footway 1.8 m wide and 20 mm thick. The tractor shovel holds around 1/3 t. Allow 20 s for loading and a further 20 s for tipping. The tractor speed is taken as 10 mph ($\equiv$ 4.5 m/s).

$$\text{Average distance travelled by the tractor} = \frac{366}{2} \times 2 = 366 \text{ m}$$

$$\text{Number of trips to move 60 t of material} = 60 \div \frac{1}{3} = 180$$

$$\text{Time per round trip} = \left(\frac{366}{4.5} + 40 \right) / 60 \approx 2 \text{ min}$$

$$\text{Time to move 60 t} = \frac{180}{2} \times 2 = 6 \text{ h total}$$

The following timetable will operate during a typical day from 7.30 a.m. to 4.30 p.m.

7.30 — 8.00	Set up site
8.00 — 10.00	Lay first load
10.00 — 10.30	Break
10.30 — 12.30	Lay second load
12.30 — 1.00	Lunch break
1.00 — 3.00	Lay third load
3.00 — 4.30	Roll off, tidy site, etc.

The timetable shows that the rate of laying depends on the tractor's ability to supply 60 t in the 6 h laying period. In reality, delays in delivery of materials, breakdowns and site problems may reduce production and it would not be unusual to finish rolling off as late as 6.00 p.m. It is also evident that, although the output per man would be rated at 10 t, since only four men are actually handling the material, the output is actually 15 t/man. It would seem, therefore, that when laying more than 366 m away from the tipping point, it is not possible to feed enough material in six hours to keep four men working. The options are to reduce the number of men or increase the quantity of material delivered.

Site B — similar to site A but with access problems.   On a site where there is no convenient tipping point or no easy access from the road to the footpath, the use of a dumper should be considered in addition to the loading shovel.

In this instance, the six-man gang would be equipped with a small vibrating tandem roller, a tractor loading shovel and a 3 t dumper. Again, consider laying 60 t of material at 40 mm thickness with one man driving the roller. Another man drives the dumper and loads it using the tractor. The remaining four men handle the material.

During the day, the dumper only has to make 20 journeys. However, it may take 360 s to load and to unload based on 20 s for each manoeuvre of the tractor or dumper. Assuming an average trip distance of 400 m, the time taken to deliver the 60 t would be 4 h whereas an average trip of 800 m would take 6 h for the same 60 t laid by four men handling 15 t/man with the only additional costs for the day being the hire charge for the dumper.

The temptation to cut laying costs by reducing the number of men in the gang from say six to four men, although showing a saving of approximately £200, would usually mean reducing the output from 60 t to 40 t. The effect of this would not be an overall saving in the cost of laying since the overhead cost of supervision, the crew bus, roller, tractor and/or dumper remain the same.

Suppose the number of men actually laying is now reduced to two, compelled to work 33% harder with each man laying 20 t. While this may be possible for a short period, it does tend to produce stress and strain, increased levels of sick leave and men seeking less strenuous employment.

The conclusions from the above are as follows.

- On sites where delivery distances exceed 800 m, an additional dumper and driver is likely to increase the delivery distance up to 1600 m from the tipping point. Employing one extra man and dumper would double the efficiency of a six-man gang working between 800 m and 1600 m from the point of tipping.
- Gang resources must be adjusted to take account of variation in site conditions. However, it is seldom economical to reduce the size of a gang from six to four men and the standard six-man gang operates most efficiently when tipping is no more than 400 m from the point of laying.

Recommendations when laying by hand include the following.

- In all cases, explore the possibility of increasing gang resources and thus increasing unit output.
- Calculate gang output on the basis of the output per man actually laying material rather than the total number of men in the gang.
- Programme the minimum resources for delivery at the point of laying with 15 t/day for each man actually handling the material being a reasonable and sustainable output.
- Try to programme work such that tipping points are available within 400 m of the point of laying.
- Use a dumper where the delivery distance is between 400 m and 800 m and two dumpers plus one extra man where the delivery distance exceeds 800 m.

Cost calculation
The calculation in Table 7.10 relates to a site where the material is delivered to the point of laying; it illustrates the false economy of reducing labour from six men to four men. Table 7.11 illustrates the benefit of using an extra dumper and driver when the point of tipping is greater than 800 m from the point of laying.

Table 7.10. Comparison of four-man and six-man gang costs

Four-man gang costs	Item	Six-man gang costs
4 × £75 = £300 1 × £30 = £30 1 × £30 = £30 1 × £30 = £30 4 × £75 = £300 4 × £75 = £300	Labour Roller Tractor Van Fuel Supervision	6 × £75 = £450 1 × £30 = £30 1 × £30 = £30 1 × £30 = £30 £60 £40
Total cost = £470		Total cost = £640
Laying 40 t/day	Output	Laying 60 t/day
470/40 = £11.75/t	Unit cost	640/60 = £10.67/t

Table 7.11. Comparison of six-man and seven-man gang costs

Six-man gang costs	Item	Seven-man gang costs
6 × £75 = £450 1 × £30 = £30 1 × £30 = £30 1 × £20 = £20 1 × £30 = £30 £30 £40	Labour Roller Tractor Dumper Van Fuel Supervision	7 × £75 = £525 1 × £30 = £30 1 × £30 = £30 2 × £20 = £40 1 × £30 = £30 £30 £40
Total cost = £630		Total cost = £755
Laying 30 t/day due to distance	Output	Laying 60 t/day
630/30 = £21/t	Unit cost	755/60 = £12.58/t

7.8.4. Machine laying

The types and method of operating pavers are covered in detail in Chapter 6 in Section 6.3. However, some simple basic rules are summarized below.

- As the paver increases in speed, the thickness of the laid mat will decrease with the converse being true.
- When laying on a radius, the inside of the screed travels slower than the outside with a corresponding increased material thickness on the inside and decreased material thickness on the outside of the mat.
- Hot material passing under the screed will lay thicker with the converse being true.
- As the speed of the tampers increases so too does the thickness of the mat. Failure of the tampers will result in the paver laying a thinner mat.
- A screed which has less material fed to it will lay a thinner mat. Bleeding out or an increased amount of material in front of the screed will result in a thickening of the mat.

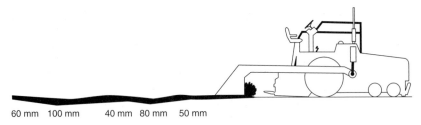

Fig. 7.13.   Time lag for screed changes to take effect

- Variations in the consistency (i.e. the workability) of the material flowing under the screed will result in a variation in the thickness laid.
- A hot screed will produce a thicker mat than a cold screed.

Generally speaking, a paver has a crew of two men. One drives the machine whilst the other operates the screws and is called, unsurprisingly, the 'screwman'. Even when the variables described above are in operation, an experienced screwman, working regularly with the same machine, will produce consistently high-quality work. Notwithstanding, these variations should be minimized as far as possible.

Whenever the screwman alters the layer thickness, he must add an amount to allow for the fact that the laid thickness reduces after compaction. It can take up to three paver lengths for a change in level to take effect, as illustrated in Fig. 7.13.

The floating screed paver is capable of taking out frequent small irregularities in the surface to be covered. Each successive layer will reduce these bumps and hollows and produce a more planar surface. It is not possible, however, to lay a mat where finished levels vary markedly over short distances as illustrated in Fig. 7.14.

Nowadays, works often consist of planing off an existing wearing course (usually 40 mm HRA) and replacing it with the same material or perhaps a thin surfacing. Surface regularity is then checked. The means of doing so is most usually by means of the rolling straight edge. This test checks whether localized irregularities exist. This test is included in the standard UK specification, the Specification for Highway Works (SHW),[1] and is discussed in detail below. It is, however, most inappropriate in these circumstances. If the original road had irregularities, these will continue to exist in the new surface (albeit to a lesser degree because of

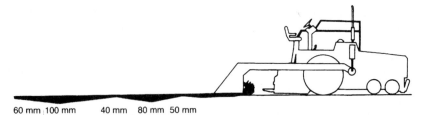

Fig. 7.14.   Overlaying an existing surface which has poor vertical geometry

the smoothing effect of the paver). If they are in excess of the specified requirements, the new surface will appear not to comply with the SHW.[1] Surfacing contractors should be wary of undertaking work where this may be the case and may request a rolling straight edge before beginning planing operations in order to establish the existing frequency of irregularities. Patching, or perhaps even application of a shaping layer, should be undertaken after the planing operation has been completed in order to ensure that any sizeable irregularities (less than 10 mm in the case of a wearing course) are eliminated. Patching tends to be expensive because of the high labour and plant element compared with the low material content. However, doing the work again is substantially more expensive.

Laying asphalts on a gradient
When laying on a gradient, laying should commence at the lowest point of the gradient and continue uphill. Doing so means that the roller driver only has to use his brakes to stop at the low end of each pass where the material is coolest and, therefore, stiffest. At the high end of each pass the roller will naturally roll to a standstill and, consequently, no braking is required which would adversely affect the appearance of the finished material. When laying a chipped HRA wearing course, any displacement causes bald patches with inadequate precoated chippings and low texture depth, as the laid mat is stretched. In heavy rain, downhill laying may be necessary to prevent water building up in front of the screed. Most specifications prohibit laying in such circumstances but it may occur when rain starts unexpectedly. This situation always requires extreme care. Figure 7.15 illustrates why asphalts should be laid uphill. Figure 7.16 shows the effect of laying a chipped HRA wearing course downhill.

Laying on crossfalls
Figure 7.17 shows why great care must be taken when laying asphalts on a surface which has a crossfall. Laying thick layers of asphalt is to be avoided if possible in the circumstances shown in Fig. 7.17(*b*). Low density and loss of level control will occur on the low-side free edge. In wet or very windy weather, laying may have to be suspended as the risk of losing the whole mat is too great.

Use of pavers
The correct use of pavers is a complex process and there are no substitutes for training and experience. It is essential to observe a number of points when using pavers and these are listed in Table 7.12.

7.8.5. Laying thin surfacings and SMAs
Recent years have seen the emergence of proprietary thin surfacings most of which are basically variants of SMA formulations. There are fundamental differences between thin surfacings/SMAs and other asphalts and these are covered elsewhere in this Book (see Chapters 3 (Section 3.7.5), 4 (Section 4.13.6) and 10 (Section 10.7)). In regard to placing these

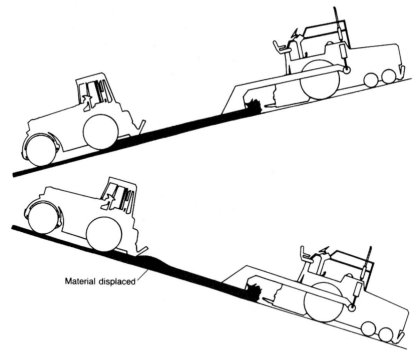

Fig. 7.15.   Laying bituminous mixtures on a gradient

materials, the same principles that apply to the laying of traditional asphalts are equally valid for thin surfacings/SMAs and should not present any problems for an experienced gang. The following basic rules summarize the main points to be considered.

- Existing surface should be clean and free from all loose material.
- Existing surface may be damp but not wet. There should be no standing water.
- Ironwork should be adjusted prior to laying.
- Minimum surface temperature should be −1 °C and rising.

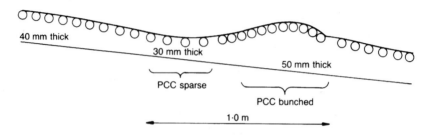

Fig. 7.16.   Results of laying chipped hot rolled asphalt wearing course downhill

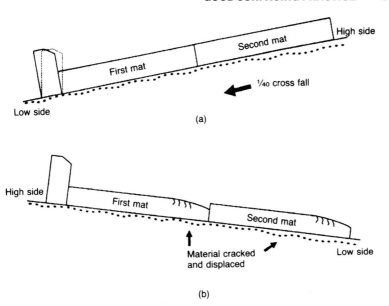

Fig. 7.17. Laying on a crossfall

- Tack coat:
  - If the material is being laid at less than 30 mm thick then it should be hot applied by tanker using a polymer modified emulsion.
  - Otherwise hand sprayed K1-40 emulsion should be adequate.
  - Binder should be applied at $0.5 \, \text{l/m}^2$.

- Paver should be in good working order. Faulty augers, flights, screws, etc. will all affect the quality of the mat.
- Great care should be exercised when reversing lorry onto paver.
- Less compaction is required compared with traditional asphalts — never over-roll and avoid flushing up bitumen at all costs.
- Usually use at least two rollers, one of which should be a tandem.
- Keep roller as close to the paver as possible.
- Vibratory compaction should be avoided.
- Longitudinal joints should be cut and painted like other asphalts.
- If gritting is required, spreading should take place before final rolling at a rate between 0.5 and $2 \, \text{kg/m}^2$.
- Hand raking should be kept to an absolute minimum with placement by fork and raking with wide toothed rake.
- Layer thickness should not exceed 50 mm. Minimum thicknesses are dependent on the nominal stone size (nss) of the mixture:
  - for 6 mm nss, the minimum thickness is 15 mm
  - for 10 mm nss, the minimum thickness is 20 mm
  - for 14 mm nss, the minimum thickness is 25 mm.

Table 7.12.   Use of pavers

Safety comes first. Observe all safety precautions
A paver should not stop when laying to obtain best results
Flow control gates should be used
The paver should go to the truck and not vice versa
Material cools on the flat surfaces of the hopper which should be folded after at least every second truck load
A large permutation of speeds is available because of variable engine speed, a large gearbox and/or hydrostatic drives
The screed need only be 3–8 °C above the temperature of the material being laid
It may take up to five tow lengths for a mat thickness correction to become fully effective. Use two tow lengths as a rule of thumb
Only two men should touch the controls of the paver, driver and screwman
The tampers should work in opposition to each other
Tampers are important and should be set correctly: 1 mm for fixed screeds, 2 mm for telescopic screeds, above the surface of the screed plate
Tyre pressures should be kept equal on both sides of the machine
The auger is for spreading material and not mixing it and the material should be correctly mixed before it reaches the paver
The floating screed will react to different temperatures in material and, therefore, these should be kept consistent
A roller will mirror an uneven contour on deep lifts due to differential compaction
The paver engine has to be checked for the correct amout of lubrication, fuel oils and water before starting
The oil levels in the hydraulic system must be correct before starting
The paver should be cleaned and lubricated at the end of the working day and not the following morning
Engine oil pressure and battery charging rates should be monitored when working
The bolts holding the screed to the side arms must be checked for tightness, especially after adjustment of the turnbuckles
The botton tip of the auger should be 50–70 mm above the finished surfaced of the mat for best results
Over-vibration is harmful to the machine and to the mat
High-amplitude, low-frequency beat can be eliminated by speeding up either the tamper or the vibrator speed will eliminate these

For the highest densities, work in the lowest traction gear or speed
The parking brake has to be applied before starting the engine
All gears must be in neutral before starting the engine
The auger clutches must be checked to ensure that they are disengaged before starting the engine
The traction clutch should be disengaged or the traction speed control lever should be in neutral before starting the engine
All control switches must be in the off or neutral position before starting the engine
If standing for a period, put gears in neutral and engage the traction clutch to reduce wear
The maximum aggregate should be less than 2/3 of the mat thickness
Vibrators should always be set in line (in phase)
Material must not be dropped in front of the paver
The flow gates must not be wide open unless necessary
Fold the hopper after every second load
Auger boxes must not be over filled
Do not pave in too high a gear or speed bearing in mind the supply of material available
The lorry must not be permitted to hit the paver
Work with the augers at the correct height if height adjustment is available
Do not overheat the screed
Do not turn the air off when extinguishing the flame in the screed heater
Support the screed adequately and evenly when work is finished otherwise it may cool down with a twist in it
Park the screed with even support
Have the hoist cables or rams supporting the screed weight except when lifting the screed from the mat
Do not make any sudden movement on controls, i.e. main clutch, pump speed control, brakes, or excessive lock on front wheels except in cases of emergency
Do not attempt to move the tampers until the screed is heated
Do not idle the engine for long periods
Do not run the engine with a faulty injector
Do not permit the fuel to become contaminated with dirt or water

Table 7.12.   Continued

Take anti-frost precautions, remembering also the tyres if these are hydro-filled
Check engine for abnormally low or high oil pressures
Keep the various reservoirs, i.e. engine fuel, hydraulic oils, etc., topped up
Examine all fuel, air and oil filters regularly
Do not run the engine with abnormally high water temperature
The screed control buttons and switches should not be operated needlessly
The screed must not sit stationary on the laid mat with the burners operating
The deckplates must not be lifted while the engine is running due to danger from moving parts
Remove electrical leads from battery and alternator before carrying out any welding on paver
No one must stand in the hopper while the conveyor is running
Fit safety pins to auger clutch levels before travelling machine
Exhaust air reservoir before travelling machine if air-assisted clutches are fitted
Switch off the sensor unit before lifting screed
Drain the air reservoir (where fitted) daily after use
Have the engine running before moving hydrostatic control levers
Collect all accessories before moving paver from working site
If tyre pressures are reduced for extra traction, the tyres must be re-inflated before the paver is travelled

7.8.6.   Repair to and reinstatement of thin surfacings and SMAs

Since one of the main reasons for the use of thin surfacings/SMAs as an alternative to chipped HRA is to avoid the problems associated with the loss of chippings, remedial works in the early life of these materials are uncommon. Indeed, major surfacing contractors report substantially less remedial costs associated with thin surfacings/SMAs than is the case with chipped HRA. If remedial works are necessary the following points should be considered.

- Infra red re-heating and rolling is not advised since binder would rise to the surface resulting in a loss of texture and, therefore, skid resistance.
- Texture can be improved by the application of a high-pressure jet of water.
- Utility trench reinstatement should be undertaken using fresh, hot-mixed materials. The use of a hot box may be of substantial assistance.

- Smaller patches and potholes may be reinstated using the same normal stone size DBM basecourse or conventional cold-mix bagged material.
- As the use of thin surfacings/SMAs increases, it may well be necessary for suppliers to produce a cold-mixed bagged thin surfacings/SMA for use in reinstatements.

7.9. The application of precoated chippings into HRA wearing course

An HRA wearing course is almost unique in having precoated chippings added to it after placement by the paver but prior to compaction by the roller. (They may also be added to dense tar surfacing and fine textured bitumen macadam (i.e. fine cold asphalt) wearing courses, but the use of such materials is very rare nowadays.) Although the use of unchipped wearing courses is now the rule rather than the exception, chipped HRA continues to be used on many roads. Indeed, in Scotland, chipped HRA wearing course remains the standard material on trunk roads and motorways (refer to Tables 2.2E and 2.2S in HD 36,[7] which is part of the standard design method used in the UK, the DMRB).[8]

Chippings are usually 20 mm single size aggregate on major roads coated with $1.5\% \pm 0.3\%$ of 50 pen binder as specified in BS 594: Part 1.[9] The chippings are spread close behind the paver before the material is rolled. They should be rolled into the wearing course as soon as possible after the mat has been placed by the paver, although it should be borne in mind that very hot HRA wearing course may be displaced by the action of the roller.

The timing of further passes should be carefully judged. The material must not be so hot that the chippings are pushed so far into the surface that the mat loses all texture. Equally, it must not be so cold as to resist chipping embedment. Such a situation will almost certainly result in substantial amounts of chippings being lost from the surface a few days after being opened to traffic (a condition described as 'fretting'). In cold windy weather, rolling is undertaken as close as possible to the paver whereas in hot still conditions, some delay will be necessary after the initial pass of the roller.

The chipping machine, which is commonly used in the UK, was jointly developed by the TRL and the manufacturer, Bristowes Machinery Ltd. Chipping machines are available in a variety of widths corresponding to common surfacing widths. Their principle of operation is illustrated in Fig. 7.18.

The machine consists of a full-width spreading hopper which is fed by a small traversing hopper. This feeding hopper runs on rails along the top of the main hopper, thus facilitating the maintenance of adequate chippings in the main hopper. A manually operated gate at the base of the traversing hopper allows operatives to control the flow of chippings from the traversing hopper into the main hopper.

The required chipping rate is defined in BS 594: Part 2[10] as either that which is necessary to give 70% or 60% (if that will achieve adequate

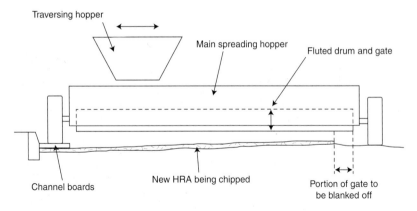

Fig. 7.18.   Schematic diagram of a typical chipping machine

texture) of the shoulder-to-shoulder cover. The rate is measured using the method prescribed in BS 598: Part 108.[11] In mechanical terms, the rate is controlled by a moveable gate and a revolving fluted drum, both of which run the full length of the machine. Calibration of the machine is checked by running the machine prior to the commencement of surfacing and using chippings trays and spring balances. The area of trays is simple to calculate and the spring balance gives the weight and therefore the mass per square metre. The method given in BS 598: Part 108[11] specifies the use of five trays to give the rate of spread on a half width of the chipping machine. However, the larger the trays the more accurate the measurement. Trays that are 300 mm square (area $= 0.09\,\mathrm{m}^2$) are specified[11] but trays having an area of $1\,\mathrm{m}^2$ are to be preferred. If contractor and engineer can agree this method then assessment is simpler, quicker and more accurate.

Figure 7.19 shows the chipping density being checked and Fig. 7.20 illustrates the cross section of the main hopper and control gate of a chipping machine.

The chipping machine will not run for long without needing some adjustment.

- Changes in gradient affect the rate of spread. Where the drum is in front of the backplate (generally the case in older machines) the machine will chip more heavily downhill and more lightly uphill. This situation is reversed with hydraulically driven machines which usually have the drum behind the hopper backplate.
- The chipper gate control often has some slack in its mechanism and this will have to be allowed for when making adjustments.
- The backplate is easily damaged when driving over rough ground or kerbing. If there is a gap between the backplate and drum, any eccentricity in the drum will allow chippings to drop through on each revolution and create a ribbed (often called a 'washboard') effect on the road surface.

Fig. 7.19. Checking the rate of spread of precoated chippings. Photograph courtesy of Wrekin Construction

- Small blockages will produce a longitudinal line of low rate of spread which can be corrected by the addition of further chippings by hand. An iron bar (a crowbar or similar) is often used to dislodge small clumps of precoated chippings and is an essential part of the chipping machine driver's equipment.
- The most important factor related to erratic chipping is the condition of the precoated chippings. Chippings should always be wet before use

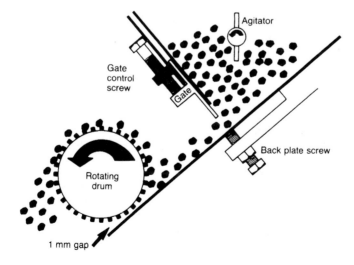

Fig. 7.20. Cross-section of a chipping machine hopper

except where the temperature is at or below 0 °C. It is not uncommon on site for stockpiles to be watered before laying begins. The chipping machine may chip well initially but then it may start to spread lightly or erratically. The reason for this is that the wet precoated chippings have been used up and dry ones have begun to stick in the machines. The solution is not to increase the chipping rate but to adjust the poor areas by hand and water the precoated chippings as soon as possible.

• Calibration trays should be checked each time before being used as precoated chippings. Excessive water or asphalt adhering to the bottom of the trays will distort the result.

• Precoated chippings can easily jam in one of the many sprockets on the drive mechanism. This will produce a lurching motion which will throw rows of precoated chippings off the drum and distort the rate of spread.

If chippings are not required for the full width, part of the gate can be closed off with one or more blanking plates. It is usual to have a channel which is not chipped. This facilitates the flow of surface water shed from the carriageway along the channel. This is achieved by placing timber boards in the channel after the asphalt is placed but before the chippings are spread. The nearside wheels of the chipping machine may run on these boards.

The chippings should be thoroughly loosened in their stockpiles and, thus, be free flowing when they are fed into the chipping machine. Chippings which are newly manufactured should be placed in stockpiles less than 1 m in height or sprayed with water to facilitate cooling in line with the recommendations contained in BS 594: Part 1.[9] Similarly, during and after warm weather, the bitumen on the chippings may soften and can result in their agglomeration. They may be loosened by being repeatedly dropped from a height by a tractor. This is unsophisticated but effective and the chippings do not shatter.

The operation of the chipping machine should be as continuous as possible if a uniform rate of spread is to be achieved. The chipping rate needs to be checked regularly since it may vary during operation of the machine, particularly if the gradient or the chipping characteristics change. Chippings should not be left in either of the hoppers overnight or for any lengthy period since there is a tendency for them to stick together and jam in the machine.

7.10. Rolling and compaction

The type and number of rollers to be used will be influenced by

• material specification
• layer thickness
• speed of laying
• compaction requirements.

The importance of using skilled, experienced roller drivers and the need to roll to a predetermined pattern for equal compactive effort across the full width of the mat cannot be over-emphasized.

At the start of rolling the temperature of the mat should be at the maximum permitted by the specification. This allows the maximum time for optimum compaction before the minimum rolling temperature is reached. It also avoids the possibility of waves in the mat which may occur if the material is too hot.

The main factors affecting the time available for rolling are

- the thickness of the mat
- the temperature of the mat at start of rolling
- the wind speed.

Special attention must be given to the rolling of joints. Transverse joints (sometimes called 'cross joints') should be rolled immediately. As indicated above, the single location where failure is most likely is in the vicinity of transverse joints against existing material or material which has cooled overnight. Longitudinal joints (sometimes called 'long joints') should be rolled progressively as the paver moves forward. They must be rolled before the adjacent mat is rolled.

The basic rules for rolling are

- follow the paver as closely as possible
- compact the joints first
- start compaction at the low edge of the mat
- turn off vibrations before reversing when using vibrating rollers
- change rolling speeds smoothly
- run forwards and backwards in the same rolling lane
- change rolling lanes on the cold side and avoid lane changes where the material is hot
- run in parallel rolling lanes and reverse at another section away from these adjacent rolling lanes
- keep the drums sufficiently wet to avoid picking up hot material but not more than necessary
- do not let the roller stand on the hot mat.

7.10.1. Rolling patterns
Figure 7.21 illustrates good rolling practice and is based on the following points.

- Rolling patterns should start at the joint and move across the mat with overlapping passes.
- Each run should advance across and forward along the mat as the paver progresses. In practice, this operation is not easy with as many as three rollers on the mat.
- The rolling pattern should be agreed between the foreman and roller drivers before laying begins.
- Rollers should approach the paver on the first run and then reverse away from the paver in a straight line on the return run.
- The move over to the second pass should be made away from the paver on the cooler compacted material to avoid undue displacement of the uncompacted material.

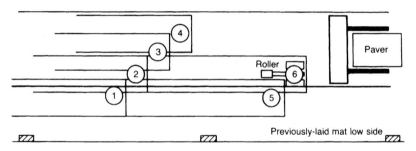

Fig. 7.21. Rolling of asphalts

- Shorter runs give better compaction with the first roller on hot material immediately behind the paver. However, short runs have the disadvantage of frequent stops and manoeuvres on hot material which produces irregularities in the finished work.
- Runs of 30 m–50 m and one roller for every 50 t per hour laid give good compaction for DBM laid in 100 mm thick layers.

The Department of Transport Standing Committee Preferred Method 8^{12} gives good guidance on all aspects of rolling, although it suggests that low air temperature substantially reduces the time available for rolling. As this and other Chapters have emphasized, the single most important variable in this respect is the wind factor. Compaction equipment manufacturers[13,14] also publish useful information on all aspects of rolling including patterns.

7.10.2. Correlation between layer thickness and density of the compacted mat

The standard UK specification for highway construction is the SHW[1] and is discussed below in Section 7.13. In respect of macadams, it requires compaction trials to be carried out on both the proposed and laid materials. This is carried out by means of both a Nuclear Density Gauge (NDG) and the Percentage Refusal Density (PRD) Test. The procedure for the former can be found in TRL Supplementary Report 754,[15] whilst the latter is detailed in BS 598: Part 104.[16] The introduction of the PRD Test and the predominance of the use of dense bitumen macadam roadbases in varying thicknesses has meant that it has been necessary to establish the optimum layer thickness at which the contract PRD can be achieved. The use of NDGs is now specified in the SHW.[1] Their principle of operation is illustrated in Fig. 7.22.

Experience has suggested that when laying 28 mm or 40 mm dense bitumen macadam roadbase or a heavy duty macadam (HDM) roadbase, failure to achieve PRD can be expected when the layer thickness is less than 70 mm and that layers of thickness greater than 85 mm usually meet the PRD specification with optimum thicknesses being 95 mm–100 mm. The reasons for this can be seen in Fig. 7.23.

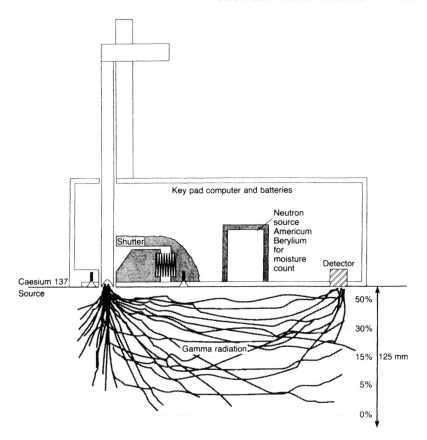

Fig. 7.22.   Principle of operation of a Nuclear Density Gauge

7.10.3.   Factors affecting rolling technique and the achievement of optimum mat density

When asphalts are placed in situations with good containment, e.g. in a trench or against a kerb, good compaction is achieved relatively easily since the material cannot spread under the compactive effect of the roller.

If the material is unrestrained then the only forces acting to prevent the material from spreading are the tensile strength of the material, which only becomes significant as the material cools, and the friction between the mat and the underlying surface. Too much compactive effort (called 'over-rolling') can crush the aggregate at points of contact. This is particularly true where the aggregate is a limestone. When this happens, the bitumen bond is broken at the point of contact. The effect is a loss of tensile strength which is not regained as the asphalt cools. The result will be that the mat will, when rolled, spread rather than compact. Once the tensile strength of the mixture is lost then the only force acting to keep the mat together is the friction between the fresh material and the underlying surface. Where this layer is a sub-base or a dense bitumen macadam using

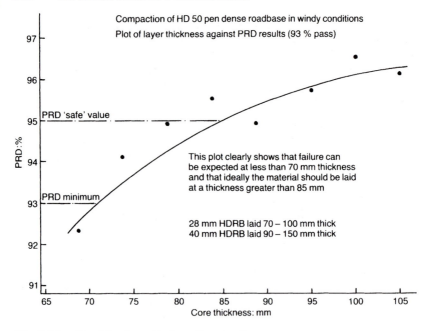

Fig. 7.23. *Densities from Nuclear Density Gauges*

a limestone aggregate which has been rolled such that the aggregate which is exposed at the surface has been crushed leaving a fine powder, there will be no bond. Similarly, there may be no bond if the underlying surface is wet or very cold. When any of these conditions exist then a crazing pattern may appear on the surface of the mat and the density of the layer will be very low. Once this has happened, the situation cannot be retrieved without replacing the mat. This situation was very common when heavy vibrating rollers were first used in road construction.

If this possibility is to be avoided then the initial compaction should be kept to the minimum until the mat has gained tensile strength when final compaction can commence. This is followed by further rolling to close up ('polish') the surface of the mat. Successful rolling is largely dictated by experience.

The following factors have an influence on how rolling should be undertaken and the resultant mat

- material characteristics
- gradient
- crossfall
- degree of edge restraint
- type and number of rollers used
- surface condition (wet or dry)
- surface temperature
- cleanliness of surface.

It is good practice for the roller driver to keep away from the low-side free edge after the first pass and to roll off at the minimum rolling temperature or lower when rolling a fragile material in adverse conditions. Since this material is subsequently removed before the adjacent lane is placed, the relatively low level of final compaction is unimportant. Compaction is also discussed in Chapter 8.

7.11. Vibratory compaction

In static mode, a vibratory roller compacts in exactly the same way as a traditional three-point dead weight roller, i.e. it compresses the mat and squeezes the air out of the material thereby reducing the air voids within the mixture. When used in the vibratory mode, the roller has a fundamentally different effect on the mat. Unless this fundamental difference is appreciated and fully understood then it is very easy to damage the mat beyond redemption.

The compactive effort is influenced by three main factors — the Static Linear Load (SLL), the frequency of oscillation and its amplitude. Other factors include the rolling speed, the number of vibrating drums and weight of the frame. The SLL is the weight of the roller divided by the rolling width of the drum expressed in kg/cm or N/mm. The higher the SLL, the fewer passes are required to compact the mat to an adequate level. The frequency is the number of drum impacts per unit time measured in Hertz (Hz) or v/min (vibrations per minute). Amplitude is the maximum movement of the drum from the axis and is usually measured in millimetres. This means that the drum movement corresponds to twice the actual amplitude. On asphalts, a frequency between 35 and 50 Hz ($\equiv$ 2000 and 3000 v/min) have been found to produce the best results. Suitable amplitudes for asphalts range from 0.4 to 0.8 mm. High frequencies result in small impact spacing (the distance between each drum impact), which helps prevent rippling of the surface.

The impact spacing is a function of frequency and speed, with low frequency at high speed giving wide impact spacing whilst high frequency at low speeds gives close impact spacing. Dual amplitude is often advantageous since it enables the driver to vary the vibratory force which is exerted by the roller. This facility permits the compactive effort to be adjusted for different types of materials and layer thicknesses. Variable amplitude is essential in asphalt compaction. Mixtures having good workability and laid in thin lifts should be rolled at low amplitude. Conversely, mixtures of poor workability laid in thick lifts require relatively high amplitude.

Modern asphalt rollers are equipped with automatic vibration control which shuts off vibrations below a prescribed speed. This prevents vibrations acting on the surface when the roller is stationary or when it slows down to change the direction of travel.

Rolling speed also influences compactive effort. Rolling at higher speed requires an increased number of passes for equivalence. However, the optimum speed for rolling asphalts lies between 3 and 6 km/h.

7.12. Quality control and checks

Numerous factors have an influence on the quality of the finished product and these can be conveniently grouped under the following headings

- the gang
- the plant
- the weather
- condition of the site
- the material
- other factors.

7.12.1. The gang

Emphasis has already been placed on the need to use operatives who are well trained and experienced and under the control and guidance of a competent and enthusiastic foreman. Paver drivers, roller drivers, tractor drivers and chipping machine operators should all be formally accredited. Examples are the schemes established by the Construction Industry Training Board (CITB), the Construction Skills Certification Scheme (CSCS), or the Quarry Products Training Council (QPTC) Certificate of Competence. Whilst the CITB Certificate is largely based on 'grandfather' rights, the QPTC Certificate includes assessment on site. These qualifications or equivalent are the minimum required by most major clients in order to provide evidence of competence. These qualifications are being replaced by National Vocational Qualifications (NVQs) and both the CITB and the QPTC are offering NVQs in the road building sector.

The morale of the gang and the attitude of the site staff are extremely important. The temptation to increase output beyond that which can reasonably be sustained should be avoided. There is little point in ordering quantities of material which exceed that which reasonably could be laid in the normal working day. The inevitable result will be lorries standing on site containing asphalt which is losing heat whilst a tired and dispirited gang try to lay cold materials in the dark. The correct approach is to assess the gang's likely output and place a reasonable order taking into account all the circumstances.

If this happens the job will run smoothly, remedial works will be eliminated or, at worst, kept to a minimum and the supervisor's time can be spent more beneficially on planning and organization rather than troubleshooting.

7.12.2. The plant

All plant should be well maintained and under the control of a competent operator. Manufacturer's operating instructions and safety rules must be followed at all times.

The temptation to minimize hire charges is understandable in view of the high costs involved. The use of an additional dumper discussed previously in relation to material which is laid by hand, the need for an extra roller when laying high tonnages or the desirability of having standby plant such as a spare paver, roller and chipping machine on a major contract are all factors which may well have a direct affect on the quality of the finished mat.

Some breakdowns are unavoidable. On a critical job, the standby plant must be checked regularly to ensure that it is available for immediate use in the event of a plant breakdown. A spare roller is of no value if it has a flat battery or seized water pump.

7.12.3. The weather

Bad weather is the cause of more disruption and poorer quality work than the combined effect of all other problems. Since the UK does not suffer from the extremes of weather to the degree found in some other parts of Europe, asphalts are regarded as materials which can be laid in all weather conditions. As a consequence, some regard it as being unworthy of the care and attention which is afforded to some other construction materials. However, much can be done to reduce the detrimental effects of poor weather.

- If frost is forecast for the next day, laying a 100 mm thick layer of roadbase is likely to ensure that the mat will be frost free and ready for the next layer the following morning.
- On a sunny winter's day, the surface temperature of the road is often three or four degrees above the reference air temperature used by the Engineer's staff. The use of thermometers at the point of laying can often mean an earlier start for operations and the avoidance of keeping material standing in lorries.
- A cross wind has a far more detrimental effect on asphalts than either low air temperature or light rain. A still, dry atmosphere with air temperatures below freezing may well be ideal laying conditions. The wind will have the effect of substantially reducing the time available for compaction.
- Laying in wet or very damp conditions should be avoided if the low side edge of the mat is unsupported. In such circumstances, the material may move sideways under rolling due to lack of adhesion to the surface below and the lack of edge restraint.
- Weather centres, such as those located at local airports, are extremely good at forecasting likely weather conditions up to 2–3 h ahead which should be enough time to plan for the delivery of materials from the supply source to the site.
- It is relatively easy to insulate the sides of the hopper of the paving machine with 40–50 mm thick hardwood. This greatly reduces heat loss from the material in the hopper. Indeed, most paver manufacturers offer this as an extra. It is not expensive and this practice is strongly recommended to all surfacing contractors. All pavers should be manufactured with insulated hoppers as standard practice.
- Applying tack coat on existing surfaces the night before laying may well avoid delays the following morning waiting for the coat to break or for frost to thaw.

- Tighter control of water sprinklers on the wheels of rollers may be necessary in cold laying conditions.

The above list is not exhaustive but it serves to support the view, expressed earlier, that time spent in anticipating problems may well reduce remedial works which are always expensive.

7.12.4. Condition of the site

As discussed earlier, a thorough site inspection is necessary prior to commencing work on site. All too often this is not done and has catastrophic results.

7.12.5. The material

Generally speaking, asphalts are manufactured to high standards and most manufacturing plants now operate in line with the Sector Scheme Document for the Quality Assurance of the Production of Asphalt Mixes[17] which adds to the requirements set out in the normal standard for Quality Assurance BS EN ISO 9002.[18] The Sector Scheme[17] recommends an organoleptic, i.e. of the senses, inspection as the first level of checking the newly manufactured material. It is suggested that delivered asphalt should be the subject of a similar check. Material which contains some uncoated aggregate due to inadequate mixing or insufficient bitumen or some other plant problem can be spotted by an experienced eye. Similarly, it is possible to smell bitumen which has been overheated. The temperature of the delivered asphalt should be taken. It should comply with the values set out in the specification. SHW[1] contains minimum temperatures but if that document does not apply then there are also values contained in BS 594: Part 2[10] or BS 4987: Part 2.[19]

There are other steps which can be taken to minimize the possibility of non-compliant material being laid.

- Only lorries that are insulated and maintained properly should be used for the delivery of asphalts. It is strongly recommended that material in the lorry should be double sheeted, particularly in the case of HRA wearing course.
- Lorries should not stand on site for long periods.
- Segregation problems with macadams can be avoided by ensuring that loose aggregate from the side of the load in the lorry is well mixed in the hopper of the paver. Similarly, a build up of material should be avoided at the ends of the feed screws in the paver.
- Cold material must not be allowed to accumulate in the corners of the paver hopper. This is achieved by the paver driver regularly raising the sides of the hopper.
- Asphalts should be laid at a thickness such that the coarse aggregate does not crush when the material is compacted. Thicknesses are specified in BS 594: Part 1,[9] and BS 4987: Part 2,[19] but a useful rule of thumb is that the layer thickness should be at least two and a half times the nominal stone size.

7.12.6.   Other factors

There are other factors which may affect the quality of an asphaltic mat.

- A surfacing gang which is well trained will take pride in its work with each member of the gang taking personal responsibility for his part of the operation. The golden rules are do it once, correctly with each man sticking to his own job.
- Attention to detail is of paramount importance. Premature failure and the resultant extensive remedial works can often be traced back to a lack of such attention.
- The foreman must be free to oversee the operation and check the newly laid asphalt.
- Constant adjustment of plant and equipment should be avoided. Regular adjustment of the paver screed should not be necessary. Once the paver has been set up, the paving train is in motion and material is arriving at regular intervals, little can go wrong.
- Frequent checking of the following will ensure that quality is maintained

  - material consistency and temperature
  - weather conditions — temperature, rain and, most important of all, wind
  - a steady flow of material without unnecessary stoppages
  - level control, mat thickness and rolling temperature
  - efficient rolling patterns — the importance of compaction cannot be overstated
  - in the case of a chipped HRA wearing course, regular use of chipping trays.

- The surfacing, and in particular the wearing course, should be the last operation on the contract. However, conflicting priorities often dictate that this may not be the case. The temptation to use asphalt layers as a convenient workbench or temporary haul road should be resisted wherever possible. Where this is unavoidable then adequate protection measures must be undertaken.

7.13.   Requirements of the Specification for Highway Works

Trunk roads are roads which have been the subject of a 'Trunking Order'. A Trunking Order makes central government the 'Roads Authority' (as defined in the Highways Act 1980 for England, Wales and Northern Ireland and the Roads (Scotland) Act 1984 for Scotland) and, consequently, responsible for the maintenance of the road. This duty is undertaken in England by the Highways Agency, in Scotland by the Scottish Executive (formerly The Scottish Office), in Wales by the Welsh Office and in Northern Ireland by the Department of the Environment for Northern Ireland. These Agencies publish the standards to be employed for road construction and maintenance in a suite which is entitled the Manual of Contract Documents for Highway Works[20] (MCHW). The

MCHW is written under the auspices of the Highways Agency with contributions from the Scottish Executive, the Welsh Office and the Department of the Environment for Northern Ireland.

The MCHW[20] is published in seven Volumes

- Volume 0: Model Contract Documents for Major Works and Implementation Requirements
- Volume 1: Specification for Highway Works (SHW)
- Volume 2: Notes for Guidance on the Specification for Highway Works (NGSHW)
- Volume 3: Highway Construction Details
- Volume 4: Bills of Quantities for Highway Works
- Volume 5: Contract Documents for Specialist Activities
- Volume 6: Departmental Standards and Advice Notes on Contract Documentation and Site Supervision

In the UK, Highway Authorities for the vast majority of roads are central government and local authorities. The motorway network is some 3300 km in length, whilst other trunk roads exceed a further 12 000 km in length together these represent some 4.2% of the total length of road network. Other principal roads are some 36 000 km in length, other classified roads are some 113 000 km in length and unclassified roads are some 205 000 km in length; together these represent some 95.8% of the total length of road network. Even given the fact that motorways and trunk roads have more lanes than most other roads, it can be seen that local authorities are responsible for the maintenance of most of the road network in the UK. Most of these authorities adopt the MCHW[20] as their standard roads specification.

The government Agencies adopt an enabling role and do not, themselves, undertake the detailed preparation of road construction and maintenance works or the supervision of same. Such duties are undertaken by local authorities or private sector civil engineering consultants. It is a requirement of the Agencies that the terms of the MCHW[20] apply to all such works.

In technical terms, the interest of surfacing contractors will focus most closely on the contents of Volume 1 and 2. The SHW[1] and the NGSHW[3] dictate the technical standards which apply to the construction and maintenance of trunk roads in the UK. Whilst the SHW[1] contains the specification, the NGSHW[3] provides guidance and amplification of the specification. Most, although not all, Clauses within the SHW[1] contain an equivalent Clause within the NGSHW[3] which provides this guidance and amplification. Consequently, the SHW[1] must never be read without reference to the equivalent area within the NGSHW.[3]

The standards related to the manufacture and laying of asphalt are contained within Series 700 and 900 of the SHW[1] and the NGSHW.[3] Clauses are numbered to match the Series number, e.g. 701, 702, etc. in the SHW[1] and NG700, NG701, NG702, etc. in the NGSHW.[3] Sub-Clauses are numbered 1, 2, 3, etc. The following section cites and comments on requirements within the MCHW and the relevant Clause numbers are given in brackets as (#702.9), for example. Where a # precedes the Clause number, then there is a change for at least one of the Overseeing Departments of Scotland, Wales and Northern Ireland and reference to these amendments should be made for the country concerned.

7.13.1.   Series 700 Road Pavements — General

The edge of the pavement shall be correct within 25 mm (#702.1).

Tolerance levels which apply to the finished surface levels of pavement layers are given in Table 7/1 in the SHW[1] and shown below as Table 7.13 (#702.2).

It is not permissible to accumulate these tolerances and they must not result in a reduction in the combined thickness of the roadbase, basecourse and wearing course of more than 15 mm or a reduction in the thickness of the wearing course of more than 5 mm (#702.3). Surface levels are checked using a grid of points located as described in Appendix 7/1 which is included in any contract let under the MCHW.[20] Pavement layers other than the wearing course are deemed to comply when not more than one in any ten consecutive measurements either longitudinally or transversely does not exceed the tolerance given in Table 7.13. In the case of the wearing course, compliance is achieved when no point exceeds the tolerance (#702.4). The grid size is decided by the person who prepares the document (the 'compiler') and stated in Appendix 7/1 of SHW[1] but it is normally 10 m longitudinally and 2 m transversely. Compliance may be waived if levels are consistently high over long lengths but clearance under bridges is not affected, drainage is not impaired, all tolerances other than those on the final road surface comply and rideability is acceptable (NG 702.3).

As far as surface regularity is concerned, the tolerances are given in Table 7/2 of the SHW[1] and shown below as Table 7.14 (#702.5).

Table 7.13.   Tolerances in finished levels of pavement courses[1]

Road surfaces — general — adjacent to a surface water channel[a]	±6 mm + 10–0 mm
Basecourse[a]	±6 mm
Top surface of roadbase in pavements without basecourse[a]	±8 mm
Roadbase other than above[a]	±15 mm
Sub-base under concrete pavement surface slabs laid full thickness in one operation by machines with surface compaction	±10 mm
Sub-bases other than above	+ 10–30 mm

[a] Where a surface water channel is laid before the adjacent road pavement layer the top of that layer, measured from the top of the adjacent edge of the surface water channel, shall be to the tolerances given in table 7/1 of SHW

Table 7.14.   Maximum permitted number of surface irregularities[1]

	Surfaces of carriageways, hard strips and hard shoulders				Surfaces of lay-bys, service areas, all bituminous basecourses and top surface of roadbases in pavements without basecourses			
Irregularity	4 mm		7 mm		4 mm		7 mm	
Length: m	300	75	300	75	300	75	300	75
Category A[a] roads	20	9	2	1	40	18	4	2
Category B[a] roads	40	18	4	2	60	27	6	3

[a] The category of each section of road is described in Appendix 7/1 of SHW

   An irregularity is defined as a variation of not less than 4 mm or 7 mm of the profile of the road surface. As can be seen, the permitted number of irregularities is a function of the Road Category, either Category A or Category B. The appropriate Category is stated in Appendix 7/1 within the contract documents. Generally speaking, only roads where the average speed is 50 km/h or less are accorded Category B status with all others being designated a Category A classification (NG 702.5). Surface regularity is measured using the rolling straight edge and this apparatus is shown in Fig. 7.24.

   The rolling straight edge has a bell which rings whenever a pre-set depth of irregularity is reached or exceeded. A gauge at the front of the machine then enables the user to read the depth of irregularity. The level at which the bell rings is normally set at 4 mm. All rings are recorded as being equal to or greater than 4 mm but less than 7 mm or as being equal to or greater than 7 mm but less than 10 mm or as being equal to or greater than 10 mm. No irregularities above 10 mm are permitted (#702.5).

Fig. 7.24.   Rolling straight edge. Photograph courtesy of the City of Edinburgh Council

Surface regularity is checked on completed lengths of 300 m, or 75 m if the total length of the pavement is less than 300 m (#702.7).

Levels across the mat are measured by use of a 3 m straight edge. The maximum allowable difference between the surface and the underside of the straight edge when placed at right angles to the centre line of the road at regular intervals as stated in Appendix 7/1 (#702.8). In addition, the straight edge is used to check the longitudinal surface regularity for lengths less than 75 m of wearing courses, basecourses and top surface of roadbases in pavements without basecourses and concrete slabs or where use of the rolling straight edge is impractical. In such circumstances, the maximum allowable difference shall be 3 mm for pavement surfaces, 6 mm for basecourses and the top surface of basecourses (#702.9).

The texture depth requirements and measurement thereof are discussed in detail in Section 9.4.6 of Chapter 9.

Rectification is required where surface levels, surface regularity, thickness, texture depth, material properties or compaction do not meet stated requirements. The full extent of the area which does not comply '*shall be made good*'. The top layer of asphalt roadbases or basecourses or wearing courses are to be replaced with fresh material for the full length and width of the non-conformity. The length which is to be replaced is to be at least 5 m long in the case of roadbases or basecourses and 15 m in the case of wearing courses and the width in all cases is to be that which was laid in one paving operation. Where rectification is due to non-compliance with textural requirements then the areas of wearing course which are replaced shall be sufficient 50 m lengths (starting with the section having the lowest texture depth values) to return the surface to compliance. Replacement consists of the full depth of wearing course being removed or 20 mm if the repave method (see Section 10.9.1 of Chapter 10) is adopted (#702.10). Note that checking of the wearing course must be carried out before the road is opened to traffic. A surfacing contractor may be rightly aggrieved if the checking for compliance is not carried out until after the road has been trafficked.

There are strict requirements relating to weather conditions and when asphalts can (or cannot) be laid. SHW[1] contains, as Figures 7/1 and 7/2, restrictions related to air temperature and wind speed and these are reproduced here as Figs 7.25 and 7.26. Note that there are similar Figures for a porous asphalt wearing course (Clause 938).

An anemometer is required by the SHW[1] (Clause 708.8). It is to be positioned either at 2 m or 10 m by implication (the previous version was much clearer on this point). Although there are questions about the relevance of the detailed requirements (discussed in Section 8.4.4 of Chapter 8), these standards must be applied by any surfacing contractor.

Except as specified below, asphalts can be laid in light precipitation providing that both the temperature of the surface to be covered and the air temperature are above 0 °C (#708.1). Light precipitation is defined as rain which is <0.5 mm/h (NG 708.6). In the absence of rain, laying may proceed if the temperature of the surface which is to be covered is ≥ 0 °C and the air temperature is ≥ -1 °C except as set out below (Clause 708.1).

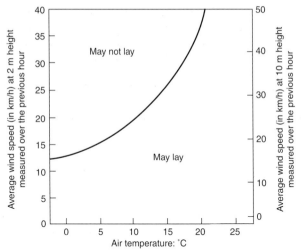

Fig. 7.25.   Wind speed and air temperature laying restraints for 45 mm thickness rolled asphalt wearing course[1]

Unless stated differently in Appendix 7/1, HRA wearing course with 30% coarse aggregate is to be laid 45 mm thick and delivered at a minimum temperature of 155 °C and can only be laid within the wind speed/air temperature envelope shown in Fig. 7.25. When an anemometer is not available (and the SHW[1] does not seem to permit same) then the minimum delivery temperature is raised to 165 °C and it cannot be laid if the temperature falls below 5 °C unless the temperature of the surface to

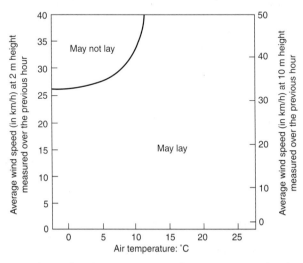

Fig. 7.26.   Wind speed and air temperature laying restraints for dense bitumen macadam wearing course or basecourse, or heavy duty macadam basecourse layers[1]

be covered is $\geq 3\,°C$ (Clause 708.3). Unless stated differently in Appendix 7/1, HRA wearing course with 35% coarse aggregate is to be laid 50 mm thick and delivered at a minimum temperature of 155 °C and can only be laid where the air temperature $\geq 0\,°C$ and the wind speed <40 km/h at 2 m or 50 km/h at 10 m anemometer height (Clause 708.4). Unless stated differently in Appendix 7/1, HRA basecourse and roadbase can be laid when the air temperature $\geq 0\,°C$ and the wind speed <40 km/h at 2 m or 50 km/h at 10 m anemometer height. The minimum delivery temperature is 135 °C (Clause 708.5).

Dense and heavy duty macadams basecourses and wearing courses can be laid at least 50 mm thick in light precipitation if the surface and air temperatures are $>0\,°C$ or, if dry and unfrozen, then if the surface temperature is $\geq 0\,°C$ and the air temperature is $\geq -1\,°C$ (Clause 708.6). Compliance with the latter restrictions is waived if compaction, i.e. PRD, is to be assessed (Clause 708.7). PRD testing is discussed more fully in Chapters 4 (Section 4.7.4) and 9 (Section 9.4.5).

7.13.2.   Series 900 Bituminous Bound Materials
In laying terms, this Series contains requirements related to good practice and most of these have been described above. However, some points are worthy of note.

Joints are to be staggered by at least 300 mm from the layer below. (This contrasts with the requirements for the widening of pavements contained in HD 27[22] which can be found in the design equivalent of the MCHW,[20] the *Design Manual for Roads and Bridges*[8] (this Manual, particularly Volume 7 thereof, is discussed in much greater detail in Chapter 4 of this book). Joints in wearing course or porous asphalt are to coincide with the lane edge or lane marking, as appropriate. Longitudinal joints are not permissible in the wheel tracks of materials which are subject to PRD testing (Clause 901.23).

The upper roadbase in pavements which have no basecourse or basecourse is to be covered within three consecutive days unless an extension is agreed with the overseeing organization. This may be due to bad weather or other good reason (Clause 901.26). No sanction for non-compliance is given.

Regulating courses are to comply with the appropriate tolerances for horizontal alignment, surface levels and surface regularity (Clause 907.2). The tolerances for horizontal alignment, surface levels and surface regularity of milled (often called planed) carriageways are those given for basecourses. Non-compliance is to be rectified by means of further milling or the use of a regulating course, as appropriate (Clause 917.3).

7.14.   Economic analysis of operations
In view of the considerable value generated by supervising squads daily, it is of paramount importance that cost/value comparisons are made on a daily basis at site level and summarized on a weekly basis for submission to management.

7.14.1. Estimate summary

An extract of the original estimate for the contract should be provided to the supervisor and/or foreman. Table 7.15 shows a typical summary of the estimate which would be issued to a supervisor. If this information is to be given to a foreman, the cost per tonne columns would normally be excluded. The two most important columns in this document are the 'Supers' and 'Output'. The former column indicates the area which must be covered by one tonne of material, whilst the latter indicates the average output which must be achieved.

7.14.2. Cost/value

At the end of each working day, the supervisor should complete the daily costs and value building up to a weekly total. This document should highlight the following

- Loss of material by not achieving the estimated 'Supers'.
  If this is the case, the dips on the underlying layer should be reviewed and a request for payment of additional tonnage should be made if this is warranted. It is normal practice for the surfacing contractor to agree a price per tonne with the main contractor for just such a contingency. Whilst the main contractor will have a tolerance for his finished work, e.g. the finished level of the sub-base, this is ignored for the purpose of calculating additional tonnage used. So, whilst the main contractor may lay the sub-base generally low albeit within tolerance, he will have to recompense the surfacing contractor for the additional asphalt which is required.
- The actual outputs which have been achieved.
  If the actual outputs fall short of those which were estimated at the time of tender then the reasons for such shortfall must be investigated. Other than on very small works, it is unusual for the surfacing contractor to be given possession of the entire site and so it is important that the areas presented for surfacing are in accordance with the agreed programme. Main contractors are keen to have sub-base protected from the elements by a layer of asphalt at the earliest opportunity and so have a tendency to fragment the presentation of the works to the surfacing contractor. This must be resisted by the latter for two reasons. The first is that output reduces dramatically when only short runs or limited areas are presented. The second is that when the surfacing contractor takes possession of part of the site he becomes responsible for that portion and will be required to rectify other works, such as sub-base, if he attempts to lay on an area which is in dubious condition and damage results.
- Detention charges.
  The asphalt supplier normally allows 30 minutes to discharge a load. Thereafter, a charge is made for the vehicle being detained on site. Laying will not commence until sufficient material is on site to ensure that the paver is fed continuously. An element of detention charges is thus inevitable and will have been allowed for in the estimate. These

Table 7.15. Summary of the estimate

ESTIMATE: Distributor Road, Anytown

Item No	Description	Area (m²)	Supers (m²/t)	Tonnage (t)	Squad code	Output (t/day)	Total duration	Cost per tonne					Bill rate
								Labour	Plant	Material	Misc	Total	
1	HMB 15 pen, 275 mm thick	15500	1.32	11742	M8	733	16 days	1.31	0.48	21.35	0.07	23.21	20.39 sq m
2	DBM 28 mm nom, 55 mm thick	16750	7.6	2204	M8	551	4 days	1.74	0.64	22.60	0.09	25.07	3.83 sq m
3	HRA 30% Sch 1B, 45 mm thick	18120	9.3	1948	M12	390	5 days	3.72	1.54	27.75	0.13	33.14	4.17 sq m
	20 mm PCC whin included in above			200		incl. above				26.00		26.00	incl. above
4	DBM 20 mm nom, regulating course			150		150	1 day	6.40	2.33	22.75	0.33	31.81	36.9 tonne

NOTES: £0.10 per tonne allowed for detention charges.

charges should not exceed a minimal amount, say £0.15 per tonne. If this figure is exceeded, this indicates either poor ordering by the supervisor or disruption to the laying operation which would be reflected in the outputs achieved.

It is almost inevitable that additional works will be instructed. These tend to be instructed verbally with all concerned being quite clear on the instruction issued. However, when this results in an additional charge, the main contractor often requests details of the instruction which was issued. If the cost/value comparison is carried out on a weekly basis then any such instruction not issued in writing by the main contractor can be identified and an appropriate record (usually called a confirmation of verbal instruction (CVI)) is issued retrospectively by the supervisor.

A worked example of a weekly report is shown in Table 7.16 and this indicates a loss in the week of £4797. It is split into two parts. Table 7.16(a) contains an analysis of the cost elements of the week's activities whilst Table 7.16(b) is the equivalent for value. The format varies from company to company and may be on an A3 form or on both sides of a sheet of A4 paper.

The information provided identifies where the loss has occurred and allows management to take corrective action before the financial situation deteriorates further. Examination of the example shows the following.

- Outputs have been achieved, but does this refer to straightforward work—i.e. do the time consuming tie-ins remain to be undertaken?
- Material loss has occurred in the roadbase layer without a compensatory gain in the basecourse. Dip sheets for sub-base should be checked to calculate and claim the additional tonnage.
- A 7% loss (25 t/345 t) in HRA is unacceptable; why has this occurred? Is the final surface within tolerance? Has the material been tipped and, if so, why?
- Material losses account for

    $104 \, t \times £23.21 = £2413$

    $8 \, t \times £25.07 = -£200$

    $\underline{25 \, t \times £33.14 - £828}$

    Net loss $= £3041$

- Detention charges amount to

    $$\frac{£228}{4000 \, t} = £0.06 \text{ per tonne}$$

    which is acceptable and indicates that there has been no undue delay in discharging the material from the vehicles.

Table 7.16(a).   A typical weekly cost analysis

WEEKLY COST / VALUE ANALYSIS

WEEK ENDING:

COST

MATERIALS

Supplier	Material	Mon	Tues	Wed	Thurs	Fri	Sat	Sun	Total	Price (£)	Cost (£)	Remarks
AS	HMB	560	1070	1120	460				3210	21.25	68,213	
BQ	28 DBC				420				420	22.50	9,450	
BQ	30% HRA					370			370	27.65	10,231	
BQ	PCC		50						50	26.00	1,300	
	Detention	75	40	38	30	45					228	
TOTAL											£89,421	

LABOUR

Operatives	Hours	Mon	Tues	Wed	Thurs	Fri	Sat	Sun	Total	Price (£)	Cost (£)	Remarks
8	9	72										
8	10		80									
8	9			72								
10	9				90							
12	10					120						
TOTAL									434	12.50	£5,425	

PLANT

Hirer	Type	Mon	Tues	Wed	Thurs	Fri	Sat	Sun	Total	Price (£)	Cost (£)	Remarks
PH	Paver	X	X	X	X	X			5	150	750	
Own	10t Roller	X	X	X	X	X			5	40	200	
Own	10t Roller					X			1	40	40	
PH	Bomag 160	X	X	X	X	X			5	90	450	
Own	Van	X	X	X	X	X			5	35	175	
Own	Trac/comp	X	X	X	X	X			5	35	175	
PH	Chipper					X			1	25	25	
TOTAL											£1,815	

Misc

Supplier	Type	Mon	Tues	Wed	Thurs	Fri	Sat	Sun	Total	Price (£)	Cost (£)	Remarks
Stores	tack coat										25	
Stores	Gas										35	
TOTAL											£60	

TOTAL COSTS   £96,721

Delays and disruption: Mon: paver ordered for 7.00 am, low loader did not arrive on site until 9.00 am. Material on site from 8.15 am, detention charges and 3 hours down time for the squad.

7.15. Remedial works

On the ideal site, inclement weather would not occur, material supplies would always arrive on time and comply with the middle of the specification, plant would never break down, operatives would always work at peak performance and promises made by others would always be kept. Unfortunately, the ideal site does not exist and remedial work is often necessary.

Typically, an allowance of between 0.5 and 1% is added to the total cost at the estimating stage to cover average remedial costs. It is important that remedial costs are clearly identified and monitored but do not form part of the cost/value analysis. For example, a contractor may have included 1% in his overhead allowance for remedial work. This value

Table 7.16(b). A typical weekly value analysis

VALUE

Item No	Description	Quant.	Rate	Value £	Material				Outputs			
					Est. supers	Est. tonnes	Act. tonnes	Gain/loss	Est. output	Est. duration	Act. duration	Act. output
1	HMB 275 mm thick	4100	20.39	83,599	1.32	3106	3210	-104	733	4.38	3.5	917
2	DBM 55 mm thick	3250	3.83	12,447.5	7.6	428	420	8	551	0.76	0.5	840
3	HRA 45 mm thick	3210	4.17	13,385.7	9.3	345	370	-25	390	0.95	1	370
										6.09 days	5 days	
	TOTAL			109,432								
	LESS O/H & PROFIT ALLOWANCE			17,509								

TOTAL NETT VALUE   £91,923

LESS COSTS   £96,720

PROFIT / LOSS   £-4,797

would then be allocated to a global remedial fund and similarly the cost of carrying out remedials would be charged to this fund.

As a guide to likely costs, typically, a HRA wearing course may be laid for between £25 and £30 per tonne whereas removal and replacement would typically cost between £100 and £150 per tonne when account is taken of traffic management charges.

Surfacing contractors report that remedial costs on works employing thin surfacings for the wearing course are substantially less than those sites where a chipped HRA wearing course is used.

Clearly, removal of laid materials must be avoided if possible. When gauging the need for remedial work, the Engineer's staff should weigh up the benefits of replacement — often the replaced material is little better but the surface is scarred and extra joint lengths are left for maintenance by the client.

The increasing use of end result and performance-related specifications will allow for more appropriate methods of dealing with minor deviations from stated tolerances such as extended guarantees or reduced payments. Faced with the current situation, however, acceptable remedial methods must be negotiated between the parties.

Infra red road heaters, when operated under the control of an experienced foreman and in conjunction with an experienced raker and roller driver, can markedly alter an area where precoated chippings are not fully embedded or where they have been lost from the mat due to chilling. Similarly, small areas of total chip loss or potholing may be treated by means of invisible patching involving a reheating of the full depth of wearing course around the suspect area, removal by shovel of the affected material and the placement of fresh material. Careful raking of the treated patch followed by hand chipping and skilful rolling completes the operation, producing a reinstatement invisible to the inexperienced eye once traffic has run on the material for a few days.

7.16. Summary

This Chapter has considered all aspects of the activities of surfacing contractors. Safety matters are of paramount importance and must not be neglected. For too long, no thought was given to the safety of the gang. Whilst this trend has, in large measure, been reversed, efforts must continue to minimize the possibility of injury to all personnel on site. Improved welfare facilities are now required by law, reflecting a greater awareness of the importance of gang motivation.

The derivation of the estimate was considered, showing how each element of cost or potential cost is combined to produce the price at which work is offered, usually to main contractors. The technicalities of the laying process were covered in detail with guidance on how different aspects of work should be tackled. The need to comply with normal specification requirements was underlined with advice on meeting the target values. Finally, the performance was considered in economic terms.

As asphalts and surfacing plant continue to become ever more sophisticated, increasing reliance will be placed on the skill of the surfacing contractor, particularly in relation to the performance of his site management and his gangs. Only those contractors who think and operate in a progressive manner will be able to meet the needs of tomorrow's clients.

References

1. HIGHWAYS AGENCY *et al. Manual of Contract Documents for Highway Works, Specification for Highway Works.* TSO, London, 1998, **1**.
2. BRITISH STANDARDS INSTITUTION. *Specification for High-Visibility Warning Clothing.* BSI, London, 1994, BS EN 471.
3. HIGHWAYS AGENCY *et al. Manual of Contract Documents for Highway Works, Notes for Guidance on the Specification for Highway Works.* TSO, London, 1998, **2**.
4. DEPARTMENT OF TRANSPORT *et al. Traffic Signs Manual, Traffic Safety Measures and Signs for Road Works and Temporary Situations.* HMSO, London, 1991, Ch. 8.
5. HEALTH & SAFETY EXECUTIVE. *Provision of Welfare Facilities at Transient Construction Sites.* HSE, Sheffield, 1997, Construction Information Sheet No. 46.
6. HUNTER R. N. *Claims on Highway Contracts.* Thomas Telford Publishing, London, 1997.
7. HIGHWAYS AGENCY *et al. Design Manual for Roads and Bridges, Pavement Design and Maintenance, Surfacing Materials for New and Maintenance Construction.* TSO, London, 1999, 7.5.1, HD 36/99.
8. HIGHWAYS AGENCY *et al. Design Manual for Roads and Bridges.* TSO, London, various dates, 1–15.
9. BRITISH STANDARDS INSTITUTION. *Hot Rolled Asphalt for Roads and other Paved Areas, Specification for Constituent Materials and Asphalt Mixtures.* BSI, London, 1992, BS 594: Part 1.
10. BRITISH STANDARDS INSTITUTION. *Hot Rolled Asphalt for Roads and other Paved Areas, Specification for the Transport, Laying and Compaction of Rolled Asphalt.* BSI, London, 1992, BS 594: Part 2.
11. BRITISH STANDARDS INSTITUTION. *Sampling and Examination of Bituminous Mixtures for Roads and other Paved Areas, Methods for the Determination of the Condition of the Binder on Coated Chippings and for Measurement of the Rate of Spread of Coated Chippings.* BSI, London, 1990, BS 598: Part 108.
12. DEPARTMENT OF TRANSPORT STANDING COMMITTEE ON HIGHWAY MAINTENANCE. *Preferred Method 8 — Road Rolling.* Cornwall County Council, Truro, undated.
13. DYNAPAC HEAVY EQUIPMENT AB. *Compaction and Paving, Theory and Practice.* Dynapac, Karlskrona, Sweden, 2000.
14. KLOUBERT H-J. *Compaction of Bituminous Materials.* Bomag-Menck GmBH, Boppard, Germany, November 1997.
15. WORKING PARTY. *Nuclear Gauges for Measuring the Density of Roadbase Macadams.* TRL, Crowthorne, 1982, Supplementary Report 754.
16. BRITISH STANDARDS INSTITUTION. *Sampling and Examination of Bituminous Mixtures for Roads and other Paved Areas, Methods of Test for the Determination of Density and Compaction.* BSI, London, 1989, BS 598: Part 104.
17. SECTOR SCHEME ADVISORY COMMITTEE FOR THE QUALITY ASSURANCE OF THE PRODUCTION OF ASPHALT MIXES. *Sector Scheme Document for the Quality Assurance of the Production of Asphalt*

Mixes, Sector Scheme No. 14. Quarry Products Association, London, 1998.

18. BRITISH STANDARDS INSTITUTION. *Quality Systems, Model for Quality Assurance in Production, Installation and Servicing*. BSI, London, 1994, BS EN ISO 9002.

19. BRITISH STANDARDS INSTITUTION. *Coated Macadam for Roads and other Paved Areas, Specification for Transport, Laying and Compaction*. BSI, London, 1993, BS 4987: Part 2.

20. THE HIGHWAYS AGENCY *et al. Manual of Contract Documents for Highway Works*. TSO, London, 1998, **0–6**.

21. BRF. *Road Fact 99*. BRF, London, 1999.

22. HIGHWAYS AGENCY *et al. Design Manual for Roads and Bridges, Pavement Design and Maintenance, Pavement Construction Methods*. HMSO, London, 1994, 7.2.4, HD 27/94.

8. Compaction of asphalts

8.1. Preamble

Laying asphalts involves a number of processes which can all profoundly affect the life of a pavement. Each is important but it is the compaction process which commands particular attention as failure by either the surfacing contractor or the engineer to understand and adequately address the key issues may lead to a substantial reduction in the life of the road.

Compaction is the densification of material by the application of pressure, initially from the screed of the paver and subsequently from the rollers. An increase in the density of an asphalt improves the resistance to permanent deformation and to fatigue cracking along with an increase in elastic stiffness, all desirable improvements.

This Chapter sets out general observations about compaction and justifies these by reference to published work on the effect of compaction on the performance of a pavement. It also covers rolling methods, compaction plant and other factors including environmental aspects. It examines the current standard UK compaction specification and considers the practical application of compaction technology.

8.2. Importance of compaction

Asphalts are nothing more than a blend of aggregates and a binding agent. Depending on the material type, the aggregates used to manufacture a particular mixture can range from coarse particles through fine aggregate to filler. The binding agent is normally a bitumen which is obtained from the distillation of crude oil.

In terms of common UK usage, asphalt mixtures have traditionally been conveniently subdivided into coated macadams and hot rolled asphalts (HRAs), which is reflected in the UK Standards for these materials.[1-4] Coated macadams are described as continuously graded since they consist of a number of aggregate sizes. HRAs are described as gap graded because the aggregate sizes within the mix are not continuous. These concepts are illustrated in Table 8.1.

Various Chapters in this Book have discussed the new materials available to highway engineers for the construction and maintenance of roads. The late 1990s were dominated by the emergence of stone mastic asphalts (SMAs) and the proprietary thin surfacing variants. Other

Table 8.1.   Comparison of typical coated macadams and hot rolled asphalts

Sieve size	Percentage by weight passing each sieve	
	Coated macadam 20 mm dense basecourse	Hot rolled asphalt Column 2/2–50/14 roadbase, basecourse and regulating course
28 mm	100	—
20 mm	95–100	100
14 mm	65–85	90–100
10 mm	52–72	65–100
6.3 mm	39–55	—
3.35 mm	32–46	N/A
2.36 mm	N/A	35–55
600 μm	N/A	15–55
300 μm	7–21	N/A
212 μm	N/A	5–30
75 μm	2–9	2–9

materials also emerged — heavy duty macadams (HDMs) and DBM50 both of which use a 50 pen bitumen compared with the traditional 100 pen or 200 pen for coated macadams. It was expected that the future would see an increase in the use of high modulus base (HMB) because roadbases constructed with this material are much thinner than those which use other mixtures. However, the Highways Agency suspended the use of HMB 15 and HMB 25 in March 2000 as a result of testing which suggested that such layers had reduced stiffness values.

Notwithstanding, the system of categorisation described above remains valid. SMAs are gap graded and HDMs, DBM50s and HMBs are all continuously graded. The compaction of these particular variants is discussed in Section 8.5 below.

'Compactability' is a concept related to the ease with which a material can be compacted. It is usually expressed in relative terms. All materials have an optimum voids content. A material with a high compactability requires less compaction to achieve the desired voids content. A number of factors affect the compactability of a mixture and these are illustrated in Fig. 8.1.[5]

Figure 8.1 illustrates a number of points about dense bitumen macadams (DBMs, traditionally the most common form of coated macadams in road construction and maintenance) and HRAs. DBMs derive their strength from internal friction and mechanical interlock between the pieces of aggregate present in the mixture. It is these mechanisms which provide the resistance to compaction. HRAs derive their strength from the sand/filler/bitumen 'mortar'. Both DBMs and HRAs are laid and compacted at relatively high temperatures. This ensures that the binder is sufficiently fluid to act as a lubricant between the aggregate particles, reducing the internal friction of the mixtures and assisting in achieving good aggregate interlock. Therefore, as the mixture temperature rises, so too does the compactability of an asphalt. In the case

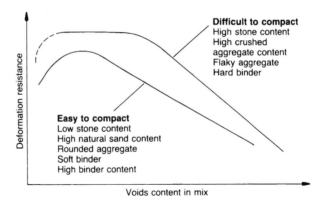

Fig. 8.1.   Variation in the compactability of bituminous mixtures due to changes in composition[5]

of DBM this is because the binder is less viscous (viscosity is defined as 'the resistance to flow') and acts as a lubricant between the aggregate particles. In the case of HRAs, this is because the bitumen has not set and the asphaltic mortar has little strength until the binder hardens.

Figure 8.2 shows a core of wearing course (HRA) and basecourse (DBM). Examination of the cross section of each material illustrates why HRAs are, generally speaking, more easily compacted than DBMs.

Figure 8.3[5] shows a typical curve illustrating the influence exerted by temperature on the compactability of a mixture. Compaction difficulties

Fig. 8.2.   Typical road core of wearing course (chipped HRA) and basecourse (DBM)

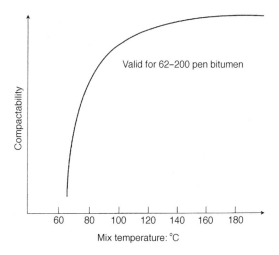

Fig. 8.3. Effect of temperature on the compactability of bituminous mixtures[5]

can often be solved by the simple expedient of raising the specified
delivery temperature of the mixture.

8.3. Published work

Proof that compaction of asphalts is a singularly important topic comes
from the large volume of published work on the topic. Most literature
concentrates on DBMs for the reason that these materials predominate in
roadbases and basecourses.

Much of the published research work relates to asphaltic concrete,
which is a material used in the USA and mainland Europe. It is
continuously graded and therefore results for this material will, in the
main, be valid for DBM.

It is convenient to split compaction technology research into four
categories

- the effects of compaction on pavement performance
- rolling methods
- compaction plant
- other factors affecting compaction.

8.3.1. Effects of compaction on pavement performance
In the 1940s in the USA, the Marshall Method[6] was developed for the
design of continuously graded asphalts. The method enjoys extensive use
in many countries throughout the world for the design of asphaltic
concretes for road surfacing. Until the early 1990s, the Marshall Method[6]
was used to design the most economic HRA wearing course mixture for a
given set of constituent materials. This was done using a cylindrical
mould containing the specimen under test. The sample was compacted by
a standard hammer for a specified number of blows and then subjected to

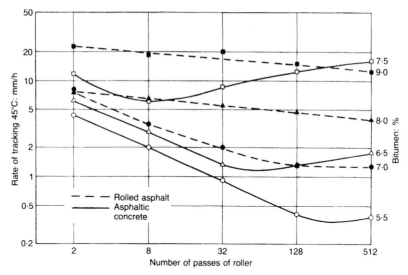

Fig. 8.4.   Effect of compaction on the resistance to deformation of asphaltic concretes and rolled asphalts[8]

various tests from which certain parameters were ascertained. (The current procedure is described in BS 594: Part 1[1] and is based on a method which involves the entire mixture whereas, previously, the evaluation was based on the bitumen/sand/filler element only.)

The Wheel-Tracking Test[7] measures the resistance of asphalts to permanent deformation, i.e. rutting. It does so by running a standard wheel backwards and forwards across a compacted sample of an asphalt. Full details of both this test and comment on the Marshall Method[6] are given in Chapter 9 (Sections 9.4.4 and 9.4.3 respectively).

Jacobs[8] carried out laboratory work on the compaction of both HRAs and asphaltic concrete using both the Marshall Test and the Wheel-Tracking Test apparatus. The relationship that he found between resistance to deformation and the achieved level of compaction is reproduced in Fig. 8.4 and illustrates how the resistance to deformation of an asphalt increases as the degree of compaction rises.

Jacobs[8] concluded that HRAs are more easily compacted than are asphaltic concretes (and therefore DBMs). This substantiated the earlier general observation. He also found that the resistance to deformation increased as the degree of compaction rose. In the case of asphaltic concretes, the deformation increased once air voids content fell below 3.5%. No such phenomenon occurred with HRA but the Jacobs' report does not consider air void contents of 2.5% or less. The change in resistance to deformation was less marked in HRAs than in asphaltic concretes. His general conclusion was that compaction control was more important in dealing with asphaltic concretes (DBMs) than with HRAs.

Lister and Powell[9] examined the relationship between compaction and pavement performance in DBMs. As part of their study, they plotted

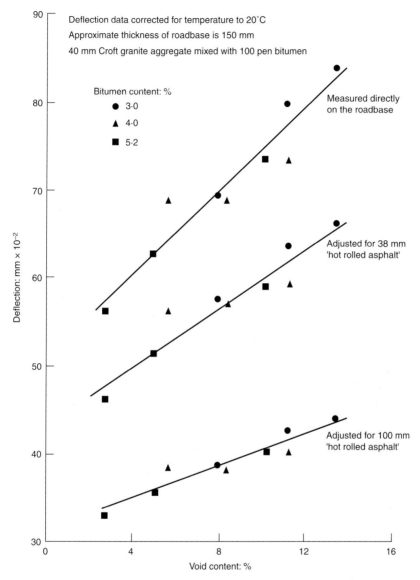

Fig. 8.5.   Variation of deflection with void content in a roadbase[9]

values of air void content and/or voids in the mineral aggregate (VMA) against deflection, with values being adjusted to include HRAs. Some of their findings are illustrated in Fig. 8.5.

Air voids are defined as 'the total volume of air between the coated aggregate particles of the compacted mixture'. This does not include air in the aggregate pores trapped beneath the binder films. VMA is defined as 'the volume occupied by the air voids and the amount of binder not absorbed into the pores of the aggregate'. Both air voids and VMA are expressed as percentages of the total volume of the compacted specimen. The methods of determination of these parameters can be found in Chapter 9 (Section 9.4.3) and the importance of air voids content is discussed in Chapter 4 (Section 4.7.3).

As a result of their work, Lister and Powell[9] concluded that worthwhile extensions in the life of the pavement could be obtained with increased compaction.

8.3.2.   Rolling methods

Leech and Selves[10] examined the rolling of DBMs. In their work,[10] they state that there is substantial variation in the achieved levels of compaction across the lane width and this is depicted in Fig. 8.6.

In any road there is a tendency for the wheels of traffic to use the same narrow strips, thus concentrating the loading on these areas. These are described as the 'wheel paths' or 'wheel tracks' and are those parts of a road that are most likely to show signs of wear or failure. Their location varies according to a number of factors including the width of the road. For a single two-lane carriageway having a width of 7.3 m, they are centred around 0.9 m and 2.7 m from the nearside kerb. The inner wheel track is narrower because vehicles tend to maintain a standard distance from the kerb and the widths of vehicles vary. (In terms of loading a pavement, the effects of cars can be ignored[11] — only commercial vehicles exceeding 15 kN (about $1\frac{1}{2}$ t) are included in any count for the purposes of traffic assessment leading to the determination of loading of a particular pavement.) It is, therefore, the inner wheel track which is stressed to the highest level at the greatest frequency. As can be seen from Fig. 8.6, the maximum compaction level, and therefore the greatest density, is achieved at the centre of the lane where there is only sporadic

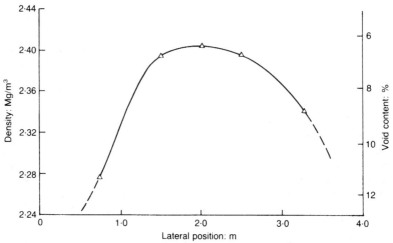

Fig. 8.6.   Variation of compaction across a lane width of DBM[10]

loading. This comes as no surprise as the traditional method of rolling is from the centre out towards the edges. Thus, the compactive effort is concentrated in the middle of the lane and this is where the density is maximized.

Leech and Selves[10] also examined existing roads to find where densification had occurred under the action of traffic. They looked at modifying the rolling pattern to equalize the compactive effort across the lane. They concluded that the level of compaction and density achieved across the width of the lane was typically 3% lower in the nearside wheel track than in the centre of the lane. They also confirmed that there was little or no additional densification as a result of the action of traffic except where there was poor compaction at the time of construction. The conclusion is that adequate compaction is essential at the time of construction. Modified rolling was found to reduce the disparity in density across the lane width.

As a footnote to the above, some concessionaires on DBFO (design, build, finance and operate) contracts have considered, as part of their contract strategy, an effective variation in the lateral location of the nearside kerb during the life of the road. The attraction is that when the pavement, i.e. the nearside wheel track, is reaching the end of its useful life but is still intact the traffic will move laterally. This avoids failure in the nearside wheel track by transferring the loading to a virtually unloaded part of the pavement and extending the life of the road before major maintenance is required.

8.3.3. Compaction plant

Initially, compaction is imparted to the asphalt by the screed of the paver. As discussed in Section 6.3.4 in Chapter 6, the screed may be a tamping screed, a vibrating screed or a combination (tamping/vibrating) screed. Little information is available on the degree of compaction imparted by screeds in roadbases and basecourses. Hunter[12] carried out some testing on chipped HRA wearing course, after laying but before rolling, which returned average air voids contents of 11%. Normally, the target for this material is $\leq$5%.[13] Accordingly, the rollers need only induce a further 6% reduction in the air voids content to achieve full compaction. However, one of the effects of the impregnation of the precoated chippings is to de-compact the HRA in the upper part of the layer. Most published work in this area relates to roller performance.

Traditionally, three-point dead weight rollers have been used to compact both DBM and chipped HRA wearing courses. A typical example of such a roller is shown in Chapter 6 (Fig. 6.26). During the last ten years or so, vibrating rollers have become standard on surfacing operations. Typical examples of tandem vibrating rollers are shown in Chapter 6 (Figures 6.28 and 6.29). Powell and Leech[14] examined the use of vibrating rollers in relation to a number of trials. The results are shown in Fig. 8.7.

Powell and Leech[14] concluded that vibrating rollers are capable of achieving 3% more compaction than conventional 8–10 t dead weight

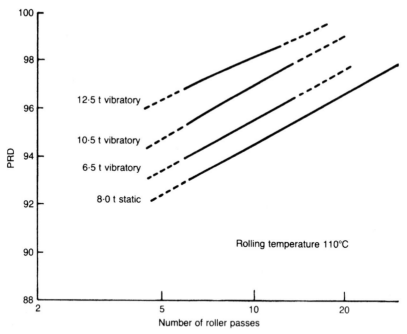

Fig. 8.7. Comparison of compaction levels achieved by vibrating and dead weight rollers[14]

rollers. This could result in up to a 30% increase in the life of a pavement. Chapter 6 discusses plant, including the available types of roller, in detail.

8.3.4. Other factors affecting compaction

Other factors that affect the degree of compaction achieved are related to the material itself and the environmental conditions under which it is laid.

In a soil, as the amount of water increases it becomes easier to compact (up to the optimum moisture content). Any soil compacted around its optimum moisture content will achieve the maximum dry density, by definition. It is convenient to think of heat in a hot asphalt as a lubricant. Similar to water in a soil, the more heat that exists in an asphalt when it is compacted, the greater will be its final density for a given compactive effort (again, up to the optimum heat content). As soon as the material is mixed, it begins to lose heat. The temperature at which it is mixed at the production plant is governed by BS 594: Part 1[1] for HRAs and BS 4987: Part 1[3] for DBMs. It is transported to site in insulated vehicles ready for laying and Chapter 6 (Section 6.2) gives more detail on this subject. It is rare for an asphalt to be over-compacted although this can occur with asphaltic concretes and, therefore, DBMs.[8]

A failure by the contractor to compact adequately may have an implication in terms of the specification or in terms of performance. The former is discussed later in Section 8.4. In performance terms, such an error may result in the pavement having a shorter lifespan than would

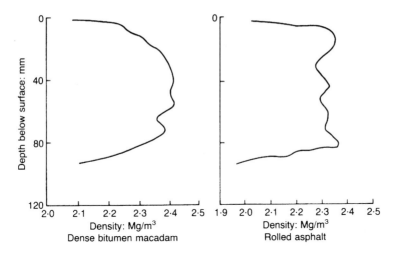

Fig. 8.8. Typical density/depth profiles[9]

otherwise have been the case. This would arise if the roadbase was inadequately compacted. If the basecourse and/or wearing course were inadequately compacted then they would require replacement earlier than would otherwise have been the case. If the wearing course is a chipped HRA then the results may be apparent very shortly after laying — it may not retain the precoated chippings. This is likely to occur soon after the road is opened to traffic if the material has not been compacted at temperatures high enough to ensure bonding of the precoated chippings with the parent asphalt layer.

Lister and Powell[9] scanned cores in order to ascertain how density varies through layers of an asphalt. Figure 8.8 shows their findings for a DBM and a HRA used in a roadbase or a basecourse. This illustrates how the temperature of an asphalt affects the final level of compaction achieved. Since the upper and lower portions cool faster than the centre of a layer, the final densities at the top and bottom of the layer are correspondingly lower. Unchipped and chipped materials cool by the same mechanisms, but the chipped analysis is much more complex due to the effects of the addition of precoated chippings to the latter.

The cooling of unchipped asphalts
Figure 8.9 illustrates how heat flows to and from a hot layer. The means by which heat is lost or gained are

- *thermal conductivity* — the transfer of heat through the hot material into the base layer and then into the layers below
- *convection* — heat is dissipated by the wind
- *radiation* — heat is transmitted into the atmosphere.

The rate at which heat flows into or out of a layer is influenced by the following parameters

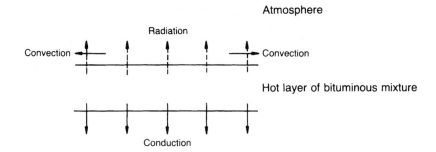

Fig. 8.9.   How heat is lost from a hot bituminous mixture as it is laid

- air (ambient) temperature
- wind speed
- emissivity of the surface, i.e. how efficiently it radiates heat
- absorption coefficient of the surface, i.e. how efficiently it absorbs heat from the Sun
- amount of solar radiation, i.e. how much heat comes from the Sun
- depth, thermal conductivity, density, specific heat and temperature of the hot layer
- depth, thermal conductivity, density, specific heat and temperature of the first underlying layer
- depth, thermal conductivity, density, specific heat and temperature of the second underlying layer — it is unlikely that there would be any effect below this level.

There have been a number of studies analysing the rate of heat loss from a layer of a hot asphalt in order to establish the time available for compaction. Four papers on this subject are related. Corlew and Dickson[15] carried out the pioneering work that was continued by Jordan and Thomas[16] and thereafter by Daines[17] and Hunter.[18]

These papers describe methods for calculating temperatures in a hot newly laid layer a short time period after placement. These values are then used to calculate the temperatures after a further equal time increment. This process is repeated until the values for the required time periods are computed. As many discrete calculations are involved such analytical methods are ideal for solution by computer. The layer is considered infinite in terms of the width and it is suggested that this is not an unreasonable assumption. Temperatures along any horizontal line have, therefore, the same value for any given time after laying. This means that the temperatures vary in one dimension.

A method which computes the temperature values at the current time by using those ascertained at the last time increment is called an 'explicit method'. It can be shown mathematically that the use of temperature values at the current time is a more accurate method. This is known as an 'implicit method'.

Corlew and Dickson[15] used an explicit method to compute temperatures. Jordan and Thomas[16] used an implicit method but obtained the first estimate of temperatures for the next time level with an explicit method, thus effectively rendering the entire analysis explicit. Their analysis was written to run on a mainframe computer. Daines[17] used Jordan and Thomas'[16] computer analysis so that, too, is explicit. Hunter[18] used a wholly implicit method to derive temperatures. This is a more accurate means of computing temperatures.

It is possible, by varying the value of each parameter, to examine the relevance of each of the parameters involved in the flow of heat within an asphaltic layer.

Daines[17] concluded that wind speed and air temperature have a substantial effect on the rate of cooling. As will be shown, the former does but the latter does not. Hunter[18] wrote a computer program using a wholly implicit method to compute temperatures in an asphalt and allowing them to be interrogated by the user. Temperatures at one-minute intervals are stored during computation. The program runs on a 'RISC OS' computer but a version has also been written to run in a Windows$^©$ environment as a module, CALT$^©$, forming part of Shell's SPDM$^©$ Pavement Design Software. (This is discussed in Chapter 4, Section 4.8.1.)

Hunter and McGuire ran a large number of tests using a variety of parameters.[19] Variation of each of the parameters in turn, allowed the relative effect of each of these factors to be ascertained. On the basis of the results, they concluded that changes in the various parameters affecting the rate of cooling of unchipped asphalts have the following effects.

- Wind speed has a major effect on the cooling rate.
- Layer thickness has a major effect — although thicker layers lose heat at a higher rate, the percentage loss is lower.
- Ambient temperature has little effect on the cooling rate.
- Thermal conductivity of the hot layer has a major effect.
- Density and specific heat of the hot layer have little effect.
- Incident solar radiation has little effect on the cooling rate.
- The underlying layer temperature, thermal conductivity, density and specific heat have little effect on the cooling rate.

Although the thermal conductivity of the material is an important determinant in the rate of heat loss, it is unlikely that there are any practical options available to alleviate this since the thermal conductivity of an aggregate is a function of its geology. The major finding is that the amount of heat loss does not vary markedly for substantially different air temperatures. Clause 6.1, in particular note (b), of BS 594: Part 2[2] goes some way to recognizing this fact. The major variable is loss through convection, i.e. incident wind. The importance of these findings will become apparent when the specification is considered later.

The cooling of chipped asphalts
The cooling of a chipped layer is much more difficult to analyse. The functions of the roadbase, the basecourse and the wearing course have

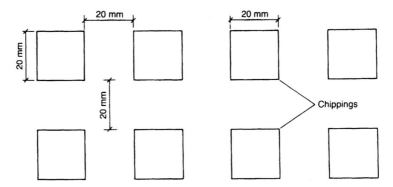

Fig. 8.10.   Plan of idealized model[18]

been covered in Chapter 3 (Section 3.6). Premature failure of any layer will manifest itself in different ways in that roadbase failure will shorten the life of the pavement as a whole whereas basecourse and wearing course failure will necessitate reconstruction of the surfacing. Nevertheless, failure of any layer in a pavement is wholly undesirable and, given the unacceptability of delays to traffic, must be avoided. One factor in minimizing the possibility of wearing course failure is effecting adequate compaction to the material. The same mechanisms which govern the cooling of unchipped layers also apply to chipped materials such as HRA wearing course.

In relation to the temperature in a chipped HRA wearing course Daines[17] states 'it is estimated that the average asphalt temperature is reduced by 15 °C'. No basis for this estimate is given and it seems highly unlikely that one value will apply for all the different sources of materials that exist and also for the vast range of circumstances which may occur during any laying operation.

Hunter[18] devised a mathematical model which accurately analyses heat loss during the laying of chipped HRA wearing course. Work by Hunter and McGuire[19] used this model to analyse temperature changes (Figs 8.10 and 8.11).

In terms of heat retention, a chipped HRA wearing course has little in its favour. It is a thin layer, normally 45 mm,[†] but specification tolerances usually allow this to be as little as 40 mm. Immediately after it is laid, cold precoated chippings are rolled into the material. It is these chippings which complicate the cooling process. Since the chippings are three dimensional, so too is the variation in temperature. Thus the analysis is also three dimensional.

Hunter[18] wrote a computer program which computes temperatures in the asphalt and allows them to be interrogated by the user. Temperatures

---

[†]Contractors are often afforded the option of employing a 50 mm thick wearing course with a consequent 5 mm reduction in the basecourse thickness — sometimes a very attractive alternative.

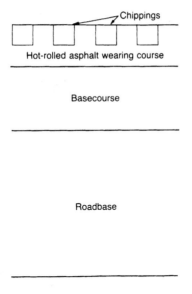

Fig. 8.11   Section through idealized model[18]

at one-minute intervals are stored during computation. The program operates on a RISC OS computer and calculates a matrix of temperatures in chipped HRA wearing course at one-minute intervals for 30 minutes after placement.

As with the unchipped layers, Hunter tested a large number of examples using a variety of parameters.[20] Again, variation of each of the parameters in turn allowed the relative effect of these to be ascertained. On the basis of the results, he concluded that changes in the various parameters affecting the rate of cooling of chipped asphalts have the following effects.

- Wind speed has a major effect on the cooling rate.
- Minor changes in layer thickness have a major effect because the nominal thickness is so small.
- Ambient temperature has little effect on the cooling rate.
- Thermal conductivity of the hot layer has a major effect.
- Density and specific heat of the hot layer have little effect.
- Incident solar radiation has little effect on the cooling rate.
- The temperature, thermal conductivity, density and specific heat of the underlying layer and precoated chippings have little effect on the cooling rate.

As in the unchipped analysis, the incident wind speed is of fundamental importance. It was noted that ambient temperature has a minor effect. Thermal conductivity is a major factor in determining the rate of heat loss but it is unlikely that this fact will be of any practical use.

Table 8.2. Temperature values in a 40 mm thick chipped hot rolled asphalt wearing course layer with different chipping densities

Distance from surface	Temperature* at 12 kg/m^2	Temperature* at 15 kg/m^2
0	70	53
5	75	57
10	80	61
15	83	65
20	85	70
25	85	72
30	83	74
35	80	73
40	75	70

* Temperature at 1 min after chipping impregnation.

Precoated chippings are added to HRA wearing course to provide frictional resistance for braking vehicles. They are normally 20 mm single size aggregate coated with bitumen of the same type and penetration as that used in the manufacture of the HRA wearing course. The required chipping rate is normally prescribed (defined in BS 594: Part 2[2]) as either that which is necessary to give 70% or 60% (if that will achieve adequate texture) of the shoulder-to-shoulder cover. The rate is measured using the method given in BS 598: Part 108.[21] This test is discussed further in Chapter 7 (Section 7.9). Examination of this method confirms that the more cubic the chipping the greater will be the required rate of spread. The chipping density is commonly around 13 kg/m^2 but values exceeding 15 kg/m^2 can occur. Hunter and McGuire[19] examined the effect of higher chipping densities on the rate of heat loss in chipped HRA wearing course. They compared the heat content in the upper 20 mm of the layer for precoated chipping densities of 12 kg/m^2 and 15 kg/m^2. The results obtained are shown in Table 8.2 and illustrated in Fig. 8.12.

Figure 8.12 assumes that chippings are fully embedded five minutes after the asphalt is deposited and shows that the heat loss in the first minute after chipping impregnation is greater for the material with the higher chipping density. The material with the lower chipping density contains more heat five minutes after laying than the material with the higher chipping density has after one minute.

Hunter[18] also modified his program to compute temperature values for a 35 mm layer. A comparison could then be made between the temperature profiles in a 35 mm thick layer and those in a 40 mm thick layer (Fig. 8.13).

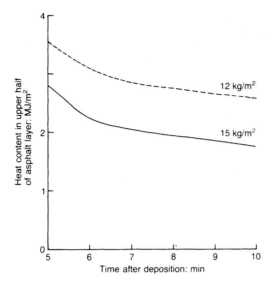

Fig. 8.12.   Heat contents in a 40 mm thick chipped hot rolled asphalt wearing course layer with different chipping densities

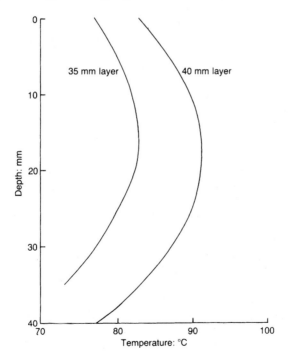

Fig. 8.13.   Comparison of the temperature profiles through a 35 mm thick and a 40 mm thick chipped hot rolled asphalt wearing course layer 10 minutes after laying[15]

The temperatures are around 10 °C lower for the 35 mm layer. Hence the reason why the thinner the layer, the more difficult it becomes to lay satisfactorily. Some years ago, the standard thickness of HRA wearing course was raised from 40 mm to 45 mm — a tacit recognition of the difficulties associated with embedding a 20 mm chipping into a layer that may be as little as 35 mm thick.

8.4. Compaction specification

8.4.1. Background

The standard specification for road construction and maintenance in the UK is the Specification for Highway Works[22] (SHW). It is part of a suite of publications entitled the *Manual of Contract Documents for Highway Works*[23] (MCHW). The System is published by the Highways Agency, the Scottish Executive, the Welsh Office and the Department of the Environment for Northern Ireland. The SHW is Volume 1 (sometimes referred to as MCHW 1). Volume 2 is entitled the Notes for Guidance on the Specification for Highway Works[24] (NGSHW and sometimes referred to as MCHW 2) and it provides indispensable guidance and further information relating to the use of the SHW.[22]

A companion suite is entitled the *Design Manual for Roads and Bridges*[25] (DMRB). It consists of some 15 Volumes numbered 1–15 which deal with all the technical issues associated with the design of highways. Volume 7 is entitled Pavement Design and Maintenance[26] and is often designated DMRB 7. Like the MCHW,[23] it is published by the Highways Agency, etc.

The use of the MCHW and the DMRB is mandatory for maintenance or construction works on trunk roads including motorways. Although the trunk road network represents some 4% of the total length of roads in the UK,[27] the MCHW and the DMRB are used for the majority of roads schemes in the UK.

As the name implies, the SHW is the standard specification. However, it should never be consulted in isolation and users are strongly encouraged to refer to the NGSHW in concert with the SHW. In addition, DMRB 7[26] provides useful information on the use of particular materials and the suitability of a wide range of maintenance techniques. These three documents often reflect the most contemporary thinking in matters associated with highway construction and maintenance.

The SHW often refers to the relevant British Standards for coated macadams and HRAs[1–4] except where its express provisions override these. The sections of the SHW which will be of most interest to those involved in pavement design and construction are designated Series 700 (Road Pavements — General), Series 800 (Road Pavements — Unbound Materials) and Series 900 (Road Pavements — Bituminous Bound Materials). The equivalent Series in the NGSHW have the same numbers and titles except that Series numbers are preceded by the letters 'NG', i.e. NG 700 etc. Clauses are numbered 701/NG 701 etc.

The NGSHW is not normally mentioned in contract documents and is, therefore, at first sight, non-contractual. It is, nevertheless, a document which dictates the way that specifications are prepared. This raises the status of this document to one which is quasi-contractual which means that it is, in effect, a contract document.

Where a Clause is preceded by a '#', this indicates that one of the 'Overseeing Organizations' other than the Highways Agency (i.e. the Scottish Executive, the Welsh Office and the Department of the Environment for Northern Ireland) has a substitute National Clause.

8.4.2.   Series 700 Road Pavements — General
Clause #702.10 deals with rectification of non-compliant work. It states that where '*any pavement area does not comply with the Specification for ... Compaction ... the full depth of the top layer*' of roadbase '*... shall be removed and be replaced with fresh material laid and compacted in accordance with the Specification*'. In the case of wearing course, basecourse and the top surface of roadbase in a pavement which has no basecourse, rectification requires the '*full depth of the course removed, or in the case of roadbase in pavements without base course, the topmost layer, and replaced with fresh material laid and compacted in accordance with the Specification*'. The area is to be the full laid width, some 5 m long if roadbase or basecourse or 15 m long if wearing course. The relevance of this Clause in relation to roadbase and basecourses will become apparent in the light of the compaction requirements discussed below. In terms of the wearing course, there are currently no end-performance requirements in terms of compaction or chipping embedment unless the material is a Clause 943 (Hot Rolled Asphalt Wearing Course (Performance-Related Design Mix)) discussed below in Section 8.4.3. However, Volume 7[26] contains, in HD 37,[28] a '*Method for determination of loss of chippings and proportion of broken chippings*', which suggests that a loss of less than 5% is considered reasonable. This may become a specified requirement at some future time. With regard to chipping embedment, it is suggested that if the asphalt is showing signs of significant distress when opened to traffic, then a surfacing contractor should recognize this as a failure and replace the material. It is further suggested that any new panel would need to be at least 15 m long in order to achieve any measure of level control on the replacement asphalt.

Clause 708 is entitled 'Weather Conditions for Laying of Bituminous Materials and Dense Tar Surfacing'. These requirements are set out in detail and discussed in Section 7.13.1 of Chapter 7.

8.4.3.   Series 900 Road Pavements — Bituminous Bound Materials
Clause 901 Bituminous Roadbase and Surfacing Materials
Clause 901.3 requires the use of insulated vehicles for the transportation of hot asphalts unless otherwise agreed by the Overseeing Organization. Asphalts are to be '*covered whilst in transit or awaiting tipping*'. Insulation on lorries normally takes the form of insulated panelling on the sides and base of the tipper body. The load is covered with a tarpaulin sheet tied to the sides of the lorry, although some vehicles incorporate a system which moves the sheeting mechanically. The sheeting is removed,

either wholly or partially as required, when the load is discharged into the paver. Experience shows that this is extremely effective although material at the tailgate (the flap at the rear end of the lorry) can be cold upon discharge. Every effort should be made to retain heat in the material and the adoption of sheeting which has greater insulation properties is strongly recommended.

Clause 901.14 dictates that compaction begins as soon as the uncompacted asphalt '*will bear the effects of the rollers without undue displacement or surface cracking*' and '*shall be substantially completed before the temperature falls below the minimum rolling temperatures given in BS 594: Part 2 and BS 4987: Part 2*'. This Clause has little meaning for two reasons. Firstly, the specification for measuring such temperatures, BS 598: Part 109[29] (specified in BS 594: Part 2[2] and BS 4987: Part 2[4]) requires measurements to be taken as near as possible to the midpoint of the layer. Even if that is achieved their relevance is questionable since the temperature varies throughout the layer, as demonstrated earlier in this Chapter. It is the lowest temperature in the mixture which is important, not the mid-layer temperature which is liable to be close to the maximum. Secondly, there is no indication of what constitutes '*substantial*' compaction. Rolling (and therefore compaction) has to continue far below the temperatures listed to eliminate roller marks.

Clause 901.15 specifies the use of

> '*8–10 tonnes deadweight smooth wheeled rollers having a width of roll not less than 450 mm, or by multi-wheeled pneumatic-tyred rollers of equivalent mass, or by vibrating rollers or a combination of these rollers. Wearing course and basecourse material shall be surface finished with a smooth-wheeled roller which may be a deadweight roller or a vibratory roller in non-vibrating mode. Vibratory rollers shall not be used in vibrating mode on bridge decks.*'

Vibrations can affect elements of a structure.[30] At particular frequencies they can, depending on the type of bridge and its configuration, break bonding of concrete, loosen bolts or cause cracking in welds.

Clause 901.16 qualifies the use of vibrating rollers by specifying that they '*may be used if they are capable of achieving the standard of compaction of an 8 t deadweight roller*'. They must be fitted with meters which indicate the frequency of vibration and the speed. Such meters must be readable from the ground. The performance of vibrating rollers is to be assessed either by site trials in accordance with the requirements laid down in BS 598: Part 109[29] or by the contractor producing evidence of independent trials which prove that the proposed vibrating roller achieves a degree of compaction which is equivalent to that which would be obtained by using an 8 t dead weight roller. In relation to the former, BS 598: Part 109[29] sets out a trial area which is split into two areas longitudinally. One area is compacted by an 8 t dead weight roller and the adjacent area is compacted by the proposed vibrating roller. Specified cores are taken and the densities obtained in accordance with the procedure given in BS 598: Part 104.[31] The mean density of the proposed vibrating roller must be at least the same as that achieved by a dead weight roller. The achieved coverage (obtained by multiplying the width of the vibrating roller by the speed of travel) must also at least equal that of the dead weight roller. In reality, most contractors would be able to obtain information from manufacturers of vibrating rollers in the form of evidence from independent trials.[32] The same Clause also states that if the degree of compaction achieved is to be determined in accordance with Clause 929 (Design, Compaction Assessment and Compliance of Roadbase and Basecourse Macadams, discussed below) then the above provisions do not apply and the contractor may use any plant to achieve the required compaction above the specified minimum rolling temperatures.

Clause 901.17 contains good advice on effecting compaction, rolling longitudinally, driven wheels nearest the paver, etc. Section 7.10 of Chapter 7 sets out and discusses good rolling practice.

Appendix 7/1 is included in almost all contracts prepared in accordance with the MCHW.[23] This Appendix contains contract-specific information related to the pavement and each of the available options for pavement materials. According to Clause 901.19, unless stated otherwise in Appendix 7/1 (although it is an option, it is difficult to perceive circumstances where it would), the degree of compaction which is actually achieved in roadbase and basecourse macadams is tested in accordance with Clause 929.

Clause 929 Design, Compaction Assessment and Compliance of Roadbase and Basecourse Macadams

Clause 929 is a most important Clause. It dictates that roadbase and basecourse macadams are tested for compaction compliance in two phases, a 'Job Mixture Approval Trial' and the 'Permanent Works'.

The Job Mixture Approval Trial requires that a trial area is compacted at least three days before macadam from each source is to be laid as part of the permanent works. The proposed material is laid in a trial length of 30 m — 60 m and at a width and thickness dictated in the contract. Four 150 mm diameter cores are then taken in accordance with BS 598: Part 100[33] at each of three locations, i.e. twelve cores in all. Two of the three locations are in the wheel tracks, the other location is agreed with the Overseeing Organization. Two cores from each location are tested for percentage refusal density (PRD) in accordance with BS 598: Part 104[31] (discussed in Chapter 9, Section 9.4.5). The maximum density is measured in accordance with BS DD 228: 1996 Issue 2.[34] At or adjacent to the core locations, the density is measured with a Nuclear Density Gauge (NDG). The in situ air voids content and the refusal air voids content are then calculated for each core as are the binder volumes for each location. In addition, compositional analysis as per BS 598: Part 102[35] (discussed in Chapter 9, Section 9.4.1) is carried out. The trial area and, therefore, the grading and binder content are acceptable if

- the average in situ air voids content of the twelve cores $\leq 8\%$

- the average in situ air voids content of each pair of cores $\leq 9\%$
- the average refusal air voids content of each pair of cores $\geq 1\%$
- the average binder volume at each location is $\geq 7\%$
- the aggregate grading and binder content comply with the contractor's suggested target
- horizontal alignments, surface levels and surface regularity comply with Clause 702.

If compliance is not achieved then a new grading and binder content are proposed by the contractor and the testing is repeated until *all* the above requirements are met.

Thereafter, the permanent works are checked by means of a NDG and the testing of cores. In each lane, pairs of cores are to be taken every 500 m and in situ air voids content and refusal air voids content obtained in accordance with BS 598: Part 104.[31] Using the NDG, readings are taken every 20 m in alternate wheel tracks starting at a location where cores are to be taken. Additional NDG readings are to be taken 300 mm from the edge of the mat adjacent to each core location. Further compaction is required if the NDG shows low densities provided the material is above the minimum rolling temperature specified in BS 4987: Part 2.[4] In fact, there is good reason to prohibit further compaction. As discussed in Section 8.5 below, over-compaction may result in crushing of the coarser aggregate fractions. If this happens then the attributes of the mixture may well be significantly reduced. Each NDG is initially calibrated against the trial area cores and then each is re-calibrated against the results from loose samples and the cores taken every 500 m if the individual gauge shows a different bias. Loose samples are taken from the paver augers in accordance with BS 598: Part 100.[33] Note that NDGs only measure down to about 80 mm. If the mat is much thicker then the core is to be inspected visually for signs of increased voidage towards the interface. NDGs are discussed in detail in TRL Supplementary Report 754.[36]

The permanent works are acceptable if, for each mixing plant

- the average in situ air voids content from any six consecutive NDG readings $\leq 8\%$, if not then cores are taken and the same criterion applied
- the average in situ air voids content of each pair of cores taken every 500 m $\leq 9\%$
- the average refusal air voids content of any three consecutive pair of cores taken every 500 m $\geq \frac{1}{2}\%$, if not then new grading and binder content must be suggested by the contractor
- the aggregate grading and binder content resulting from compositional analysis undertaken in accordance with BS 598: Part 102[35] comply with BS 4987: Part 1[3] with additional requirements for HDMs
- horizontal alignments, surface levels and surface regularity comply with Clause 702.

There are additional requirements for basecourse macadams placed below porous asphalt wearing courses.

It is not difficult to achieve compliance with the requirements of this Clause. The purpose of this Clause is to ensure that these mixtures are adequately but not over compacted. The danger of the latter is that the voids will be overfilled with binder, the result of which is discussed in Section 11.2.3 of Chapter 11.

Contractually speaking, HMBs do not require to be the subject of PRD testing under Clause 929. This will undoubtedly change with time but it is not expected to be difficult to achieve compliance. Whilst it is by no means perfect, testing by means of the PRD Test is a useful means of ensuring that macadams are properly designed and compacted. (PRD is also considered in Section 4.7.4 of Chapter 4.)

Clause 938 Porous Asphalt Wearing Course
Porous asphalt offers substantial advantages in terms of its low noise and spray generation characteristics and it is discussed further in Chapter 3 (Section 3.7.2) and Chapter 4 (Section 4.13.5). Clause 938 specifies that compaction plant shall be two 6–8 t dead weight rollers with steel drums per paver. Rubber tyred and three-wheeled rollers are not permitted. Compaction is to be 'substantially completed' before stated minimum temperatures are reached. Vibratory rolling is not advisable since it could result in crushed aggregate within the mixture and, therefore, a significant reduction in the performance of the material.

Clause 942 Thin Wearing Course Systems
In relation to thin surfacings, compaction is to be effected by means of at least two passes of a tandem roller, capable of vibration, and with a minimum dead weight of 6 tonnes, before the material cools below 80 °C, measured at mid-depth. These requirements have little relevance in practice. Surfacing contractors who supply such materials are well versed in appropriate laying and compacting techniques. Indeed, one of the HAPAS[37] requirements is that only trained personnel lay the mixtures.

Clause 943 Hot Rolled Asphalt Wearing Course (Performance-Related Design Mix)
The compaction requirements are very similar to those which are specified in Clause 929. Both the Job Mixture Approval Trial and the permanent works must return air voids contents of $\leq 7\frac{1}{2}\%$ for a pair of cores at a specific location and $\leq 5\frac{1}{2}\%$ for the mean of any six consecutive determinations.

Clause 943.34 contains an interesting requirement. Traditionally, new surfacings are not exposed to traffic until the material has cooled to ambient temperature as per BS 594: Part 2.[2] Clause 943.34 contains a novel approach. It requires that the surfacing should not be opened to traffic until the surface temperature has fallen to at least 25 °C or the temperature anywhere in the layer shall be below 35 °C. In fact, experience suggests that material which has reached around 35 °C is unlikely to be damaged by traffic — carriageways are occasionally well above that temperature in periods of prolonged sunshine. It is an old hypothetical rule of highways that roads subjected to long periods of solar heat can reach a temperature which equates to approximately twice the ambient temperature. Test runs using CALT$^©$ (part of Shell's Pavement Design Suite, discussed in Chapter 4, Section 4.8.1) demonstrate this and the wearing course reached 58 °C when the air temperature was 30 °C. This explains, in part, why south facing roads subject to prolonged solar heat can suffer permanent deformation. Thus, a temperature of 35 °C is not particularly rare. The temperature at which particular mixtures will be subject to deformation under traffic loading will vary according to the compositional details, particularly in relation to the characteristics of the binder. It is not unlikely, given the pressing financial and political need to open roads as quickly as possible, that the SHW[22] will extend this requirement to other materials. If this occurs in practice then research may well be necessary to derive a means of assessing the temperature and other parameters which control the temperature at which asphalts deform under traffic loading. The main difficulty with the current SHW requirement is that there is currently no means of measuring the temperature anywhere other than on the surface because the asphalt is so stiff that it is impenetrable to a thermometer.

8.4.4. Use of 'may lay'/'may not lay' diagrams
The SHW contains four number 'may lay'/'may not lay' diagrams. They are captioned as follows.

- *Figure 7/1: Wind Speed and Air Temperature Laying Restraints for 45 mm Thickness Rolled Asphalt Wearing Course*
- *Figure 7/2: Wind Speed and Air Temperature Laying Restraints for Dense Bitumen Macadam Wearing Course or Basecourse, or Heavy Duty Macadam Basecourse Layers*
- *Figure 9/1: Limiting Weather Conditions for Laying Porous Asphalt Containing 200 Pen Bitumen with Natural Rubber*
- *Figure 9/2: Limiting Weather Conditions for Laying Porous Asphalt Containing 100 Pen Bitumen with Natural Rubber.*

Figures 7/1 and 7/2 are reproduced in Chapter 7 as Figs 7.25 and 7.26 respectively. Figures 9/1 and 9/2 are reproduced below as Figs 8.14 and 8.15 respectively.

The purpose of all these diagrams is to dictate when the materials cited can or cannot be laid. It is understood that they have been compiled on the basis of allowing 10 minutes for 'substantial compaction' to be completed. There are a number of concerns about the applicability of these diagrams to all situations for the following reasons.

- They take no account of variations in the thermal conductivities.
- They assume that there is some fixed relationship between an anemometer reading 10 m above a particular location and 2 m above the laying operation.
- Examination of the fundamental heat flow equations shows that air temperature is not a significant criterion.

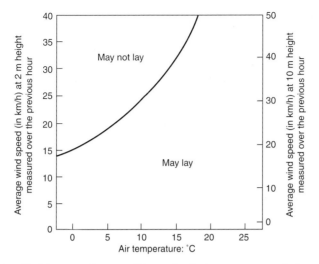

Fig. 8.14. *Limiting weather conditions for laying porous asphalt containing 200 pen bitumen with natural rubber*[22]

Perhaps the shortcomings of the SHW[22] are best demonstrated by plotting the total heat content in mixtures against time. Figure 8.16 compares heat contents with time after laying for two specific sets of circumstances. Condition A is for temperatures of air, chippings and substrate of −10 °C and a wind speed of 0 mph. Condition B is for temperatures of air, chippings and substrate of 0 °C and a wind speed of 5 mph. The paradox is that the condition which results in the material having the higher heat content would not be allowed under the SHW,[22]

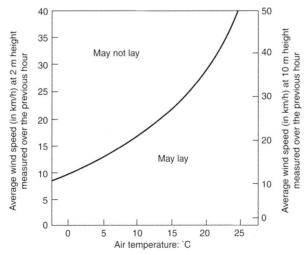

Fig. 8.15. *Limited weather conditions for laying porous asphalt containing 100 pen bitumen with natural rubber*[22]

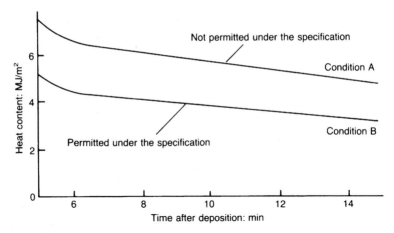

Fig. 8.16.   Heat content of chipped hot rolled asphalt wearing course under varying physical conditions[38]

while the condition which results in the lower heat content would be permissible. Such is the complexity of the interaction of all the parameters involved in heat flow from a hot asphaltic layer, it is suggested that simple diagrams such as those above do not adequately reflect all prevailing circumstances. Accordingly, the use of such diagrams is inappropriate.

8.5.   Practical application of compaction technology

8.5.1.   General

Other than in very favourable conditions, most contractors will order material such that it arrives on site close to the maximum temperature permissible. This maximizes the time available for compaction. Computerized control of coating plants ensures accurate control of the proportions and mixing times and temperatures. Modern lorries with effective body insulation and motorized sheeting minimize heat loss during transportation.

The choice of compaction equipment is a major issue and is discussed below.

Most pavers are not automatically equipped with insulated hoppers although most manufacturers make insulation available as an optional extra. All pavers used in the UK should be so equipped as standard. Having manufactured the mixture at elevated temperatures and transported it in insulated vehicles, it is rather illogical to expose the material to the atmosphere against a very effective heat sink, the metal sides and base of the paver hopper. Mixed materials must be retained in the lorry until immediately before laying.

Once an asphalt has been placed, compaction must be effected immediately behind the paver. If the material is too hot then it will displace or display significant surface cracking. (Minor cracks are not important and are usually closed up by back rolling once the material has cooled below the relevant minimum rolling temperatures. In the case of wearing courses, even if such compaction (sometimes called 'polishing') does not close them up, the action of traffic/weather often does so.)

8.5.2.   Vibrating rollers versus static rollers

The choice of roller is a major issue. The first decision which has to be taken is whether the roller should be static or vibrating. These terms describe the nature of the applied loading. The effect of the roller can be increased by vibrating the drum. This is illustrated in Fig. 8.17.

Vibrations emanate from an asymmetric weight within the drum. The effect of this can be measured by considering two parameters — the amplitude and the frequency. Amplitude is the maximum distance by which the drum deviates from its axis. Frequency is the impact rate. The values of these parameters can be varied by design within limits independently by the manufacturer. It is important to consider the effect of changing these values. The higher the amplitude value, the greater the depth to which the compactive effort will be effective. Appropriate frequencies for asphaltic compaction are from 20 Hz to 40 Hz. Changes within this range do not affect the compactive effort. Amplitude is, however, critical.

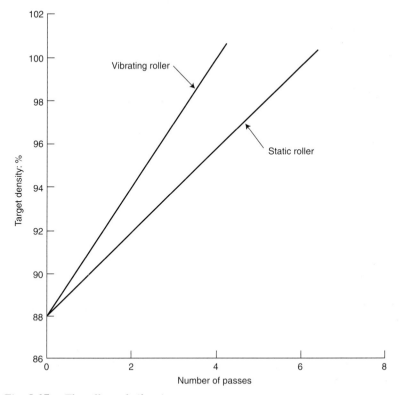

Fig. 8.17.   The effect of vibration

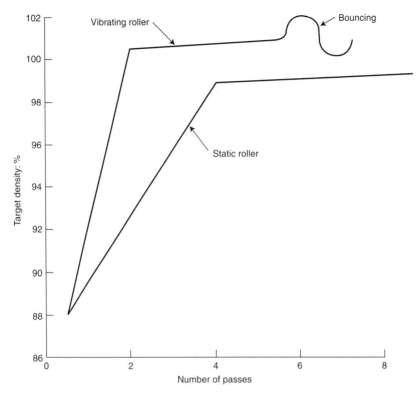

Fig. 8.18.   The effect of over-compaction

In the UK, it is not unusual for contractors to use one or more vibrating rollers on surfacing operations but to employ them in static mode only, keeping the vibratory option for occasions when areas of the mat appear to be under-compacted. This is inefficient and a waste of the potential of vibrating rollers. This is also illustrated in Fig. 8.17. By contrast, Fig. 8.18 depicts the dangers of over-compaction.

'Bouncing' occurs when a particular material has been compacted to refusal. The effect of further compactive effort is that the density of the material oscillates about a particular value. Further compactive effort will be felt through the steering of the roller with difficulty in retaining control. Bouncing starts before it is felt through the steering. What is particularly undesirable about this phenomenon is that it can result in the aggregate crushing. This means that a portion of the aggregate is uncoated and the mixture takes on the characteristics of an unbound material. In essence, vibrational compaction will result in more efficient compaction but it must be used with care. Over-compaction may adversely affect the performance of the layer quite significantly.

It is important to study not only the price of the equipment but also its performance characteristics. All manufacturers supply statistics on their rollers.

8.5.3.   The compaction of chipped HRA wearing course

The application of precoated chippings to an HRA wearing course means that there are special problems in ensuring adequate compaction of this material. The results of inadequate compaction of this material usually become apparent shortly after the surface is exposed to traffic. The processes which take place when chippings are introduced into an HRA wearing course are illustrated in Fig. 8.19.

The paver imparts initial compaction to the HRA wearing course. The chippings are dropped onto the top of the laid material. As the precoated chippings are pushed into the laid mat, they de-compact, i.e. de-densify, the upper part of the layer. This asphalt will cool rapidly due to the de-densification and its physical contact with the cold precoated chippings. As the chippings continue to be forced into the material, this interstitial asphalt is re-compacted. It is vital that the asphalt is re-compacted as quickly as possible. If the final density is inadequate, the material will not be able to resist the shearing force of traffic tyres and will abrade. This abrasion, in turn, exposes the chipping to a higher level of the shearing force than that experienced by embedded chippings and, as a result, some of them will become detached from the layer. This phenomenon is called 'fretting' (shown in Fig. 8.20). Other failures in asphalt pavements are discussed in Chapter 11.

Fretting invariably occurs shortly after the road is opened to traffic. Abrasion will stop at a level where the material itself possesses adequate density (and consequently abrasion resistance) to successfully withstand shear loading due to vehicle tyres. Fretting usually stops after a day or two of exposure to traffic. Its occurrence is serious and may require that the layer is planed out and replaced or perhaps surface dressed. Either eventuality is serious. The former means further delays for traffic while the latter means that the surface will not be of the expected quality. (Although surface dressings are an excellent maintenance measure, they are undesirable on a new or reconstructed road.) Both options often involve substantial expense.

It is relatively easy to tell whether a chipped HRA wearing course has been properly compacted. The interstitial asphalt has a smooth surface. If that area is rough then that may well indicate that there has been inadequate compaction and that the material may fail.

8.5.4.   The compaction of thin surfacings

In the case of thin surfacings, the level of compaction which is required is minimal but no less vital. Only two or three passes of a vibrating roller are required in order to achieve the required degree of compaction. A further two or three passes will complete densification and give the material the desirable finish. The uniform appearance of thin surfacings is an attractive but rarely mentioned feature of these asphalts. A key difference between these materials and more traditional asphalts is their sensitivity to the amplitude of the roller. An amplitude of 0.2 mm is suitable for thin surfacings laid up to 15 mm thick. Amplitudes of 0.3 mm or more are much more likely to result in aggregate crushing. Thicker layers optimally require amplitudes of 0.3 mm to 0.4 mm.

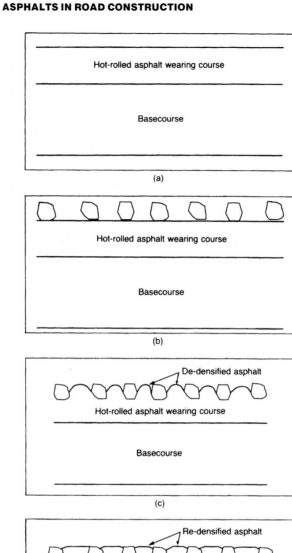

Fig. 8.19. The stages of compacting chipped hot rolled wearing course: (a) hot rolled asphalt wearing course laid on basecourse; (b) chippings placed on top of hot rolled asphalt wearing course; (c) chippings partially embedded; (d) chippings fully embedded

Fig. 8.20.   The fretting of chipped hot rolled asphalt wearing course

8.5.5.   The compaction of HMB

As suggested previously, high modulus bases made with low penetration bitumens may become the most common roadbase materials if doubts about its stiffness over time can be resolved. Assuming this is the case then it may seem difficult to compact materials with such hard binders but experience shows that this is not the case. The important factor is the delivery temperature which allows compaction at the appropriate temperature (usually around 130–140 °C).

8.5.6.   Rolling patterns

Rolling has to be carried out by skilled operatives. It is of vital importance to provide thorough training in effective compaction practice. In particular, compaction should follow a pre-planned pattern which ensures that the compactive effort is equalized across the width of the lane. Despite incorrectly stating that air temperature and base material temperature are important parameters affecting the time available for compaction, Preferred Method 8 — Road Rolling[39] gives excellent guidance for all involved in laying hot asphalts. It covers the planning of surfacing operations including evaluation and flow charts, sample calculations, methods for checking rolling patterns and a separate section specifically for roller drivers which covers all aspects of rolling from the constitution of asphalts (albeit somewhat outdated) through good rolling practice to daily maintenance. As well as the advice contained in Section 7.10 of Chapter 7, major compaction equipment manufacturers issue excellent advice on rolling patterns.[5,40]

8.5.7. Dynapac's PaveComp© Computer Program

A recent positive development has been the introduction of Dynapac's 'PaveComp'© computer program. Produced by the Company's International High Comp Centre at Karlskrona in Sweden, it is designed to make the choice of paver, screed and roller much simpler and relates to that company's own equipment. Some equipment manufacturers make their catalogues, which include performance figures, available on CD-ROM. PaveComp© is rather more useful since it contains a number of routines which calculate the time available before minimum rolling temperature is reached. These data combined with the number of passes required to achieve the specified level of compaction and the necessary speed of the roller, provide uniquely useful information to the surfacing contractor.

The calculation of time available for compaction is facilitated by solving the heat flow equations which apply in a compacted layer. Results from PaveComp© have been compared to those obtained by running CALT© (CALT© is discussed in 8.3.4 above and in Section 4.8.1 of Chapter 4). PaveComp© calculates temperatures on the basis of fewer parameters than CALT©. Nevertheless, for the range of relevant temperatures (70–120 °C) the times which were returned showed a good correlation. A screen dump from PaveComp© is shown in Fig. 8.21.

A wide range of parameters can be input, including

- mixture details
- foundation type, bound or unbound

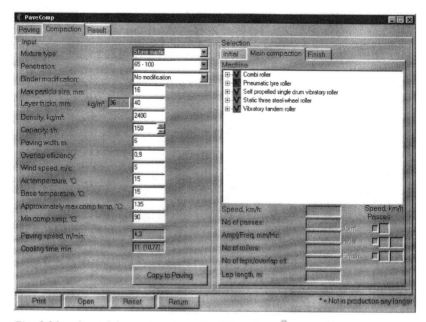

Fig. 8.21.   One of the screen views from PaveComp©

- hourly delivery quantity
- layer density, width and thickness
- bitumen penetration/modification
- environmental parameters, including wind speed, air and base temperature.

This information allows the surfacing contractor to exercise close control in both economic and performance terms on the paving and compaction process.

8.6. Summary

This Chapter has examined the subject of compaction. Published work supports the view that this process is fundamental to the successful construction of pavements and that greater care is necessary in ensuring adequate and homogeneous compaction across the full width of laid mats. The differences between static and vibrational rolling were discussed, along with the economic benefits of vibrational compaction. The importance of amplitude — with the emphasis on training including the adoption of rolling patterns — was advocated.

Surfacing contractors now have at their disposal the first of many packages which will assist in exercising greater control on the surfacing operation. These will ensure that compaction is based on knowledge and experience and not guess work. Already, some rollers have a meter in the operator's cabin which shows, by visual inspection, the degree of compaction achieved as the roller passes over the material being compacted. The same system could be extended to record detailed information about the levels of compaction which are actually achieved on a carriageway. Until these refinements become commonplace, engineers and surfacing contractors will have to continue to exercise great care when controlling the compaction of asphalts.

References

1. BRITISH STANDARDS INSTITUTION. *Hot Rolled Asphalt for Roads and other Paved Areas, Specification for Constituent Materials and Asphalt Mixtures.* BSI, London, 1992, BS 594: Part 1.
2. BRITISH STANDARDS INSTITUTION. *Hot Rolled Asphalt for Roads and other Paved Areas, Specification for the Transport, Laying and Compaction of Rolled Asphalt.* BSI, London, 1992, BS 594: Part 2.
3. BRITISH STANDARDS INSTITUTION. *Coated Macadam for Roads and other Paved Areas, Specification for Constituent Materials and for Mixtures.* BSI, London, 1993, BS 4987: Part 1.
4. BRITISH STANDARDS INSTITUTION. *Coated Macadam for Roads and other Paved Areas, Specification for Transport Laying and Compaction.* BSI, London, 1993, BS 4987: Part 2.
5. KLOUBERT H-J. *Compaction of Bituminous Materials.* Bomag-Menck GmBH, Boppard, Germany, November 1997.
6. MARSHALL CONSULTING & TESTING LABORATORY. *The Marshall Method for the Design and Control of Bituminous Paving Mixtures.* Marshall Consulting & Testing Laboratory, Mississippi, 1949.
7. BRITISH STANDARDS INSTITUTION. *Sampling and Examination of*

Bituminous Mixtures for Roads and other Paved Areas, Method of Test for the Determination of Wheel-Tracking Rate and Depth. BSI, London, 1998, BS 598: Part 110.

8. JACOBS F. A. *Properties of Rolled Asphalt and Asphaltic Concrete at Different States of Compaction.* TRL, Crowthorne, 1977, Supplementary Report 288.

9. LISTER N. W. and W. D. POWELL. *The Compaction of Bituminous Base and Base-Course Materials and its Relation to Pavement Performance.* TRL, Crowthorne, 1977, Supplementary Report 260.

10. LEECH D. and N. W. SELVES. *Modified Rolling to Improve Compaction of Dense-Coated Macadam.* TRL, Crowthorne, 1976, LR 274.

11. HIGHWAYS AGENCY *et al. Design Manual for Roads and Bridges, Pavement Design and Maintenance, Traffic Assessment.* TSO, London, 1996, 7.2.1, HD 24/96.

12. HUNTER R. N. *An Examination of the Fretting of Hot Rolled Asphalt Wearing Course.* Heriot-Watt University, MSc thesis, Edinburgh, 1983.

13. DAINES M. E. *Tests for Voids and Compaction in Rolled Asphalt Surfacing.* TRL, Crowthorne, 1995, PR 78.

14. POWELL W. D. and D. LEECH. *Compaction of Bituminous Road Materials using Vibratory Rollers.* TRL, Crowthorne, 1983, LR 1102.

15. CORLEW J. S. and P. F. DICKSON. Methods for Calculating Temperature Profiles of Hot-Mix Asphalt Concrete as Related to the Construction of Asphalt Pavements. *Proc. Tech. Sess. Asphalt Paving Tech.*, 1996, **35**, 549–50.

16. JORDAN P. G. and M. E. THOMAS. *Prediction of Cooling Curves for Hot-Mix Paving Materials by a Computer Program.* TRL, Crowthorne, 1976, LR 729.

17. DAINES M. E. *Cooling of Bituminous Layers and Time Available for their Compaction.* TRL, Crowthorne, 1985, RR 4.

18. HUNTER R. N. *Calculation of Temperatures and their Implications for Unchipped and Chipped Bituminous Materials During Laying.* Heriot-Watt University, PhD thesis, Edinburgh, 1988.

19. HUNTER R. N and G. R. MCGUIRE. Are Texture and Density in Hot Rolled Asphalt Wearing Courses Incompatible? *Journal of the Inst. of Asphalt Tech.*, 1988, **40**, 29–31.

20. HUNTER R. N. Cooling of Bituminous Materials During Laying. *Journal of the Inst. of Asphalt Tech.*, 1986, **38**, 19–26.

21. BRITISH STANDARDS INSTITUTION. *Sampling and Examination of Bituminous Mixtures for Roads and other Paved Areas, Methods for Determination of the Condition of the Binder on Coated Chippings and for Measurement of the Rate of Spread of Coated Chippings.* BSI, London, 1990, BS 598: Part 108.

22. HIGHWAYS AGENCY *et al. Specification for Highway Works, Manual of Contract Documents for Highway Works.* TSO, London, 1998, **1**.

23. HIGHWAYS AGENCY *et al. Manual of Contract Documents for Highway Works.* TSO, London, 1998, **0–6**.

24. HIGHWAYS AGENCY *et al. Manual of Contract Documents for Highway Works, Notes for Guidance on the Specification for Highway Works.* TSO, London, 1998, **2**.

25. HIGHWAYS AGENCY *et al. Design Manual for Roads and Bridges.* TSO, London, various dates, **1–15**.

26. HIGHWAYS AGENCY *et al. Design Manual for Road and Bridges,*

Pavement Design and Maintenance. TSO, London, various dates, **7**.

27. BRF. *Road Fact 99*. BRF, London, 1999.

28. HIGHWAYS AGENCY *et al. Design Manual for Roads and Bridges, Pavement Design and Maintenance, Bituminous Surfacing Materials and Techniques*. TSO, London, 1999, 7.5.2, HD 37/99.

29. BRITISH STANDARDS INSTITUTION. *Sampling and Examination of Bituminous Mixtures for Roads and other Paved Areas, Methods for the Assessment of the Compaction Performance of a Roller and Recommended Procedures for the Measurement of the Temperature of Bituminous Mixtures*. BSI, London, 1990, BS 598: Part 109.

30. MANNING D. G. *Effects of Traffic-Induced Vibrations on Bridge Deck Repairs*, National Cooperative Highway Research Program Synthesis of Highway Practice. Transport Research Board, Washington, 1981, **86**.

31. BRITISH STANDARDS INSTITUTION. *Sampling and Examination of Bituminous Mixtures for Roads and other Paved Areas, Methods of Test for the Determination of Density and Compaction*. BSI, London, 1990, BS 598: Part 104.

32. BOMAG. *The Compaction of Bituminous Materials*. Bomag 'Great Britain' Ltd, Larkfield, UK.

33. BRITISH STANDARDS INSTITUTION. *Sampling and Examination of Bituminous Mixtures for Roads and other Paved Areas, Methods for Sampling for Analysis*. BSI, London, 1987, BS 598: Part 100.

34. BRITISH STANDARDS INSTITUTION. *Methods for Determination of Maximum Density of Bituminous Mixtures*. BSI, London, 1996, DD 228, Issue 2.

35. BRITISH STANDARDS INSTITUTION. *Sampling and Examination of Bituminous Mixtures for Roads and other Paved Areas, Analytical Test Methods*. BSI, London, 1996, BS 598: Part 102.

36. WORKING PARTY. *Nuclear Gauges for Measuring the Density of Roadbase Macadams*. TRL, Crowthorne, 1982, Supplementary Report 754.

37. BRITISH BOARD OF AGRÉMENT. *Guidelines Document for the Assessment and Certification of Thin Surfacing Systems for Highways*. BBA, Watford, 1999.

38. HUNTER R. N. and G. R. MCGUIRE. Winter Surfacing — Only the Specification is out in the Cold. *Highways & Transportation, London*. IHT, London, 1986, **33**, 12.

39. DEPARTMENT OF TRANSPORT STANDING COMMITTEE ON HIGHWAY MAINTENANCE. *Preferred Method 8 — Road Rolling*. Cornwall County Council, Truro, undated.

40. DYNAPAC HEAVY EQUIPMENT AB. *Compaction and Paving, Theory and Practice*. Dynapac, Karlskrona, Sweden, 2000.

9. Standards for testing

9.1. Preamble

The results of measuring and testing procedures throughout the whole process of flexible road construction are relied on for guidance. Decisions are taken based on the data generated at virtually every stage of the operation, including initial investigation, cost estimation, production control, installation, acceptance and assessment of performance.

Confidence in the accuracy of the data being analysed is, therefore, vital. This can only be achieved by demonstrating that sufficient care has been taken by skilful, knowledgeable individuals in following prescribed methods for proper sampling and testing that yield reliable results. Errors can lead to costly, false assumptions being made about the quality of materials and products.

This Chapter looks at many of the tests which are undertaken on asphalts and their constituents. There is little point in testing unless the sample is representative of the whole and this issue is considered along with precision. Many test methods are considered, including those for bitumens, compositional analysis, mixture design and adhesion. Surfacing contractors will be particularly interested in the sections on percentage refusal density and texture depth. As the move continues towards performance specifications, knowledge of wheel-tracking, creep and elastic stiffness will become mandatory.

Intelligent interpretation of the data provided is essential in the light of the precision of the test methods and the reliability of measuring and test equipment. This can be achieved with the aid of appropriate statistical techniques, many of which involve only simple calculations. Good interpretation of reliable data leads to the correct decisions being taken and good roads being built at the right cost.

Some testing matters are covered in other Chapters. Aggregate assessment and performance testing is covered in Chapter 1. Testing of laid material is discussed in Chapter 7. Other Chapters which touch on testing issues are noted in the text within this Chapter.

9.2. Sampling

Possibly the greatest concern about the testing of asphalts is whether laboratory portions are representative of the bulk quantity of the material being examined. Unfortunately, asphalts are prone to segregation during handling, whereby differently sized aggregate particles in the mixture separate in either a random or uniform manner. The effect is exacerbated in coarser mixtures which have thin binder films coating the aggregate particles. Thicker binder films resist the relative movement of the particles within the mixture and so aid homogeneity. Where there is a tendency to segregate, strict adherence to standardized sampling practice is of paramount importance. The sampler should be properly trained, competent in implementing prescribed sampling methods and able to work safely without direct supervision.

A number of factors influence segregation of asphalts. The nature of the specified or designed aggregate grading can markedly affect the tendency of the mixture to segregate. Gap grading of the type illustrated in Fig. 9.1 can be a cause as the aggregate blend comprises high proportions of the larger and smaller particles with a lack of intermediate sizes. Gap graded mixtures, such as hot rolled asphalts (HRA), have certain advantages and these are covered in Chapters 3 and 4. Segregation in gap graded mixtures is hindered if the material has relatively high binder contents and thick binder films.

However, continuous gradings (see Fig. 9.1) are less sensitive to segregation, particularly where the blend comprises a good mix of aggregate particles changing progressively over the full size range with a relatively small maximum size. The difference will not be so evident when, for example, comparing large stone mixtures of dense bitumen macadam (DBM) and hot rolled asphalt (HRA), due to the uncertainty over the fully continuous nature of the aggregate gradings of dense macadams and the stabilizing effect of the richer sand/bitumen mortar in HRAs.

Segregation of aggregate constituents in stockpiles or feed bins can also persist in the asphalt, depending on the method of manufacture of the mixture. Inspection of aggregate stocks and bins for signs of segregation may determine whether difficulties in sampling a segregated asphalt can be traced back to the methods of handling the aggregate constituents and whether this can then be overcome by taking appropriate measures to improve the handling techniques.

The increasing use of large bins for the storage of hot asphalts brings further potential segregation problems for the sampler. Brock[1] discusses in detail the different methods used to minimize the effects and concludes that both bin charging batchers and rotating chutes (Figs 9.2 and 9.3) are adequate devices to help eliminate segregation.

Ideally, lorries transporting asphalts should be loaded evenly along their bodies to prevent larger aggregate particles rolling to the base of the conical heaps formed during loading. Brock[2] suggests loading in three different drops in the sequence indicated in Fig. 9.4 to ensure a good mix when discharging from the lorry into the paver hopper. It is therefore important that the sampler takes due account of the actual loading

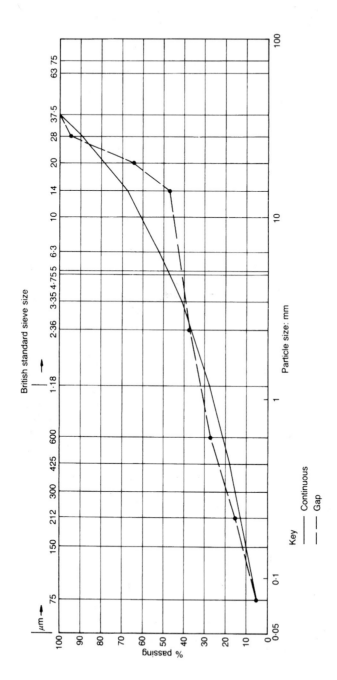

Fig. 9.1. Aggregate grading

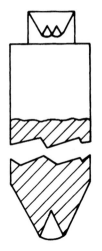

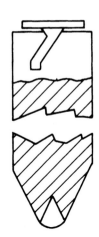

Fig. 9.2. Bin charging batcher[1]   *Fig. 9.3. Bin charging rotating chute*[1]

methods employed to determine the best sampling procedure to be followed at the site of the mixing plant.

BS 598: Part 100[3] describes methods of obtaining bulk samples, both from loaded lorries and during discharge, from the mixing plant. The minimum mass of bulk sample is given in Table 9.1. This is sufficient to provide a laboratory sample for both the supplier and the customer.

Sampling from the loaded lorry using a suitable shovel is frequently preferred by the sampler and this is easily and safely done if a sampling platform has been constructed at the correct height in relation to the height of the sides of the lorry body. Management should ensure that appropriate protective clothing is always worn by samplers and that they are familiar with the hazards associated with moving plant and vehicles and are trained in the treatment of burns in the event that skin contact occurs.

A number of increments, depending on the nominal size of the aggregate constituents, are taken from different positions across the top of the load after surface material has been removed from those positions.

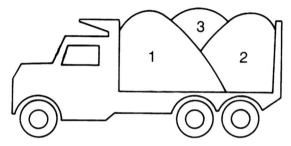

Fig. 9.4. Loading a lorry[2]

Table 9.1.   Minimum mass of bulk samples[3]

Nominal size of aggregate: mm	Minimum mass of bulk sample: kg
>20	24
≤20	16

Alternatively, a sampling pan secured by a hinged bracket to a convenient rigid section of the mixing plant can be operated so that it passes transversely through the centre part of the curtain of material being discharged. Again, a number of appropriate increments are taken to provide the required mass of bulk sample. The risk to the sampler's personal safety using this method is a deterring factor. However, a distinct advantage of the method is the ability to sample individual batches during production without possible intermingling by other batches. ASTM Standard D979[4] also describes procedures for sampling asphalts from both belt and skip conveyors.

Segregation can also occur at the laying site. It follows that the sampler should be aware of those locations in the laying process where the taking of either coarser or finer samples than are representative of the bulk quantity of material being examined is most likely to occur. Brock[1] makes a number of suggestions on avoiding excessive segregation at the laying site, including the following.

- Whenever possible, do not empty the paver hopper between each lorry load.
- Only operate the paver hopper wings infrequently.
- Keep the paver hopper as full as possible during loading from the lorry.
- Do not starve the augers or overfeed them with material.
- Run the paver and augers continuously whenever possible.

The points made in Chapters 7 and 8 about heat loss should be borne in mind in relation to these suggestions.

BS 598: Part 100[3] no longer includes the procedure for sampling from a paver hopper, which entailed withdrawing the lorry from the paver after it had discharged approximately half its load before sampling, often causing the cessation of paving operations. This practice is clearly inconsistent with a number of the suggestions. The alternatives are to sample from the material around the augers of the paver and from the material extruded from beneath the paver screed with the sample taken before rolling.

When sampling material from around the augers, two increments should be taken from each side of the paver. The sampler should be aware of a tendency for the augers to throw coarser material to the edges of the paved widths if the auger box is not filled. Samples should, therefore, only be taken after the box has been properly filled by a continuous slow flow of material from the conveyors.

Generally, there should be less chance of significant segregation actually appearing in the laid material, provided all the necessary precautions are taken upstream of the laying operations. In practice, surface defects caused by segregation and other factors do appear from time to time, frequently only becoming apparent after rolling, and these are discussed in Chapter 11. Samples may be taken from the laid material before rolling by placing two sampling trays on the ground just ahead of the paving operation and removing the trays with the aid of wire attachments after the material has been laid. The method is not favoured for sampling wearing course material because of the possibility of leaving surface blemishes at the sampling positions after infilling. It is also inappropriate for relatively thin layers of coarser material because of a real possibility of disturbance as the paver moves over the sampling positions.

For various reasons, samples may need to be taken from the completed layer, for example to measure the density of the layer. Samples may be extracted using a core cutting machine or other equipment suitable for cutting out squares of material. Normally, samples are not taken from a localized spot which is visibly unrepresentative of the whole unless there is a specific reason for examining the material, perhaps to investigate the cause of some apparent difference. Despite this, there is a good argument for devising a plan of random sampling that eliminates intentional (and also minimizes unintentional) bias on the part of the sampler.[5] In this way, there can be no criticism of the sampling regime.

ASTM Standard D3665[6] describes procedures that can be followed to determine the location of samples with the use of a table of random numbers. It is recommended that on receipt of the samples at the laboratory, note is made of their condition and any discernible variations to aid interpretation of subsequent test results. A degree of randomness is removed by dividing the whole into smaller lots within which random sampling is carried out.

It may be convenient to reduce the size of the bulk sample before dispatch to the laboratory, depending on the planned tests or on whether there is a need to divide the sample between interested parties. This may be done for hot mixtures using a sample divider or riffle box, or by quartering on a clean, hard surface as described in BS 598: Part 101.[7] The laboratory sample obtained should be accompanied by a record of the sampling details. A typical example of a certificate of sampling is shown in Fig. 9.5.

Sampling procedures for constituent aggregates are described in BS 812: Part 102[8] and the subject is discussed in detail by Harris and Sym.[9] Standard methods for sampling bituminous binders are described fully in BS 3195: Part 3[10] and ASTM D140.[11] No method exists which will guarantee that a sample is truly representative of the whole in all respects.[12] There will always be random sampling error and sampling bias, albeit small in respect of both.

Large differences between the sample and the total quantity of asphalt are often caused by segregation. Acknowledgement is made in the British Standards for asphalts[13,14] of the particular difficulties involved in obtaining representative samples of the whole in the provisions made for sampling and testing error by the application of specified tolerances.

CERTIFICATE OF SAMPLING

SAMPLE DESTINATION		SAMPLE IDENTIFICATION NUMBER
MATERIAL DESCRIPTION		TEMPERATURE
PLANT / SUPPLIER		BINDER TYPE
CUSTOMER		BINDER TARGET %
FLUX %		SPECIFIED BINDER %
LOCATION / SOURCE OF SAMPLE	/ WAGON NUMBER / TICKET NUMBER	
DESTINATION		REASON FOR SAMPLING
DATE OF SAMPLE		TIME OF SAMPLE
METHOD OF SAMPLING	B.S. No. Part No. other Method	

Remarks

Fig. 9.5.   Certificate of sampling

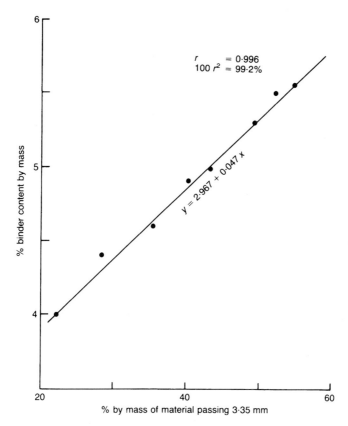

Fig. 9.6. Relationship between binder content and fine aggregate content for a particular uniformly coated bituminous mixture

However, for a uniformly coated mixture, there is a close association between the amount of bituminous binder and the aggregate grading, or more specifically, the aggregate surface area. This fact can be used to determine whether the sample is likely to be unrepresentative. This can be demonstrated by taking a large sample of mixture and intentionally dividing it into test portions having different fine aggregate contents to simulate sampling error due to segregation. Analysis of the constituent parts after the extraction of binder from each portion confirms that the binder content changes in proportion to the change in fine aggregate content. This is because the fine aggregate has a larger surface area per unit volume and so holds a greater quantity of binder coating than the same mass of coarse aggregate particles.

The relationship is illustrated in Fig. 9.6 for a particular asphalt. The percentage of aggregate passing the 3.35 mm test sieve, representing the independent variable, has been plotted against the corresponding binder content. The plot suggests a linear association between the variables which may be described by the best straight line $y = a + bx$.

Regression analysis of the data gave the following information for this material

Intercept (a)	= 2.967
Slope (b)	= 0.047
Correlation Coefficient (r)	= 0.996
Percentage fit ($100r^2$)	= 99.2%

The resultant straight line $y = 2.967 + 0.047x$ can then be drawn through the plotted points on the graph. The 'Correlation Coefficient' gives a measure of the degree of linear association between the variables, whereby values of -1 or $+1$ indicate perfect association and a value of 0 indicates the absence of a linear association. The 'Percentage Fit', indicates the extent to which the variation in binder content has been caused by the variation in material passing the 3.35 mm sieve. The remaining percentage is unaccounted for and may be caused by other independent variables. The equation describing the straight line relationship will depend on the physical and volumetric properties of the constituents of the particular mixture and, especially, the physical differences between the fine and coarse aggregate particles. For example, a steeper slope would indicate a greater capacity of the fine aggregate for holding binder in relation to that of the coarse aggregate particles.

Once the relationship between binder content and coarse/fine aggregate proportions for a particular mixture is established, subsequent plots on the graph during production deviating from the straight line would suggest either poor mixing, resulting in lack of uniformity of binder coating, or a change in constituent proportioning. Points plotted on the graph lying close to the straight line would suggest that segregation is a cause of variation and that the individual test results need not, necessarily, be indicative of an unsatisfactory product. A variation of the analysis based on a summation of each percentage amount passing the fine aggregate sieves was devised by Bryant.[15]

BS 598: Part 102[16] includes tables for adjusting the binder content found on analysis to correspond with the midpoint of the grading passing the top fine aggregate test sieve specified in the British Standard product specifications, to take account of variations due to sampling and sample reduction procedures. Similar adjustment values are included for aggregate passing the 75 μm test sieve to correspond with the same midpoint on the top fine aggregate test sieve. However, it is debatable whether the midpoint of the specified limits on this sieve should actually be taken as reference, rather than the average percentage passing the sieve from the designed aggregate grading that may be quite different, depending on the specified tolerance, from the midpoint.

Variation in test results due to sampling and testing for binder content of asphalts has been estimated to be around 65% of the total variation from studies made in the 1960s.[17] Later studies[18] confirmed that this percentage had not changed appreciably. Section 9.3 considers the variation that occurs after sampling and during testing of the sample. This variation is described by the 'precision' of the test method.

9.3. Precision

Asphalts are generally produced to a specification provided by the purchaser. Certain properties of the mixture are influenced by its volumetric composition and the nature of its constituents. Control of these influential factors is important. Knowledge of how certain properties relate to the performance of the product incorporated in the works frequently also leads the specifier to include requirements for a number of measurements to be made. Certain desirable features of the constructed pavement may also be measured and these too can be included in the specification.

During the course of the production and installation processes certain changes occur. Test results and other measured data vary due to different factors such as sampling techniques, testing, changes in the quality and nature of constituents and plant variations. Some degree of control is therefore required to ensure that the desired properties of the end product are realized. Permitted deviations from the values of the characteristics whose behaviours are to be investigated are specified.

It follows that the tolerances set should allow for errors in sampling and testing, and for plant and material inconsistencies that may be expected to occur when all reasonable care has been taken to avoid them. However, this need not always be the case. It is important, therefore, to be able to estimate the magnitude of these inherent variations so that realistic specification limits can be determined for the variable characteristics measured.

When available, precision data are frequently included in standard documents for test methods. The data may be expressed in different forms but the terms 'repeatability' and 'reproducibility' are commonly used.

There is a distinction between 'precision' and 'accuracy'. In simple terms, precision is concerned with the variability of individual values, irrespective of how close these values are to the true value. Accuracy, on the other hand, is concerned with how close the measured values are on average to the true value, irrespective of the degree of variability of the individual values. This distinction is illustrated in Fig. 9.7 with the aid of the type of target board that may be found on a firing range. It is clear that an analysis of variable characteristics must include a study of both scatter and location.

The scatter of test results may, of course, be influenced by the inherent variations or precision of the test methods. These testing variations can be determined by experiment and such a procedure is described in BS 5497: Part 1.[19] However, the accuracy of the test method, i.e. how close the overall mean value is to the true value of the characteristic, can be more difficult to determine due to an uncertainty as to the true value.

When determining the precision of a test method, every effort must be made to minimize the influence of variations from other causes. The experiment concerns both a measure of the variability due to a single

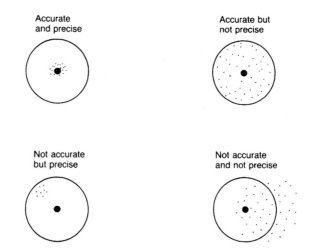

Fig. 9.7.   Precision and accuracy

operator and a measure of variability between different laboratories, all other factors being equal.

Testing by a single operator should be carried out each time in circumstances that are as similar as possible and on samples of identical material. Repeat testing on the same test specimen ensures that identical material is used. Unfortunately, most of the tests carried out on asphalts are destructive and so particular care is needed in the preparation of test portions from laboratory samples which should be derived from a mixture that is considered to be as uniform as possible in terms of both composition and structure. Repeat tests should also be carried out within as short a period of time as possible.

Single operator variation is described by a measure of 'repeatability' determined under *'conditions where mutually independent test results are obtained with the same method on identical test material in the same laboratory by the same operator using the same equipment within short intervals of time'*.[19] The variations associated with the inherent variability of constituent aggregates and the difficulties involved in reducing samples of asphalts to give representative test portions unavoidably contribute to the measure and so form a real part of it when the same test specimen cannot be used. So, in this case, both material variation and sampling error will tend to increase the magnitude of the reported precision data above that due to testing alone. Depending on the nature of the particular test method under experiment, the error may be minimized by combining measured quantities of the constituents of the mixture to provide the individual test specimens.

The 'repeatability value' *r* is defined as *'the value below which the absolute difference between two single test results obtained under repeatability conditions may be expected to lie with a probability of 0.95'*.[19] It is calculated from the following equation[19]

$$r = 2.8S_r \tag{9.1}$$

where S_r is an estimate of repeatability standard deviation and 2.8 is a factor derived from the basic statistical model used.

Variation between laboratories is described by a measure of reproducibility determined under '*conditions where test results are obtained with the same method on identical test material in different laboratories with different operators using different equipment*'.[19] This is a measure of the maximum variability in results and differs from repeatability due to different operators using different equipment of different calibration status that is used under different environmental conditions.

The 'reproducibility value' R is defined as '*the value below which the absolute difference between two single test results obtained under Reproducibility conditions may be expected to lie with a probability of 95%*'.[19] It is calculated from the following equation.[19]

$$R = 2.8S_R \tag{9.2}$$

where S_R is an estimate of reproducibility standard deviation and 2.8 is the same factor as in equation (9.1).

Harris and Sym,[9] in their study of the precision of testing aggregates, used a statistical model that included an allowance for variation between laboratory samples. This takes account of an additional variation that arises when it is not possible to obtain test results on the same laboratory sample and contributes to a higher value of R. As similar difficulties are experienced with sampling the two types of product, the model is equally applicable to asphalts. However, the magnitude of the contribution may be greater with aggregates which are particularly dry and therefore more prone to segregation.

Table 9.2 gives precision data which has been published in the UK for some of the more common test methods. Precision data published by the American Society for Testing and Materials (ASTM)[21] and the American Association of State Highway and Transportation Officials (AASHO)[22] is given in Table 9.3 for similar tests carried out under conditions described in ASTM Standard Practice C802.[23]

Some of the findings of early research work carried out in the USA[24] on the normal variation of test results of asphalts from a large number of paving projects are given in Table 9.4. Precision data for the test methods give a guide to the differences that may be expected to occur when testing the same material under the stated conditions and caution should be exercised when comparing these data with the variation of field test results that arises due to sampling, testing and materials variation.

9.4. Test methods

9.4.1. Composition of asphalts

The composition of an asphalt has a definite influence on its behaviour in service, to an extent depending on the property being measured. The

Table 9.2.   Precision of British test methods

Test	Standard	Repeatability standard deviation, S_r (Coefficient of Variation $= S_p/\bar{x}.\ 100\%$)	Repeatability, r (% of mean)	Reproducibility standard deviation, S_R (Coefficient of Variation $= S_R/\bar{x}.\ 100\%$)	Reproducibility, R (% of mean)	Unit
Penetration of bitumens < 50 pen	BS 2000: Part 49: 1983	0.35	1	1.4	4	0.1 mm
Penetration of bitumens > 50 pen	BS 2000: Part 49: 1983	(1.1%)	(3%)	(2.8%)	(8%)	—
Ring and ball softening point of paving grade bitumen	BS 2000 Part 58: 1988		1		2.5	°C
Binder content of bituminous mixtures %	BS 598 Part 102 1996		0.3		0.5	—
Adjusted filler content of bituminous mixtures < 75μm, %	BS 598: Part 102: 1996		1		1.5	—
Percentage refusal density, %	BS 598: Part 104: 1989		1.2		1.8	—

Texture depth by laser texture meter for test result levels BS 598: Part 105: 1990					
0.8 mm	0.032	0.09	0.065	0.18	mm
1.2 mm	0.047	0.13	0.078	0.22	mm
1.4 mm	0.051	0.14	0.091	0.25	mm
Texture depth by sand patch for test result levels BS 598: Part 105: 1990					
1.3 mm	0.062	0.17	0.073	0.20	mm
1.9 mm	0.103	0.29	0.129	0.36	mm
2.3 mm	0.135	0.38	0.156	0.44	mm
Laboratory design of hot-rolled asphalt mixtures[20] BS 598: Part 107 1990					
Density		17		26	kg/m³
Stability		1		2.2	kN
Flow		0.6		1.3	mm
Quotient		0.7		1.4	kN/mm

Table 9.3 Precision of American test methods

Test	Standard ASTM No. (AASHTO No.)	Repeatability standard deviation, S_r (Coefficient of Variation $= Sr/\bar{x}.100\%$)	Repeatability, r (% of mean)	Reproducibility standard deviation, S_r (Coefficient of Variation $= S_R/\bar{x}.100\%$)	Reproducibility R, (% of mean)	Unit
Penetration of bitumens $<$ 50 pen	D5 (T49)	0.35	1	1.4	4	0.1 mm
Penetration of bitumens $>$ 50 pen	D5 (T49)	(1.1%)	(3%)	(2.8%)	(8%)	—
Ring and ball softening point of bitumen	(T53)		2.0		3.0	°C
Bitumen content of bituminous mixtures, %	D2172 (T164)	0.18	0.52	0.29	0.81	—
Bulk specific gravity of compacted bituminous mixtures using saturated surface-dry specimens	D2726	0.0124	0.035	0.0269	0.076	—

Test	Designation					
Theoretical maximum specific gravity of bituminous mixtures	D2041 (T209)	0.004	0.011	0.0064	0.019	—
% air voids in compacted dense and open bituminous mixtures, %	D3203 (T269)	0.32	0.91			—
Surface macrotexture depth using a volumetric technique	E965	(as low as 1%)		(as low as 2%)		—
Effect of water on cohesion of compacted bituminous mixtures, %	D1075 (T165)	6	18	18	50	—
Viscosity of bitumen residue at 60 °C after Rolling Thin Film Oven Test	D2872 (T240)	(2.3%)	(6.5%)	(4.2%)	(11.9%)	—

Table 9.4.   Typical values of standard deviation of test results of bituminous mixtures during paving projects[24]

Test	Average standard deviation
Bitumen content, %	$0.17^a–0.42^{b*}$
Bulk specific gravity of compacted samples	$0.02–0.04$
Air voids, %	$0.8–1.8$
Marshall Stability, kg	128
Marshall Flow, mm	0.3

*[a] is a value of $\frac{1}{3}$ least variable jobs and [b] is the value $\frac{1}{3}$ most variable jobs

effect of constituents on performance is discussed in detail in Chapters 1 and 2.

Richardson[25] described methods used at the beginning of the twentieth century for the examination of asphalts for their bitumen content and mineral aggregate grading. These tests involved dissolving out the bitumen with an appropriate organic solvent and trapping the mineral matter in a filter or centrifuge apparatus. The recovered mineral aggregate was then passed through a series of sieves and the aggregate graded by the amounts passing the various aperture sizes. The preferred solvent was bisulphide of carbon (carbon disulphide) which was available cheaply and could be readily redistilled for reuse if desired.

Carbon disulphide has a boiling point slightly higher than that of methylene chloride (dichloromethane) commonly used today, but is extremely flammable and gives off very poisonous vapour. At the time, pure chloroform (trichloromethane) was found to be an ideal solvent for determining total bitumen content due to a high bitumen solubility. However, it was prohibitively expensive and could not be evaporated or burned off rapidly, as was frequently done when determining residual mineral matter in the extracted binder solution during analysis.

The early analytical methods for fine asphalts involved determining the bitumen quantity in small samples of the mixtures by subtracting the weight of recovered mineral matter from the initial weight of the sample and expressing this as a percentage of the mixture. In order to prevent the solvent with dissolved bitumen from carrying over fine particles of mineral matter also, the solution was passed through a filter paper supported in a funnel that was placed in a conical flask to collect the filtered solution. After filtering, the percolate in the flask was allowed to stand overnight to allow any sediment to settle to the bottom. The percolate was decanted and the remnants containing sediment were passed back through the filter paper. The binder solution was evaporated and burned and the bitumen incinerated to determine a correction factor for any mineral matter that passed through with the bitumen, to be added to the weight of mineral aggregates recovered from the filter paper. The used filter paper was also burned to correct for any fine mineral matter remaining trapped in the pores of the paper.

Alternatively, where a more rapid analysis was required, the mineral aggregate was recovered in a centrifuge machine rotating at a speed of 1500 r/min. This method was also more suitable for analysing mixtures containing relatively large quantities of mineral filler. After centrifuging, the binder solution collected in the centrifuge tubes was decanted leaving the sediment for further centrifuging with new solvent.

When all bitumen was removed from the aggregates, the solvent remaining in the tubes was evaporated and the residue weighed. The binder solution removed from the machine was evaporated and burned and the extracted bitumen incinerated to correct for mineral matter removed with the bitumen in the calculation of their contents from the loss in weight in the centrifuge tube.

Coarser mixtures were analysed using a specially designed centrifuge that could handle larger samples. A filter ring of roofing felt was inserted into the apparatus to catch fine mineral matter thrown out during centrifuging at between 1500 and 1800 r/min. Facilities were also included to recover solvent for reuse. The separation was considered to be of such efficiency that no correction was made for traces of mineral matter remaining in the extracted bitumen. Today, the methods of analysis have not changed greatly and the principle of the test remains the same, comprising the following basic operations

- binder extraction by dissolving in an organic solvent
- separation of mineral matter from the binder solution
- determination of binder quantity by difference or binder recovery
- calculation of soluble binder content
- grading of mineral aggregates.

This is illustrated in Table 9.5 which summarizes a selection of analytical methods used in Europe, breaking them down into the same basic operations. An additional column is included (headed insoluble binder) to cater for those products such as tar which are not totally soluble in the solvent used.

A variety of equipment is used for the various test methods and the main items are illustrated in Figs 9.8 to 9.16.

Some of the test methods include determination of binder directly by weighing the total recovered binder or a portion of it (Table 9.6). BS 598: Part 102[16] describes a method of direct determination from a portion of the total binder solution collected after extraction by roller bottle (Fig. 9.10). The portion is quickly poured into a number of centrifuge tubes which are then capped securely to prevent evaporation of the solvent and placed in a bucket type centrifuge (Fig. 9.16). The time centrifuging continues depends on the acceleration developed in the machine to ensure that fine mineral matter separates satisfactorily to the bottom of the tubes. The centrifuged solution is then transferred to a burette, care being taken not to disturb the mineral matter that is left in the tubes, and measured into a boiling flask. The solvent is then removed from the solution by boiling

Table 9.5. Analytical test methods for bituminous mixtures

Test method	Basic operation						
	Binder extraction	Separation of mineral filler	Binder quantity	Ash determination	Soluble binder	Insoluble binder	Aggregate grading
(D) DIN 1996: Part 6: 1988 Differential	(1) Hot extraction apparatus or (ii) Cold agitation or centrifugal filter press using trichloroethylene or toluene	Continous flow centrifuge	By calculation	For arbitration only DIN 52005	By subtracting mass of total mineral aggregate	By calculation as a function of filler content or by test	DIN 1996: Part 14
Recovery	As above	As above	Total recovery of binder by evaporation and rotary evaporator or distillation apparatus with fractionating column	As above	From mass of recoved binder	As above	As above
(GB) BS 598: Part 102: 1996 Extraction bottle: binder by difference	Bottle rolling using methylene chloride or trichloroethylene	Pressure filter	By calculation	—	By subtracting mass of total mineral aggregate	% by mass of binder insoluble in solvent pre-determined	BS 812: Part 103

Extraction bottle: binder directly determined	Bottle rolling using methylene chloride	Bucket type centrifuge	Recovery of binder from a portion of centrifuged binder solution by evaporation under reduced pressure	—	From mass of recovered binder residue after evaporation of aliquot portion	As above	Filler directly by pressure filter or Filler by difference after decanting through 75μ sieve BS 812: Part 103
Hot extractor	Hot extraction apparatus with filter paper lining using trichloroethylene	Filter paper	By calculation	—	By subtracting mass of total mineral aggregate	As above	(i) Filler directly by pressure filter or filter paper or (ii) Filler by difference after decanting through 75μm sieve BS 812: Part 103
(N) Strassentest Automatic binder extraction machine	By recycled solvent spray	Continuous flow centrifuge	By calculation	—	By subtracting mass of total mineral aggregate	—	Vibrating sieving unit of six sieves
(NL) Test 65.0 Soxhlet extraction	Modified Soxhlet equipment with pressed filter paper extraction case	Filter paper extraction	Total recovery of binder by evaporation	Incineration of portion of recovered binder	From mass of recovered binder with correction for ash	—	Test 6.0

Table 9.5. Continued

Decanting-jar centrifuge	Cold agitation and through sieves	Continuous flow centrifuge	By calculation or Recovery of binder from all or portion of centrifuged binder solution by evaporation	Evaporation and incineration of portion of centrifuged binder solution	By subtracting mass of total mineral aggregate or from mass of recovered binder	—	Test 6.0
Automatic decanting-jar centrifuge	Cold agitation and by solvent spray	Continuous flow centrifuge	As above	As above	As above	—	Test 6.0

Fig. 9.8.   DIN hot extractor. Photograph courtesy of Control Testing Equipment Ltd

in a water bath under vacuum to leave approximately one gramme of soluble binder.

The flask is dried and allowed to cool in a desiccator to prevent absorption of moisture from the air before weighing. The percentage binder content is then calculated by scaling up in the ratio of the total volume of solvent used to that contained in the measured portion of centrifuged binder solution.

Some methods, as indicated in Table 9.6, rely on the total recovery of the binder in the sample of mixture to determine binder content. The binder may be recovered from solution after extraction from the mixture by solvent and removal of fine mineral matter, using either a rotary film evaporator or a simple distillation apparatus with a fractionating column (Figs 9.17 and 9.18). The simple distillation apparatus may be used for fluxed or cutback bitumens, as well as penetration grades. The rotary film evaporator is used for penetration grades only.

The Rotary Film Evaporator Method[26] is possibly becoming more popular when handling mixtures containing ordinary penetration grade bitumens. The apparatus is operated under reduced pressure and a relatively large sample can be recovered quite quickly. The flow of binder

Fig. 9.9.   Centrifuge extractor. Photograph courtesy of ELE International Ltd

solution into the evaporating flask (Fig. 9.17) is controlled by the induction stopcock so that the rate is approximately equal to that at which the distillate flows into the receiving flask. The evaporating flask is placed in an oil bath at a temperature of 115 °C and rotated at 75 r/min, minimizing the possibility of overheating the bitumen.

Fig. 9.10.   Bottle rolling machine. Photograph courtesy of ELE International Ltd

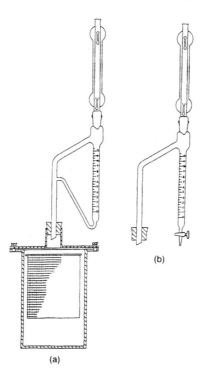

Fig. 9.11.   BS hot extractor: (a) solvent density > 1; (b) solvent density < 1

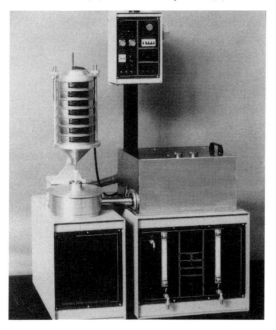

Fig. 9.12.   Automatic binder extraction machine. Photograph courtesy of
Control Testing Equipment Ltd

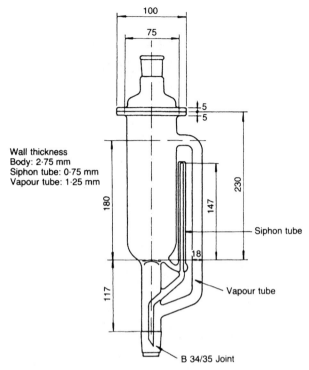

Fig. 9.13.   Soxhlet extractor

After all the binder solution has been transferred to the evaporating flask, the pressure inside the apparatus is increased and the temperature of the oil bath raised to 150 °C and above to remove all the remaining solvent. The recovered binder is weighed to give the binder content of the mixture but may also be used for further testing such as bitumen penetration[28] or ring and ball softening point,[29] both of which are discussed in Section 9.4.2 below.

The distillation apparatus with fractionating column is shown in Fig. 9.18. In this case,[27] the bulk of the distillation is carried out under atmospheric conditions with the oil bath maintained at a temperature of approximately 100 °C while the binder solution is continuously stirred to prevent bumping. The temperature of the oil bath is increased to 175 °C or 100 °C above the expected softening point of the binder, whichever is higher, as the final portion of binder solution introduced into the distillation flask reduces in volume. When the rate of distillation drops significantly, the pressure in the apparatus is reduced to remove the last traces of solvent. During this last stage, unless very volatile flux is contained in the binder, carbon dioxide gas is passed through the residue in the flask to assist the removal of solvent while hindering oxidation of the binder. Volatile oils in the binder are recovered from the distillated vapours as they pass through the fractionating column.

Fig. 9.14.   Continuous flow centrifuge. Photograph courtesy of Control Testing Equipment Ltd

Fig. 9.15.   Pressure filter. Photograph courtesy of ELE International Ltd

Fig. 9.16. Bucket centrifuge. Photograph courtesy of ELE International Ltd

The selection of a particular method of determining binder content may be largely a matter of personal preference. Factors influencing the choice include cost, exposure to solvent fumes, precision, complexity, rapidity and purpose of the test. For quality control testing during production, the time taken to produce a result can be one of the most critical factors. Table 9.7 gives an estimate for those tests detailed in Table 9.5 that can produce a result within two hours and so may be considered for quality control purposes.

Nuclear testing devices are also now available to give a very rapid estimate of the binder content of a sample of a mixture. However, although an interesting innovation, for proper assessment of conformance it is still necessary to remove the binder from the mineral aggregate, recover the fine mineral matter from the binder solution and carry out an aggregate grading. Methods of determination of binder content that eliminate the need for organic solvents are also of interest. An Ignition Test that involves burning the asphalt sample in a furnace is being

Table 9.6. Binder determination directly or by difference

Country	Standard	Binder directly	Binder by difference
Germany	DIN 1996: Part 6: 1988	Recovery method[a]	Differential method
UK	BS 598: Part 102: 1996	Extraction bottle: directly determined	Extraction bottle: by difference Hot extractor
Norway	Strassentest		Automatic binder extraction
Netherlands	Test 65.0	Soxhlet extraction[a]	Decanting-jar centrifuge

[a] Requires recovery of total binder

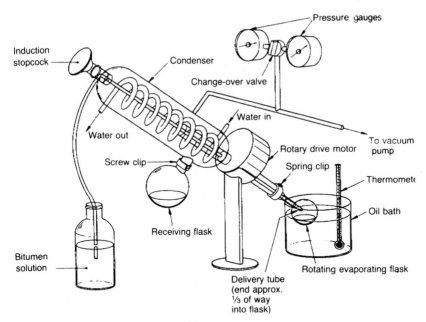

Fig. 9.17. Rotary film evaporator[26]

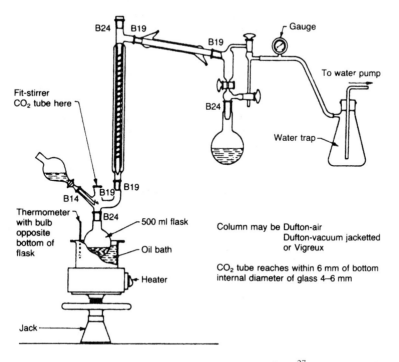

Fig. 9.18. Distillation apparatus with fractionating column[27]

Table 9.7. Duration of tests for composition

Country	Standard	Up to 2 hours	Over 2 hours
Germany	DIN 1996: Part 6: 1988	Differential method	Recovery method
UK	BS 598: Part 102: 1996	Binder by difference Binder directly determined	Hot extractor method
Norway	Strassentest	Automatic binder extraction	
Netherlands	Test 65.0	Decanting-jar centrifuge	Soxhlet extraction

developed. A standardization trial run by the TRL under contract to ETSU[30] has shown that the precision of the ignition method is at least as good as the solvent method. Calibration factors, predetermined from the mean difference between the calculated and known binder contents from a series of samples, are applied to individual mixtures to account for mass losses affecting the binder content. Certain aggregate types and sizes may be more prone to the effects of the high temperature used in the test (540 °C) and supporting data for a range of different aggregates are required before the method is published as a test standard. Further details of this proposal can be found in Section 1.13.1 of Chapter 1.

Due to the difficulty in determining the binder content of asphalt mixtures containing certain modified bitumens using standard methods, there is scope in particular for developing the test for these materials.

Various combinations of test equipment can be adopted to suit a particular need (Fig. 9.19).

The centrifuge extractor (Fig. 9.9) consists of a bowl, an apparatus in which the bowl may be revolved, a container for collecting the solvent thrown from the bowl and a drain for removing the solvent. A test portion of the asphalt mixture is placed into the bowl and covered with solvent. A filter paper disc is fitted between the edge of the bowl and cover, and clamped tightly in place. The centrifuge is operated at speeds of up to 3600 r/min and until the extract that is forced through the filter paper is virtually colourless.

The soxhlet extractor is an apparatus for extraction of the binder from the asphalt mixture with an organic solvent using a continuous reflux process. It is a glass extractor, consisting of a flask, an extraction case with tap and vapour tube, and a condenser (Fig. 9.13). The extraction case with the test portion is placed on a gauze in the extractor which has been filled with solvent. The extractor tap is then opened and a heater switched on. The extraction is stopped when the solvent collected in the extractor becomes virtually colourless.

Possibly the most common equipment used in Europe to separate out fine mineral matter from binder solution is the continuous flow centrifuge (Fig. 9.14) which can cope with relatively high proportions of filler. After

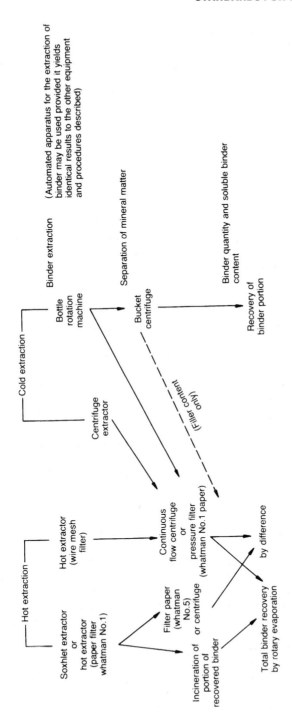

Fig. 9.19. Determination of binder content

extracting the binder from the sample of mixture with a suitable solvent in, eg a rolling bottle apparatus, the binder solution containing fine mineral particles is carefully poured through a funnel into a pre-weighed centrifuge cup that rotates inside the machine at 9000 r/min. The binder solution, due to the centrifugal effect, rises up the sides of the cup and deposits the fine mineral particles on the sides. Pure solvent is passed through the centrifuge until all binder is removed from the revolving cup. The centrifuged solution is discharged and may be collected for recovery of the bitumen or the binder content determined by difference. The fine matter collected in the cup is dried and added to the coarser mineral aggregates recovered from the rolling bottle.

If determining binder content by difference, it is especially important that due allowance is made for any water that may be present in the sample being tested, since this will affect its accuracy. If water is suspected in the sample, its content may be determined by hot extraction of the binder and measurement of the volume of moisture driven off with the solvent in a Dean and Stark Apparatus[16] (Fig. 9.11). Different solvents may be used, but trichloroethylene is preferred due to its higher boiling point (87 °C) and satisfactory bitumen solubility. The Dean and Stark Apparatus illustrated in Fig. 9.11(a) is suitable for use with trichloroethylene which is denser than water.

A different design that allows measurement of water collected at the bottom of the graduated receiver would be required for use with solvents less dense than water (Fig. 9.11(b)). The sample of mixture is placed in the extraction pot with sufficient solvent to permit refluxing in the water cooled condenser to take place during the distillation. Heat is applied to the pot until the volume of water collected in the receiver remains constant.

Standard test methods that are used by the Member States of the European Union are currently being harmonized by Comité Européen de Normalisation (CEN), the European Committee for Standardization.[31] Most of the test methods for composition of asphalt that are specified in BS 598: Part 102[8] are also included in the draft CEN documents, but the individual laboratory will have the option to adopt any of the other procedures described above.

9.4.2. Bitumen tests

The Shell Bitumen HandBook[32] discusses the properties of bitumen products and describes the various tests that are used to measure these properties. Currently, the British Standard Specification for Road Bitumens[33] includes test methods which are considered important for classification purposes in the UK.

The Penetration Test[28] is the most common control test for penetration grade bitumens and the midpoint of the range of penetration values is used to designate the particular grade. It is a measure of the consistency or hardness of the bitumen.

The penetration value is determined using a penetrometer (Fig. 9.20). This test[28] uses a standard stainless steel needle under a load of 100 g, set

Fig. 9.20.   Penetrometer. Photograph courtesy of ELE International Ltd

at the surface of the material, to penetrate a sample of bitumen which is
maintained at a temperature of 25 °C. The penetration value is the distance
(in tenths of a millimetre) that the needle penetrates the bitumen in 5 s.
The conditions of loading, time and temperature may be changed to suit
special applications.

The test method is convenient and simple to perform in site
laboratories as a means of monitoring the quality of supplies. However,
penetration has been found not to correlate well with tests used to assess
the performance of asphalts.[34] A much better linear association has been
found between bitumen softening point and mixture properties defined by
wheel-tracking rate,[35] Marshall Stability and Marshall Quotient.

The physical state of bitumen changes gradually with temperature from
solid or semi-solid to a softer more fluid material. A specific melting
point cannot be measured. The ring and ball softening point[29] is a
temperature which is measured at a specific viscosity of the bitumen at
some point during its transition from solid to liquid. (Viscosity is defined
as the resistance to flow and its measurement is discussed in detail in
Chapter 2.) The actual viscosity has been found to be around 1300 Pa s[36]
for bitumens with low Penetration Index and low wax content, equivalent
to a consistency as measured by 800 penetration units. This fact enables a

Fig. 9.21.   Ring and ball softening point apparatus. Photograph courtesy of ELE International Ltd

viscosity/temperature relationship to be readily established for those bitumens over a wide temperature range.

The softening point is found by placing a 3.5 g steel ball on the surface of a bitumen specimen that has been moulded into a tapered brass ring and allowed to steadily increase in temperature at a rate of 5 °C/min until the weight of the ball stretches the bitumen and forces the underside of the specimen to fall 25 mm onto a base plate. The temperature at which the bitumen touches the plate is the softening point. The apparatus used is illustrated in Fig. 9.21. Distilled water is the liquid used in the bath to transfer heat to the specimen for bitumens expected to have softening points up to 80 °C. Glycerol is used for more viscous materials. A mechanical stirrer is included to ensure uniform heat distribution throughout the liquid. The equivalent ASTM Method D36[37] does not include stirring in the procedure and results in a higher measured value of approximately 1.5 °C.[32]

Heukelom[36] developed a Bitumen Test Data Chart (BTDC) to describe the viscosity/temperature relationship of bitumens (Fig. 9.22). The chart is designed so that, for bitumens of low penetration Index and low Wax content, designated Class S by Heukelom, the relationship can be

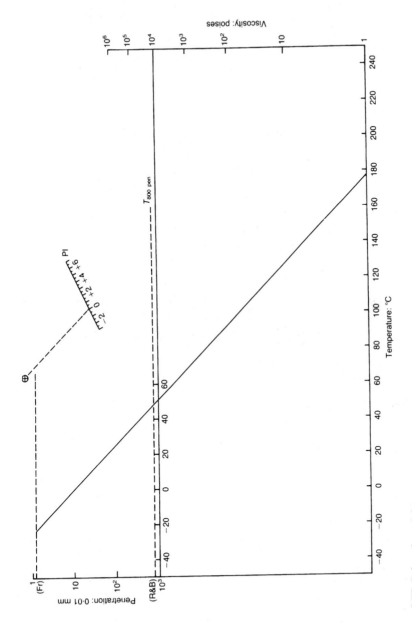

Fig. 9.22. BTDC for class S bitumens

described by a single straight line constructed by joining the plotted points for penetration and ASTM softening point. The line can then be extrapolated to give an estimate of Fraass breaking point (*Fr*), the temperature at which the bitumen cracks under controlled test conditions, and viscosities at higher temperatures. Alternatively, the line for a bitumen can be drawn using penetration values measured at two different temperatures.

Waxy bitumens give differently shaped curves and require further test data for their construction. The curves characteristically comprise two parallel misaligned straight lines joined by an ill defined transition zone over which the waxy constituents change state from crystalline to molten. The behaviour at lower temperatures may be similar to that of equivalent bitumens but for low wax contents. At higher temperatures when the wax is molten, the bitumen has correspondingly lower viscosities.

Blown bitumens are produced by blowing air through a low-viscosity grade of bitumen or heavy oil at high temperatures which results in a product of low temperature susceptibility suitable mainly for certain industrial applications.

Bitumens which have been blown also give differently shaped curves. The viscosity/temperature relationship is represented by two intersecting straight lines, where the line at higher temperatures is characteristic of unblown bitumens of the same origin and that at lower temperatures has a less steep slope, indicative of lower temperature susceptibility over that range.

The Penetration Index (PI) of a bitumen is a measure of its temperature susceptibility, varying from about −3 for highly temperature susceptible bitumens to about +7 for bitumens of low susceptibility. The chart shown in Fig. 9.22 also includes a means of determining the PI. The indices are found by drawing lines through the PI reference point parallel to the lines constructed from the bitumen test data and reading the values at their intersection with the PI scale.

In its applications, bitumen is used to coat aggregate particles in films, some of which are so thin as to be transparent. Regardless of the film thickness, bitumen makes an excellent binder because it is adhesive, cohesive, self-healing, resistant to abrasion and waterproof.

Bitumens can suffer a gradual loss of these desirable properties with the passage of time. This lessening of effectiveness is caused by hardening which is the result of continued exposure to heat, light, air and moisture, which are always present in the environment. Sparlin[38] also found that ultraviolet energy is capable of producing measurable increases in the hardness of bitumen films, characterized by their viscosity, when they are sealed from the atmosphere and held at room temperature. Hardening or ageing of the bitumen leading to brittleness has particular consequences for a wearing course as it is likely to be the cause of cracks initiating at the surface.

The effect of bitumen hardening on the performance of bituminous basecourse and roadbase has been investigated by the TRL.[39] It was shown that significant changes in stiffness due to binder hardening can occur in the early life of these layers, mainly dependent on temperature and air voids content. The gradual stiffening of the main structural layers appears to be beneficial for their long-term performance, provided that they are part of a thick, well constructed road that is designed for heavy traffic.[40] This effect is described as 'curing'.

The hardening of a bitumen binder is usually discussed in terms of the separate reactions that occur in service and result in a progressive change of properties. The reactions are given below.

- Oxidation (a process involving the loss of electrons, frequently with the gain of oxygen or loss of hydrogen).
- Volatilization (the loss of lighter constituents).
- Thixotropy (an isothermal gel/sol transformation brought about by shaking or other mechanical means, e.g. a gel on shaking may form a sol which rapidly sets again when allowed to stand). (Sol is an abbreviation of solution and is used to describe a fluid suspension of a solid in a liquid whereas a gel is a semi-solid suspension of a solid dispersed in a liquid.)
- Syneresis (the separation of liquid from a gel or jelly-like substance on standing).
- Polymerization (the formation of large molecules, comprising repeated structural units).

In general, it is agreed that the hardening of these binders in service is due in large part to oxidation and volatilization, and to a lesser degree to thixotropy, syneresis and polymerization.

During the manufacture of asphalts, the rate of oxidation can be particularly high in the mixer of a conventional batch mix plant[41] where thin films of bitumen are exposed to a large volume of oxygen in the air under high temperature conditions. If mixing is unduly prolonged, significant hardening may occur with consequential loss of the desirable bitumen properties.

Oxidation of bitumen is the combination of the hydrocarbon compounds with oxygen and results in a direct union, with the elimination of a portion of the hydrogen and carbon.

Abraham[42] characterized these reactions as follows

$$C_xH_y + O \rightarrow C_xH_yO \tag{1}$$
$$C_xH_y + O \rightarrow C_xH_{y-2} + H_2O \tag{2}$$
$$C_xH_y + 2O \rightarrow C_{x-1}H_y + CO_2 \tag{3}$$

These reactions require progressively greater amounts of activating energy (heat and light) to promote them. An oxidation reaction of type (1) occurs mainly on the exposed surface of bitumen. It forms a protective skin on the outer surface, and if this protective skin is not abraded or washed off, the rate of oxygen absorption slows, profoundly modifying further reaction with the oxygen and evaporation of volatile oils from the binder and hardening is retarded.

However, reaction in the absence of light must be considered as the main long-term cause of the deleterious hardening of bituminous binders which have access to oxygen through, say, an interconnecting network of air voids in a compacted asphalt layer.

The susceptibility of different bitumens to hardening is, therefore, important. To assess the resistance of a bitumen to hardening, it is usually subjected to the Thin-Film Oven Test[43] or the Rolling Thin-Film Oven Test.[44] In the former test method,[43] 50 g samples of bitumen are prepared in weighed containers to give a thickness of approximately 3 mm and placed in a ventilated oven maintained at 163 °C onto a shelf that rotates at 5 to 6 r/min. The sample containers are removed from the oven after 5 h and weighed to determine loss due to evaporation. The same samples may be melted carefully at a low temperature and re-mixed to determine the loss in penetration value or other test property after heating. The loss in weight and penetration is normally expressed as a percentage of the original value before heating.

The results of viscosity measurements made on bitumen subjected to the Rolling Thin-Film Oven Test[44] method have been shown to correlate well with the changes produced during mixing and laying of dense graded asphalts under normal conditions of temperature control.[45] In this case, 35 g samples are poured into glass bottles which are secured horizontally into openings in a vertical, circular carriage fitted inside the oven (Fig. 9.23).

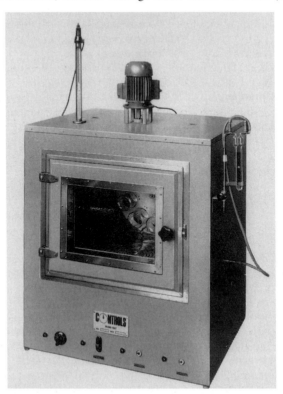

Fig. 9.23. Rolling thin film oven. Photograph courtesy of Control Testing Equipment Ltd

The carriage is rotated at a rate of 15 r/min and heated air blown by an air jet into each bottle at its lowest point of travel for 85 min. The test temperature of 163 °C should be reached within 10 min, otherwise the test is discontinued. The method of rotation during the test ensures that a fresh surface of bitumen is continuously being exposed to air. At the end of the test the percentage loss in weight and penetration may be found as before.

Bitumen is soluble in carbon disulphide and this solvent can be used to detect the presence of coke or mineral matter in the bitumen. The degree of solubility is also important information in the extraction of bitumen for analysis of composition. For health and safety reasons, trichloroethylene is normally used as an equivalent solvent to determine bitumen solubility and the test method is described in BS 2000: Part 47.[46] This involves dissolving the sample of bitumen in the solvent and filtering through a layer of powdered glass in a sintered crucible. The insoluble matter is then washed, dried and weighed.

Direct or indirect measurement of flow behaviour or viscosity of bitumens may be of interest at particular temperatures e.g. at 60 °C or 135 °C roughly corresponding to maximum in-service and layer compaction temperatures respectively. The viscosity at particular temperatures may be determined by measuring the time for a fixed volume of bitumen to flow through a capillary or standard orifice of various types of viscometers.[47] The viscosity may be reported as simply efflux time or converted into fundamental units by multiplying by the calibration constant of the particular viscometer.

An absolute measure of viscosity is the ratio between the applied shear stress and rate of shear. The cgs unit of measurement is the poise (1 g/cm s) and the increasingly more common SI unit is the Pascal second (Pa s, equivalent to 10 poise). When this ratio is independent of the rate of shear, the substance is said to be Newtonian. The viscosity of most bitumens changes with the rate of shear and so they are not strictly Newtonian, although they are sometimes treated as such by researchers.[48] However, at higher temperatures, they may be regarded as essentially Newtonian. More complex flow behaviour is termed non-Newtonian.

The ratio of viscosity to the density of bitumen gives a measure of its resistance to flow under gravity and this is called 'kinematic viscosity'. It is expressed in cgs units of cm^2/s or the Stoke. The SI unit is m^2/s and 1 centistoke $= 1 \times 10^{-6} m^2/s = 1 mm^2/s$.

The sliding plate micro-viscometer[49] permits the direct determination of the absolute viscosity of liquids and viscoelastic substances in the range of 10 to 10 billion Pa s. The operation of the viscometer is based on the principle of shearing a sample between two parallel flat plates of dimensions $2 \times 3 \times 0.6$ cm under the action of a constant shearing stress. The sample is spread and pressed between the two plates to a thickness of between 50 and 100 μm. Thicknesses out of this range are influenced by plate effects.

A uniformly thick film is necessary to ensure an accurate measurement of viscosity. The thickness may be determined by weighing and assuming a specific gravity of 1 for the bitumen without introducing any significant error. The plates are then clamped in the micro-viscometer, allowing a few minutes for the sample to reach the bath test temperature before the test begins. A load is applied and the movement of the glass plate is recorded on a millivolt recorder. Measurements are made at different rates of shear by varying the applied load and the viscosity determined by interpolation to the particular shear rate desired.

The Strategic Highway Research Program (SHRP), established in 1987 in the USA, included the objective of developing a new system for specifying asphalt binders that was based directly on performance rather than on purely empirical relationships. The result of this research is the SuperPave[TM] (Superior Performing Asphalt Pavements) Performance Graded Asphalt Binder Specification.[50] The testing requirements include procedures that simulate the ageing process during the in-service life of the binder and test methods for evaluating the behaviour of the binder over the broad range of temperatures that are experienced during its lifetime. The specification[50] is applicable to both unmodified and modified binders.

The use of the pressure ageing vessel (PAV) and the dynamic shear rheometer (DSR) for simulating binder hardening in the pavement and for measuring rheological properties at high and, in particular, intermediate temperatures respectively are of particular interest.

The simulative in-service ageing process involves spreading binder samples, that have been pre-aged in the rolling thin-film oven, onto sample pans that are fitted into a sample rack and placed inside the pressure vessel. The pressure vessel is positioned inside a forced air draft oven, preheated to 90 to 110 °C depending upon the design climate, and pressurized at 2070 kPa for 20 h. The final part of the process involves retaining the samples in an oven at 163 °C for 30 minutes in order to remove any air that may be entrapped as a result of a tendency for the binder to foam during the reduction of pressure in the vessel at the end of this procedure.

Other fundamental measuring techniques such as dynamic testing have been introduced and these take account of the time and temperature dependence of asphalt behaviour. Dynamic testing involves the measurement of the response of the material under an applied sinusoidal stress or strain wave. The DSR in the SuperPave[TM] specification uses the controlled stress mode to measure the rheological properties of complex modulus (G^*) and phase angle (δ).

A bitumen sample is spread and pressed between two parallel, circular plates. The lower plate is fixed and the upper plate oscillates at 10 rad/s (approximately 1.59 Hz). A conditioning period of oscillation at the test temperature determines the oscillating stress required to give a shear strain within a target range that depends upon the aged state of the binder. Ten additional cycles at the required stress produces a response that is computed by the rheometer software to give G^* and δ.

G^* is the ratio of total shear stress ($\tau_{max} - \tau_{min}$) to the total shear strain ($\lambda_{max} - \lambda_{min}$) or, in effect, the peak stress divided by the peak strain. δ is determined from the time lag between the applied stress and the resulting strain. This is illustrated in Fig. 9.24.

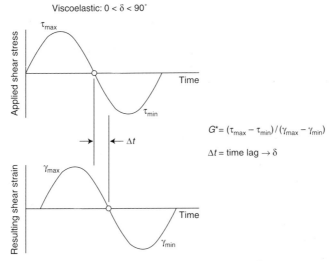

Viscoelastic: $0 < \delta < 90°$

$$G^* = (\tau_{max} - \tau_{min}) / (\gamma_{max} - \gamma_{min})$$

$\Delta t = $ time lag $\rightarrow \delta$

Fig. 9.24.   Stress/strain response of a viscoelastic material[50]

A totally elastic material deforms and recovers instantaneously under the application and release of a load. In this case, the time lag and the phase angle δ are zero. The response of a viscous material, on the other hand, is time dependent so that the time lag between applied load and strain is relatively large. Where stress and strain are totally out of phase (purely viscous), δ is $90°$. δ is the difference between the angles that describe the sine curve, corresponding to the shift of the strain wave form relative to the shear wave form. The value of δ lies somewhere between $0°$ and $90°$ for a viscoelastic material, like bitumen at normal pavement temperatures.

Once a suitable grade of bitumen has been selected, the ideal bitumen content of the asphalt may be estimated by following certain mixture design procedures.

Certain aspects of the testing of bitumen are also covered in Chapter 2.

9.4.3.   Mixture design

In the UK, a substantial proportion of the common types of asphalt is manufactured to recipe formulations that are based on observed satisfactory performance over a period of time in the field.[13,14] However, for certain types, a need was identified for a mixture design procedure that could be followed to optimize the use of available materials and to provide the necessary data that would enable the selection of an appropriate binder content for adequate durability of the mixture without sacrificing stability. Increasing traffic volumes on major roads also demanded a means of improving the resistance to deformation of wearing courses.

The Marshall Method[51] for the design and control of asphalts was developed in the USA in the 1940s and was later adopted by the UK Air

Ministry in the 1950s as a means of designing bituminous surfacings for airfield pavement construction that would offer high stability, good rideability, weatherproofing, durability, and a smooth, dense surface.[52] Marshall asphalt using continuously graded aggregates was generally favoured.

The Marshall Method involved an enhanced level of supervision and testing effort that gave rise to a more highly controlled and consistent material. In 1973, a revision of BS 594[53] included, for the first time, a laboratory design procedure based on the Marshall Method for the composition of the bitumen/sand/filler component of HRA asphalt wearing course. BS 594 included a table of adjustments of bitumen and filler contents for different percentages of coarse aggregate required in the desired mixtures. However, the current edition of BS 594[13] now specifies a laboratory design procedure for the complete mixture.

BS 598: Part 107[54] describes a method of test for the determination of the composition of designed HRA wearing course, the properties of which are discussed in Sections 3.7.1 and 4.13.4 of Chapters 3 and 4 respectively. The test procedure generally takes at least three consecutive days to complete when followed properly. A quantity of dried and blended mineral aggregate constituents is prepared for each batch of mixture that is sufficient for a compacted cylindrical specimen of 101.6 mm in diameter and approximately 63.5 mm high. The aggregates are heated for at least four hours in an oven maintained at a temperature of 110 °C above the softening point of the specified binder.

The binder is poured into a number of small containers so it can be individually heated up to the required temperature to minimize undue binder hardening. Three separate batches are mixed at each binder content for one minute at a temperature of not less than that required for compaction. Each batch is transferred to a steel cylindrical mould, the inside surface of which has been smeared with silicone grease and the base covered with a disc of paper to prevent sticking, and the mixture spaded with a spatula to prevent any bridging of aggregate. After covering the top of the mixture with another paper disc, the steel mould is secured on top of a compaction pedestal made to specified dimensions and of particular materials, including a 200 mm × 200 mm × 450 mm laminated hardwood block, and the mixture is compacted on each side at a temperature of 92 °C above the softening point of the binder by 50 blows of a 7850 g flat-footed steel hammer with a falling height of 457 mm. Before removing the compacted specimens from the moulds, they are cooled under water to prevent distortion. Each specimen is then extruded from the mould and allowed to dry. The procedure is repeated for mixtures of at least nine binder contents differing consecutively by 0.5% of total mixture. The specimens are tested for relative density, Marshall stability and flow.

The relative density is determined by weighing the specimen in air and in water and calculating the volume displaced from the difference of the two weights, i.e. assuming a relative density of 1.0 for water, and then dividing the result into the mass in air. The compacted aggregate density is also calculated by multiplying the relative density of the specimen by the percentage by mass of the total aggregate in the mixture.

The specimens are then immersed in water at 60 °C for at least 45 minutes before being individually placed in a testing head comprising two segments of specified dimensions shaped to take the curved sides of the moulded specimen. The testing head is immediately placed centrally on a compression testing machine and a load applied to the specimen at a constant rate of deformation of 50 mm/min. Both the maximum load applied and the vertical deformation of the specimen at the point of maximum resistance are recorded and reported as stability and flow, respectively.

Mean values of each property are calculated for each set of three specimens of the same binder content. In the case of stability, the individual measured values are firstly corrected for variations in specimen volume before averaging. These values are then plotted against percentage binder content, as shown in the example in Fig. 9.25. The mean of each of the binder contents at the maximum value from the

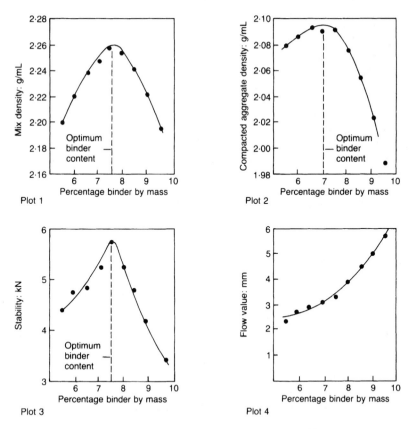

Fig. 9.25. Example of evaluation of design binder content[54]

curves of mix density, stability and compacted aggregate density is then calculated.

A factor which depends on the coarse aggregate content of the mixture is added to the mean to give the design binder content. This factor has been derived from the correction necessary in the correlation between design binder content using complete mixture and that using the binder/sand/filler component only to ensure equivalent performance in terms of stability, durability and workability. The different factors are shown in Table 9.8.

The Marshall Method[51] is also used for the design and control of asphaltic concretes produced extensively in other countries for road surfacing. These mixtures are quite similar to Marshall asphalts used as surfacings on airfields in the UK.

In asphaltic concrete mixtures used as surfacing courses, the aggregate constituents are carefully graded down through all sizes to produce a relatively fine but dense material with a controlled void content. This type of aggregate structure can be quite sensitive to changes in the percentage of added binder, but as long as due account is taken of traffic and climatic conditions in the selection of the binder content, the mixture will provide a very durable and stable surfacing product.

In designing these mixtures, the inter-relationship between aggregate packing, binder content and air voids in the total mixture is critical. The binder content should be sufficiently high for durability, without overfilling the voids in the mineral aggregate to the extent that the mixture will not develop strength from the applied loads of traffic, due to the volume of binder preventing close aggregate packing.[55]

Measurement of voids and the capacity for further densification by traffic after construction are therefore essential elements of the design procedure for these mixtures. In some cases, the specified air voids content may have to reflect an inability of the laboratory compactive effort to closely simulate the predicted final density in the field, especially where the aggregate constituents have a tendency to degrade under the impact of an increased number of blows of the Marshall Hammer.

The Asphalt Institute Manual Series No. 2 (MS-2)[56] describes such procedures. As well as Marshall, the Hveem Method of Mixture Design, developed in California, is included. An interesting feature of the Hveem procedure is the preliminary determination of an approximate binder

Table 9.8. Factors for the calculation of design binder contents of hot rolled asphalt wearing course[54]

Coarse aggregate content: %	Addition factor
0	0
30	0.7
35	0.7
55	0

content by the Centrifuge Kerosene Equivalent (CKE) Test. Dry, fine aggregate is placed in a centrifuge cup fitted with a screen and disc of filter paper. The aggregate is saturated in kerosene and then centrifuged in a machine capable of achieving an acceleration of 400 times gravity for a period of two minutes. The weight of retained kerosene is the CKE.

A surface capacity test is also carried out on the coarse aggregate. In this case, dry aggregate of 9.5 mm–4.75 mm is placed in a metal funnel and immersed in lubricating oil. The oil is then allowed to drain from the sample for 15 minutes and the weight of retained oil determined. The surface area of the aggregates is then calculated from the grading by multiplying each cumulative aggregate fraction by a corresponding surface area factor and summing. These data are then used to determine the approximate binder content from a series of charts given in the manual. A number of compacted specimens are prepared at different binder contents including those on either side of the approximate content. Tests are carried out on the specimens to determine their resistance to both deformation and to the action of water. An analysis of density and voids provides the final information necessary to determine an optimum binder content that satisfies the specified design criteria.

Both the Hveem and the Marshall Methods are used in the USA. However, it is the Marshall Method which is adopted most frequently outside the USA for the design and control of asphaltic concrete. Using the Marshall compaction apparatus, the density of the compacted specimen is increased up to a point by increasing the number of blows of the compaction hammer. MS-2[56] recommends three levels of compaction for light, medium and heavy traffic conditions, simulated by 35, 50 and 75 blows respectively on each side of the specimens. As noted earlier, an important consideration is the prediction of the amount of densification that will occur in the road layer due to trafficking after construction.

The total amount of densification is also a function of temperature and, to a large extent, the degree of initial compaction achieved during construction. Investigations[55] have shown that, for a properly designed mixture, final density should be achieved typically within a period of trafficking over three summers. In order to maximize use of the mixture design data, the density of the laboratory compacted specimen should equal the final layer density. It is not possible, owing to different variable factors, to be too precise but the laboratory density may be regarded as a best estimate.

Once acceptable laboratory compaction that simulates as closely as possible the conditions in the field is established, a voids analysis can be made for each of the compacted specimens. The volumetric proportions of interest are air voids, voids in the mineral aggregate (VMA), and voids filled with binder (VFB). Note that mineral aggregate may be defined as any hard, inert, mineral material such as crushed rock, gravel, slag or sand that has been produced in specified sizes for further processing in the manufacture of asphalts or other products. Air voids are defined as the total volume of air between the coated aggregate particles of the compacted mixture. This does not include air in the aggregate pores trapped beneath the binder films. VMA is defined as '*the volume occupied by the air voids and the amount of binder not absorbed into the pores of the aggregate*'. Both air voids and VMA are expressed as percentages of the total volume of the compacted specimen. VFB is defined as '*the percentage of the VMA filled with binder*'.

Percentage of air voids is probably the most important criterion for the evaluation of the performance of asphaltic concretes and selection of the appropriate binder content. Minimum percentage contents of VMA are recommended by the Asphalt Institute to ensure that the mixture is neither deficient in binder nor air voids. There should be enough room in the aggregate structure to take sufficient binder for durability of the mixture while still leaving a sufficient volume of air voids to avoid problems with plastic deformation (rutting). Limits placed on VFB, control the balance between the effective binder content, i.e. excluding binder absorbed into the pores of the aggregate, and air voids content.

The air voids content is calculated from

$$V_t = \frac{S_t - S_b}{S_t} 100 \qquad (9.3)$$

where V_t = percentage air voids
$\quad\quad S_t$ = theoretical maximum density of loose mixture
$\quad\quad S_b$ = bulk density of compacted specimen

The theoretical maximum density of the mixture may be determined from the individual densities of its constituents. However, the calculation of air voids content is quite sensitive to differences in aggregate particle density (specific gravity) and so the basis of the determination of aggregate density is important. Selection may be made from aggregate particle densities on an oven dried basis (bulk), saturated surface dry basis, or on an apparent basis, their values increasing in magnitude in the same order. A change in aggregate density of 0.02 corresponds to a change in air voids content of around 0.6%. The determination of oven dried particle density is based on a volume that includes the water permeable pores in the aggregate whereas that of the apparent particle density is based on a volume that excludes those water permeable pores. The volume of water is invariably higher than the volume of binder that is absorbed by the aggregate. Use of either of those aggregate densities alone does not necessarily offer an accurate basis for the analysis of the volumetric proportions of an asphalt. Generally, an approximation is made to the effective particle density (i.e. the density based on the volume of water permeable material but excluding voids accessible to binder, expressed as a ratio of this value to the density of water) by simply using one of the measured aggregate densities or an average of two.

Such a procedure may be satisfactory for many applications, provided the effect on calculated air voids is understood. The Asphalt Institute recommends a direct determination of theoretical maximum density of the asphalt. Such a procedure is described in ASTM D2041.[57] A weighed sample of loose asphalt is placed inside a vacuum container and covered with water at 25 °C. Air is removed from between the coated particles by gradually increasing vacuum to give a residual pressure of 30 mm Hg or less. The vacuum is maintained for 5 to 15 minutes and the container shaken to assist the expulsion of air. The theoretical maximum density is calculated from Equation (9.4) or (9.5), depending on whether the volume of loose mixture is determined by weighing the container in water or air.

(i) Weight in water method

$$S_t = \frac{A}{A - C} \tag{9.4}$$

where S_t = theoretical maximum density
A = dry mass of sample
C = mass of sample in water.

(ii) Weight in air method

$$S_t = \frac{A}{A + D - E} \tag{9.5}$$

where D = mass of container filled with water
E = mass of container filled with sample and water.

The test is best carried out on mixtures having a binder content at or near their optimum, to avoid breaking of thin binder films under vacuum or coagulation of rich mixtures trapping air which would otherwise be removed. Work by Kandhal et al.[58] suggested that the test method should take account of the time dependent nature of binder absorption, by conditioning the sample for, say, four hours in an oven at 143 °C before test. If this effect is not taken into account during mixture design, lower calculated air voids may lead to a lower binder content and, subsequently, higher in situ air voids contents. The effect is more marked when using more absorptive aggregates in the mixture. Such a conditioning procedure is included in the SuperPave[TM] Mix Design Method,[59] which recommends oven ageing for 4 hours at 135 °C. Other methods used in some European countries replace water with hydrocarbon solvent. In this case, a vacuum is not necessary to expel entrapped air but, because the solvent readily penetrates the binder films, a measure of binder absorption cannot be derived and higher theoretical maximum densities are determined.

ASTM D2726[60] describes a method of determining the bulk density of compacted specimens of dense mixtures. The specimen is weighed in water after immersion for three to five minutes at a temperature of 25 °C. The specimen is then removed from the water and blotted with a damp towel before weighing in air. The difference in weights gives the mass of an equal volume of water, corresponding to the volume of the specimen for a density of water of 997 kg/m.[3] The bulk density, S_b, is then calculated from

$$S_b = \frac{A}{B - C} \qquad (9.6)$$

where A = dry mass of specimen in air
B = mass of saturated surface dry specimen in air
C = mass of specimen in water.

This test method assumes a volume for all specimens that includes both internal and surface voids. The procedure described by BS 598: Part 107,[54] however, for measuring the relative density of compacted specimens of HRA assumes a volume that excludes any surface voids and any internal voids that may be accessible to water so that the volume determined may be slightly lower, thus giving rise to a slightly higher density.

The choice of methods is probably not so critical for carefully prepared laboratory specimens of dense mixtures. However, careful thought should be given to the choice for the measurement for the density of extracted cores which are unavoidably subject to greater surface voids due to coring and surface effects at the top and base of the core. The choice should be based on a consideration of what can be realistically achieved in the field.

The density of compacted specimens may also be determined by sealing any surface voids with paraffin wax. The procedure is described in BS 598: Part 104[61] and ASTM D1188[62] and is frequently the preferred method for extracted cores of coarser materials. Although it may well be that this procedure returns a density approaching the 'true' value for the widest range of cored specimen types, it would be impracticable to use for very dense materials and for those materials approaching a porous state.

In their evaluation of methods used to determine the air voids content of asphalt mixtures, Richardson and Nicholls[63] noted that some methods were more appropriate than others, depending upon the type of product and the situation in which it is used. It was concluded, since the importance of air voids content in relation to the performance of asphalt is well established, that the method used to determine this parameter should be stated when reporting the value.

Since VMA is defined as the volume occupied by air voids and the amount of binder not absorbed into the pores of the aggregate, its calculation must be based on the aggregate particle density determined on an oven dried basis (bulk specific gravity)[64,65] as follows

$$\text{VMA} = 100 - \frac{S_b.P_a}{G_{ab}} \qquad (9.7)$$

where S_b = bulk density of compacted specimen
P_a = per cent aggregate by mass of total mixture
G_{ab} = aggregate particle density on an oven dried basis

VMA determined on any other basis may be acceptable, providing its effect on calculated voids is fully understood.

The optimum binder content of asphaltic concretes may then be determined by averaging the individual binder contents corresponding to maximum compacted mixture density, maximum stability and the specified air voids content. Providing the mean meets all other design criteria, full scale production trials at this binder content may begin. In some cases, selection of a binder content without averaging but still meeting all the specified design criteria may be considered.

More recently, a need has been identified to develop a mixture design procedure that is based on fundamental engineering properties and that more accurately predicts the in-service behaviour of asphalts than do procedures such as the Marshall Method which are empirically based and include measurements of relatively poor precision.

A research programme began in the USA on a new asphalt-aggregate mixture analysis system (AAMAS), originally funded through the National Cooperative Highway Research Program and then part of SHRP.[66]

It was proposed that an initial mixture design is carried out, based on volumetric analysis of compacted specimens of asphalts prepared by using Marshall apparatus or some other preferred method to determine a suitable composition of mixture for further testing. Specimens of this composition are then prepared for measurement of its fundamental properties. At this stage, the selection of the method of laboratory compaction is important to simulate the conditions in the field as closely as possible.

Different compactors tend to produce specimens that, although of the same composition, have quite different engineering properties. Sousa *et al.*[67] have reported an investigation into the effect of different methods of specimen preparation on permanent deformation characteristics of asphalts. Three methods were selected for investigation that subjected the mixtures to shearing motions similar to those induced by site compactors. The compactors used were the Texas gyratory, kneading, and rolling wheel apparatus. Rolling wheel compaction was recommended to simulate field compaction most closely. However, gyratory compaction may be considered more convenient for the preparation of small specimens. Specimens are also prepared for conditioning for moisture damage and age hardening before testing.

The critical mixture properties examined are resistance to fatigue cracking, permanent deformation, and thermal cracking. The proposed tests used to evaluate these properties included resilient modulus, creep and indirect tensile strength. The measured properties are then used to predict performance and for comparison with the requirements of the structural pavement design.

The performance prediction models developed by SHRP are being validated in the USA and specific analytical procedures have yet to be published. A new mix design procedure, however, is described in the Asphalt Institute's SuperPave[TM] Series No. 2 (SP-2).[59] SP-2 also describes the use of two performance test devices, the SuperPave[TM] Shear Tester and the Indirect Tensile Tester. These are used for new test procedures that would be required to be followed for a complete analysis of the proposed mixture.

The SuperPave™ mix design procedure involves selecting an asphalt binder that is appropriate for the climate and traffic loading. The grading specification is based on the expected high and low pavement temperatures. The selection is upgraded for slower and heavier traffic loading. Important aggregate properties such as shape and clay content are specified. Other inherent aggregate properties such as toughness (Los Angeles Abrasion Value) and soundness are also identified as being important for design. Aggregate grading is defined using a 0.45 power gradation chart (test sieve size in millimetres to the power 0.45 on the abscissa scale), that includes a restricted zone in the area of the intermediate to coarse fine aggregate sizes discouraging the use of fine natural sands. (Shape, toughness (Los Angeles Abrasion Value) and soundness are all discussed in detail in Chapter 1.)

A significant departure from the Marshall Method[51] of design is the use of the gyratory instead of the impact form of specimen compaction. The compaction device, which is considered to simulate compaction in the field more closely, has the advantage of an inbuilt means of measuring compactability (ease of compaction). This is achieved by recording the change in specimen dimensions with time of compaction.

Compaction is achieved by the simultaneous action of a static load (P) applied to the surface of the specimen contained in a cylindrical mould and the shearing actions due to the gyratory kneading motion (Fig. 9.26). Throughout the test, the angle of the gyratory motion remains constant and the upper and lower mould plates remain parallel and perpendicular to the axis of the conical trace of the gyratory motion. The use of gyratory compaction of specimens is also discussed in Section 3.6 of Chapter 3.

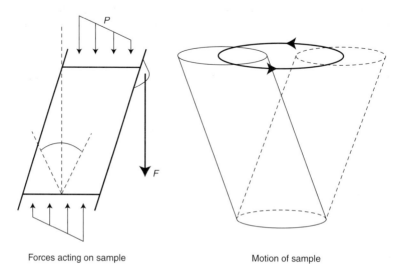

Forces acting on sample   Motion of sample

Fig. 9.26.   Gyratory compaction

A further significant development is to take account of the changing consistency and behaviour of the binder during the mixing and compaction processes. Prior to compaction, the samples of loose mixture are age conditioned to simulate the effect. This is important for both volumetric analysis and mechanical properties.

In the Marshall Method[51] of design, the aggregate grading is predetermined by specification, usually with quite small permitted deviations, and the whole mix design procedure is then based upon this grading. If the mixture fails to comply with the design criteria, the whole process is normally repeated with a slightly altered aggregate grading. The SuperPave[TM] process, however, incorporates an initial assessment of several aggregate trial blends to determine the best estimate of a design aggregate structure that will fulfil all of the requirements of the mixture. This is achieved using an estimated design binder content for each trial blend.

The optimum binder content for the selected design aggregate structure is then determined in a similar manner as before, by preparing compacted specimens with a range of binder contents and analysing their volumetric composition. However, at this stage, the results of mechanical tests are not included in the determination of the optimum binder content. These are replaced by an analysis of the compactability data. The specimen densities, or rather, their percentages of the theoretical maximum density, are determined at an initial (N_{ini}) and at a maximum (N_{max}) number of gyrations. N_{ini} and N_{max} are calculated from the design number of gyrations, which is related to climate and traffic level. An example of design criteria that may be specified for a 19 mm nominal sized mixture is shown in Table 9.9. Other properties of the mixture that may be specified are then checked at the design binder content stage.

Compactability studies can be made using other types of compacting devices. Figure 9.27 illustrates the method developed by Arand[68] using an impact or Marshall compactor. The curves take the following form

$$\rho_{(E)} = \rho_\infty - (\rho_\infty - \rho_0) \exp(-E/C) \qquad (9.8)$$

where $\rho_{(E)}$ = the density at E number of blows on each side
ρ_∞ = the theoretical maximum compacted density

Table 9.9. *Design criteria for 19 mm nominal size asphalt concrete*

Air voids: %	4.0
VMA: %	13.0, min
VFA: %	65–75
Dust proportion	0.6–1.2
$\%G_{mm}$ @ N_{ini}	<89
$\%G_{mm}$ @ N_{max}	<98

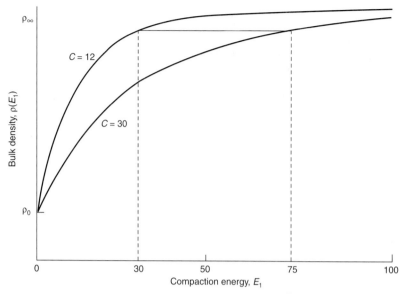

Fig. 9.27.   Compactability study by impact compaction[68]

ρ_0 = the theoretical initial density of the uncompacted mixture

$\rho_\infty - \rho_0$ = the compaction potential

C = the compaction resistance

The compaction potential of a mixture describes the scope for an increase in density to a much higher value. The compaction resistance is a function of how quickly the ultimate density is being approached. The values of ρ_∞, ρ_0 and C, are determined from the equation of the best fit curve for each mixture under test.

Assessment of the propensity of a mixture to segregate can also be a useful aid to mixture design. A test method for the determination of segregation sensitivity is being developed by the CEN.[69] The apparatus for the test is illustrated in Fig. 9.28.

The sample of hot mixture is placed in the hopper and the slide board is removed to allow the material to fall onto the platform below and form a conical pile. The inner part of the pile is then allowed to fall through the opening in the middle of the platform so that it can be collected in a container. An intermediate part is removed by widening the platform opening. The segregation value is calculated as the difference between the soluble binder contents of the inner, finer part and the remaining outer, coarser part.

9.4.4.   Adhesion

Moisture damage of asphalt layers is an important distress mode. Methods that are currently used to assess the water sensitivity of asphalts are now discussed.

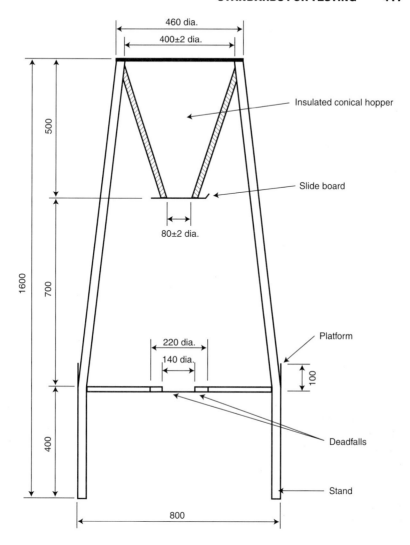

Fig. 9.28.   Segregation sensitivity apparatus[69] *(dimensions in mm)*

Sustainable adhesion of binder to aggregate in an asphalt is necessary for a long and serviceable life in a flexible pavement. Failure of the pavement can sometimes be attributed to loss of adhesion or stripping. Most aggregates have a greater affinity for water than for bitumen due to bitumen's lower surface tension and thus, inferior wetting power. It is therefore difficult for bitumen to displace water which has already formed a film over the aggregate surface, except for certain special processes. In some cases, an excessive dust coating on the aggregate will also prevent an adequate bond with bitumen and hasten the occurrence of debonding.

Adhesion between the aggregate surface and the bitumen may be broken by water forcing its way under the bitumen film and stripping it free from the surface. The risk of stripping is usually greater with softer bitumens and with acidic or high silica aggregates. Low silica aggregates such as limestone are thought to have a greater chemical attraction for the polar carboxylic acid components of bitumen. However, although properly manufactured and placed asphalts generally perform satisfactorily in the presence of water, all have a potential for damage by moisture and a number of tests have been developed in an attempt to measure their water damage susceptibility.

Static immersion tests are carried out in which dry, single sized aggregate is coated with a specified percentage of binder and, after being allowed to stand at room temperature for a specified period, the sample is immersed in previously boiled, distilled water. The temperature of the water and the period of immersion vary, but typically may be 20 °C for 48 h or 40 °C for 3 h. At the end of the immersion period, the aggregate particles are visually examined for signs of stripping and an estimate made to the nearest 5% of the residual coated area. Alternatively, the number of particles showing evidence of stripping may be counted and expressed as a percentage of the total number of particles.

Mechanical immersion tests may also be carried out. These involve measurement of the change in the mechanical properties of compacted specimens of asphalts after extended periods of immersion in water. ASTM D1075[70] describes such a procedure using statically compacted cylindrical specimens 4 in. high and 4 in. in diameter, suitable for the measurement of compressive strength. However, Marshall specimens are frequently used as an alternative. In this case, six Marshall specimens are prepared at the desired binder content and their bulk densities determined in order that they can be divided into two groups of three which have similar values of mean bulk density because the test results are sensitive to the voids content of the specimen which controls the amount of water absorbed. The first group is tested for Marshall Stability after immersion in water at 60 °C for 30 minutes and the second group tested after immersion at the same temperature but for 24 h.

An Immersion Index of retained strength is then calculated from

$$\text{Immersion index} = (S_2/S_1)\,100 \qquad\qquad (9.9)$$

where S_1 = average Marshall Stability of Group 1 (30 minutes immersion)

S_2 = average Marshall Stability of Group 2 (24 hours immersion).

A minimum index of 75% is often specified for satisfactory resistance to damage by moisture.

The immersion wheel-tracking Test[71] is a method that takes account of the role played by traffic in the process of stripping. The apparatus consists of three solid tyred wheels each approximately 5 cm wide which pass over three specimens of compacted asphalt to simulate the action of traffic (Fig. 9.29). The wheels travel with a reciprocating motion at a frequency of

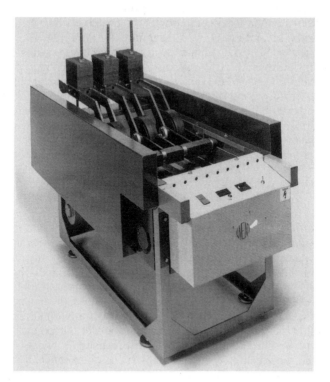

Fig. 9.29. TRL immersion wheel tracking machine. Photograph courtesy of Wessex Engineering and Metalcraft

25 cycles/min and a stroke of about 28 cm. Each wheel is loaded to give a total weight of 18 kg bearing on the specimen contained in a mould approximately 3 cm deep, 30.5 cm long and 10 cm wide, maintained horizontally in a water bath so that the level is just above the top of the specimens. The water is maintained at a temperature of 40 °C during the test.

Resistance to stripping is assessed according to the time necessary to produce failure which is defined as the point at which the rate of penetration of the wheel into the specimen sharply increases. Typically, failure within 4 h of tracking may correspond to very low resistance to stripping and failure after 20 h to high resistance to stripping. The test is normally stopped within 24 h.

Better methods of assessing the water sensitivity of asphalts were researched as part of the SHRP work. AASHO Test Method T283[72] has been recommended. This is another type of mechanical immersion test where the change in the indirect tensile strength after water conditioning is determined. Two sets of compacted, cylindrical specimens are prepared with approximately equal, average air voids contents (6% to 8% for dense graded mixtures). One set of dry specimens is tested for indirect tensile strength at 25 °C, after conditioning by wrapping in plastic bags and placing in a water bath, maintained at the test temperature for 2 h. The other set is subjected to partial vacuum saturation with water, so that the volume of water absorbed is 55% to 80% of the volume of air voids. The specimens may then be subjected to a freeze/thaw cycle, before being tested for indirect tensile strength at 25 °C. A test specimen is placed in a compression testing machine between loading strips. The specimen is then loaded until break at a constant rate of deformation of 50 mm/min. A Marshall load frame can therefore be used for the test with a test head designed to take the loading strips. The indirect tensile strength is calculated for each specimen from the peak load and the dimensions of the cylinder. A Retained Strength Index of at least 80% is considered acceptable.[59]

Other variants of the mechanical immersion test include retained stiffness or resilient modulus.

9.4.5. Percentage refusal density

The specifying authority generally prefers that a certain level of compaction of asphalts is achieved at the time of construction. This will ensure that air voids in those mixtures that depend on their density to provide satisfactory performance, will not be so high as to significantly expose the binder to the potentially damaging effects of air and water. Early distress, such as ravelling at the surface, will also be avoided. Because it is known that improved compaction of the asphalt layers leads to reduction in stresses developed in the lower pavement by traffic loading, the purchaser will wish to ensure that such a benefit is realized early in the service life of a pavement.

The percentage refusal density (PRD)[61] is a measure of the relative state of compaction of cores extracted from the pavement layers. Currently, it is only applicable to dense coated macadam roadbases and basecourses (including heavy duty macadams (HDMs)).

The PRD is the ratio, expressed as a percentage, of the bulk density of the core to its density after reheating and recompacting to refusal by a particular prescribed procedure. A description of the procedure is given in BS 598: Part 104.[61] A maximum air voids content of 9% for a typical coated macadam roadbase would correspond to a minimum PRD of around 93%, assuming an air voids content at refusal of approximately 2%.

A core of 150 mm diameter is dried in an oven to constant mass. After cooling, the core is weighed in air and then its surface is coated with molten wax. Once the wax has hardened, the core is again weighed in air and then weighed in water at 20 °C. The bulk density, G, of the core is calculated using the equation

$$G = \frac{A}{(B - C) - [(B - A)/D]} \tag{9.10}$$

where A = the mass of the dry core
$\quad\quad B$ = the mass of the waxed core
$\quad\quad C$ = the mass of the waxed core in water
$\quad\quad D$ = the density of the wax.

The wax is then removed from the surface of the core, and the core is placed into a split mould with base plate (Fig. 9.30).

The inside surfaces of the mould are coated with silicone grease and the base plate covered with a disc of filter paper to prevent sticking. The mould with the core is placed in an oven until the temperature required before compaction is reached. The mould is removed and the core is covered with another disc of filter paper. The core is then compacted with a vibrating hammer using a 102 mm diameter tamping foot around the internal circumference of the mould in a prescribed sequence for a period of two minutes. Surface irregularities are then removed by vibrating with a 146 mm diameter tamping foot. The core is then inverted and compacted on its opposite side in the same fashion as before. After cooling, the core is weighed in air and in water and its density calculated using the following equation

$$\text{refusal density } H = \left(1 - \frac{F}{E}\right)^{-1} \tag{9.11}$$

where F = the mass of the compacted core in water
E = the mass of the compacted core in air

The PRD is then determined as follows

$$\text{PRD} = 100G/H \tag{9.12}$$

The PRD Test[61] introduces a very useful measurement in the analysis of asphalts for prediction of performance. In an earlier section on mixture design, the importance of the laboratory reference density used in the Marshall Method[51] was discussed and it was noted this should simulate as closely as possible the final density in the road. However, there is no guarantee that this will actually be the case. Measurement of refusal density and calculation of the corresponding air voids content offers some additional insurance against the possibility of plastic deformation occurring, by specifying the air voids content at refusal to be sufficiently greater than zero.

The main drawbacks of the PRD Test for control purposes are the time it takes to achieve a result and the intensity of labour required. The adverse effect this also has on client/contractor relationships has been reported by various engineers since the introduction of the test for acceptance purposes.[73] Nevertheless, it seems likely that this test, or a variation of it, will be used for the foreseeable future. More discussion on PRD requirements can be found in Sections 4.7.4 and 8.4.3 of Chapters 4 and 8 respectively.

9.4.6. *Texture depth*
The need for good texture on a road surface as an aid to resistance to skidding, especially at higher speeds, is discussed in Sections 3.9 and 4.10 of Chapters 3 and 4 respectively. The UK Specification for Highway Works (SHW)[74] gives requirements for the average texture depth to be

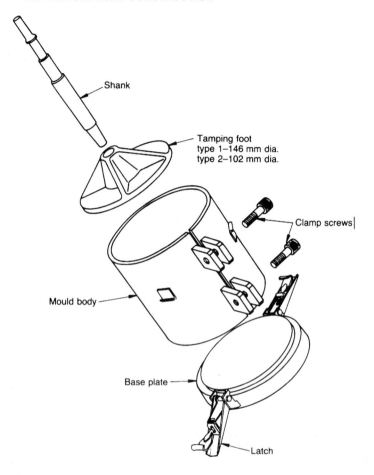

Shank

Tamping foot
type 1–146 mm dia.
type 2–102 mm dia.

Clamp screws

Mould body

Base plate

Latch

Essential dimensions (inspection) of closed mould excluding
base plate
Internal diameter 152·45 ± 0·5 mm
Length   170 ± 0·5 mm

Fig. 9.30.   Compaction apparatus for PRD test[61]

achieved over 1000 m sections of carriageway using the Sand Patch Test.[75] It is a measure of the macrotexture of the surface. The minimum specified values of average texture depth for high-speed roads in the UK (> 90km/h) are given in Table 9.10.

Corresponding values for use with the laser texture meter (called the 'sensor measured texture depth (SMTD)') may be established by undertaking calibration trials to derive the relationship between the two methods. Such a relationship was investigated by the Transport Research Laboratory (TRL) on newly laid chipped HRA.[76] Regression analysis of the results of a correlation study between the SMTD and texture depth determined by the Sand Patch (SP) Test[75] gave the following equations

Table 9.10. Minimum texture depths

Test method	Section tested	Minimum value of average texture depth: mm
Sand patch	1000 m	1.5
	Each set of ten individual measurements	1.2

$$SMTD = 0.41SP + 0.41 \tag{9.13}$$

$$SP = 2.42\,SMTD - 0.98 \tag{9.14}$$

Using these formulae, the SMTDs that are equivalent to sand patch results of 1.2 mm, 1.5 mm and 2.0 mm are 0.90 mm, 1.03 mm and 1.23 mm, respectively. The equations are applicable only to newly laid chipped HRA with sand patch texture depths in the range 1.0 to 2.0 mm. A subsequent precision experiment carried out in 1990 gave the following correlation.[77]

$$SP = 1.591\,SMTD \tag{9.15}$$

However, it must be stressed that this relationship was applicable to particular laser texture meters used on particular materials at a particular site. A calibration experiment should be carried out at each site where such a correlation is desired. The same experiment gave the precision data shown in Table 9.11, expressed as a percentage of the measured value.

On this occasion, the repeatability of the SMTD was better than that of the SP, but the reproducibility or between-operator precision was worse. It is for this reason that, at the present time, SP results are normally used for compliance purposes and SMTD results may be used only as a screening procedure.

BS 598: Part 105[75] describes the two methods of test for the determination of texture depth. The Sand Patch Method[75] may be used on any type of surface, but the use of the laser texture meter remains restricted to newly laid chipped HRA wearing course.

The advantages of measurement by the laser texture meter are ease and speed of use. The principle of the meter is that infrared light from a rapidly pulsed laser is projected onto the road surface and the reflected

Table 9.11. Relationship between laser texture meter and sand patch results[77]

Test method	Measured value, mm	Repeatability, %	Reproducibility, %
Laser texture meter	1.03	10.7	19.9
Sand patch	1.5	14.7	17.6

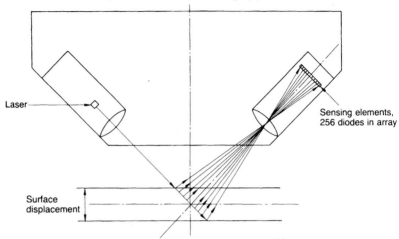

Fig. 9.31. Principle of laser displacement transducer[75]

light from the spot so formed is focused by a receiving lens onto an array of photosensitive diodes (Fig. 9.31). The position of the diode receiving most light gives a measure of the distance to the road surface at that instant and the depth of the texture is computed from a series of such measurements as the meter is propelled along the road surface. The meter is controlled by a microcomputer system and is wheeled and operated by hand. It prints the average SMTD for each completed 10 m measured, together with an overall average for each 50 m length.

The meter is operated at a speed between 3 and 6 km/h on a diagonal line across the carriageway lane width left to right in the direction of traffic flow. The meter must be calibrated by the manufacturer on a regular basis strictly in accordance with a prescribed procedure and the sensitivity of the meter must be checked before use. The sensitivity is checked by running the meter over a check mat and comparing the dropout percentage, which is the proportion of the series of height measurements from the surface which the receiver fails to detect, with that quoted by the manufacturer. The difference should not be greater than three.

Conditions to avoid when the meter is being operated include low angle sunlight illuminating the road surface, condensation on the lens, build up of material on the tyres, and foreign matter or moisture on the road surface.

Texture depth by the Sand Patch Method[75] is determined by taking ten individual measurements at approximately 5 m spacing along a diagonal line across the carriageway lane width, avoiding positions within 300 mm of the edge. The average value of these measurements gives the texture depth for the corresponding 50 m length of lane.

Before making a measurement the surface is dried, if necessary, and any foreign substances removed. A 50 ml measuring cylinder with

30 mm maximum internal diameter is then filled with rounded silica sand predominantly all passing a 300 μm test sieve which has been previously washed and dried. Care should be taken not to compact the sand by any vibration and it is finally struck off level with the top of the cylinder. The volume of sand is then poured into a heap on the road surface and spread in a rotary motion with the aid of a hard rubber disc so that the surface depressions are filled with sand to the level of the peaks. The operator should avoid being over exuberant with movement of the disc to avoid undue scatter of sand across the surface. The diameter of the sand patch is then measured with a steel rule to the nearest 1 mm at four diameters approximately 45 apart. Knee pads for the operator are recommended.

The texture depth is calculated in mm using the following equation

$$TD = 63660/D^2 \tag{9.16}$$

where $D =$ the mean diameter of the sand patch in millimetres.

The specified test method is not suitable for wet, moist or sticky surfaces. However, a test method was developed by Wimpey Laboratories Ltd in 1980[78] and which permits measurements to be made under such conditions. In this case, the Sand Patch Test is carried out with the surface being tested submerged in water. The test was developed at a time when it was considered important for the contractor to rapidly monitor the texture depth of HRA with precoated chippings to control the chipping process which markedly affects the level of texture which is achieved. As this test is unaffected by wet or hot surfaces, measurements could be made immediately after rolling.

The principle of the test is that if only sand and water are present with the exclusion of air, there can be no surface tension forces acting on the grains of sand which would otherwise cause them to cohere. A 'wet' Sand Patch Test was therefore developed whereby the road surface to be tested is submerged in water. A water-retaining ring is placed on the surface. It is then sealed to the surface by applying a plaster of Paris/sand grout mixture to its outside rim. Once the grout has set, usually after about five minutes, the ring is filled with water.

The rest of the procedure is followed in accordance with the specified standard, except that extra care may be needed in spreading the sand into a circular patch to avoid disturbing it by turbulence in the water. Results obtained may be regarded as being the same as those given by the standard test.

The need for a test that can begin rapidly after surfacing operations are completed has probably diminished due to contractors' increasing confidence in achieving the required texture by following established procedures that closely control rate of spread of chippings, laying and compaction temperatures, and rolling technique, and by the increased use of vibrating rollers.

The introduction of requirements for relatively high levels of texture depth led to localized problems with the loss of precoated chippings from

HRA wearing course due to their bunching and reduced contact with the asphalt mortar. This occurred when applying heavier rates of spread to give the cover necessary to meet the texture requirements. This and other problems experienced with HRA wearing course are discussed in Chapters 7, 8 and 11.

9.4.7.   Wheel-Tracking Test[35]

The principal function of the Wheel-Tracking Test[35] is to determine the resistance of asphalts to plastic deformation, i.e. rutting. The UK TRL has carried out studies which led to the derivation of the following relationship between the laboratory Wheel-Tracking Rate and trafficking of HRA wearing course, for deformation to be less than 0.5 mm per annum.

$$d < \frac{1400}{N + 100} \tag{9.17}$$

where d = the Wheel-Tracking Rate in mm/h at 45 °C
  N = the number of commercial vehicles per lane per day.

So, for example, a road carrying 6000 commercial vehicles/lane/day should be surfaced with a wearing course having a maximum Wheel-Tracking Rate of 2.3 mm/h at 45 °C. A maximum rate of 2 mm/h has since been recommended.[79]

Further work has been carried out to compare the Wheel-Tracking Rates of HRAs with their Marshall Stabilities and Marshall Quotients (Stability divided by Flow) which are readily determined as part of the routine laboratory mixture design procedure. Choyce et al.[80] established the following relationship for mixtures excluding those having binder contents below that corresponding to the maximum Marshall Stability

Wheel-Tracking Rate (in mm/h at 45°C)

$$= 116 \, (\text{stability in kN at } 60°C)^{-2} \tag{9.18}$$

A Correlation Coefficient of 0.88 was quoted. This suggests that a Wheel-Tracking Rate of no greater than 2 mm/h at 45 °C may be achieved with a laboratory mixture having a Marshall Stability at 60 °C of at least 7.6 kN, or thereabouts. BS 594: Part 1,[13] in turn, gives recommendations on the test criteria that may be specified for the Marshall Stability of laboratory designed HRA wearing courses that are suitable for different traffic categories expressed in terms of commercial vehicles/lane/day. The acceptable stabilities are from 3 to 10 kN depending on the traffic category.

BS 598: Part 110[35] describes a method for the determination of the Wheel-Tracking Rate and depth of cores of bituminous wearing course. The Standard is not applicable to laboratory prepared and compacted specimens, but the apparatus can be used for this purpose with modified specimen preparation procedures.

Six 200 mm diameter cores that have been marked to indicate direction of traffic flow are used to obtain a test result. Each core is bedded in a holding medium of plaster of Paris onto a flat steel base plate and held in place by plywood clamping blocks which are bolted to the base plate to provide rigid support. A mounting table is used during assembly to ensure that the surface of the core is properly level for testing. Cores of chipped HRA which cannot provide a smooth flat surface are inverted and their undersides tracked after cutting with a circular saw to form the flat surface.

The core, held in its clamping assembly, is placed in the tracking machine and maintained at the test temperature (normally 45 °C or 60 °C). The tracking apparatus consists of a wheel fitted with a 50 mm wide, solid rubber, treadless tyre and loaded to apply a force of 520 N to the surface of the core held in a table which moves backwards and forwards in a fixed horizontal plane at a frequency of 21 cycles/min over a total distance of 230 mm. The machine is set in motion and regular readings taken from a dial gauge or an automatic displacement measuring device, measuring the vertical displacement of the wheel, at no more than 5 min intervals. The wheel-tracking path should correspond to the direction of traffic flow initially marked on the core. The test is normally continued until a displacement of 15 mm is recorded or for a period not exceeding 45 min.

The rate of increase of track depth is calculated from different formulae, depending upon the number of readings taken, and gives a measure of rate of deformation over the latter part of the test. The tracking rates of each core specimen are then averaged to give the test result for the particular material in millimetres per hour.

The Wheel-Tracking Depth is the change in millimetres in the vertical displacement from the initial value to the ninth reading, when the deformation is less than 15 mm after 45 min. If the deformation reaches 15 mm, the Wheel-Tracking Depth is calculated from a formula, depending upon the time at which the deformation was reached. The Wheel-Tracking Depth is then reported as the average of the six test results.

Recommended wheel-tracking requirements for different categories of site are given in Table 9.12.[81]

9.4.8. Creep Test

The simple Creep Test[82] was developed as a better means of designing and assessing asphalt mixtures for resistance to permanent deformation than using the Marshall Test.[51] Laboratory prepared specimens or cored samples are subjected to unconfined, uniaxial loading with a constant force and the resulting axial deformation is measured with time. The reversible part of the total deformation may also be determined by removing the load and measuring the deformation after a recovery time that is usually equal to the loading time.

A disadvantage of the use of the Marshall Test (stability and flow) for assessing resistance to permanent deformation is an unrepresentative load

Table 9.12.   Limiting wheel-tracking requirements for site classifications

Classification		Test temperature: °C	Maximum wheel tracking	
No.	Description		Rate (mm/h)	Rut depth (mm)
0	Lightly stressed sites not requiring specific design for deformation resistance	Not required (Shall comply with the requirements of BS 594: Part 1)		
1	Moderate to heavily stressed sites requiring high rut resistance	45	2.0	4.0
2	Very heavily stressed sites requiring very high rut resistance	60	5.0	7.0

application compared with traffic loading. In the Creep Test,[82] a constant load is applied to the test specimen to provide an axial stress of 100 kPa for a loading time of 1 h with the test temperature maintained, usually at 30 °C. The test conditions of temperature and loading are considered to be similar to those experienced by the materials in the road.

After setting up the specimen in the test apparatus (Fig. 9.32), it is preloaded at the test temperature for 10 minutes with a conditioning load to provide a stress equivalent to 10% of the normal applied stress of 100 kPa, and any axial deformation is recorded. The load is then quickly increased to the test load and the axial deformation is measured with time from displacement readings on a dial gauge or from the output of electrical displacement transducers. The time intervals for the readings are usually predetermined. Those recommended by BS 598: Part 111[83] are 10, 40, 100, 400, 1000, 2000, 3000 and 3600 seconds after the load starts to increase to provide the applied stress for test. A graph of axial strain versus time can then be plotted to describe the deformation behaviour. A fairly accurate straight line can be drawn on a log–log plot, where a steep slope would indicate predominantly viscous behaviour. The axial strain at the completion of the test and the corresponding creep stiffness modulus are usually also determined and reported.

Care is required in the preparation of specimens for test. The end faces should be planar, smooth and perpendicular to the axis of the specimen. Frictional forces between the load platens and the end faces should be as small as possible, in order to ensure that the stress distribution in the loaded specimen is correct. A thin coating of silicone grease sprinkled with graphite flakes is applied to the specimen ends and the ends polished with a soft cloth.

Different deformation behaviour of asphalt mixtures is examined when comparing the simple Creep Test and Rutting Tests similar to that described in Section 9.4.7 above. This is due to loading statically and the lack of lateral confinement in the simple Creep Test.[82]

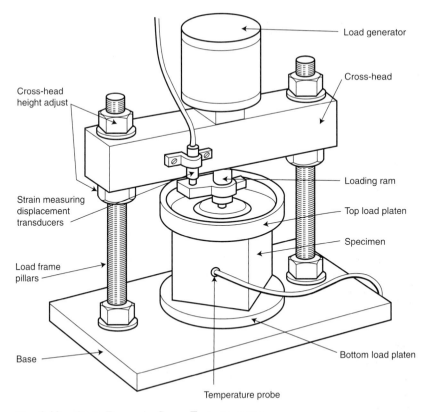

Cross-head
height adjust

Strain measuring
displacement
transducers

Load frame
pillars

Base

Load generator

Cross-head

Loading ram

Top load platen

Specimen

Bottom load platen

Temperature probe

Fig. 9.32. Static/Dynamic Creep Test apparatus

The results of the simple Creep Test have generally led to lower
predicted rutting. The difference has been assumed to be due to the
'dynamic effect'. It has been suggested that, in order to take account of the
damaging effect of dynamic loading which is more simulative of traffic
loading, an empirical correction factor should be applied. The value of such
a correction factor depends on the type of mixture under test.

The attraction of the simple Creep Test[82] is that it is relatively quick,
not too complicated and reasonably inexpensive. The development of a
simple test that included the refinement of the application of axial pulsed
stresses without being prohibitively expensive became of interest in the
late 1980s, with the introduction of the Nottingham Asphalt Tester
(NAT)[84] in the UK.

BS DD 226[85] describes a method for determining the resistance to
permanent deformation of asphalt mixtures subjected to unconfined,
uniaxial, repeated loading. In this case, the specimen is subjected to
repeated applications of load pulses of one second duration, followed by
rest periods each of one second duration. The transient axial deformations
are measured using electrical displacement transducers and a data
acquisition system. These are measured after every tenth load application

up to 100 load applications and thereafter every 100 load applications to the end of test. The axial stress, test temperature and duration of test are usually the same as for the static Creep Test.

The Repeated Load Axial (RLA) Test[85] for the assessment of the resistance to permanent deformation of asphalt mixtures is considered to be a promising performance test. However, the lack of horizontal confinement is seen as a distinct disadvantage. The unconfined test has been shown to discriminate between asphalt mixtures having similar aggregate structures with different binders, but it may not properly describe the effect of the aggregate on deformation resistance (Fig. 9.33[86]).

The fundamental Triaxial Test involves enclosing the test specimen in a membrane and testing in a triaxial cell, which is a type of pressure vessel in which the confining medium may be air, water or oil. In order to simulate traffic loading closely, both axial and confining stresses should be cycled. The test therefore requires very specialist equipment under the care of highly trained technicians. As a result, efforts are being made to develop simple methods of applying a confining pressure. Possible methods include a type of Indentation Test.[87] An axial load is applied to a cylindrical specimen with a diameter of 150 mm, through an upper platen with a diameter of 100 mm. The outer annulus of the specimen effectively provides a confining pressure, related to the properties of the asphalt mixture. The difficulty in analysing the stress state within the specimen is, however, considered by researchers to be a distinct disadvantage of the method.[88]

A limited study at the TRL[89] has demonstrated encouraging results with the development of a test that uses a partial internal vacuum applied to the specimen. This test is called the Confined Repeated Load Axial

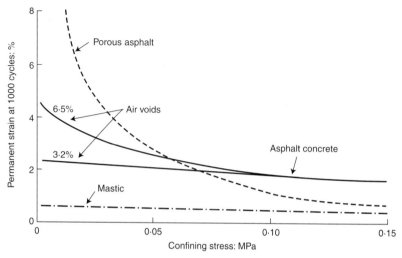

Fig. 9.33.   Effect of confining stress on the Creep Test[86]

(CRLA) Test.[90] The specimen is sealed with a rubber membrane that is secured at both ends by two O-rings which rest in grooves cut around the perimeter of two specially designed platens. The lower platen has a series of drainage holes on its top surface and an outlet pipe fitted in the base which connects to a vacuum pump and pressure regulator and gauge. So far, the test has been shown to correlate well with the Wheel Tracking Test,[35] and further studies are being undertaken to establish standard test conditions and to investigate its precision.

More information on the NAT, RLA and CRLA can be found in Section 3.10.1 of Chapter 3.

9.4.9. *Elastic stiffness*

At typical traffic speeds and pavement temperatures, asphalt may be considered to approximate to an elastic material. Its stiffness modulus under these conditions is a measure of its resistance to bending and, hence, of its load spreading ability. It is therefore considered to be an important performance property of the roadbase and basecourse.

A performance specification for these materials would ideally stipulate a target level for elastic stiffness. Measurement of this property on a routine basis has been found to be most suitably done by the Indirect Tensile (IT) Test as specified in the British Standard Draft for Development DD 213.[91] It is a non-destructive test using cylindrical specimens that may be prepared in the laboratory or sampled from site (Fig. 9.34). It involves applying several load pulses along the vertical diameter of the test specimen and measuring the resultant transient peak horizontal deformations. The measured stiffness modulus can then be determined from these values, Poisson's ratio and the thickness of the specimen. It is claimed that up to 100 specimens can be tested in a working day.

The recommended thickness of the test specimens is between 30 and 75 mm, and their diameters may be 100, 150 or 200 mm, depending on the nominal size of the aggregates in the asphalt mixture. The specimen ends are saw trimmed to ensure that they are perpendicular to the specimen axis.

As a consequence of the curing effect producing an increase in stiffness of asphalt with time (discussed in Section 9.4.2 above), particular attention is paid to the storing conditions for specimens. When the specimens are stored prior to test for longer than four days, the storage temperature should not be greater than 5 °C. In any case, details of the particular storage conditions should be recorded in the test report.

The recommended test temperature is 20 °C. This is ensured in most cases within ±2 °C by placing a dummy specimen of similar composition, with thermocouples at its surface and its centre, next to the test specimen and monitoring the difference between the two thermocouple readings until it is not greater than 0.4 °C. A factor can then be applied to the measured stiffness modulus as a correction to the target temperature. The test result is particularly sensitive to temperature and a change of 1 °C possibly resulting in 10% difference in stiffness has been suggested.[92] Storage, conditioning and testing may often take place in the same temperature-controlled enclosure.

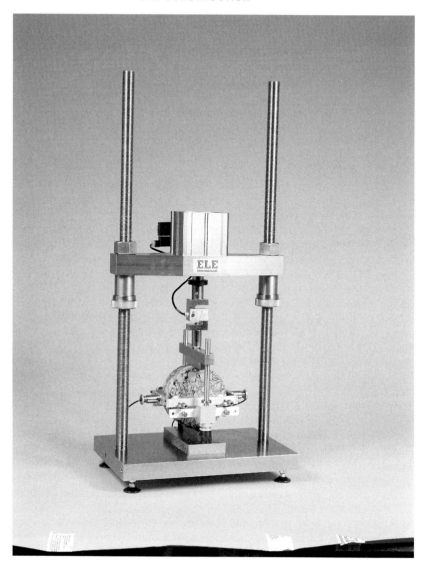

Fig. 9.34.   ELE-MATTA asphalt testing system with accessories for the Indirect Tensile Stiffness Test.[91] *Photograph courtesy of ELE International Ltd*

Typically, a pneumatic load actuator mounted on a steel load frame transmits an applied load, measured by a load cell, along the vertical diameter of the test specimen via loading platens that extend over the full width of the specimen. The transient horizontal diametral deformation is measured by two linear variable differential transformers, mounted opposite each other in a frame that is clamped to the test specimen.

The shape of the load pulse applied to the specimen has an effect on the test result. The difference in shape may be due to different causes. Nunn[93] has suggested that it may be a function of the amount of air required to pressurize the load actuator which is, in turn, related to the position of the actuator relative to the test specimen. The deformation response to a pulse shape with a faster initial rise time to peak load will be greater, resulting in a lower measured stiffness modulus. A correction factor is therefore used to account for different pulse shapes within certain limits which are functions of the area under the load pulse curve, rise time and peak load.

The recommended rise time to the peak load is 124 ms. The peak load needs to be adjusted to achieve a target peak transient horizontal deformation which depends on the diameter of the test specimen.

The measurement of the deformation response is specific to the IT Test.[91] This is illustrated in Fig. 9.35. The amplitude of the deformation is close to the maximum induced by the load pulse. Methods that specify the measurement of a different deformation response will therefore result in the determination of a different measured stiffness modulus. It has been shown that the measured deformation in the BS method may have an irrecoverable or viscous component of 14%.[93] This would reduce to only a very small portion of the total deformation when translated to the corresponding deformation at typical in-service conditions of temperature and frequency for the purpose of pavement design in the UK (20 °C and 5 Hz).

Five conditioning load pulses are applied to the test specimen to reach the specified horizontal diametral deformation and rise time. A further five load pulses are then applied. The load pulses and the corresponding deformations are recorded using a digital interface unit connected to a computer. The test is repeated by rotating the specimen through 90° and the mean value of stiffness modulus taken as the test result.

In the IT Test,[92] the effective frequency for the load pulse cycle has been found to be approximately equivalent to 2.5 Hz in the Three-Point Bending Test with a sinusoidal load pulse shape.[93] This differs from the loading frequency taken as reference for pavement design calculations in

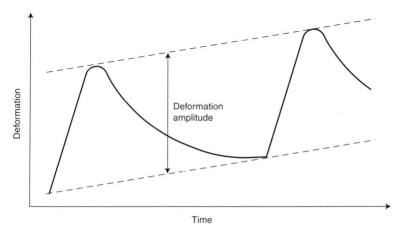

Fig. 9.35.   Indirect tensile deformation response

the UK (5 Hz). Nunn[94] suggests that the value of stiffness modulus measured at 5 Hz is typically about 25% greater than the value measured at 2.5 Hz for dense bitumen macadams. However, stiffer materials have been shown to be less sensitive to loading frequency.[95]

Other procedures that may be followed for the determination of elastic stiffness are included in a draft European Standard[96] and these include Bending Tests with beams and trapezoidal specimens, and Uniaxial Tension-Compression Tests and Direct Tensile Tests using cylindrical specimens.

9.5. Analysis of test results

9.5.1   Statistical techniques

Test results alone give only part of the story. Interpretation of the results and analysis of trends are needed to provide the full information. No single test result can be a reliable estimate of a true value because of the variations that occur due to sampling and testing. It is, therefore, preferable that any decision based on a single result is avoided whenever possible. Practically, a balance must be struck between cost and the risk associated with inferior quality of data output for satisfactory analysis. Those responsible for determining the means by which items are examined should also be continually searching for quicker and cheaper methods that produce the same or more reliable results.

The assessment of quality by an inspector or quality controller based on an analysis of test results can be made easier with the use of statistical methods. Fortunately, most of the results of tests on asphalts have a normal distribution,[97] i.e. their frequency over the range of results produced can be illustrated by the area under a bell-shaped curve. This fact means that simple statistical calculations can be made to describe their variability from a relatively small sampled proportion of the whole quantity under examination.

Those critical variables such as binder content or mixture densities determined by testing can be collated during sufficiently long production runs when steady state conditions prevail and a programme of monitoring, analysis and control can then be established.

The application of statistical techniques is limited when production runs are short due to the scarcity of data gathered and the lower probability that steady state conditions are achieved. The benefit of the techniques is therefore fully realised for major projects involving large tonnages of few different types of products.

Much of what follows is derived from a presentation by C. A. R. Harris on continuous quality improvement.

All measuring and test apparatus should be calibrated and properly adjusted before operation to ensure that variation in results is not largely due to inaccurate or misused equipment. This is discussed further in Section 9.6. All the rest of the results obtained relating to the variable characteristic of interest should be recorded on a data sheet that includes details of the nature of the data, the purpose of the test, the dates of sampling and testing, the test method, the sampling plan and the name of the person who did the test.

The test results can then be systematically recorded on a chart such as a tally sheet or histogram where the range of values is sufficiently extensive to accommodate all expected results. The construction of the tally sheet or histogram is checked for the symmetrical bell shape that is characteristic of a normal distribution. Should a normal distribution of data be uncertain, the data can be plotted on normal probability paper to confirm a straight line representation of the normal distribution.[98]

The boundaries of capability of a particular process should be known by management in order to realistically assess the risk associated with the need to comply with the customer's specified requirements. The process capability can be determined by an analysis of normally distributed data taken from a process that is under statistical control, i.e. one which is not subject to any assignable or special causes of variation. All processes have inherent variations present due to chance causes that influence measured data or test results in the same way. Variations due to common causes are therefore predictable. Variations due to special causes are unpredictable.

Special causes influence some or all measurements in different ways and produce intermittent, unusual variations that indicate specific underlying changes to the process itself. A sudden change in constituent materials or in the operation of proportioning devices on an asphalt plant affecting mixture composition or temperature would be an example of a special cause of variation. Detection of special causes is discussed later.

The mean and standard deviation of the collection of test results are calculated usually by the use of a statistical calculator or a suitable computer program. The standard deviations, s is the standard deviation, may alternatively be calculated from

$$s = \left(\frac{\Sigma(x - \bar{x})^2}{n - 1} \right)^{1/2}$$

where $x =$ the test result
$\bar{x} =$ the mean value
$n =$ the number of test results.

The standard deviation is a measure of the spread of results or their deviation from the mean. It defines the width of a normal distribution (Fig. 9.36). Another measure of spread that is commonly used is the range, which is the difference between the smallest and largest value in the group of test results. The range is convenient when looking at only a few results. Either measure of spread can be used to assess the capability of a process. The standard deviation can be estimated from the mean range of a number of sub-groups of values by dividing by a factor that depends on the number of values in each sub-group.[99]

It can be seen from Fig. 9.36 that 99.7% of values lie within three standard deviations of each side of the mean. It follows that in order to achieve substantial compliance with the specified requirements, the tolerance must be greater than six standard deviations. This is assuming that the mean equals the specification target. Measures are, therefore,

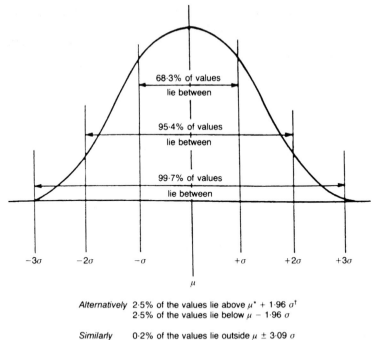

Alternatively 2·5% of the values lie above $\mu^* + 1·96\ \sigma^\dagger$
2·5% of the values lie below $\mu - 1·96\ \sigma$

Similarly   0·2% of the values lie outside $\mu \pm 3·09\ \sigma$

$^*\mu$ = Population mean
$^\dagger\sigma$ = Population standard deviation

Fig. 9.36.   Normal distribution[98]

required to indicate both spread and position of the mean in relation to the specification limits.

The following Process Capability Indices are calculated to provide those measures of capability

$$C_p = \frac{USL - LSL}{6s} \qquad (9.20)$$

$$C_{pk} = \frac{\bar{x} - LSL}{3s} \ \text{or} \ \frac{USL - \bar{x}}{3s} \qquad (9.21)$$

where USL and LSL = the upper and lower specification limits, respectively
s = the standard deviation
$\bar{x}$ = the mean value.

For a process to be capable of meeting the specification requirements, both C_p and C_{pk} should be greater than one. When only a small number of values are available, the specification target and limits may be directly compared with the mean and the smallest and largest values from the process to give an indication of capability.

When the analysis shows that the process is not capable of meeting the specification, an estimation can be made of the percentage outside specification using the following equation and Table 9.13

$$Z = \frac{(y - \bar{x})}{s} \tag{9.22}$$

where Z = standardized normal variate
   y = the upper or lower specification limit
   $\bar{x}$ = the mean value
   s = the standard deviation

Z is, in effect, the number of standard deviations between the mean and the upper or lower specification limit.

Example: use of Z, the standard normal variate

y = lower specification limit = 38

$\bar{x} = 41$

s = 2.1

∴ Z = (38 − 41)/2.1 = −1.43

From Table 9.13, substitution of Z equal to 1.43 gives a function of 0.0764. This corresponds to 7.64% of the area under the normal distribution curve that is below $\bar{x} - 1.43$ standard deviations. It can therefore be said that about 8% of product will fall below the lower specification limit.

So it can be seen how a collection of test results can give a picture of how consistent a particular process is. All too often, in other areas too, the odd test result is used as a basis of judgement and then filed and forgotten but it is clearly necessary to analyse groups of data to enable the best assessment to be made.

9.5.2.   *Control charts*

Interpretation of test results obtained during the production process is also aided by the use of control charts. In many cases, simple mean and range charts or individuals and moving range charts are suitable. However, where the detection of trends and slight changes in data is required, the CUmulative SUM (CUSUM) chart may be used to highlight the small but persistent changes. The control chart is used to distinguish between unusual patterns of variability (special causes) and natural or random patterns inherent in the process (common causes). When detected, variations due to special causes are marked on the control chart and are investigated by those responsible for the operation and control of the process and the necessary corrective action

Table 9.13. Proportions under the normal distribution curve

$Z = \frac{(x-\mu)}{\sigma}$	0.00	0.01	0.02	0.03	0.04	0.05	0.06	0.07	0.08	0.09
0.0	0.5000	0.4960	0.4920	0.4880	0.4840	0.4801	0.4761	0.4721	0.4681	0.4641
0.01	0.4602	0.4562	0.4522	0.4483	0.4443	0.4404	0.4364	0.4325	0.4268	0.4247
0.02	0.4207	0.4168	0.4129	0.4090	0.4052	0.4013	0.3974	0.3936	0.3897	0.3859
0.03	0.3821	0.3783	0.3745	0.3707	0.3669	0.3632	0.3594	0.3557	0.3520	0.3483
0.04	0.3446	0.3409	0.3372	0.3336	0.3300	0.3264	0.3238	0.3192	0.3156	0.3121
0.05	0.3085	0.3050	0.3015	0.2981	0.2946	0.2912	0.2877	0.2843	0.2810	0.2776
0.06	0.2743	0.2709	0.2676	0.2643	0.2611	0.2578	0.2546	0.2514	0.2483	0.2451
0.07	0.2420	0.2389	0.2358	0.2327	0.2296	0.2266	0.2236	0.2206	0.2177	0.2148
0.08	0.2119	0.2090	0.2061	0.2033	0.2005	0.1977	0.1949	0.1922	0.1894	0.1867
0.09	0.1841	0.1814	0.1788	0.1762	0.1736	0.1711	0.1685	0.1660	0.1635	0.1611
1.0	0.1587	0.1562	0.1539	0.1515	0.1492	0.1469	0.1446	0.1423	0.1401	0.1379
1.1	0.1357	0.1335	0.1314	0.1292	0.1271	0.1251	0.1230	0.1210	0.1190	0.1170
1.2	0.1151	0.1131	0.1112	0.1093	0.1075	0.1056	0.1038	0.1020	0.1003	0.0985
1.3	0.0968	0.0951	0.0934	0.0918	0.0901	0.0885	0.0869	0.0853	0.0838	0.0823
1.4	0.0808	0.0793	0.0778	0.0764	0.0749	0.0735	0.0721	0.0708	0.0694	0.0681
1.5	0.0668	0.0655	0.0643	0.0630	0.0618	0.0606	0.0594	0.0582	0.0571	0.0559
1.6	0.0548	0.0537	0.0526	0.0516	0.0505	0.0495	0.0485	0.0475	0.0465	0.0455
1.7	0.0446	0.0436	0.0427	0.0418	0.0409	0.0401	0.0392	0.0384	0.0375	0.0367
1.8	0.0359	0.0351	0.0344	0.0336	0.0329	0.0322	0.0314	0.0307	0.0301	0.0294
1.9	0.0287	0.0281	0.0274	0.0268	0.0262	0.0256	0.0250	0.0244	0.0239	0.0233

2.0	0.0228	0.0222	0.0216	0.0211	0.0206	0.0201	0.0197	0.0192	0.0187	0.0183
2.1	0.0179	0.0174	0.0170	0.0165	0.0161	0.0157	0.0153	0.0150	0.0146	0.0142
2.2	0.0139	0.0135	0.0132	0.0128	0.0125	0.0122	0.0119	0.0116	0.0113	0.0110
2.3	0.0107	0.0104	0.0101	0.0099	0.0096	0.0093	0.0091	0.0088	0.0086	0.0084
2.4	0.0082	0.0079	0.0077	0.0075	0.0073	0.0071	0.0069	0.0067	0.0065	0.0063
2.5	0.0062	0.0060	0.0058	0.0057	0.0055	0.0053	0.0052	0.0050	0.0049	0.0048
2.6	0.0046	0.0045	0.0044	0.0042	0.0041	0.0040	0.0039	0.0037	0.0036	0.0035
2.7	0.0034	0.0033	0.0032	0.0031	0.0030	0.0029	0.0028	0.0028	0.0027	0.0026
2.8	0.0025	0.0024	0.0024	0.0023	0.0022	0.0021	0.0021	0.0020	0.0019	0.0019
2.9	0.0018	0.0018	0.0017	0.0016	0.0016	0.0015	0.0015	0.0014	0.0014	0.0013
3.0	0.0013									
3.1	0.0009									
3.2	0.0006									
3.3	0.0004									
3.4	0.0003									
3.5	0.00025									
3.6	0.00015									
3.7	0.00010									
3.8	0.00007									
3.9	0.00005									
4.0	0.00003									

taken to eliminate the cause. Reduction in the variations due to common causes, however, would probably require a significant change in the way that the process is operated or a complete change in the process itself.

An example of the type of chart that may be used is shown in Fig. 9.37. In the case of mean/range charting, it is usual to plot the averages of, say, two to five consecutive results forming a sub-group and the corresponding ranges of each sub-group. Averaging reduces the variation, effectively smoothing out the plot and making it easier to detect changes in the process. The mean gives an indication of the general location of the variable in relation to the target. Monitoring the range will indicate changes in spread or scatter of results.

Figure 9.37 gives equations for determining appropriate control limits that can be superimposed on the chart. Points outside the control limits or unusual patterns within the limits would indicate that the process is out of control. Patterns to look out for include

- seven points in a row on one side of the mean
- seven points in a row that is ascending or descending
- less than 2/3 of points within middle 1/3 region of the chart.

In many cases, it will be more convenient to plot individual results, e.g. when the frequency of testing is relatively low. Individuals/moving range charts may then be appropriate. The same control chart shown in Fig. 9.37 may be used. The individual values are plotted in place of the averages of the sub-groups. The moving range is determined from each successive sub-group of, say, three results on a moving basis. An example of the individuals/moving range chart for bitumen content is shown in Fig. 9.38.

The control limits for individual results may alternatively be calculated from the equations

$$\text{UCL} = \bar{x} + 3s \tag{9.23}$$
$$\text{LCL} = \bar{x} - 3s \tag{9.24}$$

where ULC and LCL are the upper and lower control limits, respectively.

In the example shown in Fig. 9.38, this would give limits of 3.52 to 4.54 for a standard deviation of 0.17. If the standard deviation was to be estimated from the range, the following value would result

$$s = \bar{R}/d_2 = 0.3/1.693 = 0.177 \tag{9.25}$$

where $\bar{R} = $ the mean range
$d_2 = $ Hartley's constant (for a sample size of 3 this is equal to 1.693).

Substitution of the estimated value in Equations (9.23) and (9.24) would then give the same control limits as those shown in Fig. 9.38. The same set of test results can be used to construct a CUSUM chart which is a highly informative means of graphical presentation of data. The CUSUM plot indicates trends not necessarily revealed by other control charts. The

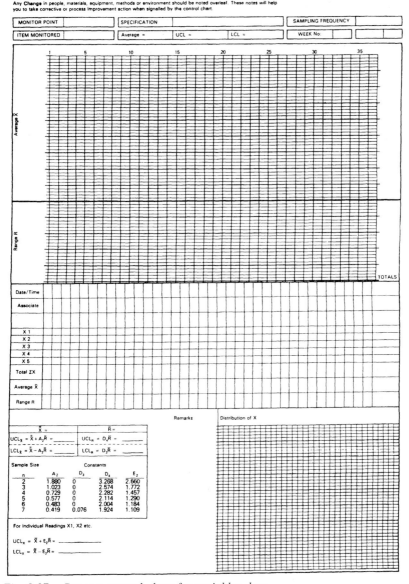

Fig. 9.37. *Process control chart for variables data*

changes in slope on the CUSUM chart indicate changes in mean value in a clearly defined manner. The essential principle is that a target value is subtracted from each observation or result and the cumulative sum of the deviations from target is calculated and plotted against its sample number. British Standard BS 5703[100] offers guidance in the use of CUSUM techniques for data analysis and quality control purposes.

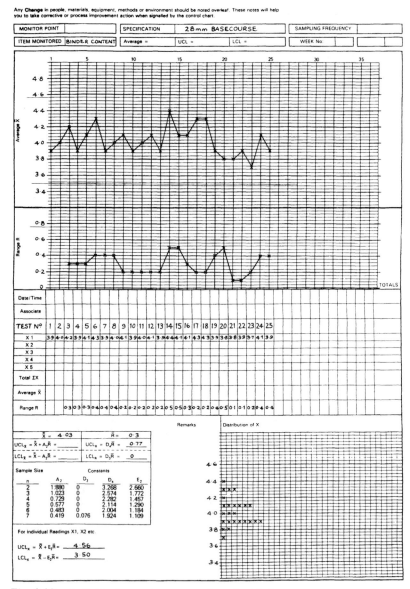

Fig. 9.38.   Individuals moving range chart for binder content

It is necessary to collect sufficient data to determine a standard error of the process for the particular variable concerned. A standard error is the *'standard deviation of an estimated statistic'* and is used in this case to fix the scaling factor in the graph. An accepted convention in scaling the CUSUM chart is that a distance on the *y* axis (CUSUM) equal to one test interval on the *x* axis should represent approximately two standard error units. The standard error is determined from

$$\sigma_e = \frac{8}{9(m-1)} \sum (Y_j - Y_{j+1}) \tag{9.26}$$

where m = the number of results taken for the estimate

     the function $\Sigma(Y_j - Y_{j+1})$ = the sum of successive differences (Table 9.14, column 6)

     Y_j, Y_{j+1} = the successive values

Table 9.14 illustrates the various stages of calculation for CUSUM charting. The deviation of each value from the target is given in column 4 and the cumulative sum of these deviations from tests 1 to 25 is given in column 5. Using Equation (9.26) yields a standard error of 0.15. Therefore, the interval of the plotted variable is 0.3. Should the calculated interval not be convenient for the graph paper used, it is acceptable to round down to the nearest convenient unit.

$$\sigma_e = \frac{8}{9} \times \frac{1}{24} \times \frac{4}{1} = 0.15$$

and

$$2 \times 0.15 = 0.3$$

In Fig. 9.39, an individual's chart is constructed above a CUSUM plot for the same test results. Although open to individual interpretation, it can be argued that the CUSUM plot clearly defines six segments where the path is either generally parallel to the x axis or at a specific angle to it. Where the path is horizontal, the mean value can be considered to be at or near to the target value. When sloping upwards, the path indicates a mean value higher than the target and lower when sloping downwards.

Comparison of the two charts shows that the CUSUM plot gives a much clearer display of changes in the process. Take, for example, test results 14 to 16 on the individuals chart. An initial high value may prompt immediate investigation to determine the cause but is it an isolated result or an indication of plant malfunction, or perhaps a change in the quality of constituent materials?

Subsequent test results show a downward trend and so may indicate that, whatever was wrong, the process is correcting itself. Reference to the CUSUM plot, though, shows a sustained upward slope, indicating that a high level is being maintained and that action is required.

A simple method of assessing approximate averages is to incorporate a slope guide or protractor on the chart. Any inclination of the CUSUM path may then be compared with the protractor as shown on the figure. The construction of the protractor is described in BS 5703: Part 1.[100]

Table 9.14.   Example of CUSUM charting of test results for 28 mm basecourse (figures in percentages)

Binder content target value = 4.0%

1 Date	2 Test number	3 Calulated value	4 Deviation from target	5 Cumulative sum of deviation	6 Successive differences
Day 1	BC 1	3.94	−0.06	−0.06	+0.06
	2	4.00	0	−0.06	+0.06
Day 2	3	4.16	+0.16	+0.10	−0.29
	4	3.87	−0.13	−0.03	+0.27
Day 3	5	4.14	+0.14	+0.11	+0.12
	6	4.26	+2.26	+0.37	−0.36
Day 4	7	3.90	−0.10	+0.27	+0.10
	8	4.00	0	+0.27	+0.10
Day 5	9	4.10	+0.10	+0.37	−0.20
	10	3.90	−0.10	+0.27	+0.08
	11	3.98	−0.02	+0.25	+0.10
	12	4.08	+0.08	+0.33	−0.15
Day 6	13	3.93	−0.07	+0.26	+0.47
	14	4.40	+0.40	+0.66	−0.27
	15	4.13	+0.13	+0.79	−0.02
Day 7	16	4.11	+0.11	+0.90	+0.16
	17	4.27	+0.27	+1.17	+0.02
	18	4.29	+0.29	+1.46	−0.36
Day 8	19	3.93	−0.07	+1.39	−0.09
	20	3.84	−0.16	+1.23	+0.01
	21	3.85	−0.15	+1.08	+0.01
	22	3.86	−0.14	+0.94	−0.14
Day 9	23	3.72	−0.28	+0.66	+0.34
	24	4.06	+0.06	+0.72	−0.12
	25	3.94	−0.06	+0.66	

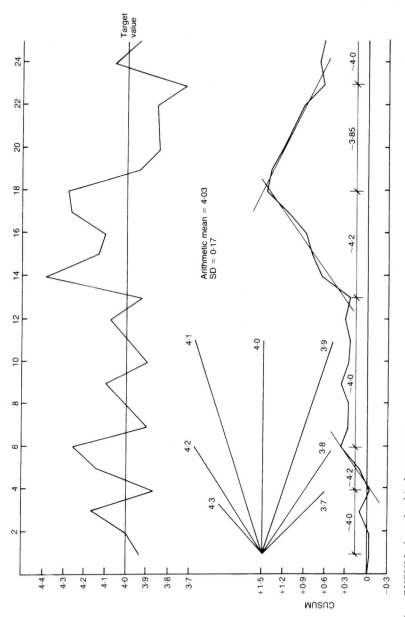

Fig. 9.39. CUSUM chart for binder content

9.5.3. *Significance testing*

It is often necessary to compare two different sets of results to try and determine whether it is likely that they belong to the same population, that is, have the same population mean and variation. A technique is available to determine whether the difference between their Means is statistically significant.

A 't test' can be performed to decide if the observed difference between small samples is due to chance only or to some real cause. The conclusion to the test has a specified probability of being correct, normally 0.95.

Take, for example, the two sets of core densities shown in Table 9.15.

Significance level $= 0.05$, i.e. 1 in 20 risk

$$\text{Standard error } (\sigma_e) = S_p \left(\frac{1}{n_1} + \frac{1}{n_2} \right)^{1/2} \tag{9.27}$$

where

$$S_p = \left(\frac{(n_1 - 1)S_1^2 + (n_2 - 1)S_2^2}{n_1 + n_2 - 2} \right)^{1/2} \tag{9.28}$$

where S_p = combined standard deviation for the 2 independent samples
  n_1 = number of results in Set 1
  n_2 = number of results in Set 2
  S_1 = standard deviation of Set 1
  S_2 = standard deviation of Set 2

Table 9.15.   Test to compare the means of two sets of core densities (small samples)

	Core density: Mg/m^3	
	Set 1	Set 2
	2.328	2.351
	2.341	2.383
	2.350	2.378
	2.382	2.396
	2.399	2.354
	2.365	2.375
Mean	2.361, $\bar{x}_1$	2.373, $\bar{x}_2$
Standard deviation	0.026, S_1	0.017, S_2
Number of results	6, n_1	6, n_2

So $S_p = 0.022$
and $\sigma_e = 0.013$

Degrees of freedom $= (n_1 - 1) + (n_2 - 1) = 10$ (9.29)

Critical value from Table 9.16 (two-sided test)

$= \pm 2.23$

$$\text{Test value} = \frac{Observed\ difference}{\sigma_e} = \frac{2.361 - 2.373}{0.013} = -0.92 \quad (9.30)$$

The test value is less than the critical value. Therefore, no significant difference between the sets of results has been detected.

The calculation of the standard error (σ_e) is specific to the test for the difference between two means when the samples sizes are small. Since, in this case, $n_1 = n_2$, the equation for the pooled standard deviations S_p may be reduced to

Table 9.16.   Critical values for the 't' test

				Significance level		
		Two-sided test			One-sided test	
Degrees of freedom	10%	5%	1%	10%	5%	1%
	(0.10)	(0.05)	(0.01)	(0.10)	(0.05)	(0.01)
1	6.31	12.71	63.66	3.08	6.31	31.82
2	2.92	4.30	9.92	1.89	2.92	6.97
3	2.35	3.18	5.84	1.64	2.35	4.54
4	2.13	2.78	4.60	1.53	2.13	3.75
5	2.02	2.57	4.03	1.48	2.02	3.36
6	1.94	2.45	3.71	1.44	1.94	3.14
7	1.89	2.36	3.50	1.42	1.89	3.00
8	1.86	2.31	3.36	1.40	1.86	2.90
9	1.83	2.26	3.25	1.38	1.83	2.82
10	1.81	2.23	3.17	1.37	1.81	2.76
11	1.80	2.20	3.11	1.36	1.80	2.72
12	1.78	2.18	3.06	1.36	1.78	2.68
13	1.77	2.16	3.01	1.35	1.77	2.65
14	1.76	2.15	2.98	1.35	1.76	2.62
15	1.75	2.13	2.95	1.34	1.75	2.60
16	1.75	2.12	2.92	1.34	1.75	2.58
17	1.74	2.11	2.90	1.33	1.74	2.57
18	1.73	2.10	2.88	1.33	1.73	2.55
19	1.73	2.09	2.86	1.33	1.73	2.54
20	1.72	2.08	2.85	1.32	1.72	2.53
25	1.71	2.06	2.78	1.32	1.71	2.49
30	1.70	2.04	2.75	1.31	1.70	2.46
40	1.68	2.02	2.70	1.30	1.68	2.42
60	1.67	2.00	2.66	1.30	1.67	2.39
120	1.66	1.98	2.62	1.29	1.66	2.36
Infinite	1.64	1.96	2.58	1.28	1.64	2.33

$$S_p = \left(\frac{S_1^2 + S_2^2}{2} \right)^{1/2} \tag{9.31}$$

The degrees of freedom is the number of independent observations or values under a given constraint. When calculating standard deviation, the constraint is that all the deviations must add up to zero. So, for a group of six results, the first five deviations calculated can take any values but the sixth value must be such that the sum of deviations is 0. Therefore, there are five independent values and so five degrees of freedom, i.e. $n - 1$.

Table 9.16 gives critical values for both two-sided and one-sided tests. A one-sided test is used only if there is interest in or there are changes anticipated in one direction. This decision and the significance level chosen defines the critical region under the normal distribution curve (Fig. 9.40). In the example shown in Table 9.15, the test value does not lie in the critical region and so no significant difference between the means has been detected at the 5% level.

The above test assumed that the two sets of results had the same standard deviation or, rather, that the sets had deviations belonging to the same population of standard deviations. It should be checked that the two standard deviations do not differ significantly as it is possible that both conditions of significance may not be satisfied. The 'F Test' is used to determine whether two standard deviations are significantly different. The 'F Value' is calculated from

$$F = S_1^2 / S_2^2 \tag{9.32}$$

where S_1 is the larger standard deviation and S_2 is the smaller standard deviation.

The same example (Table 9.15) gives an F value of 2.34. The critical value taken from Table 9.17 is 7.15, which is greater than the test value. This confirms that there is no significant difference that could be detected between the variability of the two sets of data.

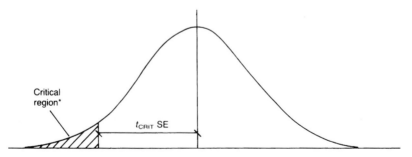

[*] The critical region lies beyond the number of
standard errors from the mean that is equal to
the critical value t_{CRIT}

Fig. 9.40.   Critical region for 't' test

Table 9.17. Critical values for the F test. Two-sided at 5% significance level

Degrees of freedom for smaller variance	1	2	3	4	5	6	7	8	9	10	12	15	20	60	Infinity
1	647.80	799.50	864.20	899.60	921.80	937.10	948.20	956.70	963.30	968.60	976.70	984.90	993.10	1010.00	1018.00
2	38.51	39.00	39.17	39.25	39.30	39.33	39.36	39.37	39.39	39.40	39.41	39.43	39.45	39.48	39.50
3	17.44	16.04	15.44	15.10	14.88	14.73	14.62	14.54	14.47	14.42	14.34	14.25	14.17	13.99	13.90
4	12.22	10.65	9.98	9.60	9.36	9.20	9.07	8.98	8.90	8.84	8.75	8.66	8.56	8.36	8.26
5	10.01	8.43	7.76	7.39	7.15	6.98	6.85	6.76	6.68	6.62	6.52	6.43	6.33	6.12	6.02
6	8.81	7.26	6.60	6.23	5.99	5.82	5.70	5.60	5.52	5.46	5.37	5.27	5.17	4.96	4.85
7	8.07	6.54	5.89	5.52	5.29	5.12	4.99	4.90	4.82	4.76	4.67	4.57	4.47	4.25	4.17
8	7.57	6.06	5.42	5.05	4.82	4.65	4.53	4.43	4.36	4.30	4.20	4.10	4.00	3.78	3.67
9	7.21	5.71	5.08	4.72	4.48	4.32	4.20	4.10	4.03	3.96	3.87	3.77	3.67	3.45	3.33
10	6.94	5.46	4.83	4.47	4.24	4.07	3.95	3.85	3.78	3.72	3.62	3.52	3.42	3.20	3.08
12	6.55	5.10	4.47	4.12	3.89	3.73	3.61	3.51	3.44	3.37	3.28	3.18	3.07	2.85	2.72
15	6.20	4.77	4.15	3.80	3.58	3.41	3.29	3.20	3.12	3.06	2.96	2.86	2.76	2.52	2.40
20	5.87	4.46	3.86	3.51	3.29	3.13	3.01	2.91	2.84	2.77	2.68	2.57	2.46	2.22	2.09
60	5.29	3.93	3.34	3.01	2.79	2.63	2.51	2.41	2.33	2.27	2.17	2.06	1.94	1.67	1.48
Infinity	5.02	3.69	3.12	2.79	2.57	2.41	2.29	2.19	2.11	2.05	1.94	1.83	1.71	1.39	1.00

9.5.4. Recommendation

Statistical techniques can be applied in a multitude of ways to analyse test results and other data. Here, only a small selection has been considered that may be found useful in interpreting the results of tests on asphalts and their regular application should contribute, in particular, to gaining a much better understanding of processes so that decisions based on the results of testing are more likely to be correct in all circumstances. In addition to the other references given on statistical analysis at the end of this Chapter, BS 2846: Parts 1 to 8[101] gives guidance on the statistical interpretation of data.

9.6. Laboratory management

The product of the laboratory is the test report. The report may be prepared for a purchaser who is interested in some property of a material or is seeking assurance that the material meets certain requirements. The purchaser may also be looking for advice that would require interpretation of test results and recommendations for further action. The report may be required for internal use only, as a source of management information. The nature and frequency of reporting will vary greatly depending upon its purpose. For quality control purposes, brief daily reports are appropriate to facilitate rapid evaluation of the ongoing quality of manufactured products.

A programme of testing may be required as part of an investigation into an identified problem or of a product development project. In such a case, a report would only be prepared once sufficient data are collected and properly analysed and would be structured in a more formal way to give the reader a clear and full explanation of the work carried out. The contents of the report may also be specifically described by the purchaser as a contractual requirement.

Whatever the purpose of the test report, it is the responsibility of the laboratory manager to establish a system that will satisfy the requirements of the customer. This means that all activities associated with the laboratory process upstream of the issue of the final report are carried out both efficiently and effectively. In other words, the manager has to be sure that everything is done correctly and, from an overall viewpoint, that the system is addressing the relevant issues to achieve its aim.

The laboratory should also be prepared to offer a follow up service after issue of the report and not simply regard reporting as the last stage in the process because, as stated at the beginning of this Chapter, test results invariably form the basis of important decision making. European Standard BS EN ISO/IEC 17025[102] specifies the general criteria for the operation of testing laboratories and offers useful guidance for assuring technical competence.

The scope of testing in the laboratory should be clearly defined to establish the overall requirements in respect of facilities, personnel, working methods, materials and any special safety measures. All laboratory personnel should be made aware of their individual responsibilities and undergo proper induction procedures.

Induction of new members of laboratory staff should include a brief introduction to the aims, objectives and organizational structure of the laboratory, explanation of rules and disciplinary procedures, health, safety and welfare policies and responsibilities. A description of the roles and duties of the individual must be given and the importance of an impartial approach stressed so that the outcome of tests may not be influenced by, for example, deliberate bias in sampling. At the same time, the policy of the organization on security of information should be made quite clear. Where necessary, an individual may be placed in the care of an experienced member of staff until he or she is able to carry out the tests competently without direct supervision.

Frequent reviews of the individual's progress should be made by the laboratory manager over the first few weeks. Periodical reviews of employee skills and knowledge should be made to assess recruitment and training and development needs. Once the needs have been identified, a training and development plan should be prepared to address objectives such as updating in new techniques or technology, improving performance, and preparation for promotion. Personal training files may be held by individuals to indicate the degree of competence achieved over a period of time in carrying out particular tests and other practices.

The working environment of the laboratory should be made as conducive as possible to the satisfactory output of reliable data. This requires that the laboratory is maintained at the required temperature and humidity conditions, devoid of excessive vibration, noise, dust and other nuisance, and is furnished to minimize the risks to the health and safety of employees, particularly with respect to the use of organic solvents to which the Control of Substances Hazardous to Health Regulations (COSHH)[103] apply.

The laboratory equipment should conform to the appropriate standards and be sufficient to complete all the tests included under the scope of testing. It should be properly maintained and instructions issued on adjustment and operation. The laboratory manager should maintain a comprehensive inventory of all equipment with details of any malfunctions, repairs and calibration status.

In order to ensure that test equipment is provided and maintained to the specified standard, a programme of regular servicing and calibration has to be established. Wherever possible, the schedule of calibration procedures for measuring and test equipment and for reference standards of measurement, e.g. standard weights, thermometers, etc. should comply with National Standards or with the manufacturers' written instructions. Table 9.18 lists some British Standards relating to the testing of asphalts that specify calibration requirements for test equipment used in a variety of laboratory test procedures.

Examples of the minimum frequency of routine calibration are given in Table 9.19. The schedule shown is not exhaustive and will vary according to the specific advice of manufacturers, purchasers' instructions, degree of exceptional use of equipment and the requirements of the particular laboratory.

Table 9.18.   Relevant British Standards for calibration of measuring and test equipment

BS	Part	Date	Title
410		1986	Specification for test sieves
593		1989	Specification for laboratory thermometers
598	102	1996	Sampling and examination of bituminous mixtures Analytical test methods
812	100	1990	Testing aggregates General requirements for apparatus and calibration
846		1985	Specification for burettes
1377	1	1990	Soils for civil engineering purposes General requirements and sample preparation
1610	1	1992	Specification for the grading of the forces applied by materials testing machines when used in the compression mode
1780		1985	Specification for Bourdon tube pressure and vacuum gauges
1792		1982	Specification for one-mark volumetric flasks
1797		1987	Schedule for tables for use in the calibration of volumetric glassware
2000	0: section 0.1	1996	Methods of test for petroleum and its products IP Standard themometers
2648		1955	Performance requirements for electrically-heated laboratory drying ovens
EN 30012	1	1994	Quality Assurance requirements for measuring equipment Metrological confirmation system for measuring equipment

Calibration reporting forms should identify the equipment and include details of the type of test, specification, test method, test results, date and place of test, and the name and signature of the calibrator. Calibration of reference standards of measurement by a qualifying calibration facility that can provide traceability to a national or international standard of measurement should not be overlooked. Calibration intervals should be appropriate to the nature and usage of the equipment.

The laboratory manager is responsible for ensuring that the relevant standards for sampling and test methods are made readily available to all technicians. In some cases, it may be necessary, for whatever reason, to follow non-standard procedures, in which case the specific requirements need to be fully documented.

Many laboratories now operate quality management systems (QMSs)[104] as a means of improving performance. The laboratory system may be part of an overall company quality system. Irrespective of how it is managed, the essential elements remain the same:

- documentation
- implementation
- internal audit
- review.

Broadly, management determines the best practical means of ensuring that it understands and satisfies the needs of its customers and documents the methods to achieve these aims. To ensure that the quality policy of the organization is being implemented, the relevant activities are audited against the requirements of the documented procedures and any deficiencies are corrected. Reviews are then made to assess whether the system is working effectively and economically and to consider ways of improving it.

Once a QMS has been established, some organizations will apply to the United Kingdom Accreditation Service (UKAS) for accreditation of their laboratories whereby verification is made by independent assessors that specified criteria for the satisfactory operation of testing facilities are being met. This seal of approval firmly establishes the credibility of the laboratory. Indeed, some purchasers now demand it for certain test requirements.

The Department of Transport has required contractors to carry out sampling and testing of asphalts for compliance purposes.[74] The contractor's testing is carried out in a UKAS accredited laboratory and therefore, audit testing by the overseeing organization is normally not then required. Guidance on the frequency of testing is given in a table in the Notes for Guidance on the Specification for Highway Works.[81] The results of such tests must be reported to the overseeing organization on official UKAS test certificates.

It is necessary to follow procedures that will ensure that samples of materials received at the laboratory are correctly identified and registered and that their integrity is preserved until ready for testing. Each sample is given a unique reference number and details of its condition recorded together with the essential details from the certificate of sampling where used (Fig. 9.5). The sample quantity should also be checked and recorded and compared with the specified requirements of the particular sampling and test method. Sample details can be conveniently kept in a register for easy reference.

The importance of good records that ensure traceability from sampling point or source to final test report cannot be exaggerated. Certain records also serve as the objective evidence of the operation of a QMS and are required to be kept to prove the effectiveness of its implementation. Every effort should also be made to reduce the amount of paper and so a procedure for the control of records that specifies the required period of retention and disposal arrangements should be established. Nowadays,

Table 9.19. Calibration guide for measuring and test equipment

Item	Method and/or reference	Maximum calibration interval
Laboratory weigh scales	By UKAS certificated weights	3 months and at least once/year by specialist firm
Thermometers (mercury in glass)	By UKAS certificated thermometer BS 593: 1989	12 months
Thermometers (bimetallic thermocouples, thermistors, electronic)	By UKAS certificated thermometer. Also check in boiling water if within scale range	6 months. Frequent, depending on use e.g. daily, weekly
Laboratory ovens	BS 2648: 1955 Temperature at mid-point of usable space of empty oven by calibrated thermometer	12 months
Volumetric glassware	Dependent on whether class A or B glassware. Refer to BS relevant to the particular glassware e.g. BS 846: 1985 and BS 1792: 1982. Also by weight of water, certificated weights and BS 1797: 1987	12 months
Test sieves	BS 410: 1986 Visual check per set. Grading check per set using either reference aggregate or reference set of sieves	1 week. 3 months
Centrifuge	BS 598: Part 102: 1996. Use UKAS certificated strobe meter, for example	6 months
Compression testing machines	BS 1610: Part 1: 1992	12 months
Proving rings	By UKAS certificated test house	12 months
Marshall test equipment	BS 598: Part 107: 1990. Mass and dimensional check. Rate of blows of automatic compactor	12 months. 6 months

Vacuum gauges	BS 1780: 1985. UKAS certificated gauge or mercury manometer corrected for air pressure	6 months
Stopclocks/watches	British Telecom speaking clock	6 months
Bottle rotation machines	BS 598: Part 102: 1996. Visual calibration of speed of rotation	12 months
Bitumen penetrometer	BS 2000: Part 49: 1993. UKAS certificated needles. Timer and loader application to be checked	At least once/year by specialist firm
Mastic asphalt hardness machine	BS 5284: 1993. Indentor, timing and force application to be checked	6 months
TRL Mini-texture meter	(i) Certification of calibration factor by the manufacturer (ii) Sensitivity on the daily check mat by user. Corrected for air pressure	12 months Daily when in use
Rolling straight edge	TRL SR 290/1977 (i) Certification by the manufacturer (ii) Calibration check by the user	6 months or 30 km Daily when in use
Nuclear density gauge	Manufacturer's instructions	

increasing use is being made of data storage on magnetic tape, computer file and microfilm. Records should be stored in an orderly manner, maintained in a clean state, protected from damage and, where necessary, indexed for ready retrieval. The test report takes many forms, but particular care should be taken in the presentation of the data and their ease of comprehension by the recipient. For the individual organization, corporate identity can be important in establishing how it is perceived by the external customer and, so, the opportunity can be taken to portray the desired image through inclusion on the report sheet of the company logo, colour, font etc.

Every laboratory test report should include, at least, the following information

• name and address of testing laboratory
• unique identification number of the report
• name of customer and project
• description and identification of the sample
• date of receipt of the sample
• designation of test method
• any deviation from the designated test method
• date of testing
• signature and title of person accepting technical responsibility for the test report
• date of issue.

Other information detailed in reports of site testing should include

• location of the site
• date of installation of materials
• weather conditions
• location of sampling point.

Should a series of test results form the basis of a larger investigation or other project, guidance may be required on an appropriate structure for interim or final reporting. It is recommended that the report consists of the following sections

• Acknowledgements
• Table of contents
• Summary
• Introduction
• Apparatus
• Procedure
• Results
• Discussion
• Tables
• Figures
• Conclusions
• References
• Appendices.

Finally, communication and teamwork are two of the most important elements of successful management and this applies right across the departmental boundaries of any testing organization. The manager needs to communicate, coordinate and facilitate.

9.7. Asphalt Quality Assurance Scheme

In the UK, a national Quality Assurance Scheme for the Production of Asphalt Mixtures[105] was launched on 27 October 1998 by the County Surveyors' Society (CSS), the Highways Agency (HA) and the Quarry Products Association (QPA) with the support of the United Kingdom Accreditation Service (UKAS) and members of the Association of British Certification Bodies and Independent International Organisation for Certification. The following emerged from the development of this scheme.

- An agreed schedule complying with BS EN ISO 9002[104] which can be used by certification bodies for the purpose of accrediting asphalt production plants.
- An agreement that the HA, the Scottish Executive, the Welsh Office and the Department of Environment for Northern Ireland Roads Service will specify accreditation to the scheme as a contract requirement and that CSS will encourage all local authorities to similarly make it a requirement.
- An implementation date of 1 April 1999.
- A joint HA/CSS/QPA/UKAS committee with a formal constitution to advise UKAS on Quality Assurance of asphalt production and to review the operation of the scheme, making amendments as necessary.

Introduced with the scheme is a system for monitoring the compliance status of an asphalt plant at any point in time. A 'Q' (Quality) Level is determined for each plant by taking the last 30 test results for all products and finding the number of these that fail to comply. For example, if it is found that four results fail to comply with the relevant specifications, the Q Level is 4. The Q Level is related to the required rate of sampling, such that, as the Q Level increases, the greater the number of samples that must be taken. However, if the Q Level exceeds six, action has to be taken to reduce it to below six within 14 days. This system has an advantage over the old system of minimum percentage compliance, by relating compliance to rate of sampling, thus encouraging the drive for continuous quality improvement. It also represents a positive move towards decision taking that is based on analysis of trends rather than individual results.

The scheme is consistent with a legal obligation of the Construction Products Directive (CPD) of the European Communities[106] for producers to meet the requirements of the process of Attestation of Conformity. Harmonized European Standards developed under mandates from the European Commission are being developed and they will include procedures for demonstrating conformity at the point of placing the product on the market — Attestation of Conformity. The whole process of attestation will include procedures relating to a producer's quality system, testing schedules and frequencies, and third party involvement dictated by the European Union for certification of the producer's conformity.

The objective is increased confidence in the quality of product and, in the case of the CPD, part of the framework being put in place for the removal of trade barriers within the European Union.

9.8. Summary

This Chapter has examined the part that testing plays in the process of road construction with asphalts. The importance of properly conducted sampling and testing procedures has been emphasized, owing to the great influence that the test result has on the success of the road construction project. The most commonly used tests for asphalt and bitumen products have been described and discussed in some detail and Section 9.4 on test methods represents a single source of information on this whole subject. The correct interpretation of test results is vitally important and some rather tentative methods of interpreting have been presented in this Chapter, although it should be noted that similar methods are used successfully in other, not dissimilar, industries. The responsibilities of the laboratory manager have been covered and the impact of increasing legislation on the running of the laboratory has also been considered.

References

1. BROCK J. D. *Segregation — Causes and Cures.* Astec Industries Inc, Chattanooga, 1986, Technical Bulletin No. T-115.
2. BROCK J. D. *Segregation — Causes and Cures.* Astec Industries Inc, Chatanooga, 1979, Technical Bulletin No. T-101.
3. BRITISH STANDARDS INSTITUTION. *Sampling and Examination of Bituminous Mixtures for Roads and other Paved Areas, Methods for Sampling for Analysis.* BSI, London, 1987, BS 598: Part 100.
4. AMERICAN SOCIETY FOR TESTING AND MATERIALS. Standard Practice for Sampling Bituminous Paving Mixtures, Construction: Road and Paving Materials; Pavement Management Technologies. *Annual Book of ASTM Standards.* ASTM, Philadelphia, 1989, D979-89.
5. NATIONAL ASPHALT PAVEMENT ASSOCIATION. *Quality Control for Hot Mix Asphalt Manufacturing Facilities and Paving Operations,* Quality Improvement Series 97/87. NAPA, Riverdale, Maryland, 1987.
6. AMERICAN SOCIETY FOR TESTING AND MATERIALS. Standard Practice for Random Sampling of Construction Materials. *Annual Book of ASTM Standards.* ASTM, Philadelphia, 1995, **04.03**, D3665-94.
7. BRITISH STANDARDS INSTITUTION. *Sampling and Examination of Bituminous Mixtures for Roads and other Paved Areas, Methods for the Preparatory Treatments of Samples for Analysis.* BSI, London, 1987, BS 598: Part 101.
8. BRITISH STANDARDS INSTITUTION. *Testing Aggregates, Methods for Sampling.* BSI, London, 1989, BS 812: Part 102.
9. HARRIS P. M and R. SYM (D. C. PIKE ed.). *Sampling of Aggregates and Precision Tests,* Standards for Aggregates. Ellis Horwood Ltd, Chichester, 1990, Ch. 2.
10. BRITISH STANDARDS INSTITUTION. *Methods for Sampling Petroleum Products, Method for Sampling Bituminous Binders.* BSI,

London, 1987, BS 3195: Part 3.

11. AMERICAN SOCIETY FOR TESTING AND MATERIALS. Standard Practice for Sampling Bituminous Materials. *Annual Book of ASTM Standards.* ASTM, Philadelphia, 1995, **04.03**, D140-93.

12. SMITH R. and G. V. JAMES. *The Sampling of Bulk Materials.* Royal Society of Chemistry, London, 1981.

13. BRITISH STANDARDS INSTITUTION. *Hot Rolled Asphalt for Roads and other Paved Areas, Specification for Constituent Materials and Asphalt Mixtures.* BSI, London, 1992, BS 594: Part 1.

14. BRITISH STANDARDS INSTITUTION. *Coated Macadam for Roads and other Paved Areas, Specification for Constituent Materials and for Mixtures.* BSI, London, 1993, BS 4987: Part 1.

15. BRYANT L. J. Effect of Segregation of an Asphaltic Concrete Mixture on Extracted Asphalt Percentage. *Journal of the Assoc. Asphalt Paving Tech.*, 1967, **36**, 206.

16. BRITISH STANDARDS INSTITUTION. *Sampling and Examination of Bituminous Mixtures for Roads and other Paved Areas, Analytical Test Methods.* BSI, London, 1996, BS 598: Part 102.

17. GRANTLY E. C. Quality Assurance in Highway Construction, Part 4 – Variations of Bituminous Construction. *Public Roads*, 1969, **35**, No. 9, 201.

18. FROMM H. Proc. American Assoc. Paving Tech. *Journal of the Assoc. Asphalt Paving Tech.*, 1978, **47**, 372.

19. BRITISH STANDARDS INSTITUTION. *Precision of Test Methods, Guide for the Determination of Repeatability and Reproducibility for a Standard Test Method by Inter-Laboratory Tests.* BSI, London, 1987, BS 5497: Part 1.

20. NICHOLLS J. C. *Precision of Tests Used in the Design of Rolled Asphalt.* TRL, Crowthorne, 1991, RR 281.

21. AMERICAN SOCIETY FOR TESTING AND MATERIALS. Road and Paving Materials, Paving Management Technologies. *Annual Book of ASTM Standards.* ASTM, Philadelphia, 1995, **04.03**.

22. AMERICAN ASSOCIATION OF STATE HIGHWAY AND TRANS-PORTATION OFFICIALS. *Standard Specification for Transportation Materials and Methods of Sampling and Testing, Methods of Sampling and Testing.* AASHO, Washington DC, 1986, Part II.

23. AMERICAN SOCIETY FOR TESTING AND MATERIALS. Standard Practice for Conducting an Inter-Laboratory Test Program to Determine the Precision of Test Methods for Construction Materials. *Annual Book of ASTM Standards.* ASTM, Philadelphia, 1990, **04.03**, C802-87.

24. NATIONAL ASPHALT PAVING ASSOCIATION. *Statistical Methods for Quality Control at Hot Mix Plants.* NAPA, Maryland, 1973 (1983).

25. RICHARDSON C. *The Modern Asphalt Pavement.* John Wiley & Sons, New York, 1908, 2nd edn.

26. BRITISH STANDARDS INSTITUTION. *Methods of Test for Petroleum and its Products, Recovery of Bitumen Binders, Dichloromethane Extraction Rotary Film Evaporator Method.* BSI, London, 1995, BS 2000: Part 397.2.

27. BRITISH STANDARDS INSTITUTION. *Methods of Test for Petroleum and its Products, Recovery of Bituminous Binders by Dichloromethane Extraction.* BSI, London, 1991, BS 2000: Part 105 (Withdrawn).

28. BRITISH STANDARDS INSTITUTION. *Methods of Test for Petroleum and its Products, Determination of Needle Penetration of Bituminous Material.* BSI, London, 1993, BS 2000: Part 49.

29. BRITISH STANDARDS INSTITUTION. *Methods of Test for Petroleum and its Products, Determination of Softening Point of Bitumen, Ring and Ball Method.* BSI, London, 1993, BS 2000: Part 58.

30. ENVIRONMENTAL TECHNOLOGY BEST PRACTICE PROGRAMME. *Solvent-Free Asphalt Testing Reduces Cost.* ETSU, Case Study, Didcot, UK, 1998, FP121.

31. RICHARDSON J. T. G. A Testing Future. *Proc. of 10th IAT East Midlands Annual Training Day on 2000 Challenges for the Asphalt Industry, Northants, 1999.*

32. WHITEOAK C. D. The Shell Bitumen Handbook, 1990. Shell Bitumen UK, Chertsey.

33. BRITISH STANDARDS INSTITUTION. *Bitumens for Building and Civil Engineering, Specification for Bitumens for Roads and other Paved Areas.* BSI, London, 1989, BS 3690: Part 1.

34. JACOBS F. A. *Hot Rolled Asphalt, Effect of Binder Properties on Resistance to Deformation.* TRL, Crowthorne, 1981, LR 1003.

35. BRITISH STANDARDS INSTITUTION. *Sampling and Examination of Bituminous Mixtures for Roads and other Paved Areas, Method of Test for the Determination of Wheel-Tracking Rate and Depth.* BSI, London, 1998, BS 598: Part 110.

36. HEUKELOM W. *An Improved Method of Characterising Asphaltic Bitumens with the Aid of their Mechanical Properties.* Shell International Co. Ltd, London, 1974.

37. AMERICAN SOCIETY FOR TESTING AND MATERIALS. Construction: Roofing and Waterproofing Materials. *Annual Book of ASTM Standards.* ASTM, Philadelphia, 1987, **04.04**, D36.

38. SPARLIN R. F. The Effect of Ultraviolet Light on Viscosity of Thin Films of Asphalt Cements. *Proc. American Soc. Testing & Materials*, 1958, **58**, 1316.

39. CHADDOCK B. *et al. Accelerated and Field Curing of Bituminous Roadbase.* TRL, Crowthorne, 1994, Project Report 87.

40. NUNN M. E. *et al. Design of Long-Life Flexible Pavements for Heavy Traffic.* TRL, Crowthorne, 1997, Report 250.

41. BROCK J. D. *Oxidation of Asphalt.* ASTEC Industries Inc, Tennessee, 1986, Technical Bulletin No. T-103.

42. ABRAHAM H. *Asphalts and Allied Substances.* Van Nostrand, New York, 1945, 5th edn, 1464.

43. BRITISH STANDARDS INSTITUTION. *Methods of Test for Petroleum and its Products, Determination of Loss on Heating of Bitumen and Flux Oil.* BSI, London, 1993, BS 2000: Part 45.

44. AMERICAN SOCIETY FOR TESTING AND MATERIALS. Standard Test Method for Effect of Heat and Air on a Moving Film of Asphalt (Rolling Thin-Film Oven Test). *Annual Book of ASTM Standards.* ASTM, Philadelphia, 1995, **04.03**, D2872-88.

45. DICKINSON E. J. *Bituminous Roads in Australia.* Australian Road Research Board, Vermont South, Australia, 1984.

46. BRITISH STANDARDS INSTITUTION. *Methods of Test for Petroleum and its Products, Solubility of Bituminous Binders.* BSI, London, 1993, BS 2000: Part 47.

47. AMERICAN SOCIETY FOR TESTING AND MATERIALS. Standard Test Method for Kinematic Viscosity of Asphalt 'Bitumens'. *Annual Book of ASTM Standards.* ASTM, Philadelphia, 1995, **04-03**, D2170-92.

48. DAVIS R. L. *Relationship between the Rheological Properties of Asphalt and the Rheological Properties of Mixtures and Pavements.* ASTM, Philadelphia, 1985, Report STP 941.

49. GRIFFEN R. L. *et al.* A Curing Rate Test for Cutback Asphalts using the Sliding Plate Micro-Viscometer. *Proc. American Assoc. Paving Tech.*, 1957, **26**, 437.

50. ASPHALT INSTITUTE. *SuperPave Performance Graded Asphalt Binder Specification and Testing*, SuperPave Series No. 1. AI, Kentucky, 1997.

51. MARSHALL CONSULTING & TESTING LABORATORY. *The Marshall Method for the Design and Control of Bituminous Paving Mixtures.* Marshall Consulting & Testing Laboratory, Mississippi, 1949.

52. PROPERTY SERVICES AGENCY. *A Guide to Airfield Pavement Design and Evaluation.* PSA, Croydon, 1989.

53. BRITISH STANDARDS INSTITUTION. *Hot Rolled Asphalt for Roads and other Paved Areas.* BSI, London, 1973, BS 594.

54. BRITISH STANDARDS INSTITUTION. *Sampling and Examination of Bituminous Mixtures for Roads and other Paved Areas, Method of Test for the Determination of the Composition of Design Wearing Course Rolled Asphalt.* BSI, London, 1990, BS 598: Part 107.

55. FOSTER C. R. *Development of Marshall Procedures for Designing Asphalt Paving Mixtures*, Information Series 84. National Asphalts Paving Association, Maryland, 1982.

56. ASPHALT INSTITUTE. *Mix Design Methods for Asphalt Concrete and other Hot-Mix Types*, Manual Series No. 2. AI, Kentucky, 1997.

57. AMERICAN SOCIETY FOR TESTING AND MATERIALS. Standard Test Method for Theoretical Maximum Specific Gravity of Bituminous Paving Mixtures. *Annual Book of ASTM Standards.* ASTM, Philadelphia, 1995, **04.03**, D2041-94.

58. KANDHAL P. S. *et al.* Evaluation of Asphalt Absorption by Mineral Aggregates. *Journal of the American Assoc. Asphalt Paving Tech.*, 1991, **60**, 207–29.

59. ASPHALT INSTITUTE. *SuperPave Mix Design*, SuperPave Series No. 2. AI, Kentucky, 1996.

60. AMERICAN SOCIETY FOR TESTING AND MATERIALS. Standard Test Method for Bulk Specific Gravity and Density of Compacted Bituminous Mixtures using Saturated Surface-Dry Specimens. *Annual Book of ASTM Standards.* ASTM, Philadelphia, 1995, **04.03**, D2726-93a.

61. BRITISH STANDARDS INSTITUTION. *Sampling and Examination of Bituminous Mixtures for Roads and other Paved Areas, Methods of Test for the Determination of Density and Compaction.* BSI, London, 1989, BS 598: Part 104.

62. AMERICAN SOCIETY FOR TESTING AND MATERIALS. Standard Test Method for Bulk Specific Gravity and Density of Compacted Bituminous Mixtures using Paraffin-Coated Specimens. *Annual Book of ASTM Standards.* ASTM, Philadelphia, 1995, **04.03**, ASTM D1188-89.

63. RICHARDSON J. T. G. and J. C. NICHOLLS. Determination of Air Voids Content of Asphalt Mixtures. *Proc. 3rd European Symposium on Performance and Durability of Bituminous Materials and Hydraulic Stabilised Composites, Leeds, 1999.*

64. AMERICAN SOCIETY FOR TESTING AND MATERIALS. Standard Test Method for Specific Gravity and Absorption of Coarse Aggregates. *Annual Book of ASTM Standards.* ASTM, Philadelphia, 1990, **04.03**, ASTM

C127-88.

65. AMERICAN SOCIETY FOR TESTING AND MATERIALS. Standard Test Method for Specific Gravity and Absorption of Fine Aggregate. *Annual Book of ASTM Standards*. ASTM, Philadelphia, 1990, **04.03**, ASTM C128-88.

66. VON QUINTUS H. L. *et al. Asphalt-Aggregate Mixtures Analysis System, Philosophy of the Concepts, Asphalt Concrete Mix Design — Development of More Rational Approaches*. ASTM, Philadelphia, 1989, Report STP 1041, 15–38.

67. SOUSA J. B. *et al.* Effect of Laboratory Compaction Method on Permanent Deformation Characteristics of Asphalt-Aggregate Mixtures. *Journal of the American Assoc. Asphalt Paving Tech.*, 1991, **60**, 533–85.

68. ARAND W. *Verdichtung Mathematisch-Analytisch Betrachtet, Bitume Teere Asphalte Peche 25*. Verlag Fuer Publicaitet, Isernhagen, Germany, 1974.

69. COMITÉ EUROPÉEN DE NORMALISATION. *Bituminous Mixtures — Test Methods for Hot Mix Asphalt, Segregation Sensitivity*. CEN, Brussels, 1995. Work Item No. 00 227 121, TC227 WG1 TG2.

70. AMERICAN SOCIETY FOR TESTING AND MATERIALS. Standard Test Method for Effect of Water on Compressive Strength of Compacted Bituminous Mixtures. *Annual Book of ASTM Standards*. ASTM, Philadelphia, 1995, **04.03**, D1075-94.

71. MATHEWS D. H. *et al.* The Immersion Wheel-Tracking Test. *Journal of Applied Chemistry*, 1962, **12**, 505–9.

72. AMERICAN ASSOCIATION OF STATE HIGHWAY AND TRANS-PORTATION OFFICIALS. *Standard Method of Test for Resistance of Compacted Bituminous Mixture to Moisture Induced Damage*, AASHO Materials, Part II Tests. AASHO, Washington, 1987, T283-87.

73. HOWE J. H. The Percentage Refusal Density Test. *J. Inst. Highways & Trans.*, 1986, **33**, No. 3, 15–18.

74. HIGHWAYS AGENCY *et al. Manual of Contract Documents for Highway Works, Specification for Highway Works*. TSO, London, 1998, **1**.

75. BRITISH STANDARDS INSTITUTION. *Sampling and Examination of Bituminous Mixtures for Roads and other Paved Areas, Methods of Test for the Determination of Texture Depth*. BSI, London, 2000, BS 598: Part 105.

76. HOSKING J. R. *et al. Measurement of the Macro-Texture of Roads, Part 2: A Study of the TRRL Mini-Texture Meter*. TRL, Crowthorne, 1987, RR 120.

77. SYM R. *Methods of Test for the Determination of Texture Depth*. BSI, London, 1990. Unpublished Report to BSI Committee RDB/36/4 on BS 598: Part 105, Statistician's Report.

78. HILLS J. F. *et al.* The Measurement of Texture Depth under Adverse Conditions. *Highways and Public Works*, 1981, **49**, Nos 1852–3, 8–12.

79. SZATKOWSKI W. S. Rolled Asphalt Wearing Courses with High Resistance to Deformation, The Performance of Rolled Asphalt Surfacings. *Proc. Inst. Civil Eng. Conf., London, 1980*, 107–22.

80. CHOYCE P. W. *et al.* Resistance to Deformation of Hot Rolled Asphalt. *Journal of the Inst. Highways & Trans.*, 1984, **31**, No. 1, 28–32.

81. HIGHWAYS AGENCY *et al. Manual of Contract Documents for Highway Works, Notes for Guidance on the Manual of Contract Documents for Highway Works*. TSO, London, 1998, **2**.

82. BOLK H. J. N. A. *The Creep Test*, SCW Record 5. Study Centre for Road Construction, Arnhem, 1981.

83. BRITISH STANDARDS INSTITUTION. *Sampling and Examination of Bituminous Mixtures for Roads and other Paved Areas, Method for Determination of Resistance to Permanent Deformation of Bituminous Mixtures Subject to Unconfined Uniaxial Loading.* BSI, London, 1995, BS 598: Part 111.

84. COOPER K. E. *et al.* Development of a Simple Apparatus for the Measurement of the Mechanical Properties of Asphalt Mixes. *Proc. 4th EuroBitume Symposium, Madrid, 1989.*

85. BRITISH STANDARDS INSTITUTION. *Method for Determining Resistance to Permanent Deformation of Bituminous Mixtures Subject to Unconfined Dynamic Loading.* BSI, London, 1996, BS DD 226.

86. BONNOT J. *Specification of Resistance to Permanent Deformation in the Fundamental Approach*, Communication to CEN TC227 WG1. LCPC, Paris, 1996.

87. BROWN S. F. and T. V. SCHOLZ. Permanent Deformation Characteristics of Porous Asphalt Determined in the Confined Repeated Load Axial Test. *Highways and Transportation*, 1998, **45**, No. 12, 7–10.

88. ULMGREN N. Functional Testing of Asphalt Mixes for Permanent Deformation by Dynamic Creep Test: Modification of Method and Round Robin Test. *Proc. 1st EAPA/EuroBitume Congress.* European Bitumen Association, Brussels, 1996.

89. COOPER K. E. Permanent Deformation. *Proc. of Workshop on Performance Related Test Procedures for Bituminous Mixtures, Strasbourg, 1996.* Boole Press Ltd, Dublin, 108–23.

90. NUNN M. E. *et al. Assessment of Simple Tests to Measure Deformation Resistance of Asphalt.* TRL, Crowthorne, 1998, unpublished Project Report PR/CE/92/98.

91. BRITISH STANDARDS INSTITUTION. *Methods for Determination of the Indirect Tensile Stiffness Modulus of Bituminous Mixtures*, BSI, London, (Final draft March 1997), BS DD 213.

92. BROWN S. F. Practical Test Procedures for Mechanical Properties of Bituminous Materials. *Proc. of ICE (Transp.)*, 1995, **111**, 289–97.

93. NUNN M. E. *The Characterisation of Bituminous Macadams by Indirect Tensile Stiffness Modulus.* TRL, Crowthorne, 1996, Report 160.

94. NUNN M. E *et al.* Towards a Performance Specification for Bituminous Roadbase. *The Asphalt Yearbook.* IAT, Staines, 1994, 72–7.

95. NUNN M. E. *Road Trials of High Modulus Base for Heavily Trafficked Roads.* TRL, Crowthorne, 1997, Report 231.

96. COMITÉ EUROPÉEN DE NORMALISATION. *Bituminous Mixtures — Test Methods for Hot Mix Asphalt, Stiffness.* CEN, Brussels, 1998, Work Item No. 00 227 132, TC227 WG1 TG2.

97. ROBERTS F. L. *et al. Hot Mix Asphalt Materials, Mixture Design and Construction.* National Asphalt Paving Association Education Foundation, Maryland, 1991.

98. OAKLAND J. S. *Statistical Process Control — A Practical Guide.* Heinemann, London, 1986.

99. NATIONAL ASPHALT PAVING ASSOCIATION. *Statistical Methods for Quality Control at Hot-Mix Plants.* National Asphalt Paving Association, Maryland, 1983, QIP95.

100. BRITISH STANDARDS INSTITUTION. *Guide to Data Analysis and Quality Control using CUSUM Techniques.* BSI, London, 1980–1982, BS 5703: Parts 1–4.

101. BRITISH STANDARDS INSTITUTION. *Guide to Statistical Interpretation of Data*. BSI, London, 1975–1977, BS 2846: Parts 1–8.
102. BRITISH STANDARDS INSTITUTION. *General Requirements for the Competence of Testing and Calibration Laboratories*. BSI, London, 2000, BS EN ISO/IEC 17025.
103. THE ROYAL SOCIETY OF CHEMISTRY. *COSHH in Laboratories*. RSC, London, 1989.
104. BRITISH STANDARDS INSTITUTION. *Quality Systems*. BSI, London, 1994, BS EN ISO 9001, BS EN ISO 9002, BS EN ISO 9003.
105. SECTOR SCHEME ADVISORY COMMITTEE FOR THE QUALITY ASSURANCE OF THE PRODUCTION OF ASPHALT MIXES. *Sector Scheme Document for the Quality Assurance of the Production of Asphalt Mixes*, Sector Scheme No. 14. Quarry Products Association, London, 1998.
106. OSBORNE J. *The Construction Products Directive of the European Communities*. Building Technical File, London, 1989.

10. Specialist surface treatments

10.1. Preamble

This Chapter describes surface dressing, a most useful and cost effective process for restoring skid resistance to a road surface which is structurally sound. Design factors, the available techniques and their applicability, the approach used by the standard UK specification, equipment, failure types and safety considerations all feature in the section.

In addition, the Chapter considers crack sealing/overbanding, slurry surfacing, high friction surfacings and mastic asphalts. The subject of thin surfacings which have become commonplace in road maintenance and improvement schemes is also included.

Nowadays, aesthetic considerations often influence the choice of material and the means of producing coloured surfacings are considered. The future will bring increased use of techniques for the recycling of existing roads and there is a section on this topic.

The recent emergence and prevalence of proprietary materials and processes and the difficulties in using these within a best value regime whilst retaining a competitive environment dictated by European Community membership led to the formation of the Highway Authorities Product Approval Scheme (HAPAS) of materials approval. The evolution, usage and future for the HAPAS are briefly discussed.

10.2. Surface dressing

In Section 3.8 of Chapter 3, the various characteristics of the surface layer of a road are described. It is essential that the Polished Stone Value (PSV) of the aggregate, the surface texture of the road and their relationship with skid resistance are clearly understood to ensure that the surface of the road provides the best possible interaction with the vehicle tyres, and thereby maximizes safety for the road user.

In a new road, the requisite surface texture is designed into the running surface by specifying requirements for both aggregate properties and texture depth. However, during its service life, the surface becomes polished under the action of traffic and the skid resistance will eventually fall below the minimum specified value as fully described in Section 4.10 of Chapter 4. A major benefit of surface dressing is that the process restores skid resistance to a surface which has become smooth under traffic.

Fig. 10.1.   Surface dressing. Photograph courtesy of Colas Ltd

Basic surface dressing consists of spraying a thin film of binder onto the existing road surface, followed by the application of a layer of aggregate chippings as shown in Fig. 10.1. The chippings are then rolled to promote contact between the chippings and the binder and to initiate the formation of an interlocking mosaic.

Surface dressing is well established, economical and a highly effective method of maintaining the surface of a road. It is the most extensively used form of surface treatment. On average about 12 000 miles of road are surface dressed each year in the UK, with local authorities spending approximately 20% of their maintenance budget on surface dressing.[1] The process can, with care, be used on roads of all types from the country lane that carries only an occasional vehicle to trunk roads and motorways carrying tens of thousands of vehicles per day. Road Note 39[2] and the *Code of Practice for Surface Dressing*[3] (the latter is published by the Road Surface Dressing Association (RSDA)) provide a sound basis for selecting the appropriate type of dressing, its design and execution.

Surface dressing as a maintenance process has three prime purposes

- to provide both texture and skid resistance to the road surface
- to arrest the disintegration and loss of aggregate from the road surface
- to seal the surface of the road against ingress of water and thus protect its structure from damage by water.

In addition, it can be used as a treatment to provide

- a distinctive colour to the road surface
- a more uniform appearance to a patched road.

However, surface dressing cannot restore evenness to a deformed road nor will it significantly contribute to the structural strength of the road structure.

10.2.1. Surface dressing specification

In addition to the British Standards, the Highways Agency, the Scottish Executive, the Welsh Office and the Department of the Environment for Northern Ireland jointly produce specifications which relate to surface dressing. *The Manual of Contract Documents for Highway Works* (MCHW)[4] is an all encompassing set of documents which is used for the preparation of contract documents for works of highway maintenance or construction. It is split into seven Volumes numbered 0 to 6.

Volume 1 is entitled the Specification for Highway Works[5] (MCHW 1) and Volume 2 is entitled the Notes for Guidance on the Specification for Highway Works[6] (MCHW 2). These two Volumes are divided into Series which cover particular site activities, e.g. Series 100 deals with Preliminaries, Series 200 Site Clearance, and so on. Volume 2 provides further information and background to the various requirements of Volume 1, the Specification. Series 900 is entitled 'Road Pavements — Bituminous Bound Materials'. Two Clauses are of particular interest to those involved in surface dressing. Clause 919 is entitled 'Surface Dressing: Recipe Specification' whilst Clause 922 is entitled 'Surface Dressing: Performance Specification'.

In the past, individual clauses have tended to call up British Standard tests and then given some guidance as to their use. However, a number of new end performance specifications have recently been developed.

The principal differences between Clause 919 and Clause 922 are that the responsibility for the design of the dressing is transferred from the Highways Agency, etc. to the contractor and performance measurement of the surface dressing is undertaken, measured at time intervals specified in the contract.

10.2.2. Factors that influence the design and performance of surface dressings

The principal factors that influence the design, performance and service life of a surface dressing are

- traffic volumes and speeds
- condition and hardness of the existing road surface
- size and other chipping characteristics
- surface texture and skid resistance
- binder type
- adhesion
- geometry of site, altitude, latitude and local circumstances.

Traffic volumes and speeds

The volume of commercial vehicles per traffic lane per day is directly related to the rate of embedment of chippings into the existing road surface. A 'commercial vehicle' is defined as having an unladen weight of 1.5 tonnes or more for design purposes. As the rate of chipping embedment is dependent on the number of commercial vehicles using each traffic lane, it is necessary to establish the number of such vehicles using the road which is to be surface dressed. Road Note 39[2] specifies various traffic categories. While cars and other light vehicles have little effect on chipping embedment, they do cause stresses when accelerating, braking and turning, particularly at high speed and, for surface dressing design purposes, it is necessary to assess the number of such vehicles using the lengths of road which are to be surface dressed. It is for this reason that Road Note 39[2] splits traffic into two groups, namely those travelling above and those travelling below 100 kph (60 mph).

Condition and hardness of the existing road surface
If the existing road surface is rutted, cracked or in need of patching, these defects must be corrected before surface dressing can be undertaken. Guidance on how this is best achieved can be found in the RSDA's *Guidance Note on Preparing Roads for Surface Dressing.*[7] Other surface conditions which require special consideration are those where the binder has flushed up, resulting in a binder-rich surface and, conversely, surfaces which have become very dry and binder lean. The design process set out in Road Note 39[2] takes these conditions into account when selecting the appropriate type of surface dressing.

The 'hardness' of the road surface affects the extent to which the applied chippings become embedded into the road surface during the life of the dressing. Choice of chipping size is directly related to hardness. The use of chippings which are too small will result in early embedment of the chippings into the surface leading to a rapid loss of texture depth and, in the worst cases, 'fatting up' of binder which may cover the entire surface of the road. The use of chippings which are too large may result in immediate failure of the treatment due to the stripping of aggregate under the applied stresses of the traffic and can also result in excessive surface texture and an increase in the noise generated between the tyre and the road surface. The hardness of the road surface also influences the rate of spread of binder required for a given size of chipping. The rate of spread must be decreased where the road surface is soft in order to compensate for the greater embedment of the chippings into the road surface under the action of traffic.

The categories of hardness of the existing road surfaces for the purposes of surface dressing design are described in Road Note 39.[2] Four graphs take account of the geographical location within the UK and the altitude of the site. The geographical location is not strictly related to latitude but also takes account of climatic conditions partly caused by the Gulf Stream. The method of measuring hardness on site is by use of the Coal Tar Research Association (CTRA) Road Hardness Probe.[8] This device measures the depth of penetration of a probe with a 4 mm semi-spherical head after the exertion of a force of 340 N for a period of 10 s. Ten or more probe tests at approximately 0.5 m intervals in the nearside

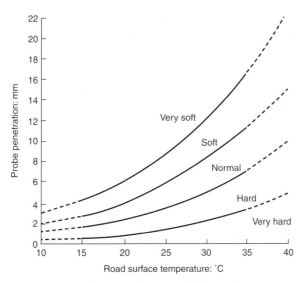

Fig. 10.2. Hardness categories determined by probe penetration and road surface temperature — south category, less than 200 m altitude[2]

wheel track of the traffic lane under consideration are taken and the average is calculated. The road surface temperature is also recorded. Experience has shown that tests are only significant within the temperature range of 15 °C to 35 °C.

Having measured the depth of penetration and the surface temperature and identified whether the site lies in the southern, central or northern region, the hardness category of the road can be established by reference to the appropriate graph, as shown in Fig. 10.2.

Size and other chipping characteristics
Gritstones, basalts, quartzitic gravels and artificial aggregates such as slag constitute the majority of chippings used for surface dressing. The guiding design principles of Road Note 39[2] are to select the size and type of chipping taking into account the following factors

- The size of chipping has to offset the gradual embedment into road surfaces of different hardness caused by traffic.
- Maintaining both microtexture and macrotexture during the life of the surface dressing by selecting a chipping of appropriate size, PSV and Aggregate Abrasion Value (AAV). Additional guidance on the selection of a suitable aggregate to provide appropriate levels of skid resistance for various traffic categories is given in TRL Report LR 510.[9]
- Relating the lane traffic category and the road surface hardness category to the size of chipping required. Figure 10.3 illustrates this relationship for a racked-in surface dressing (discussed in Section 10.2.3 below).

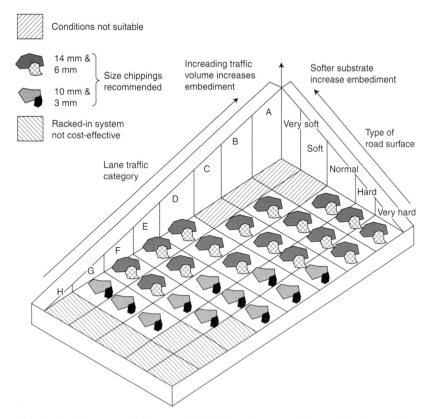

Fig. 10.3. Recommended sizes of chippings for racked-in and double surface dressings[2]

- Ensuring that the rate of spread of chippings is sufficient to cover the binder film after rolling without excess chippings lying free on the road surface.
- Binder type and viscosity (defined as 'the resistance to flow') must be appropriate to the time of year, traffic category and degree of stress likely to be encountered. Similarly, the rate at which it is applied must be appropriate for the traffic category, road hardness and nominal size of chipping.

Chippings for surface dressing should meet the requirements of BS 63: Part 2.[10] Road Note 39[2] suggests that chippings should have a nominal size of 6, 10 or 14 mm. Occasionally, 20 mm chippings are used on roads which are soft and where traffic volumes are high but such conditions are not normally regarded as suitable for surface dressing. Extreme caution should be exercised when using chippings larger than 14 mm nominal size due to the risk of premature failure.

The presence of dust can delay or even prevent adhesion and is particularly acute at low temperatures and with the smaller sizes of

Table 10.1. Typical properties of surface dressing chippings

Property	Value
Aggregate Crushing Value (ACV)	10 to 20
Aggregate Impact Value (AIV)	5 to 20
Polished Stone Value (PSV)	45 to 70
Flakiness Index (FI)	25 max

chipping. As well as being of nominal single size and dust free, surface dressing chippings need to satisfy other characteristics. They should

- not crush under the action of traffic
- not shatter on impact
- resist polishing under the action of traffic
- have a ratio of length to thickness that is not extreme (this factor is known as the 'Flakiness Index' (FI)).

Chippings are rarely cubical and when they fall onto the binder film and are rolled, they tend to lie on their longest dimension. If the chippings used for a surface dressing have a large proportion of flaky chippings, less binder will be required to hold them in place and the excess binder may 'flush up' onto the surface of the dressing. Such a dressing will have a reduced texture and life. It is for this reason that BS 63: Part 2[10] sets a maximum limit of flakiness for single size chippings for use in surface dressing. Typical requirements for surface dressing chippings are shown in Table 10.1.

It must be remembered that not all aggregates with a high PSV are suitable for surface dressing because of strength deficiency. The PSV of the aggregate defines its resistance to polishing and its determination is defined in BS 812: Part 114.[11] The relationship between PSV, traffic and skid resistance is detailed in HD 36/99[12] and discussed in Section 4.13 of Chapter 4.

Coated chippings have a thin film of bitumen applied at an asphalt plant. This bitumen film eliminates surface dust and promotes rapid adhesion to the bitumen. Coated chippings are used to improve adhesion with cutback bitumens, particularly in the cooler conditions which occur at the extremes of the season. They should not be used with emulsions as the shielding effect of the binder will delay 'breaking' of the emulsion as discussed below.

Surface texture and skid resistance
Skid resistance is defined by the Mean Summer SCRIM Coefficient (MSSC) at either 50 km/h or 130 km/h. SCRIM values are measured using the Sideways Force Coefficient Routine Investigation machine. In 1973, the TRRL[9] (now TRL) introduced the concept of 'risk rating'. This is a measure of the relative skidding accident potential at an individual site and is determined by its geometry and location. Within each category, there is a range of target Sideways Force Coefficient (SFC) values with the target value for each site depending on its risk

rating. A road will need to be surface dressed if the measured skid resistance (MSSC value) is inadequate for the risk rating of the site. The 'investigatory levels' of skid resistance for trunk roads including motorways are given in HD 28/94.[13] In this document, the Investigatory Levels are those at which the measured skid resistance of the road has to be reviewed. In the review process, it may be decided to restore skid resistance by surface dressing or, in certain circumstances, the risk rating of the site may warrant change. (This is discussed further in Section 4.10.6 of Chapter 4.)

The skidding resistance of a road surface is determined by two basic characteristics, the microtexture and macrotexture, as shown in Fig. 10.4. Microtexture is the surface texture of the aggregate. Good microtexture is necessary to enable vehicle tyres to penetrate thin films of water and thus achieve dry contact. Macrotexture is the overall texture of the road surface. High macrotexture is necessary to provide drainage channels for the removal of bulk water from the road surface.

TRL Research Report 296[14] considered the relationship between accident frequency and surface texture on roads. On asphalt road surfaces with speed limits greater than 64 km/h and probably on roads with lower speed limits the report concluded the following.

- Skidding and non-skidding accidents, in both wet and dry conditions, are less frequent if the microtexture is coarse.
- The texture level below which accident risk begins to increase is a sensor measured texture depth (SMTD, see later) of around 0.7 mm.
- All major types of surfacing contribute to texture depths across the practical range.
- Macrotexture has a similar influence on accidents whether they occur near hazards such as junctions or elsewhere.

Microtexture is the dominant factor in determining the level of skid resistance at speeds up to 50 km/h. Thereafter, macrotexture predominates, particularly in wet conditions. Traditionally, macrotexture has been measured using the Sand Patch Method.[15] A standard volume of a specified type of sand is spread in a circular motion on the road surface. On newly laid surfaces, the test is carried out on sections of length 50 m along a diagonal line across the carriageway lane width. Ten measure-

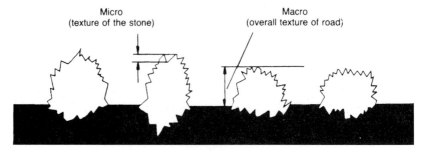

Fig. 10.4.   Microtexture and macrotexture[2]

ments are taken on each section at approximately 5 m intervals. The texture depth can then be calculated by dividing the volume of sand by the cross sectional area of the patch. The test cannot be carried out on a wet or sticky surface. Full details on method and application are given in BS 598: Part 105.[15]

Mature surface dressings in good condition typically have texture depths between 1.5 mm and 2.0 mm. Thus, a high-speed skid resistance of the surface dressing is maintained throughout the life of the surface dressing.

In recent years, there has been a move towards laser texture measurement. Technically, this is described as the sensor measured texture depth (SMTD) and is illustrated in Fig. 10.5. This method is quicker but is not directly comparable with sand patch measurements and, for this reason, the method of measurement should be stated when a texture measurement is quoted. Section 9.4.6 of Chapter 9 provides further information on texture depth, the Sand Patch Test and SMTD.

It is the petrographic characteristics of the aggregate which largely influence its resistance to polishing and this property is measured using the PSV Test.[11] This test subjects the aggregate to simulated trafficking in an Accelerated Polishing Machine. A high value of PSV indicates good resistance to polishing. The majority of road aggregates have a good microtexture prior to trafficking and, consequently, most road surfaces have a high skid resistance when new, particularly after the bitumen has worn off the aggregate. The time required for the bitumen to be abraded from the aggregate at the road surface depends upon a number of factors, including the type of mixture and in particular the bitumen film thickness. Until this has been removed by traffic the microtexture of the aggregate will not be fully exposed to the vehicle tyre. The action of tyres removes the binder and the microtexture is exposed. The exposed aggregate

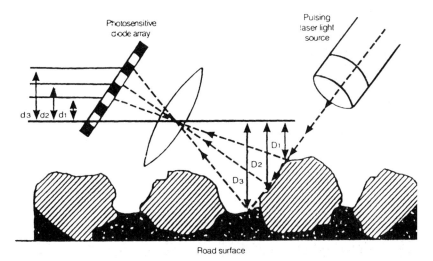

Fig. 10.5.   Laser measurement of texture depth[16]

Table 10.2. Relationship between texture depth and reduction in BFC[9]

Texture depth: mm	Percentage reduction in BFC: %
0.5	30
1.5	10
2.0	Negligible

surfaces become polished and, within a year, the skid resistance normally falls to an equilibrium level.

There are many sources of high PSV aggregate in the UK, mostly gritstones. Unfortunately, many aggregates have poor abrasion retention or a low resistance to weathering and, consequently, few aggregates which have a high PSV are suitable for surface dressing. As a result, blanket specifications for all sites are wholly inappropriate. They waste valuable resources and more thought should be given to the nature of individual sites to enable more abundant lower PSV aggregates to be used effectively. Catt[17] developed a rational approach to the selection of suitable aggregate types. The method takes into account all the factors relating to skid resistance and can demonstrate the suitability of aggregates which have previously been considered inappropriate.

The macrotexture is determined by the nominal size of aggregate used. However, resistance to abrasion is also important. Aggregates with poor abrasion resistance, determined by the AAV Test,[18] will quickly be worn away with a consequent loss of macrotexture. HD 36/99[12] specifies maximum values of AAV for various traffic levels at design life.

The rate at which skid resistance falls off with increasing vehicle speed is influenced by the macrotexture. The TRL[9] has shown that there is a correlation between macrotexture, expressed in terms of average texture depth, and skid resistance, measured by percentage reduction in Braking Force Coefficient (BFC) between 130 km/h and 50 km/h, as shown in Table 10.2.

The skid resistance which is necessary at any site will depend on the stresses likely to be induced. Sections of road with minor road junctions and with low traffic volumes are unlikely to require a high resistance to skidding whereas sharp bends on roads carrying heavy volumes of traffic will require significantly higher skid resistance characteristics. It is also important to recognize that the PSV of aggregate required to maintain any particular resistance to skidding will be higher for roads carrying high volumes of traffic than for sites which are less heavily trafficked.

Table 4[†] of Road Note 39[2] indicates the Highways Agency's assessment of the relationship between site category, traffic and PSV. It

---

[†] Table 4 purports to be a reproduction of that which is contained in the DMRB.[19] At the time of publication of Road Note 39[2] (1996), this table could be found in HD 28/94[13] but it is currently contained in HD 36/99[12] (DMRB 7.5.1) as Table 3.1. The Table has been altered and, hence, the latter differs from that in Road Note 39.[2] In most circumstances, it is likely that the version in the DMRB will take precedence. Accordingly, as stated in Road Note 39, users should check to see whether an updated version has been published. It is understood that the current version is unlikely to change in the foreseeable future.

can be seen from this that PSVs of 65 and below are sufficient for the majority of the trunk road network but, under some extreme conditions, aggregate with a PSV in excess of 70 are required. (There are very few sources of aggregate in the UK with a PSV of 70 or more — a number exist in Cumbria and Wales. Where a racked-in dressing system is used, the smaller chippings need not be of the same PSV as the larger aggregate as it is only the latter that will come into contact with vehicle tyres. Further information on skid resistance is available in Sections 3.9 and 4.10 of Chapters 3 and 4 respectively and in other publications.[20,21]

Binder type
The essential function of the binder is to seal surface cracks and provide the initial bond between the chippings and the road surface. The viscosity of the binder must be such that it can wet the chippings adequately at the time of application, prevent dislodgement of chippings when the road is opened to traffic, not become brittle during periods of prolonged low temperature and function effectively for its design life.

The subsequent role of the binder is dependent on the traffic stresses applied to the surface dressing from low-stressed lightly trafficked country roads, to highly stressed sites carrying large volumes of traffic. Clearly a range of binders is necessary if optimal performance is to be achieved under these widely differing circumstances. Three broad ranges of surface dressing bitumen binders are available

- traditional cutback and emulsified bitumens
- epoxy resin thermoset binders
- proprietary polymer modified bitumens (hot applied and emulsions).

Traditional emulsified and cutback bitumens
Bitumen emulsions. These are produced to BS 434,[22] the principal binder being K1-70 bitumen emulsion, which is a cationic emulsion containing approximately 29% water, 70% of 200 pen or 300 pen bitumen to which, approximately, 1% by weight of kerosene has been added. The relatively low viscosity of bitumen emulsion enables them to be sprayed at 75 °C to 85 °C through either swirl or slot jets (see Section 10.2.5 below). Bitumen emulsion is now, by far, the most popular type of binder used in the UK. There are two main reasons for this.

- The UK climate is such that road surfaces and chippings are often damp. Even under these conditions, it is possible to get a good bond between the road surface and the chippings in circumstances where the use of cutback binders would be unwise.
- They are sprayed at much lower temperatures than cutback binders. This reduces the cost of heating and, more importantly, lessens the risk of injury if operatives accidentally come into contact with the binder.

K1-70 emulsions will flow on road surfaces at the time of spraying because they have a relatively low viscosity. Whilst this is helpful in some respects, migration of the binder from the centre to the side of cambered roads, or downhill on gradients, requires the application of chippings to the binder film as quickly as possible after spraying. The loss of around 30% of the water from the emulsion, which is known as 'breaking', can take from as little as 10 minutes to over an hour depending on ambient temperature, humidity, aggregate water absorption, wind speed and the extent to which the dressing is subject to slow moving traffic. Humidity is one of the most important of these factors and the use of emulsions when humidity is above 80% should only be undertaken on very minor roads and where traffic speeds can be kept within the range of 10 mph to 20 mph until the emulsion has fully broken. Precoated chippings should not be used with emulsions as the coating delays breaking and serves no tangible benefit but carries additional cost.

Binder manufactures are able to control the rate at which emulsions break within certain limits, but these have to be compatible with the storage, transport and application of the emulsion. On site, breaking can be accelerated by spraying the binder film with a mist of a breaking agent before the application of the chippings. However, this usually requires specially adapted spray bars and the fitting of storage tanks for the breaking agent.

Cutback bitumens. Cutback bitumen for surface dressing is usually 100 pen or 200 pen bitumen which has been diluted with kerosene to comply with BS 3690[23] to produce a Standard Tar Viscometer (STV) viscosity of either 50, 100 or 200 seconds. Most cutback bitumen surface dressing in the UK is undertaken with 100 second material. The suffix X on cutback bitumens, e.g. 100X, indicates that they have been doped with a specially formulated thermally stable passive adhesion agent. This additive assists wetting of the aggregate and the road surface during application and resists stripping of the binder from the aggregate in the presence of water. The effectiveness of this additive can be demonstrated by either the Immersion Tray Test or the Total Water Immersion Test.[16] If surface dressing is being carried out in marginal weather conditions or if the aggregate is wet, it is recommended that an active wetting agent is added to cutback bitumens immediately prior to spraying.

Cutback bitumens are typically sprayed at temperatures in the range of 130 °C to 170 °C. During spraying, between 10% and 15% of the kerosene evaporates and a further 50% dissipates from the surface dressing in the first few years after application. At these temperatures, wetting of the road surface and the applied chippings is achieved rapidly. Cutback bitumens once applied to the road surface cool quickly and it is important to ensure that the chippings are applied as quickly as possible to ensure satisfactory wetting of the chippings.

During the surface dressing season, as temperatures rise, binder suppliers may recommend the use of different viscosity cutback binders in order to promote adhesion and avoid the fatting-up of new dressings. Both cutback and emulsion manufacturers may also make minor changes to their formulations within the ranges covered by the British Standards to meet changes in road and ambient temperatures.

Table 10.3. Advantages and disadvantages of cutback binders

Advantages	Disadvantages
Rapid adhesion to the chippings as soon as the binder cools	High spraying temperatures compared to emulsions
Reduced risk of brittle failure due to the presence of kerosene in the binder	Possibility of bleeding in early life due to the presence of kerosene in the binder

Comparison of cutback and emulsion surface dressing binders
The advantages and disadvantages of cutback binders are listed in Table 10.3. and the advantages and disadvantages of emulsion binders are listed in Table 10.4.

Modified binders
The relentless increase in the volume of all types of vehicles on roads in the UK has led to the development of modified binders with improved adhesivity and greater cohesive strengths than traditional cutback and emulsion binders. These characteristics of modified binders are particularly important immediately after chippings have been applied to the binder film, when they are less likely to be plucked from the road surface by passing traffic than when using unmodified binders. As well as greater adhesive and cohesive strength, modified binders perform satisfactorily over a greater temperature range than traditional binders. This means that they do not flush up as rapidly in hot conditions as traditional binders and are not subject to the same degree of brittleness in winter conditions.

Epoxy resin thermoset binders.   These are bitumen-extended epoxy resin binders used with high PSV aggregates such as calcined bauxite. Epoxy resin based binders are used in high-performance systems such as Spray Grip which are fully capable of resisting the stresses imposed by traffic on the most difficult sites, e.g. roundabouts or approaches to traffic lights and

Table 10.4. Advantages and disadvantages of emulsion binders

Advantages	Disadvantages
Low spraying temperature	Slow initial adhesion pending evaporation of water in the mixture, i.e. breaking
Effective on damp roads (but not wet roads)	Possibility of brittle failure during the first winter due to insufficient embedment of the chippings
Very low risk of bleeding in early life	The binder may 'skin' in hot weather due to water being trapped within the binder film*

* This problem is overcome by ensuring that only slow moving traffic passes over the newly laid dressing

designated accident black spots. It is these systems which are often described as Shellgrip but are more correctly generically described as high friction surfacings (see also Section 10.5 below). Binders used in such systems are classified as 'thermosetting', as the epoxy resin components cause the binder to cure by chemical action and harden and it is not subsequently softened by high ambient temperatures nor by spillage of fuel. The dressing thus acts as an effective seal against the ingress of oil and fuel, which is particularly important on roundabouts where spillages regularly occur. Very little embedment of the chippings into the road surface takes place with this type of binder and the integrity of the surface dressings is largely a function of the cohesive strength of the binder.

The extended life of this binder justifies the use of a durable aggregate with an exceptionally high PSV. The initial cost of this surface dressing system is high compared with conventional binders. However, its exceptional wear-resistant properties and the ability to maintain the highest levels of skid resistance throughout its life make it a cost effective solution for very difficult sites by significantly reducing the number of skidding accidents. Recent statistics show the cost to the community of a serious accident to be £124,600 and a fatal accident to be £1,042,400.[24]

Proprietary polymer modified binders.   Figure 10.6 shows a relationship between binder type and performance requirements.[25] Precise categorization is not possible, however, since there is a significant variation in severity along the length of every class of road which is not solely dependent on traffic numbers and vehicle weight.

Over 80% of surface dressing sites can be regarded as lightly trafficked and lightly stressed. Conventional cutback or emulsion binders which comply with appropriate standards are normally perfectly adequate for this category of road unless it carries high volumes of commercial vehicles.

Some 15% of sites, such as straight sections of motorways, other trunk and principal roads carrying more than 250 commercial vehicles per lane per day, are considered to be of average traffic loading. In practice, for the majority of work carried out on trunk roads including motorways and many principal roads, modified binders are now used in view of the reduced risk of early life failure, greater cohesive strength and the increased temperature range within which modified binders perform satisfactorily.

Some 4% of sites are regarded as 'difficult'. These sites include roundabouts, bends and heavy rolling sites, some of which will require the use of thermosetting binders. Approximately 0.1% of sites, such as the approaches to pedestrian crossings and traffic signals where traffic speeds exceed 40 mph, are regarded as 'very difficult'. At such locations, thermosetting or epoxy resin binders are often appropriate.

A high proportion of the road network can be classified as easy or average and these roads can largely be surface dressed using bitumen emulsion or cutback bitumen, although it may be necessary on certain highly stressed sections to consider a more enhanced treatment, i.e. at

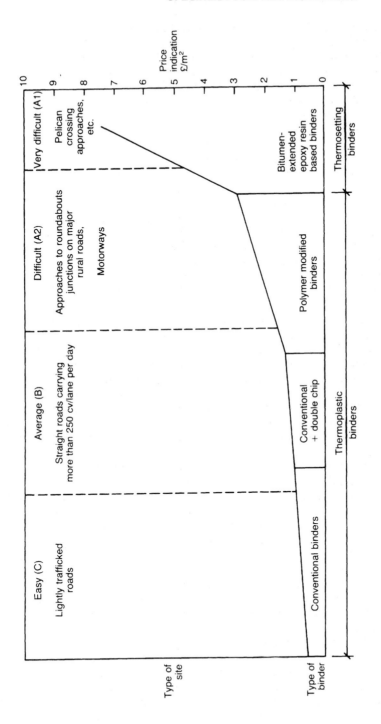

Fig. 10.6.   Relationship between binder type and performance requirement[25]

sharp bends, road junctions and in shaded areas under trees and bridges. At these locations, it is probable that the same binder can be used but the specification will be changed in respect of one or more of the following

- rate of application of binder
- chipping size and rate of application
- double chip application
- double binder, double chip application.

These latter techniques are widely practised in France on relatively heavily trafficked roads, and with correct procedures their use can significantly reduce the risk of flying chippings and consequential windscreen breakage. Conventional emulsions and cutbacks have the lowest cost and with careful workmanship, compliance with specification and appropriate aftercare, good results can be obtained.

As a further aid to consistent performance under the variable weather conditions commonly experienced in the UK, a number of proprietary adhesion agents are available which act in two basic ways. Some types ('active') modify surface tension at the stone/binder interface to improve the preferential wetting of chippings by the binder in the presence of water. Other types ('passive') improve the adhesion characteristics of the binder and give an improved resistance to the subsequent detachment of the binder film in the presence of water throughout the early life of the dressing. The effectiveness of the former type of adhesion agent reduces after a few hours at elevated temperature and so it is generally mixed with the binder a few hours before spraying. The latter type of agent is more temperature stable and is generally added to the binder by the supplier.

At the top end of the performance scale, bitumen-extended epoxy resin binders have been developed that are fully capable of resisting the very high stresses imposed by traffic at the most difficult sites. Thermosetting binders are also available with a reduced resin content and these are less costly. The curing time to achieve full performance is increased but their use can be considered where the stresses are slightly less severe.

The middle ground between lightly and severely stressed sites is the most difficult to specify with certainty and it has taken time to develop appropriate binders for such applications. Employers have been reticent to pay more for binders to use on site where, under ideal conditions, a conventional unmodified bitumen 'might just have worked'. It should be remembered that the cost of failures is high in both monetary and potential accident terms. Failures also reduce client and public confidence in surface dressing.

Two distinct considerations should be borne in mind when preparing the specification for such sites. These are to provide a surface dressing which

- is more tolerant of difficult conditions such that it will always give results at least equal to the best obtainable with conventional binders
- will enable surface dressings to be used on sites where conventional binders would normally fail.

Types of additive

Bearing these performance requirements in mind, it is possible to examine the ways in which the various polymers can be used with bitumen to modify its performance. The properties of bitumen in which improvement would be sought through the addition of a modifier are set out below.

- Reduced temperature susceptibility can loosely be defined as the extent to which a binder softens over a given temperature range. It is clearly desirable for a binder to exhibit a minimum variation in viscosity over a wide range of ambient temperatures.
- Cohesive strength enables the binder to hold chippings in place when they are subject to stress by traffic. This property can also be coupled with the elastic recovery properties of the binder which enable a surface dressing to maintain its integrity even when it is subjected to high levels of strain.
- Adhesive power or 'tack' is a difficult property to define but it is an essential one. This is especially true in the early life of a surface dressing before any mechanical interlock or embedment of the chippings has taken place.

In addition, it is desirable for the modified binder to

- maintain its premium properties for long periods in storage and then during its service life
- be physically and chemically stable at storage, spraying and service temperatures
- be applied using conventional spraying equipment
- be cost effective.

Many polymers have been used in surface dressing binders over a number of years with varying degrees of success. These are now discussed.

Rubbers.   Polybutadiene, polyisoprene, natural rubber, butyl rubber, chloroprene, random styrene/butadiene rubber and ethylene propylene diene monomer (EPDM) have all been used with bitumen and their main effect is to increase viscosity with only a small effect on elasticity. In some instances, the rubbers have been used in a vulcanized ('cross-linked') state but this is difficult to disperse in bitumen requiring high temperatures, long digestion times and can result in a heterogeneous binder with the rubber acting mainly as a flexible filler.

Thermoplastics.   Polyethylene, polyvinyl chloride, polystyrene, and ethylene vinyl acetate (EVA) are the principal thermoplastic polymers that have been tried as bitumen modifiers. Their behaviour is similar to that of bitumen, i.e. they soften when heated and they harden when cooled. They can be blended with bitumen at elevated temperature using comparatively low-shear mixers and mainly increase the viscosity of the binder and stiffness at ambient temperature. They do not enhance the elastic properties of the binder and tend to separate in storage which can lead to uneven dispersion of the polymer in the modified bitumen.

Thermoplastic rubbers. These polymers combine both elastic and thermoplastic properties. They consist of links of block copolymers of styrene and butadiene coupled in pairs to give di-blocks or star configurations. These block copolymers are more commonly known as styrene – butadiene – styrene (SBS) polymers and have been used for many years to modify bitumen to provide high-performance coatings for roofing membranes and for a range of road applications including surface dressing.

The unique properties of thermoplastic rubbers are thermoplastic behaviour at elevated temperatures, coupled with a rubbery type behaviour at ambient temperature. Their strength is provided by the hard polystyrene end blocks which associate in domains linked by butadiene chains to form a three-dimensional network. On heating to above 100 °C, the polystyrene softens, the domains disassociate and the polymer flows. The process is completely reversible and the domains reform on cooling. The viscoelastic nature of normal bitumen and its susceptibility to temperature and loading time are well known. This susceptibility is reduced by the addition of thermoplastic rubbers and the extent of the change is closely related to polymer concentration.

Substantial changes in viscosity, elasticity and temperature susceptibility can be obtained with about 3% of SBS in the bitumen providing the correct base bitumen is used. It is necessary to raise the polymer content to about 5% to 6% before a full three-dimensional network is established and the maximum improvements in elasticity are achieved.

Improved initial tack can be obtained by careful selection of polymers generally of the styrene–isoprene–styrene (SIS) type which are used in combination with SBS polymers in proprietary binders such as Shelphalt SX.

The choice of either an emulsion or cutback formulation for modified binders is not an easy one. Generally speaking, emulsions are more vulnerable during the early life of the dressing before the emulsion has fully broken. The rate at which breaking occurs is influenced by a number of factors including emulsion formulation, weather and mechanical agitation during rolling and trafficking. As discussed earlier, under certain conditions the emulsion goes through a 'cheesy' state during cure in which the bitumen droplets have agglomerated but not coalesced and the dressing is extremely vulnerable. The timing of this condition is variable and is generally worsened by the presence of high polymer contents which can cause skinning and retard breaking of the emulsion. It is essential that the emulsion is completely broken before uncontrolled traffic is allowed on the dressing. This can be a serious disadvantage on heavily trafficked roads and where work is executed under lane rental contracts. Chemical after-treatment sprays have been shown to be effective in promoting breaking.

Cutback bitumens have benefits in terms of immediate cohesive grip of the road and chippings but generally the residual binder properties are more variable due to the differences in the evolution of the diluent and its absorption into the road surface. Nevertheless, cutback formulations are the preferred choice for high-speed and heavily trafficked roads because of their predictable early life performance.

The development of polymer modified surface dressing binders to complement conventional and epoxy resin systems offers the highway engineer a range of binders suitable for all categories of site. Regardless of the improved properties of a particular binder, it is the performance of the aggregate binder system and application mode which dictate the success or otherwise of the dressing. Further details on bitumen additives are contained in Chapter 2.

Adhesion

Some aggregates adhere more rapidly to the binder than others and it is wise to avoid problems by testing the compatibility of the selected binder and chippings. When using cutback binder, this is checked using the Total Water Immersion Test.[16] In this test, clean 14 mm chippings are totally coated with cutback bitumen. After 30 minutes curing at ambient temperature, the chippings are immersed in demineralized water at 40 °C for 3 h. After soaking, the percentage of binder which is retained on the chipping is assessed visually. The extent to which the base of the chipping has been coated with the binder after 10 minutes is then assessed visually. If the coated area is below about 70%, serious consideration should be given to lightly precoating the chippings, as it is much easier to obtain a bond between the binder film and a light coating of binder on the chippings than is the case with uncoated chippings.

The amount of binder required to lightly coat surface dressing chippings varies from one type and size of chipping to another but it is typically around 1% by weight. A thick coating of binder, e.g. as used for coating chippings for rolling into the surface of low stone content asphalt such as HRA wearing course, will make the chippings sticky and will inhibit free flow through chipping machines. When lightly coated chippings are used, this free flowing property is critical to a satisfactory surface dressing. Minor pinholes in the coating of the chippings are not detrimental. Chippings should meet the requirements of BS 63: Part 2[10] prior to the application of a light coating. It is particularly important that the requirement that there shall not be more than 1% of material passing a 75 micron sieve from a sample of chippings is met. Although the British Standard does not specify any requirements concerning the grading of this fraction, it is suggested that even 0.5% of material passing 63 microns (normally regarded as silt) can represent a serious impediment to good chipping/binder adhesion for both clean and lightly coated chippings before coating. Once the chippings have been lightly coated, it is important that they are not allowed to come into contact with dust as this will impair the bond between the chippings and the applied binder film on the road surface.

Geometry of site, altitude and latitude and local circumstances

Sections of road which include sharp bends induce increased traffic stresses on a surface dressing. Two categories of bend are identified in

Road Note 39^2 — less than 250 m radius and less than 100 m radius. As discussed earlier, altitude and location within the UK are taken into consideration when determining the road hardness. Gradients steeper than 10% will affect the choice of binder rate of spread due to the fact that on the ascent side of the road traffic loading will be present for longer than on the descent side. Typically there is a difference of $0.2 \, l/m^2$ of binder between the ascent and the descent sides.

10.2.3. Types of surface dressing

The original concept of a normal single-layer surface dressing has been developed over the years and there are now a number of techniques available which vary according to the number of layers of chippings and applications of binder. Each of these techniques has its own particular advantages and associated cost and it is quite feasible that along the length of a road a variety of techniques would be used depending on the stresses at any particular feature. Figure 10.7 shows the types of technique available.

Single surface dressing

The single surface dressing system is still the most widely used and cheapest type of surface dressing. It consists of one application of binder followed by one application of single size chippings. Its advantages are that it has the least number of operations, uses the least amount of material and is sufficiently robust for minor roads and some main roads free from areas where braking and acceleration are likely to occur and where speeds are unlikely to exceed 60 mph.

Racked-in surface dressing

This type of dressing uses a single application of binder. The 10 mm or 14 mm diameter principal chippings are spread at approximately 80% to 90% cover which leaves a window of binder between the chippings. This window is filled with small chippings of 3 mm or 6 mm respectively to achieve high mechanical interlock of the principal chippings. The rate of spread of binder on this type of surface dressing is usually slightly higher than would be required for a single size dressing.

The initial surface texture and mechanical strength of a racked-in dressing are high. The configuration of the principal chippings is different from that achieved with a single dressing. The interlocking mosaic, referred to earlier, cannot be formed because the principal chippings are locked in place. In time, small chippings not in contact with binder are lost to the system without damaging vehicles or windscreens, resulting in increasing surface texture despite some coincident embedment of the larger chipping.

On a racked-in dressing vehicle tyre contact is principally on the large aggregate with little contact on the small aggregate. Thus, the micro-texture of the small aggregate is less critical and therefore consideration should be given to the use of cheaper lower PSV aggregates such as limestone for the small chippings.

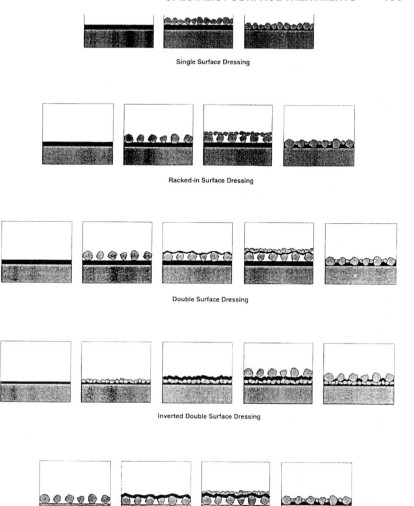

Single Surface Dressing

Racked-in Surface Dressing

Double Surface Dressing

Inverted Double Surface Dressing

Sandwich Surface Dressing

Fig. 10.7.   Schematic representation of different types of surface dressing prior to embedment[3]

The advantages of the racked-in system are

- virtual elimination of 'flying' 14 mm or 10 mm chippings, thereby reducing the incidence of broken car windscreens
- early stability of the dressing through good mechanical interlock
- good adhesion of larger size chippings
- a rugous surface texture with an initial texture depth exceeding 3.0 mm.

Since the late 1980s there has been a marked increase in the use of racked-in dressing on main and heavily trafficked roads.

Double surface dressing
In this technique, two surface dressings are laid consecutively, the first consisting of 10 mm or 14 mm chippings and the second of 6 mm chippings. It is used on main roads as an alternative to the racked-in system to provide a mechanically strong dressing with a texture marginally less than a racked dressing, and therefore quieter. The method is also suitable for minor roads which have become very dry and lean in binder.

Inverted double surface dressing
In this system, two surface dressings are carried out consecutively. The first uses 6 mm chippings and this is followed by a second dressing using 10 mm or 14 mm chippings. This technique is appropriate for use on minor concrete roads, where chipping embedment does not occur and surface texture is not an important issue, or on sections of road which have been widely patched.

Sandwich surface dressing
Sandwich surface dressing is used in situations where the road surface condition is very rich in binder. The first layer of 14 mm or 10 mm chippings is spread onto the existing road surface before any binder is applied. The binder is then sprayed over these chippings followed by an application of 6 mm chippings.

Sandwich dressings can be considered as double dressings in which the first binder film has already been applied. The degree of binder richness at the surface has to be sufficient to hold the first layer of chippings in place until the rest of the operation has been completed. The chipping sizes need to be chosen to reflect the quantity of excess binder on the surface and the rate of spread of the second binder application.

Pad coats
A pad coat consists of a single dressing with small sized chippings. It is applied to a road which has uneven surface hardness possibly due to extensive patching by the utilities. The pad coat is used to provide a more uniform surfacing which can subsequently be surface dressed. The chipping size for a pad coat is traditionally 6 mm with a slight excess of chippings. Pad coats can also be used on very hard road surfaces, such as concrete or heavily chipped asphalts, to reduce the effective hardness of the surface. However, using a racked-in system with a polymer modified binder is now the preferred option.

After laying and compaction by traffic, excess chippings should be removed before the road is opened to unrestricted traffic. Pad coats may be left for several months before application of the main dressing which may be either a single dressing or a racked-in system. Either system will embed into the pad coat and have immediate significant mechanical strength. This reduces the risk of failure. All loose material should be swept from the surface of the road prior to the application of the final dressing.

10.2.4. The Road Note 39 design approach

The design principles contained in Road Note 39[2] are as follows

- select the most suitable type of dressing for the site conditions
- identify the chipping size and basic binder rate of spread
- refine the binder rate of spread to match the source of chippings selected
- further refine the binder rate of spread on site immediately before work commences.

Road Note 39[2] is set out in such a way that the rate of spread of binder expressed in l/m^2 is only obtainable after the above stages have been completed. This is one of the major differences between the 3rd and 4th Editions of Road Note 39 and was taken in the light of experience which suggested that the refining steps described above were not being undertaken and that, in consequence, some binder rates of spread were seriously inadequate leading to premature failure. Tables in Road Note 39[2] provide coefficients for binder rates of spread for single surface dressings and racked-in surface dressings. These coefficients are then adjusted by reference to tables which take into account the type of aggregate and aggregate shape, road surface condition, gradient, the degree to which the site is shaded by trees and structures and for sites where traffic volumes are exceedingly light. The coefficients are then translated into actual target rates of spread of binder in l/m^2.

10.2.5. Surface dressing equipment

Improved design of surface dressings and advances in binder technology and aggregate quality have been major contributory factors in improving the performance of surface dressings. Advances in the design of application machinery in concert with the above have enabled the whole process to be coordinated in terms of the speed of operation, compliance with specification and continued improvements in standards of safety.

Sprayers

Surface dressing sprayers must satisfy the requirements of BS 1707.[26] This Standard sets requirements for evenness of distribution of the spray bar, heat retention, capacity of heaters and other important issues.

In the UK, it has been traditional for spray bars to have swirl jets. However, the increased use of polymer modified binders has resulted in some sprayers being equipped with bars having slot jets. The need to undertake work quickly on main roads with the minimum number of longitudinal joints has led to the use of expanding spray bars of up to 3.8 m. In the UK, the usual way of checking that the rate of spread across the bar is uniform within 15% or better, has been by the Depot Tray Test which is detailed in BS 1707.[26] This is normally checked before the start of the surface dressing season.

Figure 10.8 shows a modern surface dressing sprayer. Programmable bitumen distributors are being developed. These are extremely sophisticated spray machines using two bars to achieve differential rates

Fig. 10.8.   Modern surface dressing sprayer. Photograph courtesy of Colas Ltd

of spread across the width of the machine. Application rates can be changed for areas of patching, wheel tracks, binder-rich fatty areas and binder-lean areas. The rate of spread is automatically reduced by up to 30% in the wheel tracks and can be increased between wheel tracks or in shaded areas. Both longitudinal and transverse binder distributions can be pre-programmed. Vehicle speed, binder temperature, spray bar width and application rates are controlled by a microprocessor.

Chipping spreaders
Figure 10.9 shows an example of a self-propelled chipping spreader. This type of vehicle is now in regular use. These machines not only allow work to proceed more rapidly but also spread the chippings more evenly and accurately at the required rate of spread. They also have the advantage that the chippings are released much closer to the road surface. As a result, the likelihood of chippings bouncing either elsewhere on the road surface or off the road surface altogether and exposing binder is reduced. These chippers are also intrinsically safer because they are driven forward rather than reversing as is the case with traditional lorry-mounted spreaders.

The latest developments in self-propelled chipping spreaders include

- four wheel drive for pulling heavy tippers up steep inclines without judder
- a mechanism for breaking lumps of lightly coated chippings within the hopper
- methods for spraying additive onto the binder film
- incorporation of a pneumatic tyred roller to ensure that the chippings are pressed into the binder film at the earliest possible moment while it is still hot.

Fig. 10.9.   Self-propelled chipping spreader. Photograph courtesy of Colas Ltd

Rollers

Rubber-covered steel-drummed vibratory rollers are now regarded as the best method of establishing a close mosaic and ensuring initial bond of the chipping to the binder film without crushing the carefully selected aggregate. Pneumatic tyred rollers are also used and some rolling is still carried out using steel wheeled rollers. One of the risks of using this type of roller is that chippings can be crushed as a result of point loading. It should be noted that rolling is largely a preliminary process undertaken before the main stabilization of the dressing by subsequent slow-speed trafficking.

Sweepers

It is necessary to remove surplus chippings before a surface dressing is opened to traffic travelling at uncontrolled speed. This is normally achieved with full width sweepers which often have a full width sucking capability. Where sections of motorways or other major roads have been closed to allow surface dressing to take place, sweepers are often used in echelon in order to ensure the rapid removal of surplus chippings. Sweeping which is properly organized and implemented will do much to alleviate the damage caused to surface dressings in their early life through chipping loss.

Traffic control and aftercare

Traffic control and aftercare are vital to the successful performance of the surface dressing. Ideally, when traffic is first allowed onto the new dressing, it should be behind a control vehicle which travels at a speed of about 15 mph. If there are gaps in the traffic, it may be necessary to introduce additional control vehicles. The objectives should be to ensure that vehicles do not travel at more than 15 mph on the new dressing and to prevent sharp braking and acceleration of these vehicles.

The length of time for which control is necessary will depend largely on the type of binder used and the prevailing weather conditions. It is likely to be longest where emulsions are being used and the weather conditions are humid. The use of wet or dusty chippings or the early onset of rain delays adhesion and necessitates a longer period of traffic control. Traffic passing slowly over a dressing immediately after completion creates a strong interlocking mosaic of chippings, a result only otherwise obtained by long periods of rolling.

10.2.6.   Types of surface dressing failures
The majority of surface dressing failures fall into one of the following five categories

- loss of chippings immediately or soon after laying
- loss of chippings during the first winter
- loss of chippings in later years
- bleeding during the first hot summer
- fatting-up in subsequent years.

The type and rate of application of the binder and size of chippings has a large influence on performance. Incorrect application rates can result in failures and this is a frequent cause of premature failure. Every care should be taken to ensure that the selected application rates are maintained throughout the surface dressing operation.

Very early loss of chippings may be due to slow breaking in the case of emulsions or poor wetting in the case of cutback bitumens. The use of cationic rapid setting emulsions with high binder contents (more than 70% binder) largely overcomes the former problem and the use of adhesion agents and/or precoated chippings should negate the latter difficulty.

Many surface dressing failures are first apparent at the onset of the first prolonged frosts when the binder is very stiff and brittle. These can normally be attributed to a combination of inadequate binder application rate, the use of too large a chipping or inadequate embedment of the chippings into the old road due to the surface being too hard and/or there being insufficient time between applying the dressing and the onset of cold weather. Inadequate binder application and/or the use of excessively large chippings will exacerbate this problem. Loss of chippings in the long term usually results from a combination of low-durability binders and, again, poor embedment of chippings.

Bleeding may occur within a few weeks in dressings laid early in the season or during the following summer in dressings which were laid late in the previous year. Bleeding results from the use of binders which have a viscosity which is too low for the high ambient temperatures or from the use of binders which have a high temperature susceptibility.

Fatting-up is complex in nature and may result from one or more of the following factors.

- *Excessive binder application rate.*
- *Embedment of chippings.* Gradual embedment of chippings into the road surface causing the relative rise of binder between the chippings to a point where the chippings disappear beneath the surface. This is largely dependent on the intensity of traffic particularly the proportion of heavy commercial vehicles in the total traffic. The composition of the binder also affects the process. Cutback bitumens which have a high solvent content can soften the old road surface and accelerate embedment.
- *Crushing of chippings.* Most chippings will eventually split or crush under heavy traffic with the loss from the dressing of many small fragments. The binder-to-chipping ratio therefore tends to increase thus adding to the process of fatting-up. A dilemma here is that many of the best aggregates for skid resistance often have a poor resistance to crushing.
- *Absorption of dust by the binder.* Binders of high durability tend to absorb dust that falls upon them. Soft binders can absorb large quantities of dust thereby increasing the effective volume of the binder. This effect, coupled with chipping embedment, leads to the eventual loss of surface texture.

10.2.7.   Surface dressing season

It is clear from the above that the success of a surface dressing operation may well depend on the weather conditions during and following the application of the dressing. Generally speaking, surface dressing is undertaken from May to September in the UK. This varies throughout the country and there are areas where the period is much shorter, for example the Western Isles of Scotland. Here surface dressing is carried out using both conventional and polymer modified emulsions from mid-May until the end of June.

At the start of the surface dressing season environmental factors are critical. The weather conditions during the early life of the dressing have a significant influence on performance. For example, high humidity can significantly delay the break of a surface dressing emulsion resulting in a delay in opening the road to traffic. Heavy rainfall immediately after the installation of a surface dressing can have a detrimental effect on the bond between the binder and the aggregate which can result in stripping of aggregate early in the life of the dressing.

In the first summer, it is important that some chipping embedment into the existing road surface occurs as this will aid chipping retention during the first winter when the binder is at its stiffest and most vulnerable to brittle failure. Therefore, the later in the season the dressing is applied the more likely it is that chipping loss will occur during the first winter particularly where the traffic volumes are insufficient to promote initial embedment. This explains why, in areas such as the Western Isles, the season finishes at the end of June rather than September.

Failure analysis carried out by the RSDA has identified that most surface dressing failures are the result of work being carried out after mid-August. Overall, the analysis shows that failure rates of proprietary surface dressings are now less than 2%.[27]

10.2.8. Safety

This Section briefly highlights some of the potential hazards associated with the handling of surface dressing binders. However, it should not be used as a substitute for health and safety information available from individual suppliers or the information contained in the Code of Safe Practice[28] which is published by the Institute of Petroleum, and the Code of Practice for Surface Dressing,[3] published by the RSDA.

Handling temperatures
Cutback bitumens are commonly handled at temperatures well in excess of their flash points, i.e. the temperature at which the vapour given off will burn in the presence of air and an ignition source. Accordingly, it is essential to exclude sources of ignition in the proximity of cutback handling operations by displaying suitable safety notices.

Every operation should be carried out at as low a temperature as possible to minimize the risks from burns, fumes, flammable atmosphere and fire. Such temperatures must always be less than the maxima given in Table 10.5. Some bitumen emulsions are applied at ambient temperatures but the majority of high bitumen content cationic emulsions are applied at temperatures up to 85 °C.

Operation of surface dressing sprayers using cutback bitumens
The sprayer should be located on level ground before being heated. The flame tube heater can be used only if a minimum of 150 mm of bitumen covers the heater tubes and must not be used during product transfer or spraying. The heater should be extinguished and left to cool for a least

Table 10.5.   Recommended handling temperatures for cutback binders[28]

Grade of cutback binder (STV seconds)	Temperature °C		
	Minimum pumping*	Spraying[†]	Maximum safe handling[‡]
50	65	150	160
100	70	160	170
200	80	170	180

* Based on viscosity of 200 centistokes
[†] Based on viscosity of 30 centistokes, conforms with the maximum spraying temperatures recommended in Road Note 39[2]
[‡] Based on generally satisfactory experience of the storage and handling of cutback grades in contact with air, subject to the aviodance of sources of ignition in the vicinity of tank vents and open air operations

30 minutes before spraying. During heating, the operator should remain in attendance to ensure that the bitumen is not overheated and there are no operations in the vicinity which might release cutback vapours before the flame tube is cool.

The temperature of the cutback in the sprayer should be kept as near as possible to the optimum for the grade and should never exceed the maxima given in Table 10.5 in order to minimize fire risks and avoid excessive fuming. During spraying, clear warning signs against smoking and naked flames should be displayed at appropriate points on site where they will be visible to the general public.

Operation of surface dressing sprayers using bitumen emulsions
The same general handling precautions which apply to cutback bitumens are also relevant for bitumen emulsions. The two major differences are that bitumen emulsions are normally heated to a temperature between 60 °C and 85 °C. A maximum temperature of 90 °C is recommended because overheating to temperatures approaching 100 °C can produce 'boilover' due to the presence of water. Flammable atmospheres are not generated by bitumen emulsions at normal working temperatures. It is therefore less important that flame tube heaters are thoroughly cooled before product transfer or spraying.

Precautions, protective clothing and hygiene for personnel
When spraying either emulsion or cutbacks, all operatives should wear protective outer clothing, i.e. eye protection, clean overalls and impervious shoes and gloves, to protect against splashes and avoid skin contact. In addition, operatives working near the spray bar should wear orinasal fume masks. When working on or near live carriageways, operatives must wear clean high-visibility jackets and reflective patches. The specification for most works require compliance with Chapter 8 of the Traffic Signs Manual,[29] which recommends the use of high-visibility garments complying with BS 6629[30] '(*Class A or B both to Appendix G or better). It is recommended that Class A with sleeves to Appendix G is used on motorways and other high speed roads*'. (It should be noted that this Standard has been withdrawn and replaced by BS EN 471.[31]) Regardless of contractual considerations, it is strongly suggested that this advice be followed for all work on carriageways.

Although bitumen emulsions are applied at relatively low temperatures, the protection prescribed above is essential because emulsifiers are complex chemical compounds and prolonged contact may result in allergic reactions or other skin conditions. In addition, most surface dressing emulsions are highly acidic and, therefore, require the appropriate Personal Protective Equipment (PPE). Barrier creams, applied to the skin prior to spraying, assist in subsequent cleaning should accidental contact occur. However, barrier creams are not adequate substitutes for gloves or other PPE. If any bitumen comes into contact with the skin, operatives should wash thoroughly as soon as practicable and always before going to the toilet, eating or drinking.

10.3.   Crack sealing/overbanding

Crack sealing is one of a number of surface treatments which may be carried out to extend the life of a pavement, surface dressing being the other obvious example. It is described in Part 1 of Section 4 of Volume 7[32] (DMRB 7.4.1) of the DMRB.[19]

Where a crack appears in a road it can be sealed to inhibit the edges of the cracks deteriorating further as a result of damage by traffic and the rigours of weather and to prevent the ingress of water which will result in structural damage to the pavement.

Crack sealing is undertaken where the edges of the cracks are sound. Overbanding is applied where the edges are damaged and the top layer has to be cut back to expose a square solid running surface. The material which is used to seal/overband is specified in DMRB 7.4.1[32] and should be capable of withstanding the thermal and tensile/compressive movements which are likely to occur.

Materials used to fill cracks may not possess good skid resistance characteristics. In addition they may be mistaken for road markings. These considerations are of no consequence where the crack is ≤20 mm. Where the width exceeds 20 mm then a material with a Skid Resistance Value (SRV) >60 and a fine microtexture is required.

10.4.   Slurry surfacings/micro-surfacings

Slurry surfacing was first introduced into the UK in the late 1950s and was primarily used as a preservative treatment for airfield runways. The introduction of quick-setting cationic emulsions and the development of polymer modified emulsions for thick-film slurry applications, now termed 'micro-surfacings', have broadened their use in highway maintenance.

The development of slurry surfacings has largely been on a proprietary basis with individual companies developing their own formulations, specifications, manufacturing and laying techniques. The basic specifications are included in both British Standards BS 434[22] and the Specification for Highway Works[5] (Clause 918 and Clause 927). Clause 927 is a hybrid specification containing elements of performance-related testing of materials, quality control of the process and end-product performance. The responsibility for the design of the surfacing rests with the contractor and there is performance measurement of the surfacing, measured at intervals throughout a period specified in the contract.

The purpose of slurry surfacing is to

- seal surface cracks and voids from ingress of air and water
- arrest disintegration and fretting of an existing surface
- fill minor surface depressions to provide a more even riding surface
- provide a uniform surface appearance to a paved area.

Slurry surfacing is used for footways, carriageways and airfields.

10.4.1.   Footway slurry

Usually this is mechanically mixed and hand applied using a brush to provide surface texture. This protects the footway and is aesthetically pleasing. Modifiers can be used to give the material greater cohesive strength and coloured slurries can be used to denote cycle tracks.

10.4.2. Thin carriageway slurries
These are usually laid 3 mm thick and are applied by a continuous flow machine. This type of slurry can be laid very rapidly with minimal disruption to traffic. The texture depth which is achieved with this material is relatively low, so it is only suitable for sites carrying relatively slow moving traffic.

10.4.3. Thick carriageway slurries
These are normally single treatments using 6 mm aggregates blended with fast breaking polymer modified emulsions. They are more durable than thin slurries and are therefore suitable for more heavily trafficked locations.

10.4.4. Micro-asphalts
Micro-asphalts typically use aggregates up to 10 mm and are placed in two layers. The first layer regulates and re-profiles whilst the second layer provides a dense surface with reasonable texture.

10.4.5. Airfield slurries
Slurry surfacing has been used for runways and taxiways for many years. It is an ideal maintenance technique for airfields as the application is relatively rapid and the aggregates used are too small to present significant foreign object damage (FOD) hazard to jet engines.

10.4.6. Usage, materials and developments
The texture depth achieved with slurry surfacing is relatively low and therefore the material is not suitable for roads carrying high-speed traffic. Slurry surfacing only provides a thin veneer treatment to an existing paved area. Thus it does not add any significant strength to the road structure nor will it be durable if laid on an inadequate substrate. The process has mainly been restricted to lightly trafficked situations such as housing estate roads, footways, light airfield runways, car parks, etc. although specially designed slurry surfacings providing a higher texture depth have been used on the hard shoulders of motorways.

The mixed slurry comprises aggregate (crushed igneous rock, limestone or slag), an additive and a bitumen emulsion. All of the components need to be chosen very carefully as each affects the setting time and performance of the mixture. Accordingly, it is often the case that a dedicated emulsion is developed for each aggregate source and size. The additive is usually Portland cement or hydrated lime which is added to the mix to control its consistency, setting rate and degree of segregation. It normally represents about 2% of the total aggregate. Where rapid setting is required to permit early trafficking or to guard against early damage from wet weather, a rapid setting anionic emulsion (Class A4) or cationic emulsion (Class K3) is used. Where setting time is not critical a slow-setting anionic emulsion (Class A4) can be used. Depending on the characteristics of the aggregate, between 180 and 250 litres of emulsion are required per tonne of aggregate. The composition of the cured material is about 80% aggregate and 20% bitumen.

Slurry surfacings are normally applied by specialist continuous flow machines which meter and deliver aggregate, emulsion, water and additives into a mixer which feeds a spreader box towed behind the machine. These machines are capable of laying up to 8000 m^2/day. Light compaction with a pneumatic tyred roller may be required depending on the composition of the slurry. Rolling is applied as soon as the material has set sufficiently to support the roller.

Recent developments in binder technology utilizing polymer modification are now rapidly changing the scope and use of slurry sealing and further development work is continuing to enable the use of larger aggregate and thicker layers. Aggregate grading up to 11 mm has been tried and thicknesses up to 20 mm have been laid successfully. The use of fibre reinforcement is being developed and it seems likely that increasing the use of this technique will help to overcome some of the problems of heavily cracked surfaces.

10.5. High friction surfacings

A study carried out by the Greater London Council (GLC) in 1965 showed that 70% of all road accidents occurred at or within 15 m of conflict locations such as road junctions, pedestrian crossings, etc.[33] Les Hatherly, then the Chief Engineer of the GLC, approached Shell to develop a surfacing suitable for this type of site. Shell developed a bitumen-extended epoxy resin system and, following various road trials, Shellgrip came into being.

In recent years alternative systems have been developed including polyurethane resin, acrylic resin and thermoplastic rosin ester materials. A resin is generally a manufactured material or a natural secretion from certain plants, whereas a rosin is a hard residue from the distillation of turpentine.

10.5.1.   Bitumen-extended epoxy resins

These are produced by blending two components — one containing a resin and the other a hardener. When combined, they react chemically to form a very strong three-dimensional structure. A specially designed machine applies the material onto the road followed by the application of calcined bauxite. The curing time is usually between 2 and 4 h. The excess aggregate is swept off the surface and the road can then be opened to traffic.

10.5.2.   Epoxy resin

Such systems are a comparatively recent development. They have the advantage that they set very rapidly and adhere well to most surfaces including cement concrete.

10.5.3. Polyurethane resin

These resins are normally two- or three-part binder systems with good adhesion to most surfaces. Some systems require a primer to be applied prior to application of the resin. They are normally applied by hand.

10.5.4. Acrylic resins

These are fast setting two component systems which adhere well to most surfaces. They are clear which makes them ideal for pigmenting.

10.5.5. Thermoplastic rosin esters

These are normally blended with calcined bauxite and heated and hand screeded onto the road surface. The material cures rapidly and may be pigmented.

10.6. Mastic asphalt

Mastic asphalt consists of a mortar of bitumen and fine aggregate and a proportion of coarse aggregate. Full details of the basic raw materials are given in BS 1446[34] and BS 1447.[35] The majority of the mastic asphalt laid in the UK is specified to BS 1447.[35] This is because material to BS 1446[34] requires the import of natural rock asphalt. The fine aggregate has to be naturally occurring limestone rock with between 40% and 55% passing the 75 mm sieve and no more than 3% retained on the 2.36 mm sieve for compliance with BS 1447.[35] The percentage of coarse aggregate added varies depending upon the application. Normally the aggregate is mixed with either a 25 pen or 15 pen bitumen.

The manufacture of mastic asphalt is complex. It can be carried out in a single process only when the material is to be used immediately. In such cases, all the aggregate is combined with the bitumen in a large mixer equipped with slow moving blades. Otherwise, only the fine aggregate is mixed with the bitumen producing a mortar which is cast into 25 kg blocks which are then left to cool. These blocks can be stored and supplied to sites when required. They are then remelted in a special mastic mixer together with the required proportion of coarse aggregate.

Mastic asphalt is normally laid by hand by skilled and experienced operatives who use wooden floats to 'work' the material at temperatures between 175 °C and 230 °C until it has stopped flowing. Thicknesses may vary from 20 mm to 50 mm according to the intended end use. Its impermeability is assured by the fact that the void content of the laid material is generally below 1%. These stringent requirements along with the relatively small areas normally laid at any one time have ruled out laying by machine in the past. Large areas, such as on the Humber Bridge, have been exceptions. The routine laying of mastic asphalt by machine on German roads is not strictly comparable since the German gussasphalt mixture differs from mastic asphalt manufactured to BS 1446[34] or BS 1447.[35]

The high percentage of fines in mastic asphalt gives a smooth surface with poor skid resistance. A variety of treatments may be applied to rectify this shortcoming. Precoated chippings can be added as per HRA wearing course. The chippings are applied while the material is still plastic enough to allow partial but secure embedment and are rolled in using either hand or light power rollers. On footways and other lightly trafficked surfaces, the desired finish may be obtained by adding a special sand by float. Another method, commonly used for multi-storey car park surfaces, is to crimp the surface with an indentation roller while it is still hot.

10.7.   Thin surfacings

Thin surfacing systems were originally developed in France and introduced into the UK in 1991. Subsequently, other surfacing materials such as stone mastic asphalt (SMA) have been introduced; these can also be laid thinner than traditional materials. Thin surfacings fit into the gap between surfacing dressing and thin slurries and traditional surfacings laid 40 mm to 50 mm thick. They have a nominal thickness 15 mm to 40 mm and are used for both new road construction and maintenance. Their main use has been to improve road profiles by filling in longitudinal rutting and sealing surface cracks. They are being used as thin wearing courses to overcome level problems in road maintenance, i.e. bridges and edge detail especially in urban roads and motorways. More recently they have been used to reduce noise emanating from busy urban carriageways.

They can be categorized in several ways, but probably the simplest is to consider the material type from which they were developed.[36] Using this approach, the following categories have been identified

- thick slurry surfacing or micro-surfacings (described above)
- multiple surface dressing developed from traditional surface dressing
- ultra thin asphalt laid by paver
- thin polymer modified asphaltic concrete
- thin stone mastic asphalt.

10.7.1.   Multiple surface dressing

Multiple surface dressing was developed in Scandinavia and introduced into the UK in 1995. It consists of a 14 mm/10 mm racked-in dressing followed by a 6 mm single dressing using an emulsified binder. The 10 mm chipping is precoated with bitumen and the rate of spread of the single dressing is designed to fill the voids in the racked-in dressing. The resultant system has the skid resistant properties of a conventional surface dressing but has the added benefit of generating significantly less noise.

10.7.2.   Ultra thin hot asphalt laid by paver

This system comprises a designed porous hot asphalt manufactured with a 6 mm, 10 mm or 14 mm aggregate laid directly onto a heavy application (0.7–1.2 l/m²) of a warm polymer modified bond coat. The material is applied using either a purpose built paver (see Fig. 10.10) or a paver with an integral spray bar. When the hot asphalt makes contact with the emulsion, it breaks and the water is driven off as steam. Migration of some of the bond coat during compaction strengthens the bond between the surfacing and the substrate.

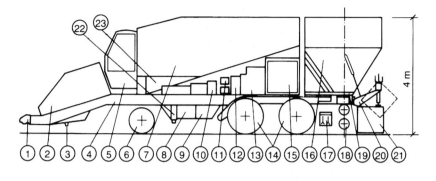

1	Towing hook	9	Fuel tank	17	Spray bar
2	Hopper	10	Electronic control panel	18	Auger
3	Tractor coupling	11	Hydraulic pump	19	Mix gate
4	Chassis	12	Gear box	20	Screed arm
5	Driver's cab	13	Engine	21	Screed
6	Front axle	14	Rear axles	22	Screw conveyor
7	Emulsion tanks	15	Engine radiator	23	Hydraulic lifting
8	Hydraulic tank	16	Holding hopper		ram

Fig. 10.10.   Diagram of a machine for laying 'Safepave'.   Diagram courtesy of Associated Asphalt Ltd

Hot mixture thin surfacing is a relatively new technique that has been used in France since 1988 and was introduced into the UK in 1991. It is the only product which fits the French classification of ultra thin hot mixture asphalt layer (UTHMAL). The original product was called Euroduit. When it was introduced into the UK by Associated Asphalt Ltd it was renamed Safepave.[37]

The main benefits claimed for this type of thin surfacing compared with traditional surfacings are

- the generation of significantly reduced traffic noise
- the generation of less spray in the wet
- texture depths in excess of 1.5 mm
- very high outputs (up to 38 000 m²/day have been claimed).

Since its development, this material has been applied in the USA, Japan, Australia and the majority of the countries of the EU with a total coverage of over 40 million square metres globally.[38]

10.7.3.   Thin polymer modified asphaltic concrete

In France, where this concept was developed, this material is termed very thin surfacing layer (VTSL). There are several proprietary VTSL systems available in the UK. The first to be introduced in the UK in 1992 was UL-M (derived from the French name ultra-mince, meaning very thin).[39,40] Now, there is a wide range of products which are encompassed in this category including Axoflex, Brettpave, Colrug, Duratex P, Hitex, Masterflex, Stratagem, Tuffgrip ULM and Viapave.

These products normally have a 10 mm nominal size, are gap graded and manufactured using a polymer modified bitumen. They are laid through conventional paving equipment onto a tack coat which, for more heavily stressed sites, may be polymer modified. The main benefits claimed for this type of thin surfacing compared with traditional surfacings are

- the generation of significantly reduced traffic noise
- the generation of significant less spray in the wet
- texture depths in excess of 1.5 mm
- significantly better deformation resistance
- similar initial stiffness but better fatigue resistance.

10.7.4. Stone mastic asphalt (SMA)

SMA was developed in Germany some 30 years ago but was not included in their national specification until 1984. It is now extensively used in Scandinavia, the Netherlands, Denmark and is quite common in the UK. The material is normally made with 10 mm or 14 mm nominal size aggregate but, in some cases, 6 mm aggregate is used. The coarse aggregate used must satisfy specific test limits for abrasion resistance, Flakiness Index and Aggregate Impact Value. The filler fraction is between 8% and 13% and the sand fraction is from 12% to 17%. The binder used is, generally, 65 to 80 pen with a bitumen content between 5% and 6%. Cellulose or mineral fibres are normally added if the binder is a penetration grade. This is effected during mixing to assist homogeneous binder distribution throughout the mix. Polymer modified binders, which are becoming more popular, do not require fibres.

SMA is usually designed with an air voids content of 3%. Compaction on site is needed primarily to orientate the aggregate exposed in the upper part of the pavement layer and little actual densification occurs under the roller. The stone skeleton of SMA must accommodate all the mastic while maintaining point-to-point contact of the large aggregate. Too much mastic will cause the coarse aggregate to separate leading to a road layer which is susceptible to deformation. Too little mastic will give unacceptably high voids leading to accelerated ageing and moisture damage. Little latitude is therefore permissible in the production control of the aggregate gradation, the binder content or, if used, the fibre content.

As the mastic receives little compaction during construction, it is rich in binder to ensure that there is a low void content which is needed for satisfactory durability. The expected life of SMA surfacings on heavily trafficked roads is 10 to 15 years.

Detailed information on thin surfacings/SMA is available in Sections 3.7.4 and 3.7.5 of Chapter 3 and Sections 4.13.6 and 4.13.7 of Chapter 4 and elsewhere.[41-45]

10.8. Coloured surfacings

The appearance of traditional asphaltic surfacings, especially when finished with the normal contrasting white road markings, is generally very pleasing to the eye. However, there are some locations where a coloured surfacing is required to act as a visible and, in some cases, audible warning to drivers. Where a coloured surfacing is desired, there are a number of ways that this can be achieved

- incorporating coloured pigments into the asphalt during manufacture
- application of suitably coloured chippings to the surfacing during laying
- application of a coloured surface treatment after laying
- using a conventional bitumen with a coloured aggregate
- using a suitably coloured aggregate with a translucent binder.

10.8.1. Incorporating coloured pigments into the asphalt

The majority of coloured asphalts are manufactured by adding pigment, usually iron oxide (ferric oxide (Fe_2O_3), i.e. rust), during the mixing process. Appropriately coloured aggregates are used to ensure that, when aggregate is exposed after trafficking, the overall appearance of the material is maintained. The main drawbacks to colouring mixtures using conventional bitumens are that the only acceptable colour which can be achieved is dark red and the cost of the iron oxide which is required is fairly high; the cost of the asphalt is thus increased substantially. It is also possible to produce a dark green colour using chromic oxide but this colour is used far less frequently than red asphalt.

Binders have been developed specifically to enable asphalts to be pigmented in colours other than red. These are synthetic binders which, because they contain no asphaltenes, are readily pigmentable in virtually any colour. They possess rheological and mechanical properties similar to conventional bitumen. Coloured asphalt manufactured using these binders require 1% to 2% of pigment to achieve a satisfactory colour whereas red asphalt may require as much as an additional 5% of iron oxide.

A common application for red asphalt is in 'sand carpet' laid as protection over waterproofing membranes laid on bridge decks. The theory is that when the wearing course is being replaced, contractors milling off the old wearing course will cease the operation as soon as the tinted sand carpet is noticed, thus avoiding damage to the expensive waterproofing membrane.

It is often the case that car parks of retail shopping complexes are surfaced in an asphalt which has been pigmented red. A significant number of such schemes exhibit large areas of potholing. This is often due to the fact that specifiers indicate the addition of a specific quantity of pigment but neglect to deduct the same quantity of similarly sized aggregate from the mixture recipe.

10.8.2. Application of coloured chippings to the asphalt

HRA and fine-graded macadam (previously termed 'fine cold asphalt') can have decorative coloured chippings rolled into the surface during compaction. HRA is suitable for most traffic conditions but fine-graded macadam is only appropriate for lightly trafficked areas, private drives and pedestrian areas. HRA is the usual choice for the top layer of footways and footpaths (see Clause 1105 of MCHW 1[5]) although coated macadams and mastic asphalts are also permitted.

Pigmented bitumen-coated or clean resin-coated chippings can be applied during laying to provide a decorative finish to both of these types of surfacing. The bitumen or resin coating promotes adhesion to the surfacing. It is not recommended that uncoated chippings are used. However, in the case of fine-graded macadam laid on areas subjected to little traffic, a light application of uncoated chippings, e.g. white spar or limestone, produces a very attractive finish if they are uniformly applied and well embedded. Decorative chippings cannot be successfully rolled into the surface of high stone content mixtures such as dense macadams or rolled asphalts.

10.8.3.   Application of a coloured surface treatment
There are four surface treatments which can be applied to an asphalt surfacing to provide a decorative finish

- pigmented slurry surfacings
- pigmented thermosetting or thermoplastic systems
- surface dressing
- coloured paints.

Pigmented slurry surfacings are available as proprietary products in a range of colours. They are applied in a very thin layer, about 3 mm thick, and are only suitable for pedestrian and lightly trafficked areas. Thermosetting and thermoplastic materials are probably the most common method used to produce a coloured contrast for traffic calming and bus and cycle delineation. Thermosetting systems are two-component epoxy or polyurethane resin adhesive with a naturally coloured or dyed bauxite applied to the surface. Thermoplastic systems are hot-applied with pigment and aggregate dispersal within the system which is screeded by hand onto the surface. A summary of the uses, benefits and disadvantages of these two systems is given in Table 10.6.

Surface dressings are suitable for most categories of road application but are less suitable for pedestrian situations. The final colour of the surface dressing will be that of the aggregate used. Proprietary coloured paints are available for over-painting conventional black wearing courses. These are only suitable for pedestrian and games areas, e.g. tennis courts.

10.8.4.   Coloured aggregate bound by a conventional bitumen
When a conventional bitumen is used, the depth of colour achieved is dependent upon the colour of the aggregate itself, the thickness of the binder film on the aggregate and the rate at which the binder, which is exposed on the road surface, is eroded by traffic. In medium and heavily trafficked situations, the natural aggregate colour will show through fairly quickly but in lightly trafficked situations, where coloured surfacings are usually specified, the appearance of the aggregate colour may take considerable time.

Table 10.6. The uses, benefits and disadvantages of thermosetting and thermoplastic coloured surfacings[46]

	Thermosetting systems	Thermoplastic systems
Uses	Suitable for coloured anti-skid surfacing and all traffic calming and delineation schemes	Most traffic calming and traffic delineation schemes
Benefits	• Suitable for high stress sites • Anti-skid coloured surfacing to CL 924 specification • Machine or hand application • Provides continuous seam-free textured surface	• Versatile material extensively used in traffic calming and low cost accident remedial measures • Can be used all year round and can be applied at temperatures as low as 0 °C • Fast setting; can normally be trafficked within 15 to 45 minutes • Good retention of colour as these systems encapsulate the pigment enhancing colour durability • Lower initial cost than thermosetting systems
Disadvantages	• Extended curing times at low ambient temperature • Loss of colour under heavy trafficking • Relatively high initial installation cost	• Not as durable as thermosetting systems and tend to wear faster at locations subjected to high densities of commercial vehicles • As it is hand screeded, butt joints and screed lines are regarded by some as aesthetically undesirable • Careful control of heating required to avoid overheating of the material resulting in degradation of the pigment and possible colour changes

10.8.5. Coloured aggregate bound by a translucent binder

Several proprietary macadams, in which the binder is a clear resin rather than a bitumen, are available. A range of coloured surfacings can be manufactured by selecting appropriately coloured aggregates. The major advantage of this type of system is that the colour is obtained immediately.

10.9. Asphalt recycling

The oil crisis in 1973 dramatically increased the costs of energy, bitumen and aggregate and made recycling processes a potentially attractive proposition. Significant research and development work produced a considerable number of technically viable techniques. These used between 10% and 100% of reclaimed materials in the final product which could be used for road maintenance and rehabilitation. During the 1980s, the economic drive was reduced as the price of oil products returned to their pre-crisis levels. Although, in real terms, the price of both bitumen and aggregate have remained fairly constant over the last decade, increasing ecological considerations have renewed the interest in recycling.

It is now becoming clear that the overriding factors in the economies of recycled versus virgin material are the penalties imposed by governments or road authorities on the disposal of old material coupled with any incentives that may be given for the reuse of asphaltic materials. It would appear that without these penalties or incentives, the use of virgin material is more economic in most countries. This is certainly the case in countries with reasonable or good access to aggregate such as the UK. Environmental considerations, such as concern about the landscape being affected by the opening of new quarries and the effects of global warming, will probably lead to the introduction of incentives for both hot and cold recycling.

There are now no technical reasons why recycling should not be utilized since the performance of recycled materials can, with correct design and operation, virtually equal that of virgin material. Recent evidence suggests that within the recycling process the emphasis is shifting from the recycling of the binder to the recycling of the aggregate. During the recycling process, the opportunity can be taken to introduce an upgraded or modified binder which in turn can enhance the performance of the recycled material.

10.9.1.   In situ recycling
In the UK, in situ recycling can be conveniently divided into three processes known as repaving, remixing and retread. Repaving and remixing are hot in situ recycling processes whereas the retread treatment is a cold process.

The repave process was introduced into the UK in 1975. It involves heating and scarifying the existing surface to a depth of about 20 mm with approximately 20 mm of new asphalt laid directly on the hot scarified material after which the layer is compacted. These operations can be completed by a single pass of a purpose built machine. As both layers are hot when compacted, a good bond is achieved between the recycled material and the new material. A report on the surface characteristics of roads using the repave process has been published by the TRL,[47] and the Department of Transport issued a Departmental Standard in 1982.[48] The requirements for this process are now detailed in Clause 926 of the Specification for Highway Works.[5]

In the remixing process the old road surface is heated and scarified. This material is then mixed with a proportion of new asphalt in the pug-mill mixer of a purpose built machine and the blended mixture is paved and compacted on the scarified surface.

A cold in situ procedure known as retread has been established practice in the UK for over 30 years. Briefly, this process consists of scarifying the existing road to a depth of approximately 75 mm, breaking down the scarified material to the required size and reshaping the road profile. The material is then oversprayed two or three times with a bitumen emulsion. Thereafter, the material is harrowed to distribute the emulsion. The material is then compacted with an 8–10 t dead weight roller and finally the surface is sealed with a surface dressing. This process has been successfully carried out on minor roads, housing estate roads, etc. for many years.

10.9.2. Off-site or plant recycling

In this process, material is removed from the surface of an existing road and transported to an asphalt plant where it may be stockpiled for future use or processed immediately. Both batch and continuous plants have been successfully converted to produce asphalt containing a proportion of recycled material. A range of methods which heat the reclaimed material prior to mixing have been developed to reduce fuming and blue smoke emissions during manufacture. Performance tests on asphalt containing a proportion of recycled material suggest that it has properties which are indistinguishable from those in an asphalt manufactured from virgin components.

Recycling in batch plants is achieved by superheating the virgin aggregate and then adding the material to be recycled either immediately after the drier or directly into the pug-mill. Heat transfer takes place during the mixing cycle. Although this method successfully overcomes the problem of blue smoke emissions, it entails keeping aggregate in the heating drum for longer and consequently the maximum output from the plant is reduced. As it is not possible to obtain adequate heat transfer with high percentages of recycled material, it is suggested that the maximum quantity of recycled material that can be added is between 25% and 40%.

Other disadvantages of using batch plants for recycling are high heating costs and accelerated wear and tear on the drum and dust collectors due to the higher manufacturing temperatures. Recycling using continuous mixers involves the introduction of the reclaimed material into the drum itself. During early trials using this type of plant, environmentally unacceptable blue smoke was produced. Blue smoke is produced when vaporized bitumen condenses. The condensate takes the form of particles which are too small to be removed by conventional emission control equipment allowing them to escape through the stack.

Bitumen vaporizes at about 450 °C and, as gas temperatures in the drum of a continuous mixer can reach 2000 °C, the introduction of reclaimed material has to be carefully controlled. Continuous mixers have been designed which, by making various modifications, can take up to 60% of recycled material without exceeding statutory pollution standards. The entry point for the recycled material is approximately half way down the drum and the flights in the drum are modified to produce a homogeneous mix of virgin and recycled material prior to the addition of the bitumen and filler. The redesigned flights shield the recycled material from the intense radiant heat from the heating unit.

Clause 902 of the current Specification for Highway Works[5] allows up to 30% of recycled asphalt to be incorporated in some mixtures, providing the resultant material complies with the required specification. A significant amount of research on the recycling of asphalt has been carried out by the TRL and much of this has been published.[49,50] The current Highways Agency specification requirements and advice on asphalt recycling is given in Volume 7 of the *Design Manual for Roads and Bridges*.[19] Further detailed information on recycling is available.[51]

10.10.   HAPAS—the Highway Authorities Product Approval Scheme

The Highway Authorities Product Approval Scheme (HAPAS) is an initiative to formalize the approval process for introducing new products and innovations into the road construction and maintenance market. The objective of HAPAS is to provide a means by which engineers can choose and specify new and innovative proprietary products with the knowledge that they have been thoroughly assessed and tested, that they meet the mandatory requirements and will perform satisfactorily in use.[52]

Specialist Groups (SGs) have been established for a number of product areas to devise guideline documents for the assessment and approval of materials, for example

- SG1 — high friction surfacing systems
- SG2 — overbanding systems
- SG3 — thin wearing course systems
- SG4 — modified bitumens.

The guideline documents detail the process and the properties to be tested in order to obtain certification for products and systems. The British Board of Agrément (BBA)/HAPAS certificate will bear a strong resemblance to the format of the French Avis Technique (Technical Notices) and will essentially provide product and systems with a 'passport' so that after undergoing the testing and assessment programme they can be specified and used with confidence.[53] In relation to the standard UK specification, the SHW,[5] HAPAS was discussed earlier in this Book (Section 4.13.3 of Chapter 4).

10.11.   Summary

This Chapter has considered a number of specialized techniques which are used for the maintenance of roads. It contains much information on surface dressing, which is probably the most cost effective treatment available to highway engineers for the maintenance of roads — particularly in respect of the restoration of acceptable levels of skid resistance. Other specialised techniques have also been described, many of which have derived significant benefit from improved knowledge of the behaviour of materials and the development of new constituents, mixtures and techniques.

Perhaps above all, this Chapter serves to underline the continuing improvements in road maintenance techniques — improvements which are necessary to meet the continually increasing demands of a road network subjected to higher traffic levels and greater performance requirements.

References

1. DEPARTMENT OF TRANSPORT. *Local Road Maintenance in England and Wales 1994/95.* DOT, London, 1996, Transport Statistics Report.
2. NICHOLLS J. C. *Design Guide for Surface Dressing,* Road Note 39. TRL, Crowthorne, 1996, 4th edn.
3. ROAD SURFACE DRESSING ASSOCIATION. *Code of Practice for Surface Dressing.* RSDA, Bristol, 1995, 2nd edn.
4. HIGHWAYS AGENCY *et al. Manual of Contract Documents for Highway Works.* TSO, London, 1998, **0–6**.
5. HIGHWAYS AGENCY *et al. Manual of Contract Documents for Highway Works, Specification for Highway Works.* TSO, London, 1998, **1**.
6. HIGHWAYS AGENCY *et al. Manual of Contract Documents for Highway Works, Notes for Guidance on the Specification for Highway Works.* TSO, London, 1998, **2**.
7. ROAD SURFACE DRESSING ASSOCIATION. *Guidance Note on Preparing Roads for Surface Dressing.* RSDA, Matlock, UK, 1995.
8. BRITISH CARBONISATION RESEARCH ASSOCIATION. *The CTRA Road Hardness Probe.* BCRA, Chesterfield, 1974, Carbonisation Research Report 7.
9. SALT G. F. and W. S. SZATKOWSKI. *A Guide to Levels of Skidding Resistance for Roads.* TRL, Crowthorne, 1973, LR 510.
10. BRITISH STANDARDS INSTITUTION. *Road Aggregates: Specification for Single-Sized Aggregate for Surface Dressing.* BSI, London, 1987, BS 63: Part 2.
11. BRITISH STANDARDS INSTITUTION. *Testing Aggregates, Method for the Determination of Polished Stone Value.* BSI, London, 1989, BS 812: Part 114.
12. HIGHWAYS AGENCY *et al. Design Manual for Roads and Bridges, Pavement Design and Maintenance, Surfacing Materials for New and Maintenance Construction.* TSO, London, 1999, 7.5.1, HD 36/99.
13. HIGHWAYS AGENCY *et al. Design Manual for Roads and Bridges, Pavement Design and Maintenance, Skidding Resistance.* TSO, London, 1994, 7.3.1, HD 28/94.
14. ROE P. G., D. C. WEBSTER and G. WEST. *The Relation between the Surface Texture of Roads and Accidents.* TRL, Crowthorne, 1991, RR 296.
15. BRITISH STANDARDS INSTITUTION. *Sampling and Examination of Bituminous Mixtures for Roads and other Paved Areas, Methods of Test for the Determination of Texture Depth.* BSI, London, 2000, BS 598: Part 105.
16. WHITEOAK C. D. *The Shell Bitumen Handbook.* Shell Bitumen, Chertsey, 1990, 293–5.
17. CATT C. A. An Alternative View of TRRL Research into Skidding Resistance. *Journal of the Inst. Asphalt Tech.,* 1983, **33**.
18. BRITISH STANDARDS INSTITUTION. *Testing Aggregates, Methods for the Determination of Aggregate Abrasion Value (AAV).* BSI, London, 1990, BS 598: Part 113.
19. HIGHWAYS AGENCY *et al. Design Manual for Roads and Bridges.* TSO, London, various dates, **1–15**.
20. HOSKING J. R. *Road Aggregates and Skidding.* TRL, Crowthorne, 1992.

21. HUNTER R. N. A Review of the Measurement of the Skidding Resistance of Roads with Particular Emphasis on New Surfacings. *The Asphalt Yearbook*. IAT, Stanwell, 1996.
22. BRITISH STANDARDS INSTITUTION. *Bitumen Road Emulsions (Anionic and Cationic), Code of Practice for use of Bitumen Road Emulsions*. BSI, London, 1984, BS 434: Part 2.
23. BRITISH STANDARDS INSTITUTION. *Bitumen for Building and Civil Engineering, Specification of Bitumens for Road Purposes*. BSI, London, 1989, BS 3690: Part 1.
24. BRITISH ROAD FEDERATION. *Basic Road Statistics*. BRF, London, 1998.
25. HOAD L. and T. W. S. HOBAN. *A Cost Performance Relationship for Surface Dressing*. Road Note 39 Symposium. Road Surface Dressing Association, Bristol, March 1982.
26. BRITISH STANDARDS INSTITUTION. *Specification for Hot Binder Distributors for Road Surface Dressing*. BSI, London, 1989, BS 1707.
27. WOOD J. Surface Dressing – Still the Best Value in Road Maintenance. *Proc. CSS Conference on Value Engineering, Leamington Spa, 1999*.
28. INSTITUTE OF PETROLEUM. *Bitumen Safety Code, Model Code of Safe Practice*, Part 11. IP, London, July 1990, 3rd edn.
29. DEPARTMENT OF TRANSPORT *et al. Traffic Signs Manual, Traffic Safety Measures and Signs for Road Works and Temporary Situations*. HMSO, London, 1991, Ch. 8.
30. BRITISH STANDARDS INSTITUTION. *Specification for Optical Performance of High-Visibility Garments and Accessories for use on the Highway*. BSI, London, 1985, BS 6629.
31. BRITISH STANDARDS INSTITUTION. *Specification for High-Visibility Warning Clothing*. BSI, London, 1994, BS EN 471.
32. HIGHWAYS AGENCY *et al. Design Manual for Roads and Bridges, Pavement Design and Maintenance, Maintenance of Bituminous Roads*. TSO, London, 1994, 7.4.1, HD 31/94.
33. HATHERLY L. W. and D. R. LAMB. Accident Prevention in London by Road Surface Improvements. Reprint of a paper given to the *International Road Federation 6th World Highway Conference, Montreal, October 1970*. Shell International Petroleum Company, London.
34. BRITISH STANDARDS INSTITUTION. *Specification for Mastic Asphalt (Natural Rock Asphalt and Fine Aggregate) for Roads and Footways*. BSI, London, 1973, BS 1446.
35. BRITISH STANDARDS INSTITUTION. *Specification for Mastic Asphalt (Limestone Fine Aggregate) for Roads, Footways and Pavings in Buildings*. BSI, London, 1988, BS 1447.
36. DAWS D. F. (J. C. NICHOLLS ed.). *Asphalt Surfacings, Thin Surface Course Materials*. E. & F. N. Spon, London, 1998, Ch. 10.
37. HEATHER W. P. F. Safepave–Two Years On–Thin Surfacing Process. *Journal of the Inst. Highways & Transportation*, 1994, **41**, 17–20.
38. HEATHER W. P. F. *Flexible Pavements and Bituminous Materials, Surface Treatments*. Section H, Residential Course at the University of Newcastle upon Tyne. Quarry Products Association, London, September 1998.
39. PARKINSON D. and P. J. LYCETT. A New Wearing Course from France. *Journal of the Inst. Highways & Transportation*, 1993, **40**, 5–10.
40. PARKINSON D. Five Years of UK Thin Surfacing. *Journal of the Inst. Highways & Transportation*, 1997, **44**, 27–28.

41. QUARRY PRODUCTS ASSOCIATION. *Stone Mastic Asphalt (SMA) and other Thin Surfacings, New Topics in Asphalt.* Quarry Products Association, London, 1997.
42. EUROPEAN ASPHALT PAVEMENT ASSOCIATION. *Heavy Duty Surfaces — The Arguments for SMA.* EAPA, London, 1998.
43. NUNN M. E. *Evaluation of Stone Mastic Asphalt (SMA): A High Stability Wearing Course Material.* TRL, Crowthorne, 1994, PR 65.
44. BELLIN P. A. F. Stone Mastic Asphalt in Germany. *The Asphalt Yearbook.* IAT, Stanwell, 1998, 61–74.
45. LOVEDAY C. A. and P. A. F. BELLIN (J. C. NICHOLLS ed.). *Asphalt Surfacings, Stone Mastic Asphalt Surface Courses* E. & F. N. Spon, London, 1998, Ch. 9.
46. LILES P. A. Delivering Value — Coloured Surfacing. *Proc. CSS Conference on Value Engineering, Leamington Spa,* 1999.
47. COOPER D. R. C. and J. C. YOUNG. *Surface Characteristics of Roads Surfaced using the Repave Process.* TRL, Crowthorne, 1982, SR 744.
48. DEPARTMENT OF TRANSPORT. *In Situ Recycling: The Repave Process.* DOT, London, 1982, HD 7/82.
49. EDWARDS A. C. and H. C. MAYHEW. *Recycled Asphalt Wearing Courses.* TRL, Crowthorne, 1989, RR 225.
50. CORNELIUS P. D. M. and A. C. EDWARDS. *Assessment of the Performance of Off-Site Recycled Bituminous Material.* TRL, Crowthorne, 1991, RR 305.
51. STOCK A. (J. C. NICHOLLS ed.). *Asphalt Surfacings, Recycling Materials.* E. & F. N. Spon, London, 1998, Ch. 13.
52. GARNER N. K. The Highway Authorities Product Approval Scheme (HAPAS). *The Asphalt Yearbook,* IAT, Stanwell, 1998, 37, 38.
53. KELLY S. J. BBA and the Highway Authorities Product Approval Scheme (HAPAS), *Proc. CSS Conference Modern Road Surfacing Materials — Latest Developments, Leamington Spa, 1996.*

11. Failures in asphalts and how to avoid them

11.1. Preamble

There is an old Dutch proverb to the effect that good judgement comes from experience, and experience comes from bad judgements. It is therefore a good idea to learn from the experience of others whenever the opportunity arises. Whilst there are many definitions of a failure, this Chapter takes the view that a failure has occurred when an asphalt does not perform as was intended.

Asphalts are used in well over 95% of all UK road pavement construction sites and to an even greater extent in maintenance works. This is because they are so much more convenient to handle and lay for small to medium sized jobs, and as cost effective as concrete, if not more so, on the very largest projects.

However, occasionally, they do not perform as well as anticipated. These occasions are relatively few and far between and, more often than not, they can be ascribed to a lack of care either in the placement of the asphalt or in the preceding site activities. In this respect, asphalts do not differ from in situ concrete pavements or small element block paving.

For quite understandable reasons, people are usually reluctant to discuss failures on sites with which they have been associated, and particularly so if they still have aspirations of career advancement. This is unfortunate, since others are unable to benefit from these experiences and often, therefore, the same mistakes are repeated unnecessarily.

This Chapter examines different modes of failure in asphalts, with possible reasons, layer by layer, in flexible pavements.

11.2. Roadbases

The function of a road pavement is to distribute the traffic loading stresses onto the underlying sub-base and hence onto the subgrade in such a way as to ensure that the subgrade does not deform, is not overstressed and the roadbase itself is not subjected to repeated levels of critical tensile stress which results in it failing in fatigue. This is illustrated in Fig. 11.1, which is taken from TRL Report 1132.[1] Although it was published in 1984, it is strongly recommended to all involved in either the design or the maintenance of roads.

The roadbase is the principle load bearing component of the pavement but before examining roadbase failures it is worth considering both the

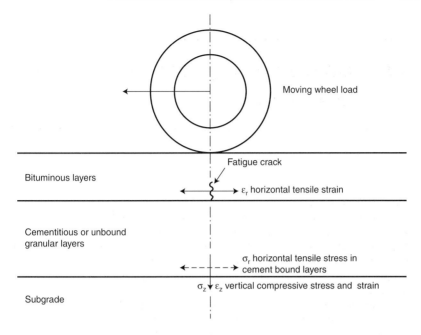

Fig. 11.1   Critical stresses and strains in a bituminous pavement[1]

subgrade and the sub-base. This is because the long term behaviour of the roadbase, sub-base and subgrade are all intimately interlinked and the failure of any one of these layers will inevitably result in the overstressing and subsequent failure of the other two layers.

11.2.1.   The subgrade
Some years ago, the Department of Transport's Engineering Intelligence Division, now superseded by the Pavement Engineering Section of the Highways Agency (HA), commissioned a risk analysis of the entire road pavement design and construction process which concluded that the greatest single area of risk was in assessing the design strength of the subgrade. This is usually expressed as its California Bearing Ratio (commonly called the CBR) since the Californian State Highways Department introduced the test into its own pavement design procedures in the 1930s. The current UK CBR tests are described in BS 1377: Part 4[2] for testing in the laboratory and BS 1377: Part 9[3] for testing in situ.

The CBR Test was subsequently adopted by the US Army Corps of Engineers in the early 1940s and was used by them in World War II throughout the world in constructing military roads and, particularly, military airfields. This widespread use of the CBR Test at that time, some sixty years ago, is probably why it is still so widely used in many countries despite frequent criticism. At the start of the second millennium, it is perhaps time that engineers involved in pavement design moved towards the more frequent use on sites of the Falling Weight Deflectometer (FWD) and other dynamic tests to determine the elastic moduli of formations.

Since the strength of the subgrade or subsoil can vary enormously between its wet and dry conditions, it is clearly desirable to construct the sub-base on a dry subgrade with adequate provision for the drainage of any water which eventually reaches it and which would tend to reduce its dry strength. This was a fact well understood by both Telford and Macadam but in an era where every effort is often made to reduce construction times to the absolute minimum, and sub-base is placed at the minimum practical thickness to reduce initial cost, it is worth re-emphasizing.

It is also tempting for the relatively inexperienced designer to assume that the condition of the subgrade as excavated will remain unaffected by later events in the life of the pavement. Depending on the possibility of the road being disturbed by excavations which have been poorly backfilled (often associated with work on utilities) or affected by defective drainage, or subject to inadequate maintenance of perfectly sound drainage systems (all of which occur all too often), the pavement might judiciously have been designed on the assumption that the subgrade strength will inevitably be weakened in its considerable service life. Indeed, experience suggests that all these types of defect (with the exception of defects resulting from inadequate backfilling on motorways where undertaker's equipment is usually not allowed) will almost certainly occur within two or three decades, at most, of a road having been constructed. It should be borne in mind that the life of a particular route may well exceed one hundred years.

Viewed against this background, it is not unrealistic to assume that at some stage the subgrade strength will be reduced to its soaked CBR value. Wherever possible, therefore, it is suggested that, for both new works and the reconstruction of existing roads, the sub-base and any capping which will support the roadbase should be designed on this basis. This is strongly recommended on heavily trafficked routes where the indirect costs to road users, or to DBFO (Design, Build, Finance and Operate) consortia, of early life strengthening or even reconstruction will make the initial additional costs of excavation, sub-base or capping pale into insignificance.

It is not unknown for a roadbase which had apparently been designed properly to fail in service. Some do so because the actual ground water conditions (established in subsequent investigations) were not anticipated at the design stage. In at least one instance, failure of the roadbase occurred because a particular undertaker laid plant which severed the gully connection pipes. Whilst such works should be routinely monitored by the Engineer's staff and should not occur, in real life things do go wrong and defeat all manner of well intentioned systems. As a result of reduced support from the soaked subsoil, the sub-base deformed providing less support for the roadbase materials which then cracked as did the basecourse and wearing course.

Even when the moisture content is reasonably constant, it is still possible to overstress the subgrade and there have been many cases of pavement failure due to contractors using the sub-base as a haul road without adequate regard for their temporary enhancement. These have then become seriously deformed and the subgrade overstressed. In such cases, some pavement failures have occurred very soon after the new road has opened to traffic when, in addition to the extra expense of the repair works, it has been embarrassing to the Highway Authority politically and to the design and construction staff professionally. On other occasions, the pavement has failed prematurely after being trafficked for some years with a reduced, but still very real, potential for embarrassment of the parties involved.

Similar unfortunate results have occurred when the inexperienced designer has assumed CBR values for the subgrade which were based on an inadequate site investigation or, worse still, in the complete absence of site investigation, on what custom and practice in that design office usually dictated. This tends to occur in smaller design offices where there has often been no significant use made of site investigations or laboratory testing of civil engineering materials.

Nowadays, engineers have the advantage of being able to refer not only to LR 1132[1] but also to HD 25[4] (part of the *Design Manual for Roads and Bridges*,[5] (DMRB) — a full description of its use can be found in Section 4.6 of Chapter 4). This enables an assessment to be made of a reliable value for the subgrade CBR upon which the thickness of the sub-base and any capping will be based. Further information on the importance of proper subgrade and foundation design can be found in Sections 3.4 and 3.5 of Chapter 3.

11.2.2. The sub-base

In the UK, the standard specification is contained within the *Manual of Contract Documents for Highway Works* (MCHW).[6] Volume 1 is entitled the Specification for Highway Works (SHW)[7] whilst Volume 2 is called Notes for Guidance on the Specification for Highway Works (NGSHW).[8] Sub-base is normally specified as Clause 803 ('Granular Sub-base Material Type 1') of the SHW.[7] According to Clause 803 it shall be crushed rock, crushed slag, crushed concrete or well burned non-plastic shale. Whilst gravels are specifically excluded, a Type 1 material can contain up to 12.5% by mass of natural sand passing a 5 mm sieve.

However, sometimes, despite the guidance given in HD 25,[4] particularly if the designer has insufficient experience, a low value for the CBR may be disbelieved since it is not in line with others in the same geographical area or within the experience of the designer or his colleagues. The danger is that a higher value of CBR is assumed. This results in either a sub-base of inadequate thickness or a capping of inadequate thickness or no capping at all, despite circumstances which warrant its inclusion.

Although not complex, sub-base design requires input from personnel with appropriate experience of soils and sites. Hence the usual UK requirement for a minimum CBR value of 30% (Sub-clause 2 of NG 803 and 804[8]) on top of the sub-base is, unfortunately, not always achieved simply because of poor design practice. This requirement is seldom checked since the NGSHW[8] (Sub-clause 3 of NG 803 and 804) states that if more than 10% of a sub-base material is retained on a 20 mm sieve, the whole material can be assumed, without test, to have a CBR value of 30% or more.

Performance testing on top of the sub-base is also rarely carried out in the UK because the CBR Test itself is unsuitable for use on layers of granular material in which the maximum particle size is greater than 20 mm. Even if the material was suitable, the in situ CBR Test[3] has unquantified but suspected very poor precision and its contractual use would be fraught with problems.

During the major road building era of the 1970s and 1980s, some of the County Highway Authorities and their associated Road Construction Units in the north of England used 300 mm diameter plate bearing tests on the completed sub-base, to ensure that the 30% CBR was achieved. Since the demise of the major road building programmes, this commendable approach to the performance testing of sub-bases appears to have largely passed out of use.

More discerning site personnel recognize that FWD testing enables the rapid assessment of the state of compaction of a formation or capping or sub-base and even indicates whether each layer is strong enough to adequately support the rest of the road structure. However, such recognition needs to be backed by the provision of a budget to undertake FWD testing and, despite the potentially high costs of failure, such funds are rarely available.

The recent introduction into the UK of the manually operated German Dynamic Plate Test (Dynamisches Plattendrückgerät or Portable Dynamic Plate Bearing Test (PDPBT)), enables effective in situ testing of both the formation and the top of sub-base or capping to be carried out both very quickly and very cheaply but even so, testing agencies which operate this equipment are not overwhelmed with requests for their services.

Type 1 sub-base which is laid and compacted at its optimum moisture content and of sufficient thickness will give adequate support to the roadbase. However, if the material becomes too wet whilst being placed, its bearing capacity will be substantially reduced. This is particularly true if the material has a high fines content. Material which is obviously too wet can be removed or loosened to enable it to dry out in better weather, if time permits. However, any material which is marginally too wet may well remain undetected in the works until the next construction operation.

Thus, when the roadbase is being laid, the resulting soft spots or areas in the sub-base may be revealed either by deformation under the wheels of the loaded delivery lorries, or by the paver itself losing traction as its wheels or tracks settle into the sub-base which has insufficient strength. Failure to detect and remedy the problem at this stage simply results in the roadbase being under-compacted since there is inadequate reaction from the sub-base to the rollers' compactive efforts and full mechanical interlock of the roadbase aggregate cannot occur.

Type 2 sub-base material, to Clause 804 of the SHW,[7] which includes natural sands, gravels and well-burned non-plastic shales, is even more prone to damage after rain. It gives very poor support to the delivery vehicles which are supplying the job, allowing severe deformation to occur in both the sub-base and subgrade with consequent permanent weakening of the latter. It is for this reason that Type 2 material is not permitted for use in the UK on trunk road sites where the design pavement loading is greater than 5 million standard axles (msa) at opening (Clause 3.16 of HD 25[4]) although it is permitted in Scotland subject to certain safeguards which include its comprising crushed, well graded gravels.

Both the capping and sub-base materials are the cheapest layers in a pavement. However, failure to ensure their correct construction will affect the entire pavement at some stage and any problems in these layers are subsequently the most expensive to rectify since all the overlying materials have to be excavated to gain access to them before any remedial works can be undertaken.

The advent of partnering and other forms of contractual arrangements where there is a less adversarial stance between the designer and the constructor means that it should be possible to construct very strong sub-base and capping layers. It is possible to assess their strengths using the FWD and possibly the PDPBT and then redesign the roadbase using one of several computer programs now available which take account of the CBR of the sub-base being well in excess of the 30% required in the DMRB.[5] Such design might take the greater strength into account either to reduce the initial roadbase thickness, to reduce initial cost, or to confidently extend the design life of the pavement. It is encouraging to note that TRL Report 250[9] examines this concept, albeit only briefly.

11.2.3.   The roadbase
Whilst dry-bound macadam, crusher run, wet-mix and the relatively weak cement bound granular base (CBGB) roadbases generally have limited lives under heavy traffic, many which are well over 30 years old still give good service in less heavily trafficked pavements, as long as they have adequate sub-base support. These materials deserve much wider use in lightly trafficked roads. However, they no longer feature in HA documents and most local authorities have no specifications of their own. Consequently, most young engineers have no knowledge of these materials and since they are not called for, producers do not include them in their range of mixtures.

Failures in asphaltic roadbases are thankfully relatively rare but again fall into two categories, during or soon after the construction period and in service.

During the construction period, small areas of failure of the exposed roadbase, usually indicated by its 'alligator' or 'crocodile' cracking, are symptomatic of its severe overstressing. Many of these are caused by hitherto unsuspected soft spots in the formation which has consequently failed to provide a uniform support for the roadbase. Other distressed areas might be indicated by deformation in wheel tracks due to the overstressing of the local subgrade. The completed pavement might have carried the early life traffic loads without difficulty but this latter problem is seen most commonly with staged construction, particularly when the roadbase itself is incomplete and carrying substantial loadings from construction traffic when it is not yet at its designed thickness.

Occasionally, in coated macadam roadbases, serious deformation occurs under construction traffic even when the pavement is nearly at its full thickness, particularly in warmer weather. This is the result of a combination of the higher stresses produced by the slowly moving delivery vehicles, and high ambient temperatures, coupled with voids in the roadbase material being overfilled with binder. Thus the stone-to-stone interlock which is so essential to the optimum performance of coated macadams is prevented by the presence of an excessive volume of bitumen and consequently, the trafficked material deforms rapidly.

This was a problem on some large trunk road sites in the early 1990s when recipe mix roadbase and basecourse materials were being used, and both had been very well compacted in accordance with the version of Clause 929 (Design, Compaction Assessment and Compliance of Roadbase and Basecourse Macadams) of the SHW[10] which was current at that time. Since there was a dearth of suitable crushed rock aggregates in the vicinity of the sites and the coated macadams were being produced using crushed gravel aggregates, it was initially considered that these were, in some way, the cause of the problem, despite their widespread and successful use in that area for very many years on lower profile sites. However, subsequent investigations revealed the significance of the very low void contents in the failed materials caused by the very high levels of compaction. Guidance then issued by the HA to consultants and con-tractors involved in these works eventually resulted in the revision of Clause 929. This now requires roadbase macadams to be compacted to a maximum of 8% air voids, but when compacted to refusal in the Percentage Refusal Density (PRD) Test,[11] the same macadams must have a void content of at least 1%. This is the practical safeguard against the problem of voids becoming overfilled with bitumen ever occurring again in such macadams.

Mixtures which are being so heavily compacted that their voids are approaching the overfilled stage usually start to become somewhat lively under vibrating rollers (sometimes called 'bouncing'). More experienced site staff will realize what is happening and arrange for the binder content of the mixture to be reduced by an initial half per cent for subsequent deliveries. This usually solves the problem.

Whether or not that overfilled material stays in or is taken up depends on judgements made at the time. A view needs to be taken on the risk of the material becoming very heavily trafficked. Normal rates of stiffening would be considerably reduced because the lower void content of the material in the layer involved will reduce the rate of binder oxidation.

It is interesting to compare this revised Clause 929 concept with that used for the design of asphaltic concretes, or Marshall asphalts, used on UK airfields, where the design process is carefully prescribed and requires the asphalt supplier having to calculate both the voids in the mineral aggregate and the voids filled with bitumen in his material. The upper limit on the latter is set at 77% for basecourse and 82% for wearing course mixtures.

In total contrast, the HA has adopted the CEN philosophy of giving the supplier complete responsibility for designing his material, using whatever design process best suits him, but checking the performance of the finished mixture in service. Under the recently introduced national Quality Assurance Scheme for Asphalt,[12] even the compositional consistency of the materials, which has always been the producer's responsibility, should be checked only by him and not subjected to previous frequent audit testing by the Agency's consulting engineers.

Other in-service failures of roadbase which are frequently noticed are caused by the inadequate backfilling of openings in roads usually by statutory undertakers. The result is lateral movement of the subgrade towards the backfilled excavation, thus resulting once again, in the loss of adequate support to the roadbase and the layers above it.

Other areas of failure, although rarer, are due to an inadequate thickness of roadbase relative to the traffic being carried. The symptoms might be either structural cracking of the asphaltic layers or, if the subgrade has been overstressed, deformation of the traffic lane width of the pavement. It is important to note that it is the numbers of heavy commercial vehicles on the length of road which count, since cars have a negligible damaging affect. These failures may be the result of substantial increases in the number of heavy goods vehicles due to the development or redevelopment of a local site, or a local traffic management scheme which has caused a significant increase in the numbers of such vehicles using a length of road or even an inadequately designed thickness of roadbase.

Fortunately, the use of a dense bitumen macadam (DBM) or a hot rolled asphalt (HRA) roadbase may well have saved many roads which would have failed if a cement bound, dry-bound or a wet mix roadbase had been employed. It is now realized that the inevitable oxidation and stiffening of DBM roadbases and, to a lesser degree, HRA roadbase is to be welcomed, as long as the total thickness of asphalt exceeds 200 mm, since it results in an increase in their dynamic stiffness. Rather than these layers getting weaker in service, they are now seen to be getting stronger. Traditionally, oxidation and the resultant stiffening were regarded as symptoms of potential embrittlement leading to a possible fatigue failure. This may still be the case if the total asphalt thickness is inadequate.

The design thickness of the roadbase will almost certainly be based on the recommendations contained in HD 26[13] (found in Volume 7 of the DMRB[5]) as amended by Interim Advice Note 29 (IAN 29).[14] (The use of IAN 29 has not been adopted in Scotland for use on motorways and other trunk roads but it does contain the latest commendable thinking on pavement design and is recommended for use wherever possible.) However, these documents both contain design curves based on the 95% confidence limits of the dynamic stiffness of roadbases produced with differing penetration grades of bitumen but some 12 months after being made and laid on site. In the meantime, and particularly when they are only a few weeks old and not yet matured, i.e. oxidized, to any extent, they will have significantly lower stiffnesses and consequently be more easily damaged.

The critical period, therefore, for a DBM roadbase is in its very early life. This will be less of a problem in building new routes which might take some years to reach their design traffic loadings than it is in reconstructed or overlaying existing, very heavily trafficked routes where the full loading is carried as soon as the road is opened to traffic.

Indeterminate life roadbases constructed with a strong lean concrete lower roadbase and a DBM upper roadbase, can often be identified after a few years service by the characteristic transverse cracks in the wearing course which lies above them. These used to be considered as indicating continuous cracking from the surface down to the concrete. However, work by the Transport Research Laboratory (TRL) and a number of County Councils[15] demonstrated that in these thick pavements, the cracking started in the wearing course above the discontinuities in the cement bound material, i.e. construction joints or transverse shrinkage cracks, and then progressed down into the basecourse and roadbase. The maintenance of such roads is well covered in HD 30[16] (found in Volume 7 of the DMRB[5]).

Failure of roadbases in fatigue
Recent research work by the TRL suggests that no fatigue failure occurs in a bituminous roadbase if the total thickness of bituminous pavement material, including the surfacing, exceeds 200 mm.[9] This accords with experience in both the Netherlands and Germany, where a sample of existing heavily trafficked roads has been the subject of research into ageing and fatigue failure for more than 20 years.

Thus, the conventional assumption in analytical pavement design computer programs that one mode of failure of a roadbase is fatigue cracking (as depicted in Fig. 11.1) starting at the bottom of the roadbase and progressively working its way to the top of the pavement, is now seen to be incorrect for these thicker roadbases. There has, therefore, been a consequent reduction in interest in progressing the development of test methods to monitor the fatigue strength of asphalts because it is now commonly believed, within the UK, that fatigue failure in roads does not occur. This view is not, however, universally shared. The French road administration certainly consider that it does occur as demonstrated by the inclusion of the Trapezoidal Fatigue Test[17] in their road material mixture design processes for roadbases and other pavement mixtures. This method involves testing at constant stress. However, for a thinner construction, say less than 150 mm thick, a better test is at constant strain and behaviour can be monitored with much less complex test equipment.

In this context, one particular example is of interest. This instance, thought to be quite rare, was where a thick non-composite roadbase had cracked and the failure was not due to subgrade or sub-base failure. This roadbase was 200 mm thick and had been laid in two 100 mm lifts. It was suspected from inspection of cores that one of these layers had been severely overheated in production. This was confirmed when the road was subsequently excavated to repair the failing pavement, since this layer was very brittle and as brown as chocolate cake. The material had cracked into multiple slabs within the roadbase which meant, in effect, the roadbase was 100 mm short of the required thickness and hence severely overstressed.

The TRL has been involved in research on the UK's motorway and other trunk road network which has been substantially built or rebuilt since the 1960s and designed either to Road Note 29[18] or to LR 1132,[1] with relatively thick bituminous roadbases. Local authorities manage the remaining 96% of the road network.[19] Their budgets are the subject of substantial financial pressures to meet a wide variety of demands and few have the resources to finance research on their own road networks. Consequently, most of the more recent experience of these local road networks has remained in the minds of those charged with their care and few have had time to contribute papers to professional journals or conferences. As a result, this experience is largely unrecorded in recent technical journals. The reality is that the great majority of these roads were never designed but rather, have become incrementally constructed with thin overlays or even surface dressings and in most, if they have a recognizable roadbase at all, it will be granular.

Some of these roadbases are of crushed brick, or slag, or stone, or even hoggin, or some might even be of 'tarred rejects' — 2 inch oversize aggregates coated with tar and once sold largely as regulating materials for use in road widening and improvement schemes. Commonly, cores in these roads reveal a variety of materials, with large variations in thickness from one road to the next, or even along the same road. Thus, the thickness of what might be considered as bituminous roadbase on these local roads is considerably less than that which is commonly found on the trunk road and motorway network. On these roads, cracks which start at the bottom of the bituminous material and progress to the top are by no means unknown. Examples of such a road with a thin bituminous pavement in the north east of England prior to its reconstruction are shown in Figs 11.2 and 11.3.

It is worthy of note that in the early 1960s, before the onset of the motorway building era, there were no DBM roadbases such as those which can be found in the current edition of BS 4987: Part 1.[20] The early motorways were built with cement bound roadbases overlaid with 100 or 150 mm of HRA basecourse and wearing course materials. The edition of Road Note 29,[21] published in 1960 was used to introduce a specification for DBM which was even ordered as 'Road Note 29'.

Fig. 11.2.   Core taken from thin bituminous construction cracked from top of sub-base upwards but stopping short of the surface. Photograph courtesy of Derek Pearson, G3 Consulting

Other roadbase problems
Some roadbase problems are due to errors in production, laying or compaction. Production problems now occur far less frequently than they did even ten years ago since asphalt producers are much more aware that poor quality materials cost them money. This is particularly true if they are dealing with a client who employs competent site staff since any production problems are more likely to be noticed and the requisite remedial actions taken without delay to the works. Additionally, virtually all UK asphalt plants are now microprocessor controlled and any material abuse, such as consciously overheating materials to meet the requirements of their customers who lay their materials, needs more conscious effort. It is worth noting here that an overheated material cannot knowingly be sold as complying with a British Standard for that material, if its temperature exceeds that specified in the British Standard.

Problems on sites are sometimes due to low binder contents which can result in the segregation of the material, particularly at the edges of the mat, since there is inadequate binding of aggregates to form the required dense homogeneous layer. Whilst this same material will have a higher indirect tensile strength since this is reduced by higher binder content if its fatigue strength was monitored, this would be lower than that of the nominally identical material made with the correct binder content. Very

Fig.11.3.   Core from the same road but with crack now extending upwards to full depth. Photograph courtesy of Derek Pearson, G3 Consulting

occasionally, defective batches do slip through their producer's quality systems and an example is shown in Fig. 11.4.

What is the correct binder content? In the UK, the binder content of recipe mixtures is specified in the relevant British Standard with manufacturing tolerances which are 0.5% for coated macadams to BS 4987: Part 1[20] and 0.6% for HRAs to BS 594: Part 1.[22] The tolerances should be the same and this anomaly is likely to be corrected when the Standards are next revised. For a designed roadbase, such as those which comply with Clause 929 (discussed in Sections 4.7.4 and 8.4.3 of Chapters 4 and 8 respectively) of the SHW,[7] the target binder content will be determined by the producer and stated by him, but will also have a manufacturing tolerance.

Oversized aggregates can also be a problem. An excess of such aggregate will result in a tendency for the material to segregate and compact to a higher refusal void level. The British Standard gradings allow a sensible degree of oversized material, usually 5% excess on the maximum nominal aggregate size, and this has been the case for many years. When the oversized content is excessive, it will cause both laying and compaction problems. This may occur due to faults on quarry screens, or aggregates overflowing into adjacent bins which should contain smaller aggregates. In

Fig. 11.4.   Poorly manufactured basecourse showing segregated material with no binder

this latter case, the computer controlling the plant will instruct draws from the bin containing the overflowed aggregate as if it was correctly filled, hence it is still necessary to have technicians monitoring the operation of the plant. This includes not only seeing what is happening within the plant but also inspecting aggregate stockpiles and the discharge of the mixture into the delivery lorry from the mixer. All these are common practices in a well managed production unit.

If the problems of oversized aggregate and low binder contents are missed at the plant, for whatever reason, the symptoms can sometimes be identified by the experienced observer on site as a delivery lorry discharges its load into a paver, but they are more commonly observed later, as cracking of the roadbase under the roller. Since this is also a symptom of the lack of adequate sub-base/subgrade support, immediate action is required to determine the layer within which there is a problem. Problems must be remedied without delay. An example of a roadbase which has cracked when being rolled is shown in Fig. 11.5. This was due to a combination of an excessive degree of oversized aggregate and a low binder content which was outside the specification.

Occasionally, despite good sub-base support, the roadbase cracks under the roller, especially at the edges of material. This is often difficult to fully understand but is exacerbated by the use of vibrating rollers. Free edges of material should be fully rolled without vibration and as soon as possible. This is essential initially to crimp the edges and then to increase their stiffness prior to them being subjected to the lateral stresses resulting from the compaction of the adjacent material, when the rest of the width of the same mat is being compacted.

Fig. 11.5. DBM roadbase cracking whilst being rolled

Problems with mixture composition will exacerbate this type of defect, hence the need to sample for possible analysis later. Too often, samples are too small to comply with BS 598: Part 100[23] and so cannot be properly tested. This is a sampling failure. If in doubt, make the sample size 28 kg. It must be placed in a clean sample bag or other container, with all relevant details. Any competent test laboratory will supply both bags and labels in the hope of being paid to test suspect or routine samples.

If a mat does crack when being rolled, note that a high filler content in the mixture will absorb an excessive proportion of the available binder, resulting in there being insufficient binder coating on the coarser aggregate for the mixture to adhere to the underlying layer. In these situations, cracking of the upper mat as it is rolled is therefore not prevented by its bonding to the layer beneath. This cracking is commonly seen as the edges of the new mat are being rolled. The remedy is more careful control to the filler content of the mixture in the plant to ensure that less filler is added.

The larger the nominal size of the aggregate, the greater will be the tendency for it to segregate, causing lack of homogeneity and voiding with consequent low tensile stiffness. Within BS 4987: Part 1[20] and BS 594: Part 1,[22] the largest nominal sized dense macadam and HRA roadbases are 40 mm but 28 mm materials are used by more experienced designers if they have had problems on site with segregation at the edges of mats.

Note here that the standard HA pavement design curves in HD 26,[13] do not differentiate between maximum nominal size materials in terms of their implied tensile stiffness. However, examination of test results often shows a 40 mm nominally sized material to be stiffer than a 28 mm nominally sized material made using identical aggregate and bitumen, due to the higher binder contents in the 28 mm materials. The differences are, however, not significant. Roadbases are usually at such depths in the pavement that the historic impermeability of much of the overlying material has protected them from water damage. Whilst the ingress of water can cause stripping of the binder from the aggregate, this is not usually a problem, as long as they are not temporarily trafficked in wet weather during construction. This is rarely tolerated by the thin binder films on coated macadam roadbases.

It should be noted that the stripping tendency is greater for some aggregates, such as flint gravel, than for others, particularly carboniferous limestones. Any tendency for binder stripping to occur can be inhibited by the addition of an anti-stripping agent during manufacture. At this juncture it is worth recording that this Chapter refers primarily to UK practice where the soundness of the aggregates used in roadbases is taken very much for granted. This is probably due to a combination of the experiences gained on both the production side and the HA purchaser side of the industry over the period since the 1900s during which asphalts have been in daily use.

Most of today's producers are very technically competent and, as in France over the last twenty years, there has been an increasingly obvious exchange of skills between client and supplier and a further enhancement of these skills as asphalt technology has progressed.

In the DMRB,[5] HD 30[16] gives good guidance on the examination of pavements in order to determine the causes of deterioration which includes failure of roadbases. Since these are not visible, reliance needs to be placed on non-destructive testing using such devices as the High Speed Road Monitor, the deflectograph and the Falling Weight Deflectometer, all of which detect changes in surface profile or deflection under load to identify weaker areas of pavement. This work is carried out in conjunction with visual surveys. (The High Speed Road Monitor is discussed in Section 4.11.2 of Chapter 4. The Falling Weight Deflectometer is discussed in Section 3.10.2 of Chapter 3.)

Sometimes, reliance has been placed on visual surveys alone, particularly in the smaller urban authorities, but also, more surprisingly, in much larger organizations. However, visual surveys can only identify those structural defects which can be seen in the wearing course and these are usually either deformation or cracking. These two are rarely seen together as is explained in Section 11.4 below. It is now realized that many of the past instances in which the cracking of wearing courses was regarded as indicating problems at basecourse and roadbase level were incorrectly diagnosed. This is particularly true on roads where the total thickness of asphalt exceeded 200 mm. Misplaced trust had been given to the mantra that *all* roads crack from the bottom up, not just some of them.

11.3. Basecourses

Until the 1960s, most roadbases in the UK were either granular, unbound or cement bound to a greater or lesser degree. Basecourse was therefore the first bituminous layer to be laid and was used both to increase pavement strength and provide a relatively even layer upon which the thin wearing course could be placed.

Since the basecourse is much closer to the top of the pavement, it is more prone to failure. Problems include reflective cracking due to non-uniform roadbase support, usually from cement bound lower roadbases in what is termed 'composite construction', and for many of the other roadbase problems discussed above. However, there is now a better understanding of the phenomena of wearing courses cracking initially, and how in thick pavements these cracks propagate downwards to also cause cracking of the basecourse and, perhaps, roadbase layers.

Deformation of wearing courses occurs relatively frequently and TRL Report 250[9] cites 250 examples of road pavements in which noticeable deformation has occurred in the vehicle wheel tracks. These have subsequently been investigated by coring and testing and this demonstrated that the deformation was almost always confined to the upper 100 mm of pavement construction, with basecourse failure due to deformation being itself a prime cause of failure at wearing course level. Thus, the basecourse and the wearing course are mutually interdependent, with a problem in one layer frequently manifesting itself in the other layer.

Sometimes, in basecourse production, sand fines have been used. If the sand has been wind blown, and is consequently rounded, the resultant mixture may lack stability and roller marks can only be eliminated when the binder cools sufficiently. These materials readily deform beneath stable wearing courses. Such a material is shown in Fig. 11.6.

Fig. 11.6.   HRA basecourse made with rounded sand fines which could not be compacted until the binder was relatively cold

Since maintenance operations often involve replacement of both the basecourse and the wearing course, basecourses are frequently trafficked whilst works are in progress. This is particularly true in urban areas where the scope for overlaying a road to improve its strength is severely limited by the levels of adjacent properties. Particularly on complex works, trafficking may last several days before the wearing course is laid and this often results in problems, the most common of which is fretting of the basecourse in the wheel tracks. This is exacerbated in wet weather, particularly when it is cold, to such an extent that areas of basecourse will need to be replaced before the wearing course may be laid. In such situations, it is recommended that an HRA basecourse is used (BS 594: Part 1[22] Column 2/3 material) unless there is concern about in-service deformation as, for instance, on a heavily trafficked route. In these cases, a 60 mm thickness of 14 mm nominal size stone mastic asphalt (SMA) should be used. The SMA may be a proprietary thin surfacing as discussed in Sections 3.7.5, 4.13.6 and 10.7 of Chapters 3, 4 and 10 respectively, or it might be a non-proprietary material. It will have a high binder content which will make it resistant to fretting. It should have a maximum surface texture of about 1 mm, when measured using the Sand Patch Test (BS 598: Part 105[24] discussed in detail in Section 9.4.6 of Chapter 9). It is most important that its maximum void content should be 6% when sampled from the laid mat. Otherwise, if it is surfaced with a thin wearing course which is permeable by intention or default (due to voided material caused by compacting at a temperature which was too low) at any time in the life of the basecourse, then rainwater will penetrate the road structure. It is important to avoid use of a DBM basecourse in a temporary trafficking situation for more than a few days, and even then, to be prepared to repair it before the wearing course is laid.

In the UK, there is a greater tonnage of 20 mm DBM basecourse produced than any other asphalt material. It is generally laid to a nominal thickness of 60 mm, i.e. following the $3D$ rule, where D is the nominal maximum size of the aggregate. At this thickness, it can be laid in weather conditions which would be too risky for wearing course work, such as in the rain and particularly in very windy conditions. However, sometimes surfacing contractors will seek to reduce their own perceived risks in laying basecourse by instructing the supplier to deliver the material at a temperature which means that it was overheated during production. This can result in the bitumen being degraded to a greater or lesser degree, depending on both the temperature to which it was overheated and the period of time for which this temperature was maintained. The result is that the basecourse is embrittled prematurely and will readily crack in use in fatigue in response to cracks propagated at wearing course level or reflected up from any discontinuities in the roadbase. Even a basecourse which has not been overheated will steadily oxidize and become embrittled with consequent loss of fatigue strength and eventual cracking. Such fatigue cracking is exposed when the wearing course is planed off prior to laying a new one, and the basecourse, instead of simply being overlaid and continuing in service, must also be replaced. Thus, it is

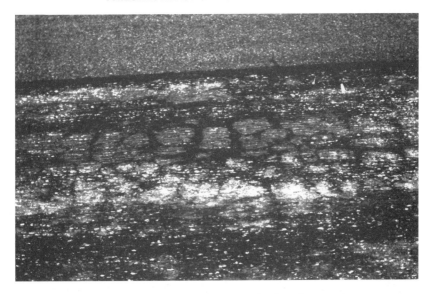

Fig. 11.7.   Basecourse fatigue failure revealed when the wearing course was planed off

categorized as a failure. An example of such a basecourse is shown in Fig. 11.7.

As stated above, the traditional thickness of basecourse is 60 mm, using a 20 mm nominal size aggregate, with its permitted degree of oversize. Specifiers are sometimes tempted to use a 55 mm thickness and whilst this is acceptable for use beneath a relatively impermeable surfacing such as HRA, the practice should be discouraged under an open graded mixture such as many of the new thin surfacings. This is because such a basecourse may be permeable in some areas since it is almost inevitable that some of the oversized aggregate will be dragged by the paver screed wherever the roadbase levels are high, but within tolerance. These dragged areas are frequently segregated and a bucket of water poured onto them will rapidly disappear. Thus, the combination of a relatively permeable thin surfacing and a possibly heavily voided area of basecourse will enable rainwater to enter the road structure with ease.

Beneath a porous asphalt (PA, discussed in Sections 3.7.2 and 4.13.5 of Chapters 3 and 4 respectively), the thickness of the basecourse should be maintained at 60 mm unless the client is prepared to seal the top of the basecourse using a very thick tack coat, some two to three times thicker than normal or even, in extreme cases, to slurry seal the basecourse before applying thick tack coat. Otherwise, surface water will drain through the basecourse and the roadbase.

11.4. Wearing courses

Wearing course problems manifest themselves as polishing, fretting, stiffening and cracking, deformation and rutting, fatting-up with resultant loss in texture and lack of bond with the basecourse or a combination of these. There is, however, another failure mode — permeability.

A wearing course will not usually exhibit cracking and deformation at the same time. This is because cracking usually indicates a weathered, hardened binder whilst deformation occurs most readily in binder-rich materials. In the latter category, the void content is very low, at or below the critical 'voids over filled with binder' stage and the rate of oxidization and weathering of the binder in these materials occurs very slowly as a result.

Recently, wearing course materials which are derived from French and German mixtures have become commonplace within the UK. Traditionally, UK wearing course mixtures have either been HRA with precoated chippings, some form of coated macadam or, particularly in the Midlands and North East, high stone content asphalt to BS 594: Part 1[22] Table 3, Column 3/4 or 3/5, made with 100 pen bitumen for machine laying or 200 pen for areas which are to be laid by hand. (This material is sometimes called 'medium temperature asphalt'.)

Although chipped HRA is not as widely used as previously (except in Scotland where it remains the default wearing course[25]), it is worth looking at the sort of problems which are experienced with chipped HRA since many can occur in other surfacing materials and they inevitably result in failure of one form or another in the finished surface.

11.4.1. Common HRA wearing course failures
Faults which are relatively common in HRA surfacing work include the following.

- Loss of precoated chippings, as illustrated in Fig. 11.8, with subsequent loss of texture and formation of pockets within the mat which hold water, freezing in winter with disruptive affects on the parent material. It is usually associated with poor compaction of the mat, which therefore has a higher than normal void content which, in turn, results in premature oxidation of the binder. Figure 11.9 illustrates a mat laid in very poor conditions which has virtually fretted to half its full depth within three years after chipping loss as soon as it was laid.
- Poorly distributed precoated chippings. This gives a non-uniform texture and can cause safety hazards to motorcyclists, a safety failure.
- Low areas which hold standing water which might freeze in winter causing a safety failure.
- High areas which, when combined with the low areas, give a poor ride, a quality failure.
- Tell-tale signs of the asphalt being too cold for adequate compaction, causing durability failures.
- Poorly finished channels cause ponding of surface water and possible freezing as well as a possibility of durability problems.
- Poor finishing around manhole covers, and the like. This asphalt should always be placed at the time of the main paving work and the ironwork must be raised before the paving commences. Failure to do

Fig. 11.8.   Loss of precoated chippings from HRA wearing course

this requires laying small areas of asphalt by hand at a later stage in the works. Small scale areas are always difficult to lay because the material cools rapidly and may well be inadequately compacted. Examples of such substandard work are shown in Figs 11.10 and 11.11.

Fig. 11.9.   HRA wearing course which has fretted to half its thickness in three years due to poor compaction

Fig. 11.10. Failing wearing course surround to gully frame, placed by hand

- Poor finishing around gully frames which would leave the frame too high for water to enter the gully is an obvious failure since the undrained water may well freeze in winter and constitute a safety hazard.
- Roller marks which not only appear as unsightly blemishes but can also intercept the free flow of water across a surface to the channel with the consequent risk of ice formation in winter.
- Poor jointing of the asphalt between adjacent mats.

In respect of the joints the following faults are common.

- Level differences across the joints causing safety hazards to motorcyclists and cyclists in particular and thus constituting safety failures.
- Excessive bitumen on the surface at the joint possibly causing a safety hazard to motorcyclists and cyclists and thus constituting safety failure.
- Poor compaction of the asphalt on one side of the joint. This is usually the second of the two mats which abut at the joint and is due either to

Fig. 11.11.   Total loss of wearing course to gully frame

heat loss from the edge of mat 2 into mat 1 or to having been laid too low for effective compaction to be applied by the roller drums which are substantially supported by the adjacent mat. This is likely to cause premature failure because of poor durability.

- Poorly cut joints, usually because the asphalt has become too cold. Once again this is likely to affect durability and result in early fretting failure at the joints. (Many SMA type mixtures have been laid with uncut joints, and very many have soon fretted at the joints.)

- The loss of precoated chippings is confined to chipped HRA. It can arise from the use of near-frozen chippings from an unprotected stockpile in the middle of winter. More commonly it occurs because the mat was too cold to permit full embedment of the chippings or because the interstitial asphalt between the chippings, having been de-compacted during the embedment phase, was not fully compacted subsequently probably because the material was below the minimum rolling temperature. This type of failure is also discussed in Section 8.5.3 of Chapter 8. If chippings are not fully embedded then the most cost effective way to deal with the problem is, if operational considerations permit, to remove the suspect area immediately and replace it with fresh material. Alternatively, judicious re-heating and re-rolling is often sucessful if done before the mat is trafficked.

11.4.2.   Other failures

Even the newly introduced materials are not without their problems, many of which mirror those encountered with HRA. For this reason they are discussed here, together with other failures encountered with more established materials. In extreme cases any of these problems could result in the failure of the wearing course, but in most cases, the problems occur over time and lead to the gradual failure of the layer. Each is discussed in turn.

Whatever skills have been used in the design and construction of the lower courses, the wearing course is the only part drivers see and they have always expected it to be safe and well drained as well as even, to produce a smooth ride. The public today now also expects road surfaces which produce lower levels of road noise compared with traditional materials, but safety is paramount.

Ideally, there should be a contractual test for measuring safety or at least for measuring the skid resistance of a surface. Unfortunately, no foolproof test exists so it is necessary to use surrogate tests which examine the resistance to polishing of the aggregate used in the surface and the surface texture of the completed work.

Polishing

A polished road is a dangerous road and hence a failure in road safety terms. An extreme example of polishing is shown in Fig. 11.12. The material used was manufactured with a limestone aggregate which has polished severely. Fortunately, the site is outside the UK. The UK enjoys one of the safest road systems in Europe; only Norway has lower accident figures. Given that all vehicles and tyres are generally similar and that road layout geometry does not differ significantly between countries, what makes the UK different?

A major factor is the use of aggregates with good microtexture (defined as 'a surface roughness which has a size of up to 0.2 mm') in the surface layer. This property is not easy to measure and, therefore, for all practical purposes a surrogate measurement is used—the Polished Stone Value (PSV) which is measured by means of the PSV Test. This determines the degree of resistance to polishing of an aggregate after being subjected to accelerated polishing with emery grit and then emery flour for six hours in purpose built test equipment. The PSV Test is described in BS 812: Part 114[26] and results of the test enable aggregates to be ranked in terms of their resistance to polishing, i.e. PSV.

Current HA surfacing policies dictate that the PSV of an aggregate used in a new road, where there are obviously no previous accident records, shall comply with table 3.1 of HD 36[25] (this table is reproduced as Table 1.10 in Chapter 1). The table relates the PSV to the risk rating of the specific site and also to the likely traffic loading in commercial vehicles per day at the design life. Note that on established sites where the accident records show that the PSV of the aggregate used previously was adequate, HD 36[25] states that an aggregate with a similar PSV can be used when the site is resurfaced even when this value is below that listed in Table 3.1.

Sections 4.10 and 9.4.6 of Chapters 4 and 9 respectively consider the importance of adequate surface texture in roads and it is the combination of both PSV and texture which gives the UK very good accident records.

Fig. 11.12. Polished limestone aggregate in a wearing course

The UK is fortunate in having adequate sources of high PSV aggregates, particularly in South Wales where most of the UK's 68 + PSV sources are located.

What is important is the microtexture of the aggregate which is more easily polished away on some aggregates than on others. Even the same aggregate will give different skid resistance measurements on different sites, and aggregates with the same PSV on the same site also give differing skid resistance.

Polishing can be particularly severe on roads carrying large volumes of heavy commercial vehicles. TRL Reports 322[27] and 367[28] provide useful further information on the skid resistance of road surfaces and are recommended reading to all involved in highway works.

In time, all roads will polish, whatever the PSV of the aggregate used initially. If the wearing course has already polished, what can be done to remedy the situation? The usual treatment, particularly on less heavily trafficked roads, is to apply surface dressing. This restores their skid resistance and also protects the bitumen in the wearing course against

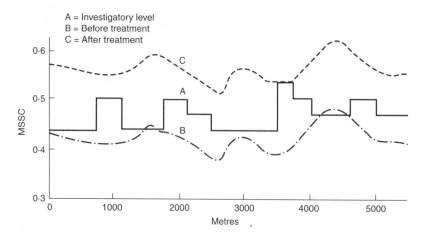

Fig. 11.13. Remedial treatment survey of polished wearing course. Figure courtesy of Klaruw Ltd

oxidation for a few years. Alternatively, there are several patented systems for restoring the microtexture and even the macrotexture of a polished surface, varying from multiple bush hammers suspended from frames beneath purpose built vehicles to shot-blasting and high-pressure water jetting. HD 37[29] (part of the DMRB[5]) considers the various techniques which can be used to restore skid resistance, but an example of a report showing the improvement resulting from treatment by bush hammering is shown in Fig. 11.13.

Surface texture
Figures 11.14 and 11.15 illustrate the concept of positive and negative texture.

For many years it has been HA policy to require a sensor measured texture depth (SMTD) of 1.05 mm on high-speed roads, equivalent to a Sand Patch Test value of 1.5 mm measured in accordance with BS 598: Part 105.[24] High-speed roads are defined as those on which the 85 percentile speed is in excess of 90 km/h (55 mph). These include all the UK motorway network and most of the trunk road network.

It is important to understand that the surface texture of the wearing course provides a means for surface water to drain from beneath vehicle tyres in wet weather and also protects the microtexture of the aggregates. There is no direct relationship between skid resistance and surface texture, although it has long been know that urban roads are safer with a minimum texture of 0.7 SMTD. This was reported in TRL Research Report RR 296[30] several years ago and has recently been reaffirmed in TRL Report 367.[28] Thus, a high-speed road on which the microtexture has been lost will be a failure in safety terms although the same surface would not be a failure on, say, a car park where traffic speeds could be expected to rarely exceed a fast walking pace.

Fig. 11.14.   HRA wearing course with a positive texture

Loss of texture due to fatting-up and/or deformation
Loss of texture is perhaps, with polishing, the most serious defect in a road surface since it deprives the motorist of a safe, skid resistant road—features which are taken for granted in the UK. On occasions, chipped HRA is compacted whilst the material is too hot and the consequence is that the precoated chippings, rather than being embedded

Fig. 11.15.   SMA with a negative texture

into the sand-filler-binder mastic in such a manner as to provide the new mat with a surface texture, are pushed too deeply into the mat, and in rare cases, almost out of a sight. Thus, the mat does not provide the expected degree of skid resistance and it is a safety failure. Sometimes, a high pressure water jet can be used on the cold mat to remove some of the excess binder and at least partially remedy the situation. Otherwise, if the provision of the texture is critical, the mat will have to be replaced.

High stone content HRAs typically contain 55% of high PSV coarse aggregate and are not chipped. They are specified in table 3, columns 4 and 5 of BS 594: Part 1[22] wearing course mixtures. These and coated macadam are used far less frequently and have characteristically far lower textures than that which results from using traditional chipped HRA. Nevertheless, high stone content mixtures can be rolled at too high a temperature and lose what little texture they might have had due to 'fatting-up'. This is the process whereby any excess binder in the mixture is brought to the surface by the rollers at laying stage, or the pumping action of vehicle tyres, especially in hot weather, when the mixture is in service. The excess binder fills the surface interstices and hence there is a marked reduction in whatever texture the material should have provided. This flushing to the surface of the binder is particularly likely to occur if the material was made to a recipe rather than having a binder content which had been optimised by a formal design procedure. This loss of texture is another instance of a safety failure.

The recently developed UK thin surfacing wearing courses, which are based on German stone mastic asphalts (SMA), being gap graded with a high coarse aggregate content, are very sensitive to minor changes in mixture composition or even changes in the aggregate characteristics.

In Germany, their SMAs are specified to have a maximum in-situ air void content of 6% which is carefully monitored, and these resultant dense mixtures are very durable because they are so resistant to oxidation and weathering of the constituent binders. They are the only European mixtures which have anything like the durability of the UKs HRAs but it should be emphasized that the Germans have no specified requirements for surface texture and hence their SMAs usually have about 0.7 mm to 1 mm texture measured by the Sand Patch Test.[24]

Here in the UK, any material producer wanting to sell into the high speed road market has had to develop a material which will virtually guarantee the achievement of the required 1.5 mm texture depth, but at the present state of mixture design with these materials, it has not yet been possible to satisfy this and at the same time, produce a mixture which has always less than 6% air voids. However, the mixtures are environmentally advantageous since they generate less road traffic noise than the traditional HRAs which they have substantially displaced, and HA has therefore accepted certain of these mixtures, which have satisfied their own acceptance criteria, despite their relatively high void content.

Those thin surfacings which are being laid on the high speed roads today are all undergoing acceptance procedure by the British Board of Agrément, as part of their HAPAS[31] accreditation, and are being laid, at least on the HA network, to Clause 942 ('thin wearing course systems') of the SHW.[7] This requires, *inter alia*, that the texture depth of these mixtures in the slow lane wheel track, immediately after laying, shall have an average value of 1.5 mm, and when checked two years later, shall be at least 1 mm (by the Sand Patch Test).

The use of Clause 942 eliminates a practice which developed within some parts of the laying industry of laying these gap graded mixtures, and then, if there was any possibility of their being rolled whilst still too hot to achieve the specified surface texture, of leaving them to cool. When rolled at their lower temperatures, there has been no difficulty in achieving the required surface texture but the mats have, to a degree, been under-compacted. Subsequently, when they have been trafficked in hot weather, secondary compaction has occurred, with resultant deformation in the wheel tracks and consequent loss of the surface texture causing them to be safety failure by virtue of their reduced skid resistance. There is no remedial action which can be used to remedy this defect. The material needs to be replaced or overlaid.

Even some thin surfacings, which can generally produce an initial texture depth of at least 1.5 mm (sand patch) and are well compacted, can subsequently suffer loss of texture due to fatting-up without deformation. Sometimes, an indication of this impending problem is apparent at the laying stage since part of the laid material will fat-up under the action of the roller. In these circumstances, the problem is usually due to lack of care in the production processes, and there is very little that the laying crews can do to avoid the problem. They are simply dealing with a mixture which is 'over sensitive', i.e. one in which the relationship between the binder content and the voids in the compacted mat is too critical. This should have been avoided in the mixture design stage but plots of binder content against voids in both a critical mixture and a non-critical mixture are shown in Fig. 11.16 to illustrate the problem.

This type of failure is exacerbated in the SMA type mixtures in which drainage of the binder which would otherwise inevitably occur in a material with a low fines content, is avoided by the use of cellulose or mineral fibres. These fibres hold the binder which cannot be held by the coarse aggregate. An alternative is to use a polymer modified binder which effectively increases the texture depth by about 0.3 mm for the same aggregate and binder content composition, and in these mixtures, loss of texture due to fatting-up is a very rare occurrence.

A wearing course which has been subject to fatting-up is illustrated in Fig. 11.17. Providing there is no deformation, surfaces which have excess binder may be remedied by high-pressure water jetting, air blasting with steel shot carried in high pressure air jets or simply using heated air at high pressure to remove the excess bitumen.

Deformation
This can be a serious problem on heavily stressed areas of carriageways such as in the wheel tracks of slow lanes on motorways and other trunk roads which carry a high proportion of heavy goods vehicles. It is also a

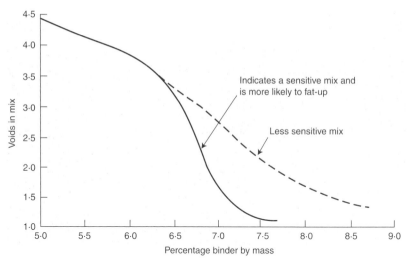

Fig. 11.16. Binder content against voids for a good SMA type mixture and an over-sensitive mixture

problem in urban areas at bus stops and the approaches to controlled junctions but is not generally a problem elsewhere on local authority roads.

The advent of the super-single tyre, three-axle articulated trailers and close-coupled super-single tyres, together with a gradual climatic warming in the 1990s has resulted in very severe deformation of HRAs, especially on the most heavily trafficked parts of the motorway and other

Fig. 11.17. An SMA wearing course which has fatted-up

Fig. 11.18.   Severely deformed HRA wearing course showing rutting in the wheel track

trunk roads network. A typical example is shown in Fig. 11.18. (These 'super-single' tyres operate at high pressures and replace two traditional tyres on the axles of heavy goods vehicles. The combination of smaller area ('footprint') and higher pressure results in significantly higher loadings on roads. They became very common on such vehicles during the 1990s.)

Chipped HRA depends for its strength entirely on the stiffness of the sand–filler–binder mortar into which the coarse aggregate is mixed but in relatively low volumes, typically 30% by mass. Thus, there is no interlock of the coarse aggregate, and with some sands, the mortar has poor resistance to wheel track deformation despite satisfying the mixture design criteria contained in BS 594: Part 1.[22]

Deformation in the wheel tracks is hazardous in dry weather, tending to upset the steering of vehicles driven across the deformed zones at acute angles, especially lighter vehicles. In wet weather, they fill with water and can result in vehicles aquaplaning. Thus, a deformed HRA presents a safety hazard and this constitutes a failure.

Deformation can also occur in thin surfacings when laid at greater thicknesses and meeting the 1.5 mm texture depth requirement but with no control on the value of maximum air voids, as described earlier in this Section.

When remedying a deformed wearing course, a common course of action is to mill (i.e. plane) out the deformed material and replace it with either more of the same or, preferably, something which is known to have very good resistance to deformation, such as a thin surfacing. It is not unknown for deformation to appear in the newly laid, supposedly superior, material. Why should this occur? The reason is that the solution has addressed the symptoms of the problem but not the basic cause.

Fig. 11.19.   Badly deformed wearing course resulting from deformed basecourse having an excessive binder content

Sometimes, an apparent deformation of the surfacing is the result of the basecourse having excessive binder content, i.e. the voids are overfilled with binder and/or low filler content and is thus easily deformed. An example of this is shown in Fig. 11.19.

The remedial works were, therefore, a failure simply because the reason for the original deformation was not established and appropriate corrective measures applied. Three cores, one in each wheel track and one in the centre of the mat, would have revealed the thickness of the basecourse at each position and with a little thought, demonstrated what was happening within the basecourse. In support of this point, the TRL has recently pointed out that deformation has occurred within the uppermost 100 mm of the pavement in almost every case that they have examined with virtually none having been observed below this depth.[9]

Deformation also occurs in coated macadams which are made to recipes and where there is no pretence of design to optimize binder contents. A more serious example is shown in Fig. 11.20, which resulted from only three months of trafficking on a motorway service area slip road used by large numbers of heavy goods vehicles. Again, it is the result of the voids being overfilled with binder although the same material on a less severely stressed site might have been extremely durable.

Permeability
Ideally, most wearing course mixtures would be totally impervious. Generations of highway engineers have been conditioned to consider a dense, impervious wearing course as the 'roof' to their road structures. Thus, any cracks in wearing courses or joints which have opened in the

Fig. 11.20.   Severe pushing of a 3-month old recipe mix wearing course

life of a wearing course have been sealed as quickly as possible by overbanding to ensure that water was not able to enter the pavement but continued to be drained from the surface into the channels and hence safely away into the road drainage system.

Whilst many have considered HRA to be waterproof, it is not. The only bituminous mixture which is absolutely waterproof is mastic asphalt. However, HRA has a very low permeability, a typical value being 10^{-11} to 10^{-10} m/s.[32] DBMs have always been regarded as more permeable because of the capillary nature of their voids and authorities sometimes surface dress these after one or two years simply to seal the road crust thus making them impervious to both water and oxidation. (For the purposes of comparison, typical permeabilities for asphaltic concrete and PA are 10^{-10} to 10^{-8} m/s and 10^{-4} to 10^{-2} respectively.[32])

More recently, the introduction of thin surfacings in the UK has very largely displaced traditional HRAs and coated macadams.

It is important to understand that in Germany, the void content of SMA is derived using the Rice SG Method, described in a draft European Standard prEN 12697-5,[33] but with a solvent instead of the water used in the UK. For some aggregates, therefore, the UK 6% corresponds to a possible maximum of 5% voids if the Germans use de-aired water in the Rice pot instead of their solvent since this also releases air trapped in the aggregate beneath the binder film where it can do no harm at all.

However, as has been seen, in order to qualify as a suitable material to be laid on the UK trunk road network, all these UK-produced derivatives must have a minimum surface texture of 1.5 mm (sand patch) and consequently those which are not produced with modified binders are rather more permeable than the HRA they are replacing and certainly more than SMAs used in Germany. In terms of forming a water-resistant seal to the road pavement, they may be regarded as relatively permeable and particularly so at the end of loads. This emanates from the standard industry practice of using uninsulated hopper sides and floors in pavers which results in the cooling, and hence stiffening, of the binder in that part of the load which is in contact with these uninsulated areas of sheet steel. This material receives no more than the usual degree of compaction so it is very much more permeable than the bulk of the remainder of the load which did not come into contact with the hopper floor or sides.

Figures 11.21 and 11.22 show successive thermographic images of laid material containing both the pre-cooled part of the mixture and some of that which has lost no heat respectively. The colder areas are the result of part of the material having been cooled by contact with the uninsulated sides or wings of the paver hopper prior to being passed through the screed. Such areas will be less well compacted and, thus, have a higher air void content than the areas of mat where the hotter material was laid. The areas with higher voids will suffer from earlier binder degradation, be more permeable and, in some cases, will be quite porous. On a very wet road where the surfacing is one of these mixtures which are enriched with fibres, it is possible to see odd, relatively dry patches. These show where the rain water has drained into the basecourse. It is not clear to where it drains. Water runs downhill so it seems inevitable that it will eventually seep through the basecourse and into the roadbase, and even into and through the sub-base. Thus, in terms of providing a waterproof seal to the road, or even one as good as HRA, they may be considered to be, at least,

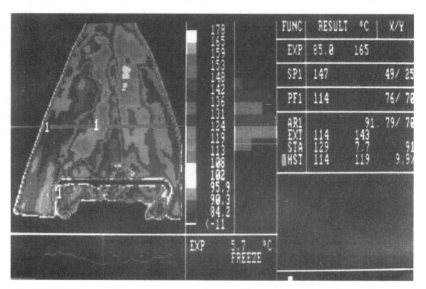

Fig. 11.21. Thermographic image showing large areas of relatively cold material in the centre of a mat

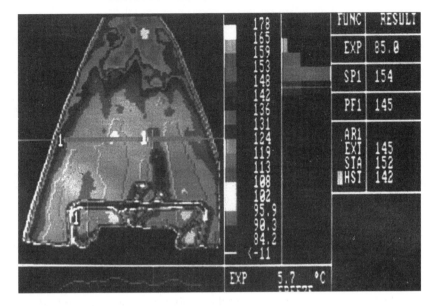

Fig. 11.22. Thermographic image showing uniformly hot-laid material being laid in the foreground

partial failures. The longer term consequences which are, as yet, difficult to anticipate, and when considered alongside the traditional practice of overband sealing open joints and cracks in HRA, the logic is not easily understood. However, there is a solution to this problem — the use of enhanced tack coats (or 'bond coats') is explained in the next Section.

In contrast, porous asphalt (PA) is a gap graded material, which is designed to be very permeable and have a void content of 20% or so, after being fully compacted but it can suffer from impermeability problems. The SHW[5] requires the finished mat to have minimum values for permeability as measured by in situ 'hydraulic conductivity' tests which accord with a modified version of the test contained in BS DD 229.[34] Any mixture design or quality control problems at the mixing plant can severely diminish the ability of the mat to drain the water from the surface down onto its specifically designed basecourse, which is as near impermeable as it can be short of having its voids overfilled with binder. Thus, this lack of permeability can constitute a failure to perform the basic function of the material. In 20 mm PA, this lack of permeability has sometimes resulted from the combined use of a 14 mm aggregate with an excessive degree of undersize and 6 mm aggregate with an excessive degree of oversize. Thus, instead of there being a gap in the grading at the 10 mm sieve, the material is almost a continuously graded mixture with substantial reduction in voids.

Occasionally, concern about the possible failure to achieve the specified degree of hydraulic conductivity might cause inexperienced staff on site to consciously under-compact the PA as it is being laid, in

Fig. 11.23.   A trial porous asphalt which was intentionally under-compacted, failing after being trafficked for one hour

the hope that the lack of compaction will assist in achieving the target. This certainly produces more voids, so the immediate aim is satisfied. However, in PA, its ability to withstand the traffic and other stresses to which it will be subjected in its relatively short life (since it usually fails by binder degradation within ten years or so) depends very substantially on the degree of mechanical interlock of the aggregate. The binder film only assists in giving the mixture tensile strength where these aggregate particles actually come into contact with each other. Therefore, any degree of under-compaction simply results in the premature failure of the mat. Such a failure is shown in Fig. 11.23. This developed after 1 h of trafficking. The photograph was taken at a national PA trial which used a material manufactured with a very strong binder and laid with greater than normal paver compaction but without any roller compaction. This approach was adopted simply to see what happened. The mat failed by total disintegration within about two days of being trafficked.

The need for enhanced tack coats
In France, engineers have come to terms with their own semi-porous mixtures, some of which are gap graded surface dressing systems laid by machine. It is normal, when laying such mixtures, to use a K1-60 emulsion at two to three times the normal rate of spray for a K1-40 tack coat. This enhanced tack coat (often referred to as a 'bond coat') is also vital for the durability of these surfacings which are substantially thinner than the HRAs and coated macadams which were used previously. The minimal thickness means that they are unable to survive the braking and

Fig. 11.24.   A micro-surfacing which was not bonded to the basecourse. Figure courtesy of Acland Ltd and Nynas UK AB

other traffic induced stresses to which they are exposed unless they are firmly fixed to the underlying road pavement layers.

Some opinion in the UK suggests that these enhanced bond coats should comprise polymer modified emulsions, but others have demonstrated extremely good bond which has usually been achieved with unmodified emulsions. This is only possible if the substrates to which they have been applied are thoroughly cleaned before the emulsion is applied. Beneath a partially permeable thin surfacing, these bond coats are the only effective seal to the basecourse. They are, therefore, essential to safeguard the pavement from the ingress of water.

Figure 11.24 shows a micro-surfacing (discussed in Section 10.4 of Chapter 10) which, as a membrane, has virtually no strength to resist breaking forces and which has been forcibly sheared off a layer to which it was clearly not bonded.

Fretting and cracking
Fretting and cracking are often symptoms of either poor compaction of a new material, perhaps because it was rolled when it was too cold, or the inevitable age-related hardening of the binder in a surfacing material. This could have been partially countered and delayed by means of a thicker binder film using either a modified bitumen or fibres to enable the aggregate to carry a greater volume of bitumen when it was mixed.

The rate at which the binder hardens in a mixture will depend on a number of factors. These include

- the characteristic void content of the mixture
- whether or not the specific example was compacted properly
- how well the binder was treated in the mixing plant

- whether it was overheated in storage
- whether it was stored for long periods resulting in a drop in penetration (this can occur even at very low temperatures)
- whether it was overheated during production.

All these factors contribute to the premature hardening of the binder and the failure of the surfacing, in which the binder can no longer withstand the fatigue stresses caused by traffic and weather. In the case of overheated materials, the mixture can fail within a very short time, sometimes as little as two years. However, even if there has been no overheating, the mixture eventually fails by cracking, sometimes into small slabs which may need to be temporarily held in place by overbanding. An example of a surface which has failed in this way is shown in Fig. 11.25.

Fretting that occurs very early in the life of a material is the result of inadequate compaction at the laying stage, there being insufficient development of the aggregate interlock which is so vital to its survival if it is to withstand normal in-service stresses.

In older materials, as the binder deteriorates and embrittles, it loses its binding powers and particles of aggregate are simply displaced by passing traffic after initial loosening by rain, pore pressures generated by tyres running over the surface and frost action.

In an HRA, and to a lesser degree in a coated macadam, at the first sign of any aggregate loss, the mat should be surface dressed or otherwise sealed as soon as possible, to arrest the rate of decay of the binder and hence the surfacing itself.

Sometimes, premature fretting failure is a symptom of a disorganized site. An example of this is shown in Fig. 11.26, where one side of a road

Fig. 11.25. An overheated load of wearing course which has cracked in three years and is held together by its overbanding

Fig. 11.26.   A fretted and potholed DBM wearing course, laid too cold due to poor site organization. The right-hand lane used identical material and was laid properly

has failed by fretting and potholing after only three years whilst the other side, laid using the same material on the same day, is in very good condition. The materials on both sides of the road were sampled and tested and found to be identical and within specification.

11.5.   Summary
The examples of failures which are given and discussed in this Chapter will allow engineers to learn from the experience of others, rather than at the expense of the client or the employer. They will also assist in gaining a fuller understanding of hot-laid bituminous materials. Although the term 'failure' might mean different things to different people, the examples described in this Chapter provide those involved in road construction or maintenance some insight into both why some failures occur, and equally important, how they might be avoided or remedied.

References
1. POWELL W. D., J. F. POTTER, H. C. MAYHEW and M. E. NUNN. *The Structural Design of Bituminous Roads*. TRL, Crowthorne, 1984, LR 1132.
2. BRITISH STANDARDS INSTITUTION. *Methods of Test for Soils for Civil Engineering Purposes, Compaction-Related Tests*. BSI, London, 1990, BS 1377: Part 4.
3. BRITISH STANDARDS INSTITUTION. *Methods of Test for Soils for Civil Engineering Purposes, In-Situ Tests*. BSI, London, 1990, BS 1377: Part 9.
4. HIGHWAYS AGENCY *et al*. *Design Manual for Roads and Bridges, Pavement Design and Maintenance, Foundations*. HMSO, London, 1994, 7.2.2, HD 25/94.

5. HIGHWAYS AGENCY *et al. Design Manual for Roads and Bridges*. TSO, London, various dates, **1–15**.

6. THE HIGHWAYS AGENCY *et al. Manual of Contract Documents for Highway Works*. TSO, London, 1998, **0–6**.

7. HIGHWAYS AGENCY *et al. Manual of Contract Documents for Highway Works, Specification for Highway Works*. TSO, London, 1998, **1**.

8. HIGHWAYS AGENCY *et al. Manual of Contract Documents for Highway Works, Notes for Guidance on the Specification for Highway Works*. TSO, London, 1998, **2**.

9. NUNN M E *et al. Design of Long-Life Flexible Pavements for Heavy Traffic*. TRL, Crowthorne, 1997, Report 250.

10. HIGHWAYS AGENCY *et al. Manual of Contract Documents for Highway Works, Specification for Highway Works*. TSO, London, 1991, **1**.

11. BRITISH STANDARDS INSTITUTION. *Sampling and Examination of Bituminous Mixtures for Roads and Other Paved Areas, Methods of Test for the Determination of Density and Compaction*. BSI, London, 1989, BS 598: Part 104.

12. SECTOR SCHEME ADVISORY COMMITTEE FOR THE QUALITY ASSURANCE OF THE PRODUCTION OF ASPHALT MIXES. *Sector Scheme Document for the Quality Assurance of the Production of Asphalt Mixes*, Sector Scheme No. 14. Quarry Products Association, London, 1998.

13. HIGHWAYS AGENCY *et al. Design Manual for Roads and Bridges, Pavement Design and Maintenance, Pavement Design*. TSO, London, 1994, 7.2.3, HD 26/94.

14. HIGHWAYS AGENCY *et al. Long Life Flexible Pavements: Revised Pavement Design Graphs*. Interim Advice Note 29 (IAN29/00), 2000.

15. NUNN M. E. An Investigation of Reflection Cracking in Composite Pavements in the United Kingdom. *Proc. Conf. on Reflection Cracking in Pavements, Liege, March 1989*, 146–53.

16. HIGHWAYS AGENCY *et al. Design Manual for Roads and Bridges, Pavement Design and Maintenance, Structural Assessment of Road Pavements*. TSO, London, 1999, 7.3.3, HD 30/99.

17. BONNOT J. *Asphalt Aggregate Mixtures*. Transportation Research Record No. 1096. Transportation Research Board, Washington DC, 1986, 42–51.

18. DEPARTMENT OF THE ENVIRONMENT. *A Guide to the Structural Design of Pavements for New Roads*. HMSO, London, 1970, Road Note 29.

19. BRF. *Road Fact 99, 1999*. BRF, London.

20. BRITISH STANDARDS INSTITUTION. *Coated Macadam for Roads and other Paved Areas, Specification for Constituent Materials and for Mixtures*. BSI, London, 1993, BS 4987: Part 1.

21. DEPARTMENT OF THE ENVIRONMENT. *A Guide to the Structural Design of Flexible and Rigid Pavements for New Roads*. HMSO, London, 1960, Road Note No 29.

22. BRITISH STANDARDS INSTITUTION. *Hot Rolled Asphalt for Roads and other Paved Areas, Specification for Constituent Materials and Asphalt Mixtures*. BSI, London, 1992, BS 594: Part 1.

23. BRITISH STANDARDS INSTITUTION. *Sampling and Examination of Bituminous Mixtures for Roads and other Paved Areas, Methods for Sampling for Analysis*. BSI, London, 1987, BS 598: Part 100.

24. BRITISH STANDARDS INSTITUTION. *Sampling and Examination of Bituminous Mixtures for Roads and other Paved Areas, Methods of Test for the Determination of Texture Depth*. BSI, London, 2000, BS 598: Part 105.

25. HIGHWAYS AGENCY *et al. Design Manual for Roads and Bridges, Pavement Design and Maintenance, Surfacing Materials for New and Maintenance Construction*. TSO, London, 1999, 7.5.1, HD 36/99.
26. BRITISH STANDARDS INSTITUTION. *Testing Aggregates, Method for the Determination of Polished Stone Value*. BSI, London, 1989, BS 812: Part 114.
27. ROE P. G. and S. A. HARTSHORNE. *The Polished Stone Value of Aggregates and In-Service Skidding Resistance*. TRL, Crowthorne, 1998, Report 322.
28. ROE P. G., A. R. PARRY and H. E. VINER. *High and Low Speed Skidding Resistance, The Influence of Texture Depth*. TRL, Crowthorne, 1998, Report 367.
29. HIGHWAYS AGENCY *et al. Design Manual for Roads and Bridges, Pavement Design and Maintenance, Bituminous Surfacing Materials and Techniques*. TSO, London, 1999, 7.5.2, HD 37/99.
30. ROE P. G., D. C. WEBSTER and G. WEST. *The Relation between the Surface Texture of Roads and Accidents*. TRL, Crowthorne, 1991, RR 296.
31. GARNER N. K. The Highway Authorities Product Approval Scheme (HAPAS). *The Asphalt Yearbook*. IAT, Stanwell, 1998, 37, 38.
32. NICHOLLS J. C. (J. C. NICHOLLS ed.). *Asphalt Surfacings*, Summary. E. & F. N. Spon, London, 1998, Ch 14.
33. COMITÉ EUROPÉEN DE NORMALISATION. *Bituminous Mixtures, Test Methods for Hot Mix Asphalt, Determination of the Maximum Density*, Part 5. CEN, Brussels, 1997.
34. BRITISH STANDARDS INSTITUTION. *Method for Determination of the Relative Hydraulic Conductivity of Permeable Surfacings*. BSI, London, 1996, DD 229.

Index